A TEXTBOOK OF
PHARMACEUTICAL CHEMISTRY

Dr. JAYASHREE GHOSH
M.Sc., Ph.D.
Head
Department of Chemistry
Anna Adarsh College for Women, Chennai

S Chand And Company Limited
(ISO 9001 Certified Company)

S.CHAND
PUBLISHING

S Chand And Company Limited

(ISO 9001 Certified Company)

Head Office: D-92, Sector–2, Noida – 201301, U.P. (India), Ph. 91-120-4682700

Registered Office: A-27, 2nd Floor, Mohan Co-operative Industrial Estate, New Delhi – 110 044, Phone: 011-49731800

www.**schandpublishing.com;** e-mail: **info@schandpublishing.com**

Marketing Offices:

Chennai	:	Ph: 23632120; chennai@schandpublishing.com
Guwahati	:	Ph: 2738811, 2735640; guwahati@schandpublishing.com
Hyderabad	:	Ph: 40186018; hyderabad@schandpublishing.com
Jalandhar	:	Ph: 4645630; jalandhar@schandpublishing.com
Kolkata	:	Ph: 23357458, 23353914; kolkata@schandpublishing.com
Lucknow	:	Ph: 4003633; lucknow@schandpublishing.com
Mumbai	:	Ph: 25000297; mumbai@schandpublishing.com
Patna	:	Ph: 2260011; patna@schandpublishing.com

First Edition 1997
Second Edition 1999
Third Revised Edition 2003;
Reprints 2008, 2010, 2012, 2013, 2014, 2017, 2018 (Twice), 2020, 2021, 2022, 2023

Reprint 2024

ISBN : 978-81-219-1508-3 **Product Code :** H6PHC68PHRM10ENAC03O

PRINTED IN INDIA

By Vikas Publishing House Private Limited, Plot 20/4, Site-IV, Industrial Area Sahibabad, Ghaziabad – 201 010 and Published by S Chand And Company Limited, A-27, 2nd Floor, Mohan Co-operative Industrial Estate, New Delhi – 110 044.

Dedicated to

My Parents
Mrs. Manju Mitra
&
Late Dr. R.N. Mitra
(Retd. Joint Director of U.P. Medical & Health Services)

Foreword

There has always been much interest in the field of pharmaceutical chemistry by practitioners in the health care field. Recently, the growth of the field has led to interest by other, non-medical fields. The emphasis of this textbook by Jayashree Ghosh, is to broadly review the field of pharmaceutical chemistry. This textbook is designed to provide basic information on pharmaceutical chemistry to non-medical science students. Given the broad spectrum of the topic, it is challenging to cover all aspects of pharmaceutical chemistry thoroughly, and Jayashree Ghosh has not only managed to provide a text that is thorough but also clear and understandable.

This textbook covers topics ranging from defining basic biomedical terms to classifying and explaining the mechanism of action of various drugs. In addition to a review of drugs, Jayashree Ghosh has also included a review of common diseases and their diagnosis and management. This book will be an excellent reference for anyone interested in the broad topic of pharmaceutical chemistry. This textbook on pharmaceutical chemistry is practical, informative and enlightening. It presents a topic that is new to many people in a way that is both simple and thorough. It will hopefully bring this important topic out for people other than in the health care field to see. This book gives new meaning and new insights to those interested in understanding pharmaceutical chemistry and the health care field. It has been a pleasure reading this book and I hope the reader and student will find it as informative as I did.

Dr. Juan R. Avila,
Pharm.D.
Assistant Professor
Duquesne University
School of Pharmacy
Pittsburgh, PA,
U.S.A.

Preface to Third Revised Edition

I am happy to present the third revised edition of the book. In this edition necessary corrections have been made and three more sets of Madras University Question papers (pharmaceutical) have been added.

I hope this edition will also prove a useful textbook for undergraduate students.

Jayashree Ghosh

Preface to First Edition

In recent years Pharmaceutical Chemistry has gained immense importance among the related sciences as it deals with the chemical and medicobiological disciplines. This book has been complied and written to provide a textbook of pharmaceutical chemistry to the students of undergraduate levels, pharmacy courses and to those who are interested in the field of medicinal chemistry.

The present text is designed to provide a comprehensive account of various topics of pharmaceutical chemistry in a single textbook, which are otherwise not included in most of the books available in pharmaceutical chemistry *viz.* concise account of common diseases, their causes, prevention, sustained release of drugs, Clinical Chemistry, Haemotology, AIDS etc. This textbook also emphasises various principles, concepts of pharmaceutical chemistry and the chemical structure of various drugs from pharmaceutical point of view. A great care has been taken to discuss the topics in such a simple and lucid way that it should be an asset even to those students who do not have any biological background. A glossary of all the medical terms have been included.

An enterprise such as this would not have been successfully completed without consulting many standard books on the subject. References of the books have been given at the end.

A special feature of the book is the appendix given at the end of the book, where the summary of various drugs, their chemical structure and therapeutic uses are discussed with the intention to help the students summarize their knowledge or refresh their understanding quickly and easily.

I hope this book will prove of real value to students interested in the field of pharmaceutical chemistry. I will be particularly greatful for suggestions to make this book more useful.

Jayashree Ghosh

29th August 1996.

Acknowledgements

I wish to express my gratitude to the management and Principal Dr. (Mrs.) Sarojini Sherry of Anna Adarsh College for Women, Madras for their encouragement and help in bringing out this book.

I sincerely convey my appreciation and thanks to all my colleagues Mrs. Jayavathana Samuel, Mrs. Sai Sathyavati, Mrs. S. Shanthi, Dr. (Mrs.) Amita Bishnoi, Mrs. Usha Rani, Mrs. P. Shanthi, Miss Mary George for their encouragement, help and co-operation. My thanks are also due to all my students for their valuable suggestions and help.

I would also like to thank Dr. (Mrs.) Rajeswari Ramchandran and Dr. (Mrs.) Prema Gurumurthy of Institute of Indian Council of Medical Research (ICMR) for their guidance and suggestions. My thanks are also due to Dr. Phalguni Gupta, Associate Professor, Deptt. of Infectious Diseases of Microbiology and Dr. (Mrs.) Chanda Gupta, Associate Professor, Endocrinology, School of Medicine, University of Pittsburgh.

I would like to convey my thanks to Mr. G. Ashokan and Mr. O.V. Chandra Mouli of 'Presto Pharmacy' for providing latest reference books of Pharmacy and Indian Pharmacopeias.

I will take this opportunity to express my deep love, gratitude and respect to my father late Dr. R.N. Mitra and my mother Mrs. Manju Mitra for their inspiration. I would further like to acknowledge my heartfelt gratitude and thanks to my beloved sister Shubha and my brother-in-law Dr. P.K. Mullick, Director of Clozapine Programme and Supervising Psychiatrists, St. Francis Hospital in Pittsburgh, U.S.A, for their valuable suggestions and help, without which this book would never have been completed. I owe my thanks to my husband Mr. Samojit Ghosh and my children Sujoy and Tanushree for their patience, understanding and co-operation.

I express my sincere thanks to Mr. Ravindra Kumar Gupta, Director, Mr. A.M. Arunachalam, Manager of Madras Branch and Mr. Chandra Babu, Senior Sales Officer of S. Chand and Company Ltd. for publishing this book promptly.

JAYASHREE GHOSH

Madras
28.8.1996

Contents

Chapter 1

Introduction

Pharmaceutical chemistry is "the chemistry of drugs" which utilizes the general laws of chemistry to study drugs. Hence, it can be summarized that pharmaceutical chemistry is the study of physical and chemical properties of drugs, their preparation, chemical composition, nature, behaviour, structure and influence on an organism conditions of their storage and the therepeutic uses of the drugs.

Pharmaceutical chemistry is a specialised science and occupies an important place among the related sciences. It depends on its organic, inorganic, analytical, physical and colloidal chemical properties and also on its medico-biological properties i.e. pharmacological, physiological and biological properties.

Spectacular developments in physiology, biochemistry and organic chemistry have greatly accelerated the advances in pharmaceutical chemistry and in its turn the pharmaceutical chemistry has helped to elucidate many basic physiological and pathological mechanisms in health and disease. A knowledge of the biological sciences is essential to understand the complex physiological processes and mode of actions of drugs and their effects on various body systems and the probable adverse effects. These scientific data in both animals and humans, form the basis of rational drug treatment.

The word drug is derived from the French word 'drogue' which means a dry herb and it is defined as "any substance used for the purpose of diagnosis, prevention, relief or cure of a disease in man or animal", or according to World Health Organisation it is defined as: A drug is any substance or product that is used or intended to be used to modify or explore physiological systems or pathological states for the benefit of the recipient. The action of a drug is not only determined by its chemical structure alone but also by its physico-chemical properties. That is why pharmaceutical chemistry which is the chemistry of drugs is closely related to physical and colloid chemistry. A knowledge of organic and analytical chemistry helps in the study of the molecular structure of a drug and for synthesizing and analysing the drug. The study of pharmaceutical chemistry helps in identification, preservation, in finding the various combination of drugs in prescription, the production method, the expiry date, storage conditions, technology of drugs and organisation of the pharmacy.

Pharmaceutical chemistry is closely related to physics and mathematics, as the physico chemical methods of analysis of drugs depend on the general laws of these sciences.

The Nature and Sources of Drugs

The various sources of drugs are:

1. **Plant.** Various medicinal compounds have been extracted from plant resources e.g., **morphine, digoxin, quinine, reserpine,** etc.
2. **Synthetic.** Many drugs are made synthetically, e.g. **sulphonamides, aspirin, procaine,** etc.
3. **Animals.** Drugs manufactured from animal sources are **insulin, thyroid, heparin** etc.

4. **Minerals.** Some examples of the medicines made from minerals are liquid paraffin, kaolin, magnesium, trisilicate etc.
5. **Micro-organisms.** Bacteria and fungi isolated from soil are important sources of antibacterial substances like antibiotics e.g., penicillin and bacitracin.
6. **Drugs produced by genetic engineering.** (DNA recombinant technology) e.g., human insulin and human growth harmone.

Majority of the drugs used in therapeutics are synthetic, but many plant products also provide important therapeutic drugs.

Pharmacologically active principles in plants are:

(*a*) **Alkaloids.** These are basic substances containing cyclic nitrogen and form water soluble salts with acids, e.g., morphine, atropine etc.

(*b*) **Glycosides.** An organic substance linked with sugar in an ether like combination is known as **glycoside.** It can be hydrolysed with mineral acids into sugar and non-sugar component called **aglycone.**, e.g., digoxigenin.

(*c*) **Oils.** Oils, too, have pharmacological and edible uses different types of such oils are as follows:

(*i*) **Fixed oils.** These are glycerides of oleic, palmilic and stearic acid. Many fixed oils are edible and have food value and some of the oils have pharmacological actions, such as, coconut oil, olive oil, castor oil, etc.

(*ii*) **Volatile oils.** These oils are volatilised by heat and have a distinct aroma. These are also known as essential oils and have immense medicinal importance as (*a*) carminatives e.g., oil of eucalyptus. (*b*) antiseptics, e.g., mouth wash (*c*) pain relieving agents e.g., oil of clove.

(*iii*) **Mineral oils.** e.g. Liquid paraffin. This is derived from hydrocarbon obtained from petroleum and is used as a lubricant purgative.

(*d*) **Gums.** These are secretory products of plants. These are used in pharmacy as suspending and emulsifying agents. Example: Agar (a gelatinous substance prepared from seaweed) etc. which is used as a bulk purgative.

(*e*) **Resins.** These are found in plants. These are formed by oxidation and polymerisation of volatile oils and are insoluble in water.

Examples: Jalap and colocynth used as purgatives.

(*f*) **Oleoresins.** These are mixtures of volatile oils and resins. These are used for tapeworm infestation. These are obtained from the bark of pine trees.

(*g*) **Tannins.** These are non-nitrogenous plant constituents which have astringent action upon mucous membrane, for example, tincture of catchu (extract of leaves and shoots of plant Accacia), used in the treatment of diarrhoea.

(*h*) **Antibacterial substances.** These are derived from moulds and fungi, e.g., Pencillin, Streptomycin.

Study of Drugs

Drugs can be studied under two categories:

(*i*) Inorganic Pharmaceutical Chemistry

(*ii*) Organic Pharmaceutical Chemistry

Inorganic Pharmaceutical Chemistry deals with the study of drugs made from inorganic elements like Al, S, etc. Organic Pharmaceutical Chemistry is the study of drugs made of organic substances. The organic drugs may be of natural or synthetic in origin.

Some important terminologies used in Pharmaceutical Chemistry are:

1. **Pharmacy.** It is a science of identification, selection, preservation, standardisation, compounding and dispensing of medical substances. The study of pharmacy helps in formulation of an active chemical entity known as active principle. The drug so formed is either in form of tablets,

capsules, powders, suppositories or aerosols. Tablet form is preferred as it . manufacture; easy to transport and suitable for long term storage. The active princi a small proportion of the whole tablet and the bulk is composed of fillers and binders to ho. tablet together and suitable agents are added to break up the tablet efficiently in the patient gastro-intestinal tract.

2. **Pharmacology.** It is a science that deals with drugs. The word is derived from Greek word *Pharmacon* = active principle; *logos* = discourse. It involves a detailed study of drugs particularly their action on living animals, organs or tissues. The action may be beneficial or harmful. The science of pharmacology is not restricted to therapeutic agents or drugs but it is also applicable to all active agents, such as, fungicides, insecticides, etc., which affect the living body. The object of pharmacology, which is a branch of biology, is mainly to provide scientific data for the rational treatment of disease, mode of action of a drug and its effect on various body systems and probable adverse effects which forms the basis of rational therapeutics.
3. **Medicinal chemistry.** The medicinal chemistry deals with the application of principles of chemistry and biology to introduce new therapeutic agents. Hence a medicinal chemist should have good knowledge of organic chemistry and biological sciences like bio-chemistry and pharmacology.

   The knowledge of medicinal chemistry helps to improve the therapeutic or beneficial effects of drugs and at the same time helps to minimise undesirable side-effects.
4. **Pharmacodynamics.** Pharmacodynamics is the study of the biological and therapeutic effect of drugs. This study helps to determine the mechanism of action of a drug and also correlates the drug action with its chemical structure. Pharmacodynamic agents are those which can stimulate or depress various functions of the body to provide relief from symptoms of discomfort, for example analgesics, sedatives, anaesthetics etc. These agents have no action on infective organism which cause the diseases.
5. **Pharmacokinetics.** Pharmacokinetics is the study of the absorption, distribution, excretion and metabolism of drugs and their relationship to pharmacologic responses.
6. **Molecular Pharmacology.** The study of action of a drug at molecular level is known as molecular pharmacology. The knowledge of molecular pharmacology helps to define the site of drug action, known as receptor. The action of drug evoked by direct contact of the drug with receptor. The selective binding of drugs to receptors is specific. The degree of binding is the measure of the ability of the drug to stimulate the tissue. The study of molecular pharmacology helps to get more consistent data on drug tissue interactions which in turn helps to ascertain more precisely the structure activity relationship.
7. **Pharmacophore.** It has been found that a particular structural unit or functional group is responsible for the physiological activity of the drugs. Such a part of the drug which is responsible for the actual physiological effect is known as pharmacophore.
8. **Antimetabolites.** The substance involved in cellular metabolic reaction is known as metabolite. Any chemical agent which has close structural similarity to the metabolite and can block the metabolism that is known as antimetabolite.

   An antimetabolite is mainly bacteriostatic in action, i.e. it can arrest the growth of the harmful micro-organism. It acts either by preventing combination of the metabolite with its specific enzyme or it can form metabolically inactive compound by combining itself with the enzyme. Sulphonamides are good examples of antimetabolites. Their antibacterial activity is due to its close structural resemblance with P-aminobenzoic acid (PABA), which is an essential metabolite. Bacteria takes up sulphonamide by mistake due to its close resemblance to PABA and thus the metabolism is blocked and the growth of bacteria stops.

   Some important antimetabolites are methotrexate, mercaptopurine, purine antagonist. These are used in the treatment of leukaemia.

NH_2 NH_2

PABA COOH SO_2NH_2 Sulphonamide

9. **Actinomycetes.** These are mould like bacteria of 'actinomyectacea' family and is made up of slender branching filaments, only a micron in diameter and without crosswall. Many species show a fragmentation of the filaments into round or rod-shaped segments in the earlier stages of their growth which are almost the same size as ordinary bacteria. Several varieties of actinomyces have been cultivated in the culture media from the human mouth and throat. This family differs from the moulds that produce penicillin or the bacterium that yields tyrothricin.
10. **Bacteria.** They belong to the class of plants called Schizomycetes. These are unicellular organisms and reproduce by binary fission. The cell wall is a complex rigid structure which gives bacteria their definite shape. The rigidity of the cell wall is due to the presence of nucleopeptide. Bacteria cells may be spherical, rod-shaped, curved or spiral in shape. Bacteria do not contain chlorophyll and may be free living, saprophytic (i.e. organism that lives on decayed organic matter), parasitic (i.e. plant or animal that lives upon another living organism) or can be pathogenic (i.e. any diseases producing micro-organism) causing diseases of plants, animals including man.
Bacteria are of two types i.e. gram positive and gram negative depending upon their staining characteristics. Bacteria which retain the violet stain's of gram's reagent (it is a solution made up of crystal violet or methyl violet or gention violet and dilute solution of iodine) are known as gram positive bacteria whereas bacteria which do not retain the violet stain of gram's reagent are known as gram negative bacteria. Some gram positive bacteria contain teichoic acid which is absent in gram negative bacteria. The chemical composition of gram negative bacteria is more complex as they contain many amino acids and high concentration of lipids.
Some soil bacteria which grows in human and animal foods render them highly lethal and cause food poisoning. Bacteria cause several infections like tetanus, diptheria, cholera, scarlet fever, etc. in man.
11. **Virus.** These are very minute micro-organisms, infectious, lack independent metabolism and are parasites within living cells. These are so small that they are barely visible under the highest power of a light microscope. It can be seen with the help of electron microscope. In its simplest form, a virus consists of a protein and a nucleic acid. The nucleic acid has either DNA or RNA. Some DNA viruses are those that cause polio, rabies, influenza and mumps, etc. The nucleic acid is the primary genetic material and it is the nucleic acid which is the infective part of the virus. The most dreaded disease Acquired Immuno Deficiency Syndrome (AIDS) is caused by HIV (human immuno deficiency) virus.
The most important differences between viruses and bacteria are that viruses are much smaller than bacteria and cannot be cultivated on an inanimate laboratory media, both these characteristics make the study of viruses much more difficult than bacteria.
12. **Fungi.** The fungi comprises of a large heterogeneous group of higher protists, that are non-photosynthetic and generally immotile (incapable of motion) and is characterized by a distinctive growth pattern. Micro-organisms which cannot be properly classified as either animals or plants are classified separately, a third kingdom, the 'Protista' which is further divided into higher protists like algae, protozoa, fungi and lower protists as bacteria and blue green algae.

It has a rigid cell wall and does not contain chlorophyll. Fungi like ringworm cause n... hair and nail infections. Certain fungi can penetrate the lung alveoli and can cause respira... allergy.

13. **Mutation.** Mutation can be defined as "a permanent transmissible change in the genetic material." There is a sudden alteration of genes, which may be spontaneous or induced and is inherited by subsequent generation. Mutation is caused by an alteration in the nucleotide sequence at some point of the DNA of the cell.
Mutation is a natural event and sometimes can bring appreciable change in a particular function of the body. For example, a streptomycin resistant mutant of the tubercle bacillus develops in a patient under treatment with the drug, multiplies. Selectively and ultimately it replaces the original drug but in a patient who is not on treatment the mutation confers no survival advantage and preferential multiplication of the mutant does not occur. Another example is the neck of giraffe that becomes longer by constant stretching in order to reach leaves on higher branches, the characteristic of a longer neck is then passed on to its offspring.
Sometimes mutations are induced by certain types of radiations like ultra-violet, x-rays, etc.

14. **Chemotherapy.** The specific treatment of a disease by the administration of chemical compounds is known as chemotherapy. It is used in the treatment of parasitic infections due to insect, worms, protozoa, spiroclaetaceae, etc. The first great advancement in the field of chemotherapy was made by Ehrlich who stated that these drugs acted on the organism and not on the host because they had an affinity for specific cell receptors. Further progress in chemotherapy was achieved with the development of dye stuff industry and the first anti-bacterial drug sulphanilamide was prepared from dyes. Later, Fleming discovered penicillin, an antiobiotic which is bactericidal. The modern concept is that chemotherapeutic agents compete with cell metabolites for enzyme systems essential to the metabolism of the cell.

15. **Pharmacopoeia.** It is an official code containing a selected list of the established drugs and medicinal preparation, their physical properties and tests for their identity, potency and purity. It also defines the standards which these preparations must follow and their average doses for an adult. A few well known pharmacopoeia are the Indian Pharmacopoeia (I.P.) the British Pharmacopoeia (B.P.), the United States Pharmacopoeia (U.S.P.) and European Pharmacopoeia (E.P.).

16. **Pharmacognosy.** It is a combined form of two Greek words viz. Pharmakon means 'a drug' and gignosco means 'to acquire the knowledge of'. Hence Pharmacognosy is defined as a subject of crude drugs obtained from the plant, animals, and mineral origin. The crude drugs mean simple drugs as they exist in their natural forms. Pharmacognosy is the scientific and systematic study of structural, physical, chemical and sensory characters of crude drugs along with their history, method of cultivation, collection and preparation for the market such as.
    (*i*) drugs from vegetables—example: cinnamon, digitalia, clove etc.
    (*ii*) drugs from animals—example: Bees wax, cod liver oil, etc.
    (*iii*) drugs from minerals—example: chalk, talc, kaolin etc.

17. **Toxicology.** It is the study of the toxic or harmful effects of chemicals, as well as of the mechanisms and conditions of occurence of these harmful effects.

18. **Pharmacotherapeutics.** It is the study of the application of drugs in the prevention, treatment or diagnosis of disease and their use in purposeful alteration of functions. For example, prevention of pregnancy or in the use of anaesthetics for surgical procedures.

Chapter 2

Classification and Nomenclature of Drugs

There are several criteria by which the drugs can be classified. The classification of the drugs is an arbitrary procedure. There are approximately 5500 chemical entities which can be used to treat diseases in humans in suitable formulations. Hence, these have to be classified into groups to facilitate comprehension and assimilation. Drugs are regarded as biologically active chemical compounds mostly with a therapeutic purpose. Drugs can be broadly classified into:

I. Biological classification
II. Chemical classification
III. Classification of drugs according to commercial consideration
IV. Classification by the lay public.

I. Biological Classification

Biological classification is based on pharmacotherapeutic and chemotherapeutic agents i.e, this is a disease oriented classification of medicinal agents used in various diseases and hence this classification is based on medicinal agents which are overall descriptive but not very accurate on scientific grounds. For example, the term pain-killer, heart drug, antiarthiritis medicine, diuretic, local anaesthetic, chemotherapy, etc. In each of these categories, sometimes compounds with entirely different chemical structures and different bio-chemical modes of actions are grouped together. The broad classification of drugs based on gross overall biological effects is as follows:

1. **Drugs acting on central nervous system and peripheral nervous system.** These drugs act on central nervous system of the human body which consists of brain and spinal cord, and controls and regularises the special functions like circulation, digestion and respiration and it also modifies the psychic reactions such as feeling, attitude, thoughts and memory. The ability to think logically, to remember the past experiences, communication are unique qualities of man which can be attributed to the development of highly specialised central nervous system.
   The drugs which affect the central nervous system can either stimulate or depress the central nervous system. The CNS drugs can further be sub-divided into the following groups:
   (*a*) *Non selective central nervous system depressants*: These depressants are non-specific in action, for example, general anaesthetics, hypnotics and sedatives, narcotic analgesics, antipyretics, antirheumatic analgesics etc.
   (*b*) *Selective modifiers of CNS*: These drugs are specific or restricted to certain areas within the CNS. The action of the specific CNS depressants may also be modified by the internal environment. Certain examples of these type of drugs are psychotherapeutic agents, anticonvulsants, antitussives, central intraneural blocking agents, etc.
   (*c*) *Central nervous system*: *Stimulants (CNS)*: In therapeutic doses the stimulant action is usually confined to specific areas of the CNS; however, in toxic doses convulsions may

be produced. Stimulants acting directly on CNS are Xanthine alkaloids, carbon-di Nicotine, pipradrol etc. Other type of stimulants act on CNS reflexly, for example:, veratrum, lobeline, etc.

(*d*) *Drugs acting on Peripheral Nervous System*: Peripheral Nervous System consists of somatic and autonomic nerve fibres. Drugs act on peripheral nervous system by either stimulating or blocking it. These have the following classes of drugs:

(*i*) Local anaesthetics like procaine, chlorpromazine, etc.

(*ii*) Cholinergic and anticholinergic agents. Examples: Anticholinesterases, cholinomimetic alkaloids.

(*iii*) Histamine and Antihistamine agents. Examples: Betazole, Betahistine hydrochloride, Diaryl-carbinylethers, Dimenhydrinate, etc.

(*iv*) Adrenergic stimulants and blocking agents and inhibitors of catecholamine—biosynthesis and metabolism like Adrenaline, Isoperanaline etc.

2. **Chemotherapeutic drugs.** Chemotherapy can be defined as the "use of chemical compounds in the treatment of infectious diseases". Various drugs used in chemotherapy are known as chemotherapeutic drugs, which have important therapeutic use in the treatment of parasitic infections due to insects, worms, protozoa, viruses, bacteria, etc. These drugs destroy offending parasites or organisms without damaging the host tissues. Paul Ehrlich, an Organic Chemist, is called 'the Father of Modern Chemotherapy'; who was awarded Nobel Prize in 1909; he successfully used for the first-time certain vital dyes like methylene blue, etc. for the treatment of certain bacteria. The next major contribution to chemotherapy came from Domagk et. al. who used azodyes like Prontosil (a dye with a sulphonamide side chain) for inhibiting the growth of Streptococci infections. Later in 1928, Sir Alexander Fleming discovered Penicillin which inhibited the growth of gram positive organisms.

A chemotherapeutic agent may act by destroying the organism i.e., it is bactericidal or by inhibiting the growth i.e., it is bacteriostatic. The efficiency of the chemotherapeutic agent depends upon the natural body defence mechanisms and the development of resistance by the micro-organisms to the particular agent. Host plays an important part in the curative action of most of the specific remedies.

Chemotherapeutic drugs may be further sub-divided into the following types:

(*a*) **Antiviral agents.** The drugs used for viral infections are known as antiviral agents. Satisfactory treatment of virus infections, however, still remains difficult. Some of the drugs which have shown beneficial effect in human infections are Idoxuridine (5 - Iodo 2' deoxyuridine) , Methisazone etc.

(*b*) **Antituberculosis and antilepral agents.** The drugs used for the treatment of tuberculosis infection, caused by several species of mycobacteria collectively called tubercle bacilli are known as antituberculosis drugs, for example, streptomycin, isonicotinic acid hydrazide etc.

Leprosy is a disease caused by mycobacterium leprae. Antilepral drugs are used in the treatment of leprosy. Some antilepral drugs are sulfones (derivative of 4-4' diamino diphenyl sulfones), diamino-diphenyl thiourea, etc.

(*c*) **Anticancer, antineoplastics.** The drugs commonly used in the treatment of malignant diseases or cancer are known as antineoplastic agents or anticancer drugs, which can be classified according to their mode of action as follows:

(*i*) Alkylating agents—examples: cyclophosphamide, triethylene, thiophosphoramide, busulfan, etc.

(*ii*) Antimetabolites—examples: methotrexate 6-mercaptopurine, cytosine, arabinoside, etc.

(*iii*) Radioactive isotopes—examples: rdiogold, radiophosphorus, etc.

(*iv*) Antibiotics—examples: ariamycine, mithramycin.
(*v*) Miscellaneous agents—examples : vinca alkaloids, procarbazine, etc.
(*vi*) Hormones—examples: oestrogens, androgens, etc.

(*d*) **Antibiotics.** Antibiotics can be defined as a "Chemical substances produced by micro-organisms having the property of inhibiting the growth or destroying other micro-organisms in high dilution". Some important antibiotics are chloramphenicol, penicillins, tetracyclines, etc.

(*e*) **Antifungal and antibacterial agents.** The drugs used for fungal infections are known as antifungal drugs. The important category of antifungal substances belong to the antibiotic class. Some antifungal drugs are effective when applied topically (i.e., external application) in the form of ointments, creams and lotions. For example nystatin ointment, amphotericin, B lotion. Other antifungal antibiotics can be used orally. For example, griseofulvin, hamycin suspension, etc.
The most important antibacterial agents are antibiotics like tetracyclines, chlorampnhenicol, etc. It has been found that, in general, the antibacterial antibiotics have little effect on fungal infection.

(*f*) **Sulphonamides.** The antimicrobial synthetic compounds containing sulphonamide (So_2NH_2) group are called sulphonamides, *e.g.*, sulphadiazine, sulphadimidine, etc.

(*g*) **Antiprotozoal agents.** The protozoan infections of human beings include malaria, trypanosomiasis, leishmaniasis, amoebiasis, giardiasis etc. Some important antiprotozoal drugs are urea-stibaminie, isothinate, etc.

(*h*) **Organometallic compounds.** Organometallic compounds comprise of arsenic, antimony, bismuth, mercury and gold.
Examples: (*i*) Arsenic group—Arrhenal, Stovarsol, etc.
(*ii*) Antimony group—Tarta-metic, Stibophen etc.
(*iii*) Mercury group—Mercurial diuretics, Mersalye etc.
(*iv*) Gold group—Krysolgan, Aurothio glucose etc.
(*v*) Bismuth group—Neocardyl etc.

(*i*) *Anthelmintic agents:* Diseases caused by worms is one of the major problems in tropical countries. The common worms are hookworms, threadworms, tapeworms etc. Anthelmintics are drugs used to kill or remove the parasite worms. An anthelmintic drug which kills the worms is called vermicidal while that which affects the worm in such a way that it is easily expelled is known as vermifuge. Examples: Chloroquine, Piperazine etc.

3. **Pharmacodynamic Agents.** These drugs affect the normal processes of the body like blood circulation, hemodynamic processes, i.e., blood pressure, cardiac outputs etc. Pharmacodynamic agents include mainly the drugs which affect the heart and blood circulation. It also includes a wide variety of drugs used in allergic and gastro intestinal diseases. The pharmacodynamic agents include the following groups of drugs.

(*a*) *Antihypertensive agents:* When walls of arteries and arterioles harden, the resistance to blood flow increases and the heart has to contract more forcefully to cope with the peripheral resistance which enhances both systolic and diastolic pressure. This is called hypertension or high blood pressure. The drugs used in the treatment of hypertension are known as anti-hypertensive drugs. A few examples of these drugs are Aldomet, Chloridine etc.

(*b*) *Antiulcer drugs or gastro intestinal agents:* These are drugs used in gastro-intestinal diseases. Some of the common disorders of gastrointestinal tract are loss of appetite (example of some drugs—Chirata. IP, aromatic bitters, etc), obesity (example of drug—Fenfluramine etc.), atherosclerosis (a disease which affects the blood vessels and example of drug-clofiberate etc.), vomiting (drug used—Apomorphine, etc), diarrhoea (example of drug—Bismuth salts, pectin-IP), constipation (example of drugs—castor oil, IP etc.), peptic ulcer (antacid gels like Aluminium hydroxide gel, aluminium phosphate gel etc.).

(*c*) *Antiarrhythemics drugs:* These drugs are used to correct the disturbances of cardiac rhythm. (Example of drug used—Quinidine procainamide, etc).

(*d*) *Vasodilator drugs:* These drugs are used for the treatment of peripheral vascular diseases. Few drugs used for the treatment of peripheral, vascular diseases are phenoxybenzamine, tolazoline, nicotinic acid etc).

(*e*) *Antianginal drugs:* Antianginal drugs are used for angina pain, which is not a disease, but a symptom of myocardial ischemia (inadequate circulation of blood to the myocardium) which is produced in a variety of ways. It develops as a result of an imbalance between oxygen supply and the oxygen demand of myocardium. The oxygen consumption of the myocardium rises with increasing heart size, heart beat rate, systemic blood pressure and myocardial contractibility (velocity at which the muscle shortens). The myocardial oxygen consumption increases significantly during exercise and other states like the emotional excitement, exposure to cold with increased sympathetic activity. If for any reason, the increase in coronary blood flow is unable to match this increased oxygen demand, angina dcvelops. It is generally produced by exertion and is due to temporary stoppage of blood supply to heart due to constriction of the coronary artery. The drugs used for reducing angina pain are Nitrites like amyl nitrite etc. Aminophylline, Glyceryl trinitrate.

(*f*) *Antiallergic agents:* The hypersensitivity reactions caused by allergens are termed as allergy. The allergens are generally foreign proteins. Reactions of allergens and the specific antibody or antigen (produced by the allergens) produces allergy. The allergic reactions are caused by the release of histamine mainly which is an amino acid and has the following structure:

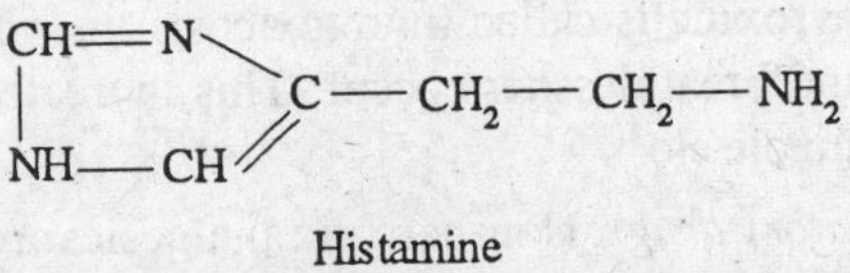

```
CH══N
|      \
|       C — CH2 — CH2 — NH2
|      //
NH — CH
```

Histamine

Antihistamines block the action of histamine at the site of action. Examples of antihistamines used in the treatment of allergy are:

(*i*) Diphenhydramine like Benadryl

(*ii*) Corticosteroids like Triamcindone etc.

(*g*) *Haematological agents:* These drugs act on blood and compromises of antianaemic drugs (example of drug—Iron dextrn (Imferon), Iron-sorbitol, Citric acid complex (Jectoferes) etc.) coagulants (fibrin, thrombin) and anticoagulants (example of drug-heparin etc.) plasma expanders etc.

4. **Metabolic diseases and endocrine function.** It includes variety of drugs which are not conveniently classified in the other groups. It includes the following groups of drugs.

(*a*) *Antirheumatic and autoimmuno disease drugs:* Among the rheumatic disorders, rheumatic arthritis is a formidable disease which can produce severe crippling deformities and functional disability. The drugs used are:

(*i*) Anti-inflammatory drugs and analgesics like aspirin, phenylbutazone etc.

(*ii*) Anti-inflammatory without analgesic action like glucocorticoids etc.

(*iii*) Immuno suppressants like azathioprine, etc.

Immunity can be defined as the ability of the body to neutralise and eliminate the pathogenic micro-organisms and their toxic products. Immunological agents which produce active or passive immunity are used to prevent or to modify certain infectious diseases. For example, vaccines like bacterial vaccines, typhoid vaccine, cholera vaccine etc., and viral vaccines like smallpox vaccine, measles vaccine, poliomyelitis vaccines, rabies vaccine etc., are used to prevent or modify certain infectious diseases.

(*b*) *Dermatologicals (used in skin diseases)*: Skin diseases can be due to infection by bacteria, parasites, fungi etc. Dermatologicals are the drugs used for various skin diseases. Example—Ascabiol (benzyl benzoate emulsion), used in scabies which is due to parasite infection.

(*c*) *Drugs used in endocrine disorders:* The endocrine glands consist of groups of highly specialised cells located at various sites in the body. These glands synthesise and discharge their specialised secretion directly into the blood or lymph without a duct and hence these are also called ductless glands. These secretions are called hormones. These hormones control metabolic processes. The endocrine or ductless glands are pituitary, thyroid, parathyroids, adrenal glands, islets of langarhans (pancreas) testes in male, ovaries in female. They pour their secretion directly into the blood stream.

The hypothalamus, consisting of a group of nerve cells in the brain, controls the whole endocrine system. The hypothalamus with its regulatory hormones controls the anterior pituitary which, in turn, controls other endocrine glands.

The deficiency of a hormone causes physiological diseases, which can be cured by giving that particular hormone. For example, Thyroxine (secreted by thyroid) hormone controls normal metabolism of all tissues of the body.

When thyroid gland is enlarged due to deficiency of Iodine in food, a disease called 'Goitre' develops which can be treated by giving iodised salt. Myxoedema or hypothyroidism is due to hypofunction of thyroid, i.e., due to lesser secretion of thyroxine. The symptoms of these diseases are mental dullness, dry skin, slow pulse, etc. It can be treated by giving thyroid tablet B.P, 30 mg. in each dose. Hyperthyroidism i.e., excessive secretion of thyroxine is called thyrotoxicosis, in which eyes may be pushed forward, nervousness and breathlessness occur. This is treated with antithyroid drugs. For example, carbamizole etc.

(*d*) *Anti-inflammatory drugs*: Non-narcotic analgesics are used as anti-inflammatory drugs. Example—salicylates like aspirin etc. Ibuprofen is another example which is a non-steroidal anti-inflammatory drug.

II. Chemical Classification of Drugs

Chemical classification is based on their chemical structure. According to this classification, drugs may come under one or more of these categories such as quinones, semicarbazides, phenols, lactoness, azo compounds, amides, alcohols, acetals, ketones, hydrocarbons, halogenated compounds, guanides, enols, esters, etc.

The implication of this classification is that similar chemical structures should yield similar biological activities. Thus, structurally close analogous compounds are grouped together with some overall activity but sometimes this classification is not very accurate as a few compounds with similar structure lack the activity test. Chemical structural classification of drugs is of advantage for study of methodology (the science dealing with principles of procedure in research and study) and structure activity relationship. This classification was prevalent before 1950 when the study of the therapeutic agents was according to their structure disregarding their bio-chemical and biological relationship, but, later on, it was found that not only the chemical structure but the bio-chemical and biological structure also plays an important role in determining the biological activity. For example, all amino alcohols do not have the same biological activity. For these reasons classification of drug is considered more accurate on the basis of its therapeutic purpose and it is widely accepted.

III. Classification of Drugs According to Commercial Consideration

The manufacturers and distributors of therapeutic agents classify the drugs according to their operational expenses, research investment and profit margins. Medicines for relatively rare diseases are called orphan drugs and such drugs lack patient protection as production costs are high and demand is less.

Another classification of commercial consideration depends on the way of administration of drug, i.e., drugs are given orally, parenterally (meaning — introduced subcutaneously, intravenously or by any route other than by way of the digestive tract) by inhalation, sublingually, rectally etc. On the whole, orally active drugs are preferred by physicians and patients.

IV. Classification of Drugs by Lay Public

The few broad public classification depends on the action of the drug, for example, antiseptic and disinfectant, anthelmintics, expectorants, cough mixtures, laxative and purgative, analgesics, tonics, ointments for skin diseases, etc. but this classification is scientifically inadequate and public should be well informed about all aspects of the chemistry, biological action and fate of medicinal agent by experimental biologists and medicinal chemists.

Nomenclature of Drugs

Various drugs used as therapeutic agents may be divided into two main categories:

1. **Prescription drugs.** These drugs can be sold only by the prescription from licensed medical practitioners.
2. **Non-prescription drugs.** These drugs are judged safe and they can be sold without a prescription from medical practitioner. These drugs are also known as OTC (over the counter) drugs.

Drugs may also be given a common name which shows a chemical or pharmacologic relationship among a group of drugs such as sulphonamides, sedatives, local anaesthetics, etc.

Every drug has three or more names, i.e., a chemical name, a non-proprietary name, a proprietary or a trade name and in some cases an official name. The official name of newer drug is usually synonymous with their non-proprietary name.

The chemical name is the first name and describes the chemical structure of the drugs accurately. For example, the chemical name of common antibiotic chloramphenicol is D (-) threo2dichloro acetamino 1-p-Nitrophenyl propane 1,3 diol. Although chemical names give a full account of the arrangement of atom or atomic groups but it is very complex and inconvenient to use. The non-proprietary name of a drug is the name given to it if any therapeutic usefulness is found in it. Now a days non-proprietary names are given by "United States Adopted Name (USAN)". The non-proprietary name usually becomes the official name of the drug as well when the drug is finally accepted to official compendia. The United States Pharmacoepia or the National Formulary. The names selected by the USAN Council for newer drugs may also be adopted for use on a worldwide basis through World Health Organisation. Sometimes, non-proprietary names are difficult to pronounce and also difficult to remember. Hence, trade name is given by the pharmaceutical company which is easier to remember. For example, Chlordiazepoxide is the non-proprietary name given to a drug used for treatment of anxiety. The trade name of the some drug is Librium. Another example of an antianxiety drug is diazepam (non-proprietary name) having the trade name as calmpose or valium. Proprietary name is the trade name of the drug which is shorter and easier to remember than non-proprietary name. Hence, the trade name is more frequently used. The proprietary name for oxytetracycline is clinimycin. Now the USAN Council is trying to select non-proprietary names which are easy to pronounce and to remember and also to suggest the nature of the drug. A single drug may have different trade names which is registered and its use is restricted to the owner of the patent right.

For example, the non-proprietary name of a local anaesthetic is known as Procaine in U.S.A. but in various parts of the Worlds the same drug has the non-proprietary name as Novocain, Paracain, Sevicaine, Cetain, Ethocaine, Allocaine, Naucaine, etc.

The chemical name of the organic drugs are based on the IUPAC system (International Union of Pure and Applied Chemistry). According to IUPAC system the name of the various organic groups which are used for naming organic drugs are as follows:

Acetamido	$CH_3CONH—$
Acetoxy	$CH_3COO—$
Acetyl	$CH_3CO—$
Acrylolyl	$CH_2{=\!=}CHCO—$
Allyl	$CH_2{=\!=}CH—CH_2—$
Allyloxy	$CH_2{=\!=}CH—CH_2O—$
Amidino	$H_2NC\ ({=\!=}NH)—$
Amino	$H_2N—$
Anilino	$C_6H_5NH—$
Benzamido	$C_6H_5CONH—$
Benzhydryl	$(C_6H_5)_2CH—$
Benzhydryloxy	$(C_6H_5)_2CHO—$
Benzoyl	$C_6H_5CO—$
Benzoyloxy	$C_6H_5COO—$
Benzyl	$C_6H_5CH_2—$
Bis (2-Chloro ethyl amino)	$(ClCH_2CH_2)_2N—$
Bromo	$Br—$
Butoxy	$CH_3CH_2CH_2CH_2O—$
n-Butyl-(butyl)	$CH_2CH_2CH_2CH_2—$
Sec-Butyl	$CH_3CH_2CH(CH_3)—$
Tertiary-Butyl	$(CH_3)_3C—$
Butylamino	$CH_3CH_2CH_2CH_2NH—$
Butyryl	$CH_3CH_2CH_2CO—$
Carbamoyl	$H_2NCO—$
Carbamoyloxy	$H_2NCOO—$
Carboxy	$HOOC—$
Chloro	$Cl—$
Cyano	$NC—$
Diethylamino	$(CH_3CH_2)_2N—$
2-Diethylaminoethyl	$(CH_3CH_2)_2NCH_2—CH_2—$
Diethylaminoethyl	$(CH_3CH_2)_2NCH_2—$
Diethylcarbamoyl	$(CH_3CH_2)_2NCO—$
Dimethylamino	$(CH_3)_2N—$
2-Dimethylaminoethyl	$(CH_3)_2NCH_2CH_2—$
Dimethyl carbamoyloxy	$(CH_3)_2NCOO—$
Ethoxy	$CH_3CH_2O—$
Ethoxy carbonyl	$CH_3CH_2OCO—$
Ethyl	$CH_3CH_2—$
Ethynyl	$CH—C—$
Fluoro	$—F—$
Formyl	$OHC—$
Glycyl	$H_2NCH_2CO—$
Guanidino	$H_2NC({=\!=}NH)NH—$
Hydroxy	$HO—$

2-Hydroxy ethyl	$HOCH_2CH_2$—
Hydroxy mercuri	HOHg—
Iodo	I—
Isobutyl	$(CH_3)_2CHCH_2$—
Isopentyl	$(CH_3)_2CHCH_2CH_2$—
Isopropyl	$(CH_3)_2CH$—
Mercapto	HS—
Methoxy	CH_3O—
Methyl	CH_3—
Methylamino	CH_3NH—
1-Methylbutyl	$CH_3CH_2CH_2CH(CH_3)$—
Methyl Sulphonyl	CH_3SO_2—
Methylthio	CH_3S—
Nitro	NO_2—
Phenethyl	$C_6H_5CH_2CH_2$—
Phenoxy	C_6H_5O—
2-Phenoxy ethyl	$C_6H_5OCH_2CH_2$—
Phenoxy methyl	$C_6H_5OCH_2$—
Phenyl	C_6H_5—
Phenyl acetamido	$C_6H_5CH_2CONH$—
Propionyl	CH_3CH_2CO—
Propionyloxy	CH_3CH_2COO—
Propoxy	$CH_3CH_2CH_2O$—
n-propyl	$CH_3CH_2CH_2$—
Sulphamoyl	H_2NSO_2—
Sulphanilamido	P—$NH_2C_6H_4SO_2NH$—
Tosyl	P—$CH_3C_6H_4SO_2$—
Trifluoromethyl	F_3C—
Vinyl	CH_2==CH—

Nomenclature of Heterocyclic Systems

Many drugs have heterocyclic systems. The heterocyclic compounds are those organic compounds which have a hetero atom in the ring, i.e., heterocyclic rings contain in addition to carbon one or more atoms of Nitrogen, Oxygen and Sulphur or Phosphorus, Silicon etc. though it is very rare. The heterocyclic system may be a fused or bridged structure forming a ring between carbon atoms known as carbocyclic system. In complex molecules there may be one similar or different heterocyclic molecule. The heterocyclic rings may be having different degrees of unsaturation or it can be fully saturated.

Many drugs have heterocyclic systems and a few examples of such drugs form natural resources are alkaloids (example—morphine, quinine, etc.), xanthines (example—theophylline, caffeine, etc.), antibiotics (example—penicillins, etc).

There are many synthetic heterocyclic drugs. Heterocyclic rings can be three, five or six membered ring. In one and two heterocyclic nuclei the hetero atom is numbered as 1, except in isoquinoline, where the hetero atom is numbered as 2. If there is more than one hetero atoms in the ring the numbering is done in such a direction that hetero atoms get the lowest possible number. Among the hetero atoms, their order of citation starts with the hetero atom of as high a group in the periodic table and as low an atomic number in the group, thus the order of naming will be O, S, N, P, Si, etc.

A saturated three membered ring having a nitrogen is called aziridine. The five and six membered rings are the most common heterocyclic systems present in the drug. Some of the common monocyclic heterocyclic systems found in drugs are as follows:

Pyrrole e.g. Viprynium

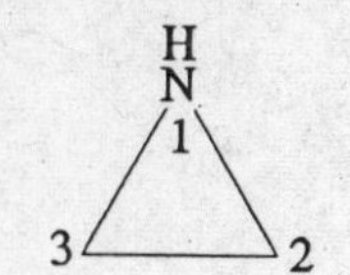

Aziridine e.g. thiotepa

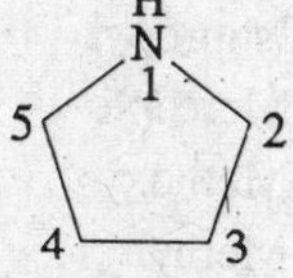

Pyrrolidine e.g., Tripro Lidine Poldine, Pentolinium

3-Pyrazoline e.g., Analgin, Phenazone

Pyrazole e.g, sulphaphenazole

Pyrazolidine e.g, Oxypnenbutazone, Phenylbutazone

Imidazole e.g., Azathioprine, Metroni-dazale, Pilocarpine

Imidazolidine e.g, Clonidine

2-Imidazoline e.g., Antazoline, Naphazoline, Pentolamine

4-Imidazoline e.g., Carbimazole

Furan e.g., Frusemide, Nitrofurantoin

Thiophene e.g., Cephalo-ridine, Cephalothin

Hydantoin e.g., Phenytoin

Isoxazolidine e.g., cycloserine

Isoxazole e.g., Cloxacillin, Sulphafurazole, Sulphamethaxazole

Thiazole e.g., Sulpha-thiazole, Thiobendazole, Thiamine

1, 3, 4 Thiadiazole e.g., Acetazolamide, Sulphamethizole

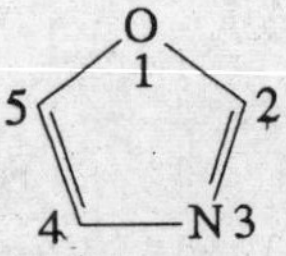

Oxazole

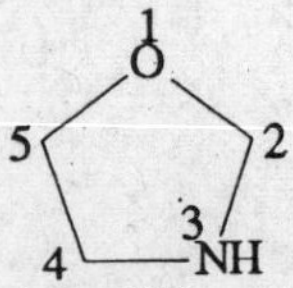

Oxazolidine e.g., Furazolidone Troxidone

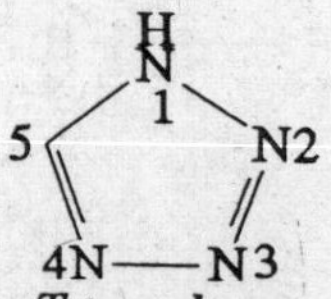

Tetrazole e.g., Pentetrazol

Examples of the six membered heterocyclic ring found in the drugs are as follows:

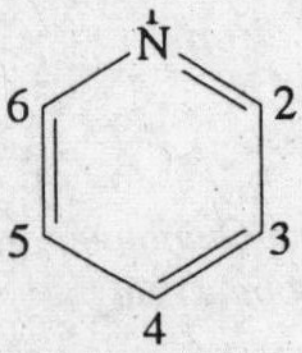

Pyridine e.g., Isoniazid, Nicotinamide, Nikethamide

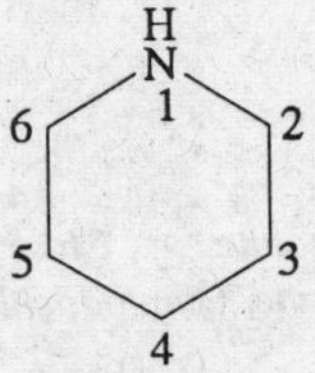

Piperidine e.g., Benzhexol, Haloperidol, Pethidine

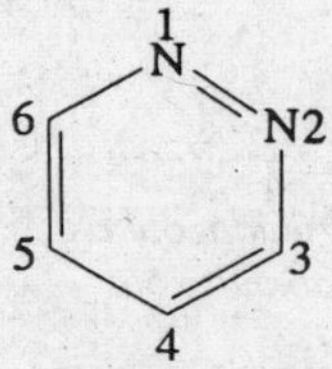

Pyridazine e.g., Sulphamethoxy, Pyridazine

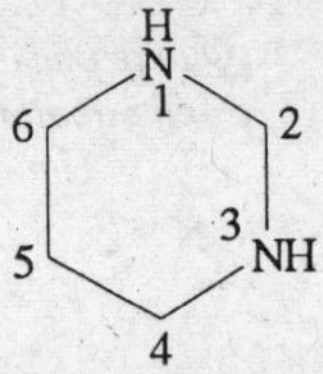

Perhydropyrimidine e.g., Primidone

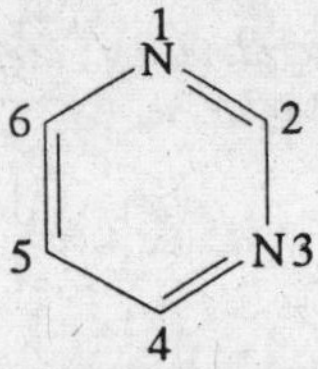

Pyrimidine e.g., Sulphadiazine, Thiamine, Pyrimethamine

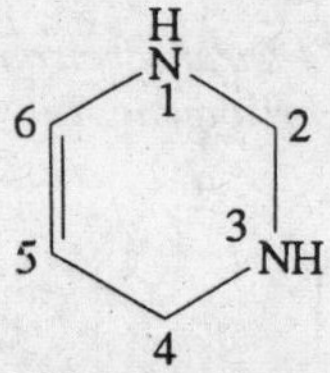

1, 2, 3, 4 tetrahydropyrimidine e.g., Idoxuridine, Propylthiouracil

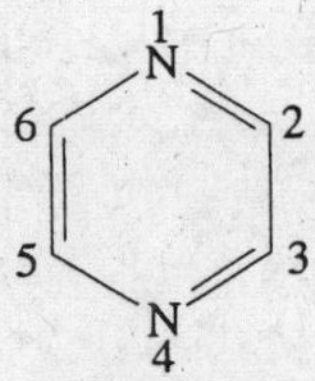

Pyrazine e.g., Amiloride, Pyrazinamide

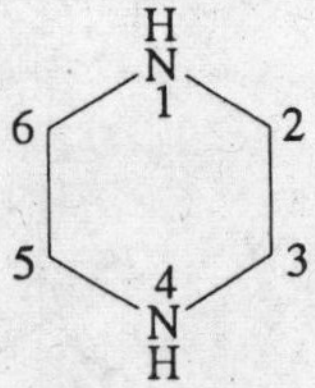

Piperazine e.g., Cyclizine, Diethylcarbamazine, Fluphenazine

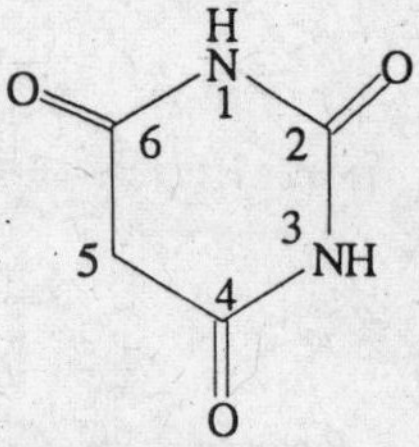

Barbituric Acid e.g., Amylobarbitone, Quinalbarbitone

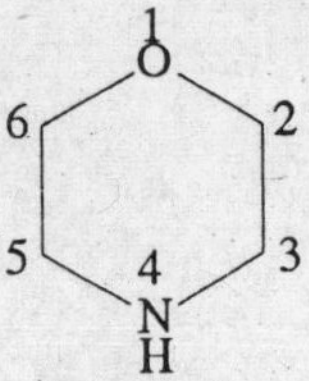

Morpholine

The drugs with fused carboxylic rings are as follows:

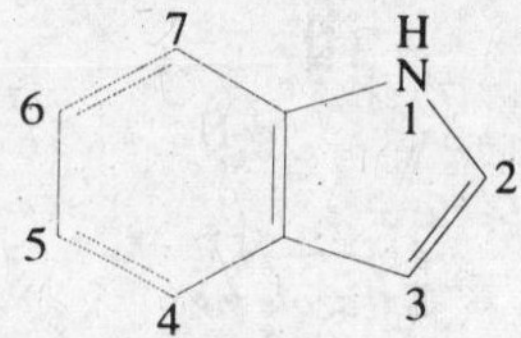

Indole e.g., Indomethacin

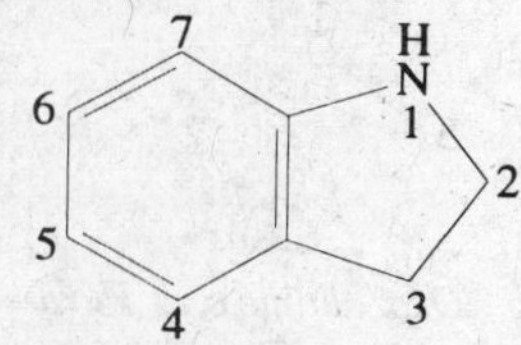

Indoline e.g., Methisazone, Indigo Carmine

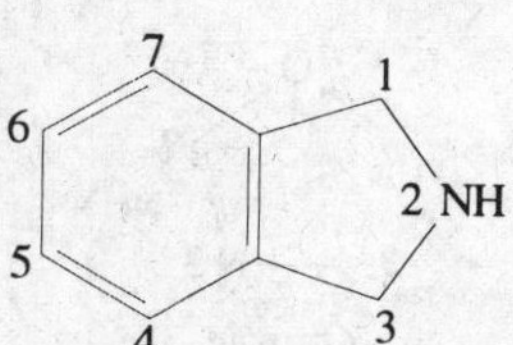

Isoindoline e.g., Chlorthalidone

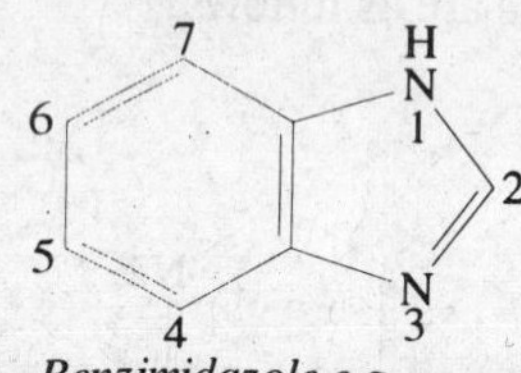

Benzimidazole e.g., Thiabendazole

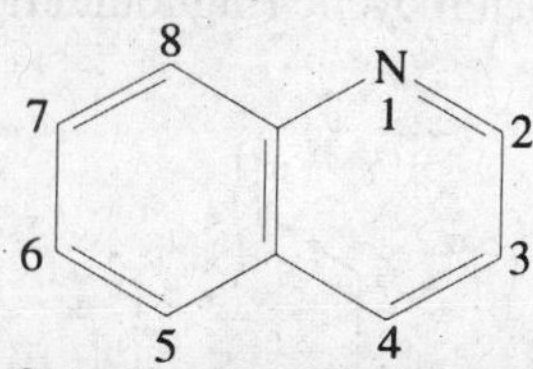

Quinoline e.g., Chloroquine Quinine, Quiniodochlor

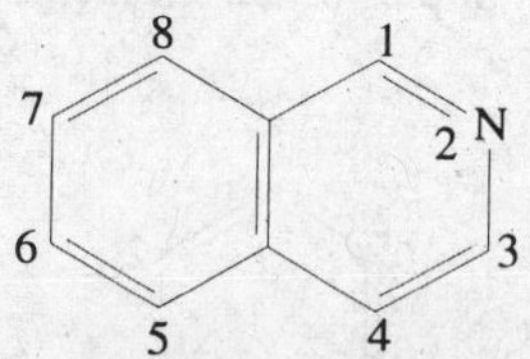

Iso-Quinoline e.g., Papaverine

1, 2, 3, 4 tetrahydroisoquinoline e.g., Debrisoquine, Tubocurarine

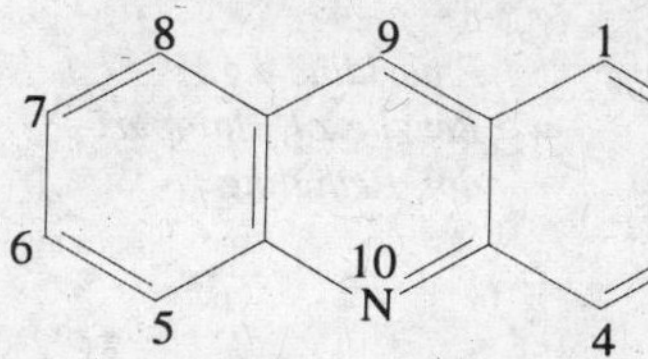

Acridine, e.g., Mepacrine, Proflavine

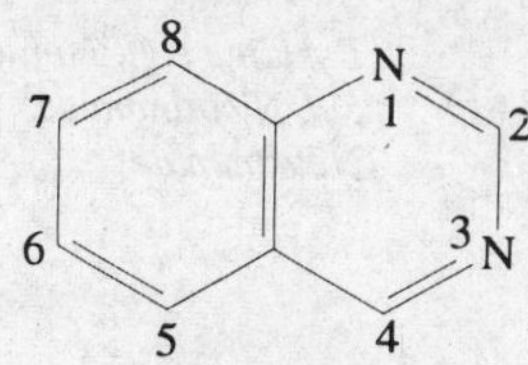

Quinazoline e.g, Methaqualone

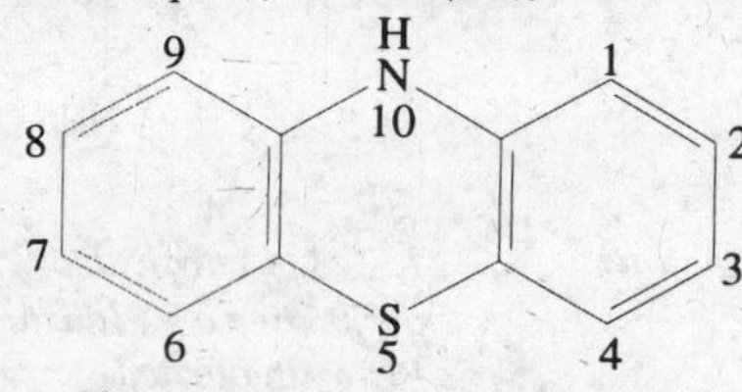

Phenothiazine e.g., Chlorpromazine, Methylene Blue, Promethazine

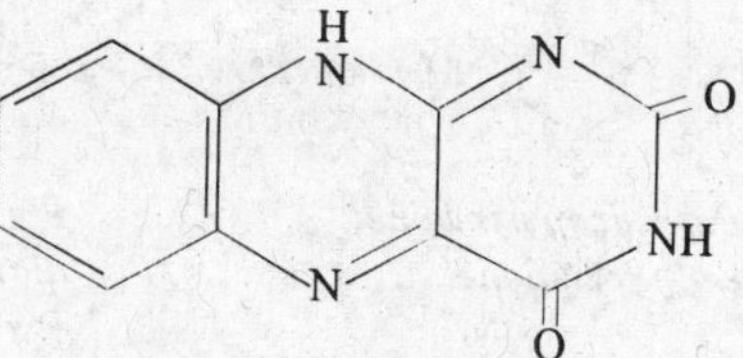

isoalloxizine

There are a few drugs which have two heterocyclic systems fused together..

Purine e.g., Azathioprine, Mercaptopurine

Xanthine e.g, Caffeine, Theobromine

Pteridine e.g., Folic Acid, Methotrexate, Triamterene

Tropane e.g., Atropine, Cocaine, Hyoscine

When one heterocyclic ring is present this is chosen as the parent compound. If more than one heterocyclic ring is present the order of preference is given to the nitrogen containing component. The naming of fused heterocyclic compounds can be explained with the help of an example like Purine which has a bicyclic system. Purine has the following structure, which has both pyrimidine and imidazole heterocycles fused together.

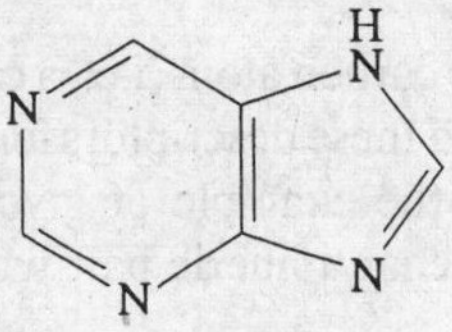

Purine 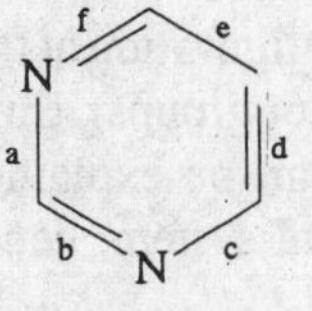 *Pyrimidine* 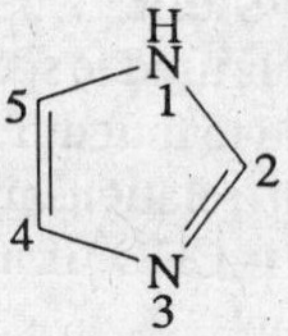 *Imidazole*

Among the two hetero systems, the pyrimidine is considered as base component due to its larger size. Purine is described as imidazopyrimidine and not as pyrimidoimidazole. But this description is not complete unless the positions of fusion is mentioned. The fusion of imidazole is from peripheral side 4,5. The peripheral sides of pyrimidine are distinguished by distinguishing the sides as a (for 1,2) b (for 2-3), c (for 3-4), d (for 4-5), e (for 5-6), f (for 6-7) and so on. It is evident from the structure of Purine that 'd' side of pyrimidine is involved in fusion with imidazole to form Purine. The systematic name for purine is imidazo (4,5 d) Pyrimidine. There is a possibility of different isomeric forms, e.g., Purine can have two isomeric forms I and II differing in the position of H atom.

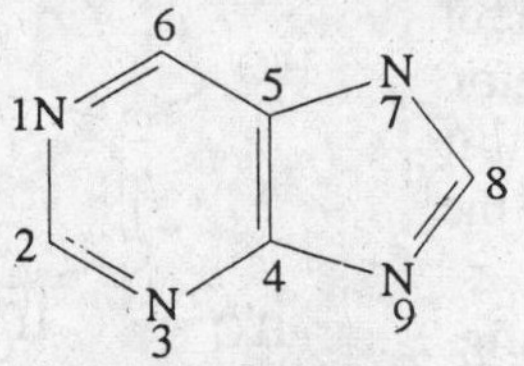

(I) 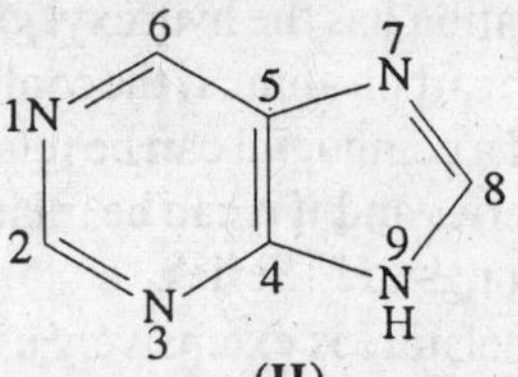 (II)

The systematic name of purine is further modified by mentioning the position of H atom; such symbols are written in the beginning. Hence, the structure (I) can be named as 7H-imidazo (4,5 d) pyrimidine and (II) as 9H-imidazo (4,5 d) pyrimidine.

If the two fused heterocyclic systems are of same size that is same numbered ring, then the ring having greater variety of hetero atom is considered as the base e.g., Levamisole, an anthelmintic has the 2,3,5,6 tetrahydroimidazo (2, 1-b) thiazole bicyclic system. Imidazole and thiazole have five membered rings and each possesses two hetero atoms. But thiazole having greater variety of hetero atoms is considered as the base and the drug is named after it.

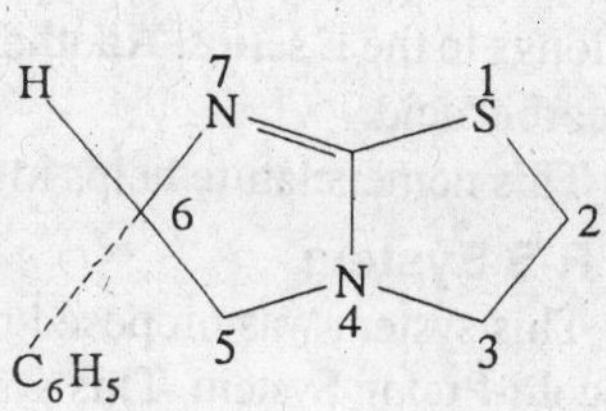

Levamisole

If the fused heterocyclic system has same size, and same kind of hetero atoms then the component or the heterocyclic system having lower numbers for the hetero atoms comes before fusion. For example, pteridine which has the adjoining structure.

It is named as pyrazine (2, 3d) pyrimidine.

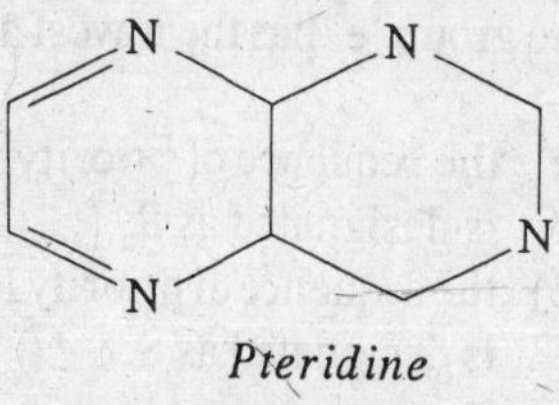

Pteridine

Stereochemical Nomenclature

Stereochemical nomenclature is used to define the stereo isomerism of a chemical compound, i.e., when there is isomerism due to the different arrangements of atoms or groups in space. The following methods are followed to explain the configuration of the isomers and these systems are followed for naming these isomers.

1. D and L System

This system is used to specify the configuration at the asymmetric carbon atom (i.e., a carbon atom which is bonded to four different atoms or groups). Emil Fischer used these descriptors for indicating configurational relationships which can be explained by taking the example of glyceraldehyde $HOCH_2CHOH—CHO$. It has one chiral system, hence it has one enantiomeric pair which can be written as follows:

```
        CHO                      CHO
         |                        |
   H —— C —— OH            HO —— C —— H
         |                        |
       CH2OH                    CH2OH
```

D (+) Glyceraldehyde *L (–) Glyceraldehyde*

Among these one component is dextrorotatory (+) whereas the other is laevorotatory (–). When the —OH or hydroxyl group is on the right hand side of the asymmetric carbon of the glyceraldehyde molecule, it is represented by the plannar projection formula as shown in the above diagram. It is known as D-configuation whereas the L-configuration has the hydroxyl group on the left hand side of the asymmetric carbon atom. If the configuration at the asymmetric carbon atom of a compound can be related to D-glyceraldehyde, it belongs to D series and if it can be related to L-glyceraldehyde the compound belongs to L-Series.

This nomenclature is extensively used to designate the configurations in carbohydrate and amino acids.

For example, natural alanine, an aminoacid, is related to L(+) lactic acid which is related to L-glyceraldehyde, hence alanine belongs to the L series. Another example is Vitamin-C which is L ascorbic acid.

```
                    O
                    ||
     HO            C
        \       /     \
          C              O
          ||            /
          C       C
        /    \  /
     HO       H C
                |
       HO  ——  C  —— H
                |
              CH2OH
```

L ascorbic acid

This nomenclature helps to define the stereochemical nature of the drug.

2. R-S System

This system was proposed by R.S. Cahn, C.K. Ingold, V. Prelog and it is also known as Cahn-Ingold-Prelog System. This system is based on the three dimensional or tetrahedral structure of the compound. In order to specify configuration about an asymmetric carbon C abde, the groups abde are first assigned in order of priority, determined by sequence rules is assigned. Thus the order of priority may be stated as

$$a > b > d > e$$

Here group 'e' has the lowest and the group 'a' has the highest priority. The conversion rule says that if—

(a) the sequence of priority follows clockwise direction or right hand direction, the configuration is designated as R.

(b) the sequence of priority follows left hand direction or anticlockwise direction, the configuration is designated as S.

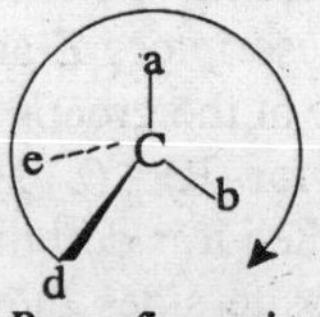

R- configuration
Clockwise direction

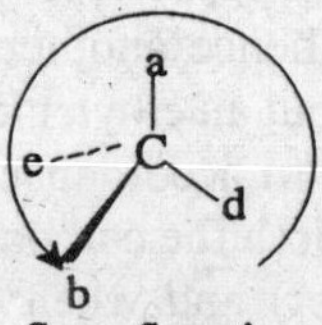

S-configurtion
Anticlockwise direction

The sequence rules or Rules of priority are as follows:

1. The atoms or groups bonded to the asymmetric carbon are arranged in order of decreasing atomic number, e.g., chlorobromofluoromethane (CHClBrF) where the order of priority is

   Br > Cl > F > H

   (At. No 35) (At. No 17) (At. No 9) (At. No 1)

2. When two or more groups have identical first atoms which are attached to an asymmetric carbon, the priority order is determined by considering the atomic number of the second atom and if the second atom is also identical, the third atom is considered.

   For example:

Methyl	Ethyl	N-propyl
1 2	1 2 3	1 2 3
$-CH_2-H$	$-CH_2-CH_2-H$	$-CH_2-CH_2-CH_3$

   In methyl and ethyl the first atom is carbon and hence atomic number of the second atom is considered in methyl and ethyl, which is H (At.no.1) and C (At.no.6) respectively. Hence ethyl > methyl and priority sequence is ethyl > methyl. Between ethyl and n-propyl the second atom is also identical, hence the third atom is considered which is H in ethyl and C in propyl and hence n-propyl > ethyl and n-propyl is given greater priority than ethyl.

3. If the first atoms of the two groups have some substituents with higher atomic number, the one with more substituents gets the priority. For example, $CHCl_2$ gets a higher priority than $CHCl_2$ as $CHCl_2$ has two chlorine atoms i.e., it has more number of the substituents with higher atomic number.

4. A doubly or triply bonded atom present in a group attached to an asymmetric carbon is considered equivalent to two or three singly bonded groups respectively.

   For example:

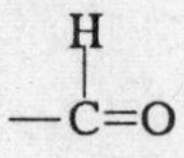

   —C(H)=O equals to —C(H)(O)(O) (O, O, H)

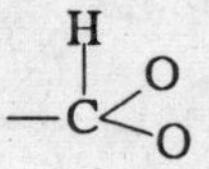

   —C≡N equals to —C(N)(N)(N) (N, N, N)

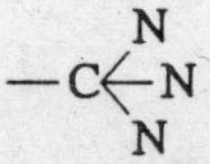

The (R) - (S) system is used for specifying the configuration in chemical names of drugs, *e.g.*, adrenaline. It is an adrenergic drug and is named as (R)-1-(3,4 dihydroxy phenyl)-2-methyl amino ethanol.

(Z) - (E) Isomerism

The *cis* and *trans* prefixes are commonly used to specify the configuration of geometric isomerism; i.e., which is shown by alkenes or their derivatives in which two different atoms or groups are attached to each carbon containing double bond. If the similar groups are on the same side, it is known as cis-isomer and if they are opposite to each other, it is known as trans-isomer. Sometimes, these relations are confusing and hence another system is introduced with descriptors Z and E. In specifying configuration of about an ethylenic double bond, first precedence of the groups attached to a double bond is determined in accordance to the sequence rule devised for (R) - (S) system. If the higher priority groups on both the carbon atoms are on the same side, then it is designated as Z (German; Zusammen, together) and when these groups are on the opposite sides they are designated as E (German) entegen opposite). These descriptors Z and E are used for designating geometrical isomerism in the chemical names of drugs.

Chapter 3

Mechanism of Drug Action and Metabolism of Drugs

The action of a drug is the process by which the drug brings about a change in some pre-existing physiological function or biochemical process of the living organism. The effects produced by a drug can be defermined and expressed only in terms of an alteration of some known function or process that maintains the existence of the organism. Any alteration that is brought about by the action of a drug may be one that either returns a function or process to normal operating levels or changes a function or process in a direction away from the normal levels. A particular effect observed as a change in a definite physiological process may be brought about in several ways. The action of the drug may be specific i.e., aimed directly as the agent responsible for the disease or it may be nonspecific i.e., improving a symptom of the disease, such as fever without getting to the basis of the disorder that can be controlled. The effects of many drugs or their actions can be characterised and identified only in terms of the knowledge of their physiological function and processes.

The part of the body namely organ, tissue or cell where a drug acts to initiate the chain of events leading to an effect is known as site of action of the drug. For example, the site of action of quinine, a drug used for malaria is the plasmodium, the cell invading the host whereas in non-malarial fever the drugs act at the discrete areas of the brain involved in the regulation of body temperature. The site of action and the effect of a drug both are mutually dependent on each other. For example, the effects of drugs on the pupil of the eye, the muscles of the iris control the size of the pupillary opening in response to light i.e., in bright light the opening is made smaller and in dim light it is larger.

Some drugs like atropine when applied to the external surface of the eye ball, cause the pupil to become larger than it would be normally under the same light conditions. The opthalmologists use atropine or similar agents to enlarge the pupil which helps them to examine the eye.

On the other hand solutions of morphine when applied externally to the eye ball, there is no change in the size of the pupil, but when morphine is given internally it produces marked pupillary constriction; in fact, this effect is so characteristic that the presence of small or "pinpoint pupils" leads to the presumption that an individual has taken morphine. The site of action of morphine is within the brain; thus, when morphine is administered internally, it reaches this site of action within the central nervous system.

Name of the drug	*The site of action*	*Effect*
Atropine	Pupil of eye	Dilated or enlarged pupil
Morphine	Brain	Constricted or pinpoint pupil

Fig. 1. *Schematic Representation: Site of action of morphine and atropine showing their effects on pupillary diameter.*

Generalising about the site of drug action is easy and conclusion can be arrived at by process of elimination, but the precise determination of the specific site and the mechanism of action of the drug is very difficult and to determine the site at which a drug acts to produce a specific effect requires detailed experimental studies. Fore example, it is easier to show that the iris of the eye is not the site at which morphine acts to constrict the pupil. But to determine the site where morphine acts requires extensive investigation to find out in which manner individual areas of the brain influence physiologic functions.

Even after determining the site of action of a drug to a certain tissue or organs it does not give sufficient idea about the effects of different drugs since different drugs, producing same or different effects may have different sites of action within the same tissue or organ. For example, caffeine, histamine, nitrites like nitroglycerine, these all act on the muscle of blood vessels to increase their diameter whereas cocaine produces constriction. Hence it is necessary to study the separate components of the tissue or organ and it has to be determined that which of them is the specialised, functional element on which the drug acts. These studies help to show that the site of action is a particular kind of cell within an organ and that the drug acts either at the surface of this cell or within the cell itself or the site of action can be extracellular, i.e., site of action can be outside the cell. The site of the action of each drug is very specific.

Mechanism of Action

When the site of action of a drug is determined, which can be either at the surface of the cell, extracellular or intracellular, the next step is to find out how the drug acts, i.e., the means by which a drug in the immediate vicinity of its site of action initiates the chain of events and which can be measured as an effect is known as its mechanism of action. The mechanism of action of most drugs involves a chemical interaction between the drug and a functionally important component of the living system. A drug may act:

(1) Extracellularly. Example: Osmotic diuretics, plasma expanders.
(2) On the cell surface. Example: digitalis, penicillin, insulin.
(3) Inside the cell. Example: antimalignancy compounds and steroid hormones.

1. Actions at extracellular site

The drug reactions occur outside the cell. One of the simplest example is the neutralisation of excessive gastric acid by antacid drugs. In this reaction a base, such as sodium bi-carbonate reacts chemically with the hydrochloric acid of the stomach and neupralizes the acid by formation of new products, i.e., salt, water and carbon dioxide.

$$\underset{}{NaHCO_3} + HCl \longrightarrow \underset{\text{(Salt Sodium Chloride)}}{NaCl} + \underset{\text{Water}}{H_2O} + \underset{\text{Carbondioxide}}{CO_2}$$

Another example is the specific antidote, Calcium disodium edetate used in the treatment of lead poisoning. This drug removes lead that is free in blood and tissue by a reaction in which the calcium of the drug is displaced by the lead. This drug has much more affinity for lead than calcium, hence this drug forms an inert complex with lead and to excreted it from the body as the soluble lead edetate complex.

The study of these examples makes it clear that the action of drug at the extracellular sites is due to a chemical reaction. These chemical reactions can be easily explained as drug involved and the other reactants are specifically identifiable. The mechanism of action of drug at extracellular sites can be readily explained since the drug involved and the functionally important molecule of the living system is easily identified.

2. Action at cellular sites

Most functions of living organisms do not take place extracellularly but they take place cellularly i.e., most drugs combine with some molecular structure on the surface of or within the cell. This molecular structure is called a receptor. The combination of the drug with the receptor results in a molecular change in the receptor, like altered configuration or charge distribution and thus triggers a chain of events leading to a response.

Drug Receptors and Biological Responses

Receptors are actually macromolecules present in tissues that combine chemically with the drug. Most drugs have considerable degree of selectivity or specificity in their actions and hence the receptors with which they interact must be equally unique, i.e., the drug acts at some sites to produce its characteristic biologic effects whereas its presence at other cells, tissues or organs have no biologic response.

The receptors are highly specialized macromolecules which recognize the drugs which are highly specific and selective in their action and trigger a chain of events which ultimately result in biologic response; hence it suggests that there are some specific physicochemical properties of the biologic molecules at the site where the responses are produced by drug interaction and these specific properties are missing at other sites. There is also a definite relationship between the chemical structure of a drug and the biological effect it produces at a particular site. Even a very slight modification in the chemical structure of the drug may yield a compound which is no longer effective at the site as the parent compound. Sometimes it is found that there are changes in the effect produced at the site, although there is no change in the chemical formula of the drug but there are only changes in the way, a few atoms are arranged in the space or in the three dimensional structure when the compounds are stereoisomers.

A drug can produce an effect only when it fits with its receptor as a special key fits a well designed lock as shown in the Fig. 2.

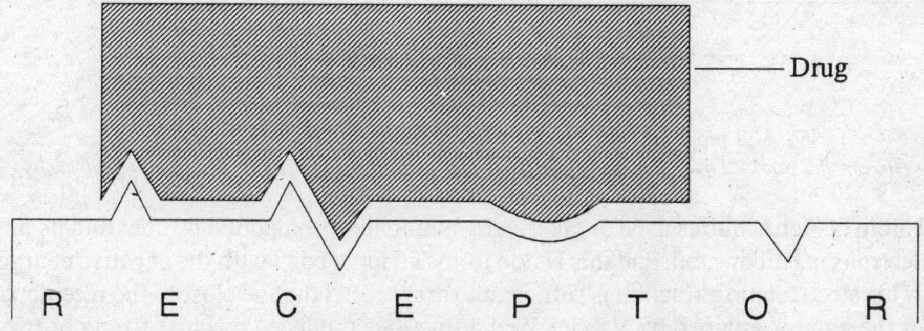

Fig. 2. *Schematic drawing of receptor site and substrate.*
(Source: "Pharmaceutical Dosage Forms and Drug Delivery Systems" Howard. C. Ansel, Nicholas G. Popovich, University of Georgia Purdue University, School of Pharmacy & Pharmacal Sciences: Lea & Febiger Williams and Wilkinson Phildelphia Baltinove London, 1995).

The drug receptor interaction can be explained through a specific example. The end plate region of a skeletal muscle fibre which contains a large number of receptors which have a high affinity for the transmitter acetylcholine (ACh). Each of these receptors, known as nicotinic receptors, are part of a post synaptic membrane (membrane pertaining to the area on the distal side of a synaptic cleft) that controls the inward movement of sodium ions. At rest the post synaptic membrane is relatively impermeable to sodium, and stimulation of the nerve leading to the muscle results in the release of acetylcholine from the nerve fibre in the region of the end plate. The acetylcholine combines with the receptors and changes them so that channels are opened and sodium flows inward. The more acetylcholine there is in the end plate region, the more the receptors that are occupied and the more channels are opened. When a large number of channels are open then after some time they reach a critical number when sodium which enters rapidly and disturbs the ionic balance of the membrane and it depolarises, *i.e.*, an action potential develops. Due to depolarization there is release of calcium from intracellular binding sites. This calcium then interacts with the contractile proteins resulting in shortening of the muscle cell. The whole process can be represented as follows:

$$\text{Ach + receptor} \longrightarrow \text{Na}^+ \text{ influx} - \text{action potential} \longrightarrow$$

$$\longrightarrow \text{increased free Ca}^{2+} \longrightarrow \text{contraction}$$

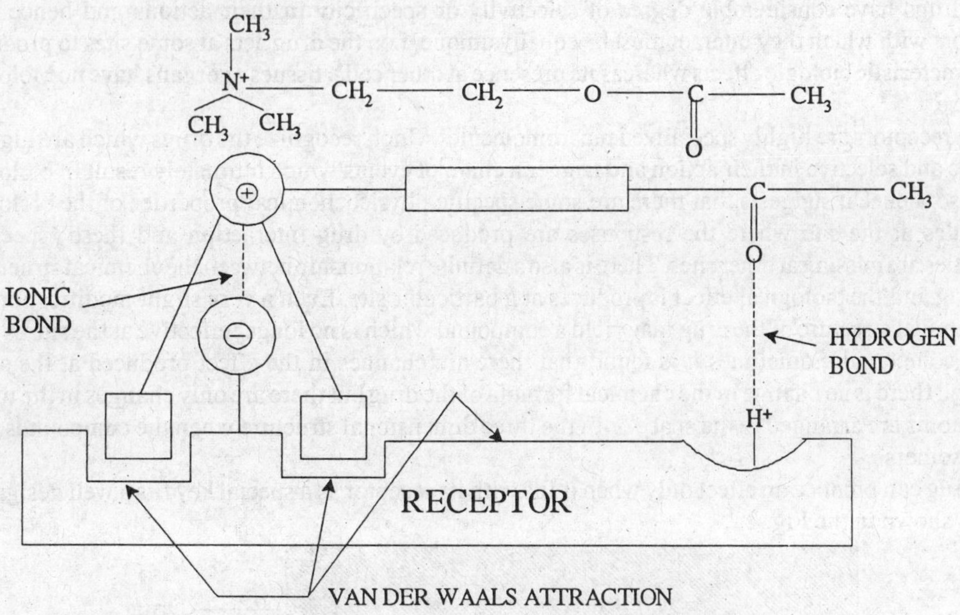

Fig. 3. *A model for the combination of acetylcholine with a receptor.*
(Source : From the book "Drug Actions and Reactions" Ruth R. Levine Ph. D. Boston University, Little Brown Co. London, Boston, Toranto, 1990)

The N atom of Ach acquires a N^+ or quartenary ammonium compound by donating its unshared pairs of electrons to carbon atom and this N^+ ion forms an ionic bond with the negative charge of the receptor. This electrostatic attraction is sufficient to draw acetylcholine close to the receptor and the stability of the bond is attained by VanderWaal attractions produced by close fitting of two of the methyl groups into the cavity in which the charged site of the receptor is embedded. Another hydrogen bond with oxygen draws the other end of the molecule close to the receptor and further increases the stability of the entire acetylcholine receptor combined.

The sequence of events by which acetylcholine brings about muscle contraction through receptors, the same way drugs can be designed to interact. A chemical with a molecular structure very similar to that of acetylcholine might be attracted to the receptor in place of acetylcholine and changes the receptor's molecular orientation in a manner which is similar to that of acetylcholine and this initiates the chain of events that causes muscle contraction. Thus, the drug shows the same actions as that of acetylcholine at the motor end plate. Examples of the drugs that have such an effect are nicotine and carbomylcholine. Chemicals that interact with a receptor and thereby initiate a cellular reaction are known as *agonists.* Thus acetylcholine as well as the drugs, nicotine and carbamylcholine are agonists for the receptors in the skeletal muscle end plate. On the other hand if a chemical is less similar to acetylcholine it may interact with the receptor but it is uanble to induce the exact molecular change necessary to allow the inflow of sodium. This chemical does not cause contraction at the same time, because it occupies the receptor site and inhibits the interaction of acetylcholine with its receptor. Such a drug is termed an *antagonist.* Example of such a compound is d-tubocurarine; this drug is antagonist of acetylcholine at the end plate receptors. Hence d-tubocurarine results in muscle relaxation.

The specificity and selectivity of drug receptor interaction is due to the number and types of bonds formed and also on the spatial configuration of the sites for bond formation at the surface of the receptor.

For example, L-epinephrine and D-epinephrine are optical stereo isomers (*i.e.* compounds with same molecular formula, same physical and chemical properties, but differ from each other in their optical properties.)

These two structures cannot be superimposed but are mirror images of each other. Though L-epinephrine and D-epinephrine are identical molecules, their biologic effect is different from each other as these are optical isomers.

L-Epinephrine D-Epinephrine

The difference in their biologic effects is due to the fact that the receptor surface can distinguish between the two optical isomers. Both L and D epinephrine are bound to the receptor by hydrogen bonding of two hydroxyl groups (OH) at one end of the molecule and by ionic bond formation at the other end where nitrogen becomes positively charged by coordinate covalent bonding with hydrogen ion. The entire spatial arrangement of all the atoms in L epinephrine has a three point attachment to the receptor and in D-epinephrine it is two point attachment as shown in the Fig. 4.

The three point attachment has maximum biologic effect.

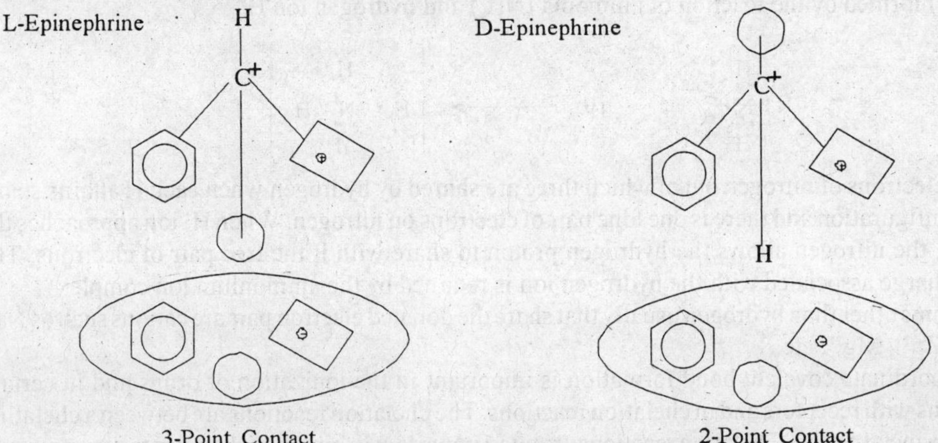

Fig 4 : *Combination of epinephrine with its hypothetical receptor.*

(Source "The Drug Actions and Reactions Ruth R. Levine Ph. D. Boston University Little Brown & Co. Boston, Toront, London, 1990).

The positively charged Nitrogen forms an ionic bond with an anionic portion of the receptor site, the hydrogen of the hydrosyl group attached to the asymmetric carbon forms hydrogen bonding with an atom of the receptor site which can donate a pair of electrons. In L-epinephrine the spatial arrangement of the hydrogen atom and the three radicals bound to the asymmetric carbon (*) permit a three point attachment with the complementary receptor.

The Chemistry of Drug Receptor Binding

Many biological receptors are composed of proteins, nucleic acids or other macromolecules. These macromolecules combine with drugs in a number of ways and the forces by which a receptor combines with drugs are fairly strong and long lasting so that the sequence of biological response could be completed. These forces are known as chemical bonds and a number of different types of bonds participate in the formation of the initial drug-receptor complex.

1. Covalent Bond

The bond formed by sharing a pair of electrons is called a covalent bond. It has a bond energy of 100 Kcal-mole and therefore it is a strong and stable bond and is responsible for the stability of most organic molecules. The example of covalent bond can be seen in methane molecule whose electronic structure is as shown below:

$$\cdot\dot{\underset{\cdot}{C}}\cdot \quad + \quad 4H^{*} \longrightarrow H^{*}\,\cdot C \cdot{*}\,H \ \text{(with H above and below)}$$

Carbon — Methane

· = Outermost electron of carbon

* = Outermost electron of hydrogen

Since covalent bonds are stable at physiological temperatures it results in the formation of a long lasting complex and it can be broken only if sufficient energy is added or by a catalytic agent such as enzyme. Covalent bond formation is found in antineoplastics or antibiotics where long lasting inhibition of cell replication is required.

If both the shared electrons that form the covalent bond are derived from only one of the two atoms, the bond is called a co-ordinate covalent bond. The atom contributing the electron pair is called the donor atom and in biologic systems it is usually nitrogen, oxygen or sulphur. Example—ammonium ion (NH_4^+) formed by the reaction of ammonia (NH_3) and hydrogen ion H^+ :

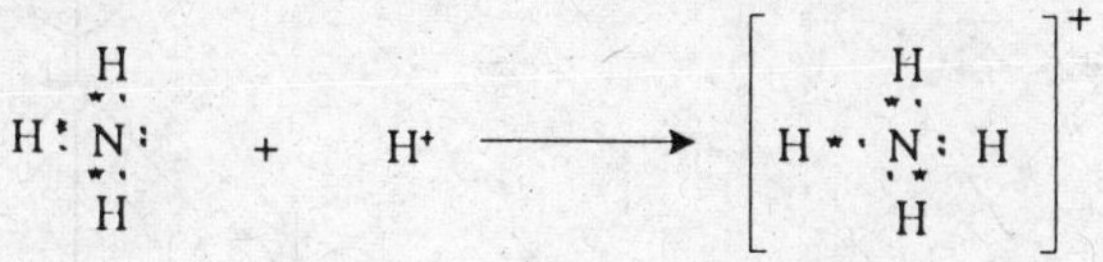

$$H{:}\overset{H}{\underset{H}{N}}{:} \quad + \quad H^{+} \longrightarrow \left[H \cdot\cdot \overset{H}{\underset{H}{N}} : H \right]^{+}$$

(• = 5 electrons of nitrogen out of which three are shared by hydrogen when each H attains stable helium configuration and there is one lone pair of electrons on nitrogen. When H^+ ion approaches the ammonia, the nitrogen allows the hydrogen proton to share with it the free pair of electrons. The positive charge associated with the hydrogen ion is retained by the ammonium ion complex).

The atoms other than hydrogen usually that share the donated electron pair are cations such as Na^+, K^+, Cu^{2+}, Zn^{2+}, Mg^{2+}, etc.

This coordinate covalent bond formation is important in the ionization of drugs and in certain interactions with receptors and in chelation reactions. The chelation reactions are between a chelating agent and a metal cation. Chelation reactions are often important in binding substrates to their enzymes and they form the basis for a number of pharmacological and toxicological effects produced by the administration of metal containing or metal chelating drugs.

2. The Hydrogen Bond

The hydrogen atom is strongly electro positive when it lies between two atoms having strong electronegativities like oxygen, nitrogen or fluorine, shows a unique property to form a bond or bridge between them, holding one by convalent bond and the other by purely electrostatic forces, known as hydrogen bond. For example, water, hydrogen fluoride, ammonia etc. whose structures are as follows:

H—F---H—F   H—O-----H—O (each O also bonded to H)   H—N-----H—N (each N also bonded to two H)

Hydrogenfluoride   Water   Ammonia

Hydrogen bonding between two atoms is generally indicated by dotted lines. The hydrogen bonding does not stop at two molecules but may extend throughout the entire mass.

H H O --- H O --- H H --- O H H ; H H O

This association between molecules of water in the liquid state makes it more resistant to disruption by heat. Hence, water has a higher boiling point than organic solvents such as chloroform ($CHCl_3$) which cannot form hydrogen bond, as carbon is not a strong electronegative element.

van der Waals Forces

These are very weak attractive forces between any two neutral atoms or atomic groups, these can operate only at short distances. These forces decrease rapidly with a slight increase in interatomic distance.

The formation of these different types of bonds gives sufficient stability to initiate the events which lead to the pharmacologic effect. These reactions are mostly reversible and the specificity and selectivity of drug receptor interactions are due to the number and types of bonds formed and also from the spatial configuration of the sites for bond formation at the surface of the receptor.

The majority of the drugs produce pharmacologic effects by mechanisms involving a drug receptor interaction. However, certain drugs can produce their characteristic effects without combining directly with a receptor and an example, e.g., volatile general anaesthetic agents.

These actions may occur at cellular sites and may involve macromolecular tissue components but the biological effects produced are non-specific, such as germicides. They are used to kill bacteria, are non-specific and show mechanism of action without involving receptors. Other examples are detergents, oxidizing agents such as hydrogen peroxide and phenol derivaties like lysol, all these agents act by destroying cellular membranes or constituents such as nucleic acid and proteins of the bacteria or germs, irreversibly.

Synopsis of the Mechanism of Action of Drugs

The mechanism of drug action is summarised as follows:

The part of the body in which the drug acts to initiate the chain of events leading to the response or the effect is known as the site of action. It may be close to the affected organ or it may be at a distance from the organ that ultimately responds. The site of the action may be at the surface of the cell, inside the cell or it can be extracellular.

The manner in which a drug acts at its site of action and produces its effect in the vicinity is known as its mechanism of action.

The drugs act by involving a chemical interaction between the drug and a functionally important component of the living system. When the drugs act at extracellular sites the living component can be identified as a distinct entity and the mechanism of action can be easily explained. Majority of drugs do not act extracellularly and their mechanism is explained with the help of the receptors.

The receptors are macromolecular tissue constituents with which the drugs combine reversibly by means of ionic bonds, hydrogen bonds and Vander Waals forces. These different types of bonds give the drug receptor complex sufficient stability to initiate the functions that ultimately lead to pharmacologic effects. The drug receptor interactions are very specific and selective and are formed from the number and types of bonds formed, and also from the special configuration of the bond formation at the surface of the receptor.

So far the drug effects have been discussed in qualitative nature to describe that drugs either stimulate or depress to indicate the magnitude of responses observed.

Given below are the 'time-response' and dose-response relationships.

Time-Response Relationships. This relationship involves three basic parameters:

(*i*) *Onset action* is the length of time of a drug to produce desirable effect.

(*ii*) *Maximal effect* is the time required for a drug to exert its maximum effect.

(*iii*) *Duration of action* is the period of time a drug continues into response after a single dose.

Dose-Response Relationships. It involves the parameters of potency, therapeutic index and maximum effect.

Potency. It is the absolute amount of a drug required to produce a given pharmacological effect.

Therapeutic index. It refers to the relative margin of safety a drug possesses. Therapeutic index $= LD_{50}/ED_{50}$. LD = lethal dose, ED = effective dose capable of producing a specific effect in 50% of the subjects. The therapeutic index must be greater than 1 if the drug is to have any clinical utility. The greater the ratio, the safer the drug.

Maximum effect. It is the greatest response produced by a drug regardless of the dose employed.

The sites of drug action, the mechanisms of their actions and their effects are collectively called 'Pharmacodynamics' and the ultimate practical application is termed therapeutics.

Mechanism of Different Types of Drug Action

Various drugs produce their effects by different processes such as: (*i*) by depression, (*ii*) by stimulation, (*iii*) as replacement, (*iv*) as anti-infective agents, (*v*) as metabolite antagonist, (*vi*) by irritation, (*vii*) by modification of immune system.

These drugs mainly produce their action only quantitatively and do not produce any qualitative change in the function of the affected organ.

(*i*) Depression

The action of these drugs is to bring depression of some specialized cells and decrease their activity. Examples are barbiturates which depress the central nervous system and are used as sedatives, hypnotics, as anticonvulsants and analgesic to reduce anxiety, pain and also to control convulsions in epilepsy.

Another example is Quindine which depresses the myocardium it is used in cardiac Arrhythmias.

(*ii*) By stimulation

These drugs stimulate the activity of specialised cells and thus increase their activity. Excessive stimulation produces changes in the protoplasm of the cell which may lead to depression of certain portions. Thus these drugs stimulate certain portion and depress other system., *e.g.*, Morphine which stimulates the vagus nerve situated at medulla oblongata and chemoreceptor trigger zone, but depresses the vomiting and the cough centres.

(*iii*) Replacement

These drugs are used as replacement when the production of endogenous substances is reduced. These drugs are of immense importance in the treatment of diseases of hormonal deficiencies. Example—insulin in diabetes mellitus, hypocortisone in the treatment of Addison's disease etc.

(*iv*) As anti-infective agents

These drugs are used for prevention and treatment of infection. They either stop the growth of the micro-organisms or ultimately destroy them. Examples of such drugs are antibiotics like pencillin, tetracycline etc. Antibiotics are chemical substances produced by living cells which are capable of inhibiting or destroying other micro-organism.

(*v*) As metabolite antagonists

These drugs are bacteriostatic in action and act as metabolite antagonists *i.e.*, it interfere with the metabolism of the bacterial cells and thus inhibit their growth. Examples are sulpha drugs (Sulphonamide). These drugs structurally resemble p-aminobenzoic acid (PABA) which is a normal cellular constituent or metabolite of the bacteria. Bacteria take up sulphonamides due to their structural similarity to that of PABA. The sulphonamide cannot perform the essential action of PABA metabolite and hence the growth and reproduction of bacteria ceases.

The inhibition or antagonism of essential metabolite by structurally similar compounds is known as metabolite-antagonism and the inhibitor is known as metabolite-antagonist or antimetabolite.

In few cases the antagonists do not resemble the corresponding metabolite structurally. Examples are as follows:

Histamines and Antihistamines

Antihistamines antagonise an essential amino acid known as histamine, but these are not structurally similar compounds.

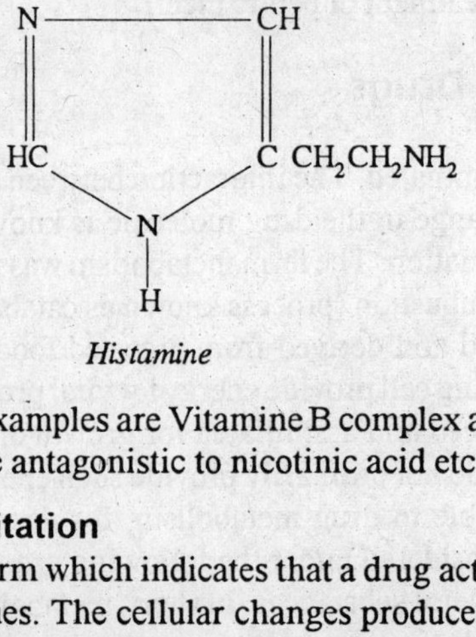

N — CH
HC — C $CH_2CH_2NH_2$
N
H

Histamine

CH_2 — C_6H_5
N
$CH_2CH_2\ NR_3$

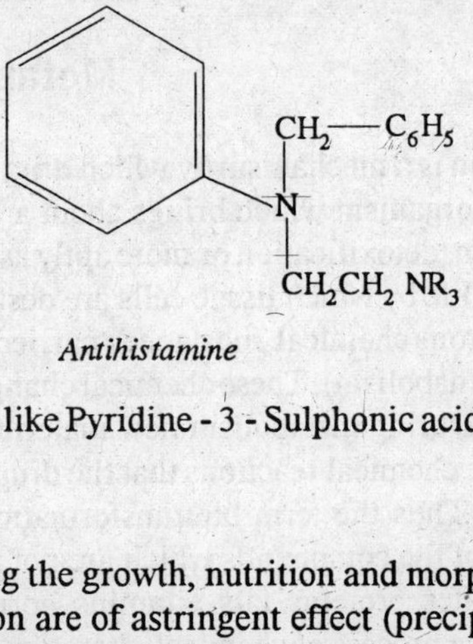

Antihistamine

Other examples are Vitamine B complex anatagonists like Pyridine - 3 - Sulphonic acid, which is found to be antagonistic to nicotinic acid etc.

(*vi*) By irritation

It is a term which indicates that a drug acts by affecting the growth, nutrition and morphology of living tissues. The cellular changes produced by irritation are of astringent effect (precipitation of protein), corrosive effect (damage to the tissue e.g. acids or alkalis), dehydration, enzyme neutralisation, cytotoxic action (damage to the cell wall). Heavy metals like mercury, silver are irritants.

Mercuric oxide is used as eye ointment in blepharitis (spasm of eyelids). Other examples are volatile oils like Turpentine oil which is used as irritant.

(*vii*) Modification of immune status

When the infected organisms enter the body, they start manufacturing toxic antigens in the body. The human body's biological defences start producing antibiotics, whih destroy these antigens.

Immunological agents produce active or passive immunity which are used to prevent or modify certain infectious diseases. These agents are known as vaccines and when these are administered in the body, it stimulates the body for manufacturing antibiotics in defence against the disease. Some infectious diseases like cholera, smallpox, typhoid, diphtheria, tetanus etc., have been prevented successfully by their respective vaccines.

A drug acts by viurte of its physical as well as chemical properties.

Physical Properties

(*a*) *Colour*: A pleasant colour may produce a psychological effect, *e.g.* tincture of cardamom.

(*b*) *Surface active agents*: Cationic surfactants which reduce the surface tension and thus damage the cell membrane of the bacteria *e.g.* cationic surfactant like cetrimide used as detergent and antiseptic for wound.

(*c*) *Electric charge*: *e.g.*, heparin which is anticoagulant and produces its anticoagulant effect by virtue of its negative charge.

(*d*) *Smell*: Volatile oils like peppermint oil etc., are used as therapeutic agents.

(*e*) *Physical mass*: Laxatives like agar etc. when taken orally absorb water and swell in size.

(*f*) *Osmosis*: Diuretics like mannital etc. produces their effect due to osmosis.

(*g*) *Taste*: Some compounds with bitter taste increase the flow of hydrochloric acid in the stomach and improve appetite, *e.g.* gentian, chirata etc.

(*h*) *Radioactivity*: *e.g.*, Drugs used in hyperthyroidism etc.

Chemical Properties

(*a*) *Chelation*: Certain harmful metals can be eliminated from the body with the help of chelating agents. These chelating agents form a non-toxic ring structure with the molecule of the metal and makes it more water soluable and thus help its excretion in the urine. Example: Dimercaprol used in arsenic poisoning. This drug combines with arsenic and forms water soluble complex which is eliminated in the urine.

Another example is anticoagulants like sodium citrate which removes calcium ions by chelation.

(*b*) *Acidity or alkalinity*: *e.g.*, antacids used in the treatment of peptic ulcer.

Metabolism of Drugs

Metabolism is a mechanism by which drug action is terminated. The interaction between a drug and the living organism which brings about a chemical change in the drug molecule is known as drug metabolism, detoxification or more aptly as biotransformation. The term metabolism was used as the process of life by which tissue cells are destroyed by combustion (process known as catabolism) and renewed from chemical substances carried in the blood and derived from digested foods (process known as anabolism). These chemical changes in the living cell provide energy for vital processes and activities. During metabolism new materials are produced and assimilated for growth and maintenance. The chemical reactions that the drug undergoes do not ordinarily provide such energy or new materials. Thus the term biotransformation is preferable to drug metabolism, for describing the functions of the compounds which are not normally considered under the following groups namely carbohydrates, proteins, fats, vitamins, hormones or mineral substances. In short, metabolism can be stated as the process of chemical alteration of drug inside the living organism. Almost all the drugs that undergo metabolic transformation are converted to products known as metabolites and these are more polar than the parent compounds.

After absorption drugs can undergo three possible types of alterations, namely, they could be metabolised by enzymes, they could change spontaneously into other substances without the intervention of enzymes or they could be excreted unchanged. Some drugs, such as thalidomide, change spontaneously into other compounds simply because of appropriate pH of body fluids. A few compounds which do not have any biological activity or have very little activity can be converted into a pharmacologically active species by matabolism. Such type of drugs are known as Prodrug, for example, the anticancer drug cyclophosphamide which is biologically inactive, but is converted into an active cytotoxic drug in the liver. Many chemical carcinagers such as benzopyrene found in cigarette smoke are examples of this group.

The matabolism of drug usually tends to make the less polar, lipid soluble substances into more polar and water soluble substances and thus facilitates their excretion by kidneys. If a drug is already highly polar and water soluble then it may not get metabolized and may get excreted as such, for example, decamethonium, methotrexate drugs, etc.

In recent years, the specific information on the metabolism of drugs have become essential for the safe and proper use of the drugs. These metabolic data are required in preclinical, clinical testing and evaluation of new drugs. Most drugs are metabolized in the liver through the action of microsomal enzymes. Liver is the dominant organ in drug metabolism because of its richness in enzymes. The other sites of drug metabolism are intestines, brain, kidneys and lungs.

Majority of the drugs are metabolized by enzymes resulting in their activation, inactivation or modification. The reaction which brings about these changes are:

(1) (*i*) Oxidation
    (*ii*) Reduction
(2) Hydrolysis
(3) Conjugation

The first two reactions are collectively referred as Phase-I reactions while the third one, *i.e.*, conjugation is termed as Phase II reaction. Oxidation-reduction enzymes are generally found in the endoplasmic reticulum, hydrolytic enzymes are mostly located in the cell cytoplasm or in the plasma, while conjugated enzymes are found associated with cytoplasm and endoplasmic reticulum.

Oxidation, reduction and hydrolysis introduce polar groups such as hydroxyl, amino, sulphhydryl and carboxyl groups into drugs which are consequently made water soluble and pharmacologically less active. The metabolism of a drug is essentially a detoxication process.

Enzymes catalyzing drug oxidation are referred to as drug-metabolizing enzymes. But these enzymes not only catalyse drugs, but they also catalyse xenobiotic substrates (*e.g.* food additives, pesticides) or endogenous substances (such as steroid hormones, fatty acids, etc.).

The enzymes which metabolize the drugs are located in the liver microsomes and these are obtained as a sediment when the liver cells are subjected to high speed centrifugation. Among these enzymes some important enzymes are esterases, amidases, glucuronyl transferases and others that catalyze a variety of oxidative and reductive reactions. These enzymes alter drugs mainly to make them more water soluble so that they can be excreted by the kidneys.

Drugs are also metabolized by enzymes which are non-microsomal in origin and these are present in liver, plasma and tissues including placenta, etc. One of the important examples of such a type of enzyme is alcohol dehydrogenase which catalyses the oxidation of ethanol to acetaldehyde.

Chemical Pathways of Drug Metabolism or Biotransformation

Drug metabolism takes place in two phases. Phase I, also known as non-synthetic reaction or pathway consists of oxidation, reduction and hydrolysis process or a combination of all these processes which alter the drug chemically. This represents first stage of biotransformation. Phase II also known as synthetic reactions and can be defined as a synthetic process by which a drug or its metabolite combines with an endogenous substance or a molecule provided by the body, resulting in various conjugates such as :

(*a*) Glucuronides conjugates
(*b*) Aminoacid conjugates
(*c*) Ethereal sulphates conjugates
(*d*) Methylated conjugates
(*e*) Acetylated conjugates.

Conjugation results in inactivation of the compound.

Phase I or Non-synthetic Reactions

1. (*i*) Oxidation or oxidative reactions

Microsomal oxidation may involve the following:

(*a*) Hydroxylation (*b*) Oxidative de-alkylation

(*c*) Oxidative deamination (*d*) N-oxidation.

(*a*) Hydroxylation

It is the introduction of a hydroxyl group in the drug molecules; both aromatic and aliphatic compounds get readily hydroxylated. For example, aromatic compound like benzene is converted to phenol or phenobarbital into p-hydroxy phenobarbital or cortisol into 6-β hydroxycortisol, etc., (represented in Fig. 5).

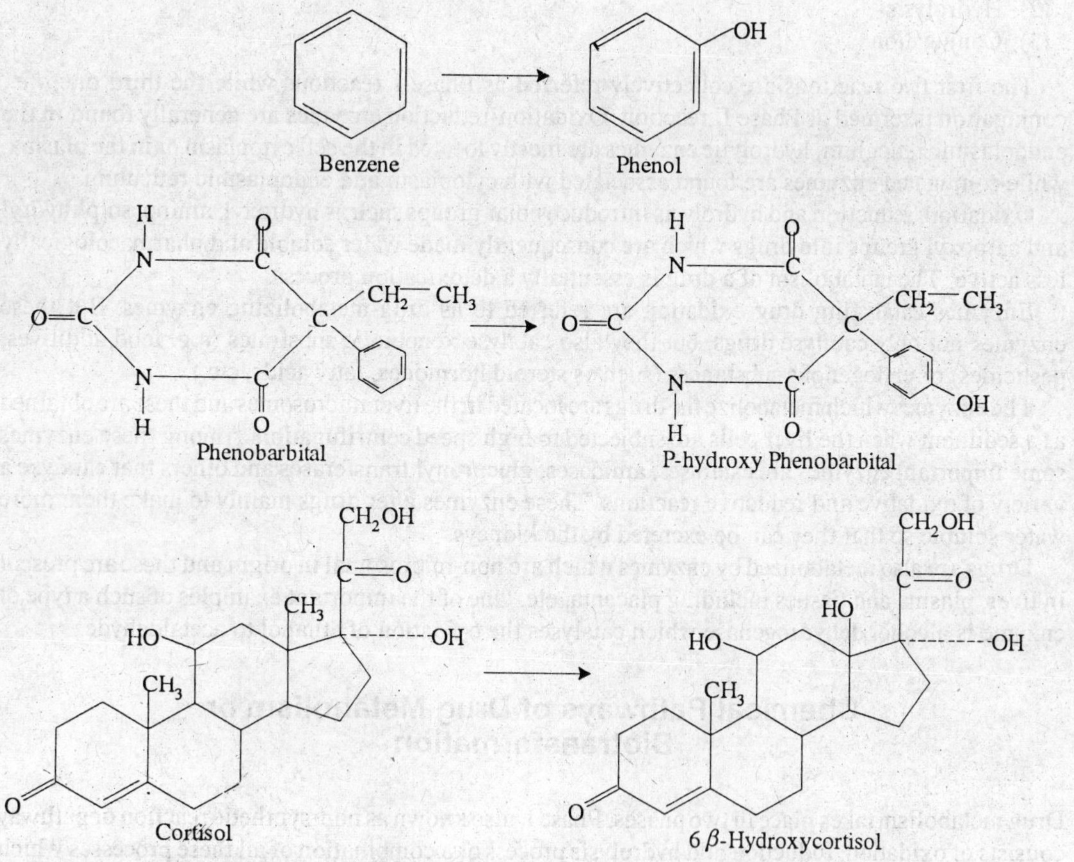

Benzene → Phenol

Phenobarbital → P-hydroxy Phenobarbital

Cortisol → 6 β-Hydroxycortisol

Fig. 5. Examples of Aromatic hydroxylation.

The aromatic hydroxylation takes place via a highly reactive epoxide intermediates (Figure 6).

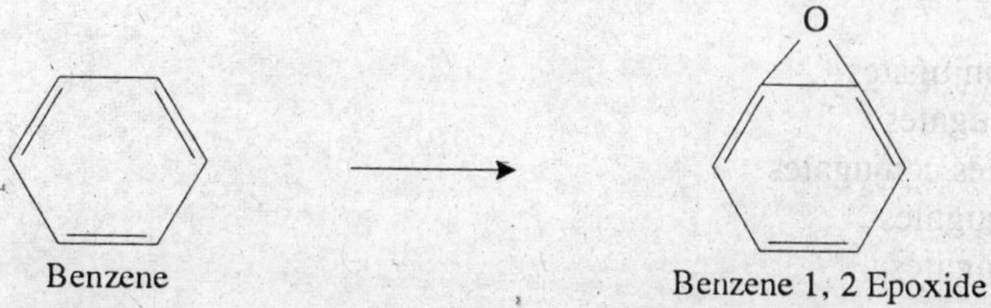

Fig. 6. *An Example of epoxidation.*

Aliphatic hydroxylation is also common, for example, oxidation of allylic, primary, secondary and tertiary carbon-hydrogen bonds like conversion of pentobarbital into hydroxy-pentobarbital (as shown in Figure 7).

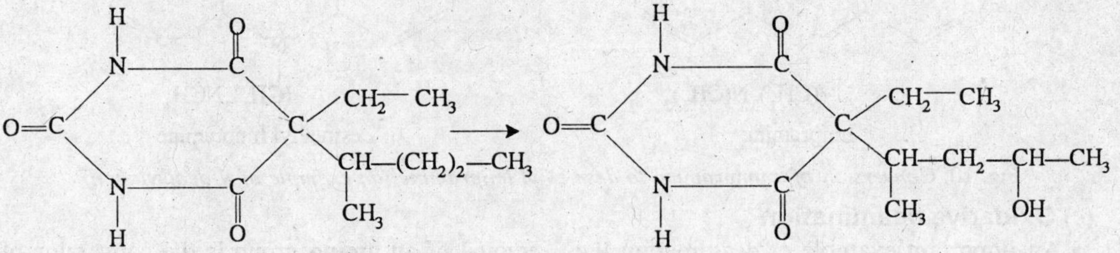

Fig. 7. *Examples of aliphatic hydroxylation.*

(*b*) Oxidative dealkylation

Both N- and O-dealkylation reactions are known where an alkyl group is removed. Example of O-dealkylation is the conversion of phenacetin to the active compound p-acetaminophenol as shwon in Figure 8.

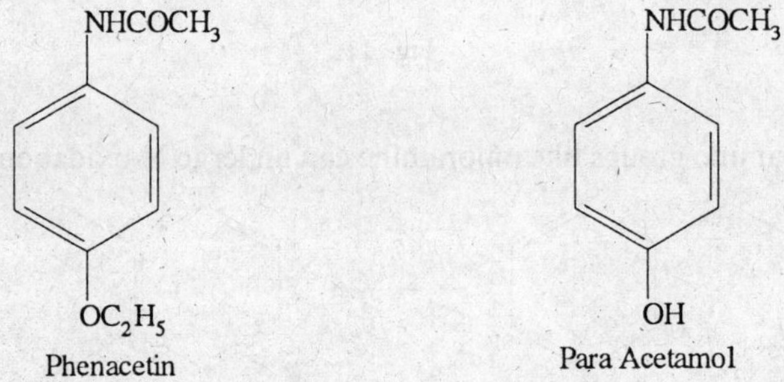

Phenacetin Para Acetamol

Fig. 8. *Examples of O-dealkylation.*

Another example of O-dealkylation is conversion of codeine into morphine as shown in the Figure 9.

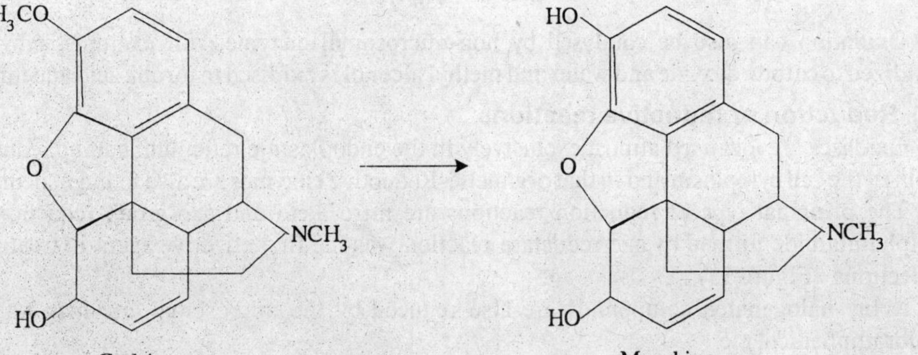

Codeine Morphine

Fig. 9. *Conversion of Codeine into Morphine (Example of O-dealkylation).*

N-dealkylation is common in secondary or tertiary amides and amines *e.g.* dealkylation of imipramine to desmethyl imipramine.

(CH$_2$)$_3$N(CH$_3$)$_2$
Imipramine

(CH$_2$)$_3$NCH$_3$
Desmethyl Imipramine

Fig. 10. *Conversion of a imipramine to desmethyl imipramine (an example of N-dealkylation).*

(*c*) Oxidative deamination

An important example of deamination, *i.e.*, removal of an amino group is the conversion of amphetamine into phenyl propane-2 (Figure 11).

CH$_2$—CH—NH$_2$ (CH$_3$) → CH$_2$—C(=O)—CH$_3$ + NH$_2$

Amphetamine Phenylpropanone-2

Fig. 11.

(*d*) N-oxidation

Drugs containing amino groups like imipramine can undergo N-oxidation. (Figure 12).

(CH$_2$)$_3$N(CH$_3$)$_2$ → (CH$_2$)$_3$—N(→O)—(CH$_3$)$_2$

Imipramine—N-oxide

Fig. 12.

Oxidation can also be catalysed by non-microsomal enzymes, for example, ethyl alcohol is oxidized to carbon dioxide and water and methyl alcohol is oxidised to formic acid and formaldehyde.

(*ii*) Reduction or reductive reactions

Oxidases are localised almost exclusively in the endoplasmic reticulum but the reductases occur both in the cell cytoplasm and in that organelle. Reductive enzymes are also found in ileum and colon.

The principal type of reduction reactions are nitro, keto and azo-group reduction. The drug sulphanilamide formed by azo reductase reaction was the first effective agent in treating bacterial infections. (Figure 13).

Many halogenated compounds are also reduced by the microsomal enzymes e.g. halothane, chloramphenicol etc.

A few drugs are reduced by non-microsomal enzymes also e.g., Chloral hydrate etc.

(2) Hydrolytic Reactions (Hydrolysis)

Most of the hydrolytic enzymes are found outside the endoplasmic reticulum. Although virtually all organs and tissues contain hydrolytic enzymes, they are found in high concentration in kidney, plasma and liver. Hydrolysis is usually carried out by enzymes 'esterases' that hydrolyse the esters forming an alcohol (phenolic or straight chains) and an acid (Figure 14).

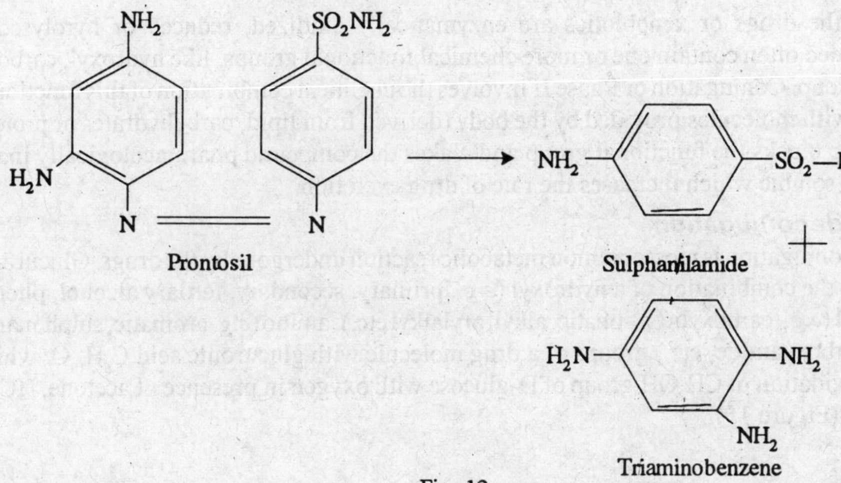

Prontosil → Sulphanilamide + Triaminobenzene

Fig. 13.

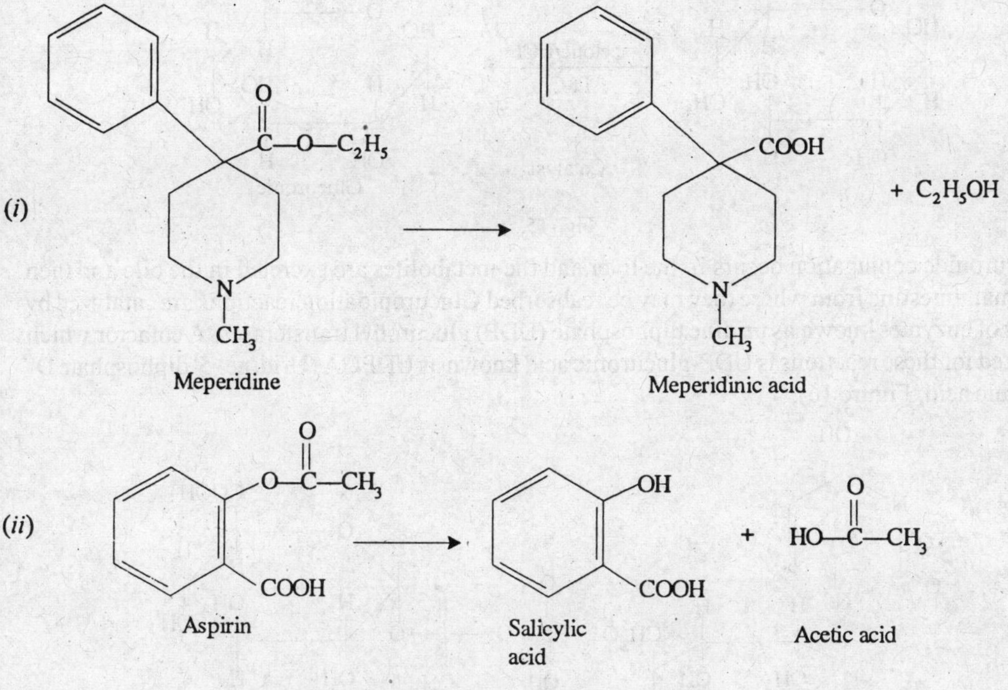

(*i*) Meperidine → Meperidinic acid + C_2H_5OH

(*ii*) Aspirin → Salicylic acid + Acetic acid

Fig. 14. *Examples of hydrolytic reaction catlysed by esterases.*

Another enzyme amidase hydrolyses amides forming an amine and an acid as products of the reactions.

The enzyme esterases are microsomal, non-microsomal and microfloral in origin. These enzymes are less specific in nature and hydrolyse many drugs *e.g.*, procaine, pethidine, acetylcholine, phenytoin, etc. Digitalis glycosides become inactive by hydrolysis. The amidases are non-microsomal enzymes and are found in liver.

Phase II–Conjugation Reaction (Synthetic Reaction)

Conjugation is a synthetic process by which a drug or its metabolite is combined with an endogenous substance forming glucumide, ethereal sulphates, methylated and acetylated compounds and amino acid conjugates.

In phase I, the drugs or xenobiotics are enzymatically oxidized, reduced or hyrolysed, the metabolites formed often contain one or more chemical functional groups, like hydroxyl, carboxylic acid or amino group. Conjugation or Phase II involves the chemical combination of this functional or reactive group with molecules provided by the body (derived from lipid, carbohydrates or proteins). The added group blocks the functional group and makes the compound pharmacologically inactive and more water soluble which increases the rate of drug excretion.

(a) Glucuronide conjugation

This type of conjugation is most common metabolic reaction undergone by the drugs. Glucuronides are produced by the combination of a hydroxyl (*e.g.*, primary, secondary, tertiary alcohol, phenolic, enolic), carboxyl (*e.g.*, carboxylic, aliphatic, alkyl, arylalkyl etc.), amino (*e.g.* aromatic, sulphonamide, heterocyclic, carboxyamide, etc.) groups of a drug molecule with glucuronic acid $C_6H_{10}O_7$ which is formed by the oxidation of CH_2OH group of D-glucose with oxygen in presence of acetone, HCl and Pt-C as catalyst (Figure 15).

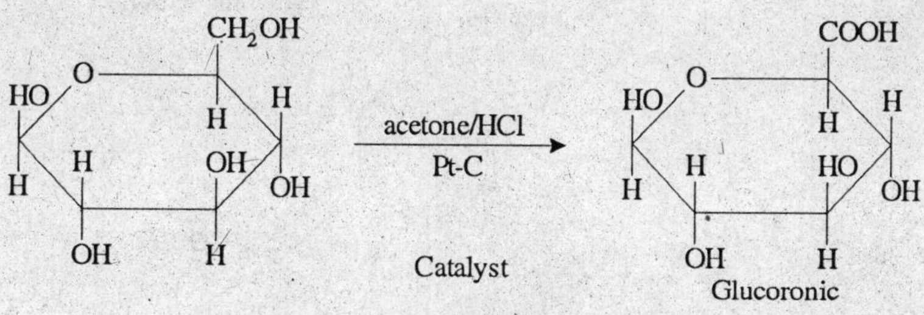

Fig. 15.

Glucuronide conjugation occurs in the liver and the metabolites are excreted in the bile and then in the small intestine from where they may be reabsorbed Glucuronidation reactions are catalyzed by a family of enzymes known as uridine diphosphate (UDP) glucuronyl transferases. A cofactor which is required for these reactions is UDP-glucuronic acid known as UDPGA (Uridine- 5' diphosphate D-glucuronic acid, Figure 16).

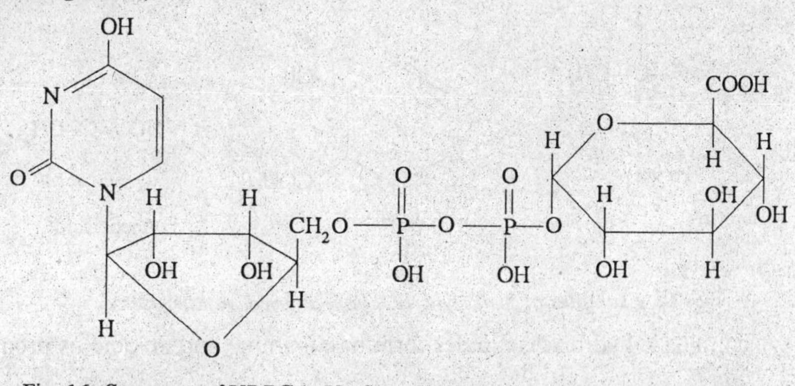

Fig. 16. *Structure of UDPGA (Uridine, 5' diphosphate D-glucuronic acid).*

These enzymes are mainly found in liver, but are also present in intestine, kidney, lung and other organs. UDPGA reacts with groups having active hydrogens, for example, thiols, amines, acids, etc., in presence of UDP-glucuronyl transferase. The sequence of the reaction is represented as follows:

$$R-OH-UDP \xrightarrow{\text{Glucuronyl Transferase}} R-O-\text{glucuronide}$$

$$R-O-\text{Glucuronide} \xrightarrow{\beta\text{-glucuronidase}} R-OH + \text{Glucuronic acid}$$

β-glucuronidase is present in the intestinal mucosa or in the intestinal bacteria which can hydrolyze the newly formed conjugate. The re-formed drug is lipid soluble and gets reabsorbed in gastro-intestinal tract.

Salicylic acid where both hydroxyl and carboxyl groups are present can combine with glucuronic acid in two ways (Figure 17).

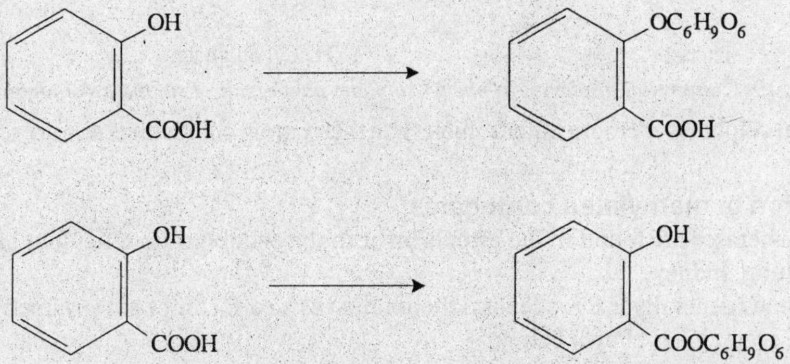

OH, COOH → $OC_6H_9O_6$, COOH

OH, COOH → OH, $COOC_6H_9O_6$

Salicylic acid Ester glucuronide of salicylic acid

Fig. 17.

The glucuronides are rapidly eliminated in the urine because these are highly ionized and water soluble. The enzymes of the bacteria normally present in the intestine can remove the glucuronic acid from the parent compound and the re-formed parent compound is more lipid soluble, it is reabsorbed.

(*b*) Amino acid conjugation

A variety of carboxylic acids which are not structurally related get conjugated with naturally occuring amino acids before excretion. Conjugation with glycine is most common, but other amino acids like taurine and glutamine are also used. The combination with glycine can be represented by taking the example of salicylic acid as shown below.

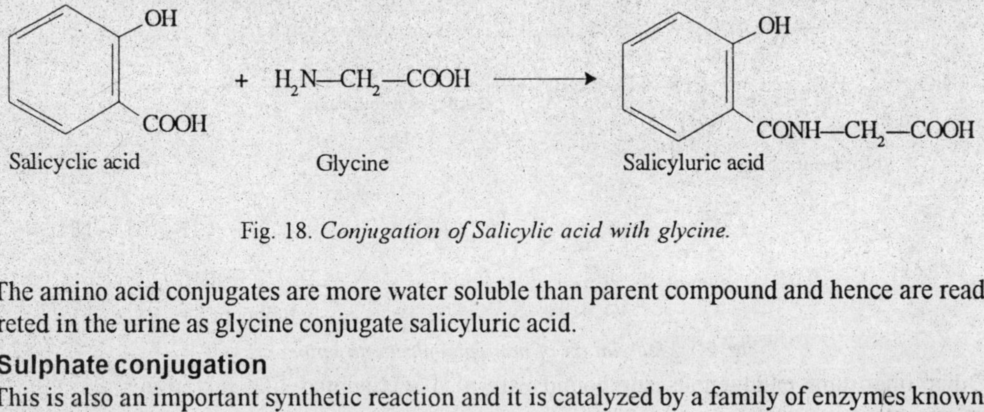

OH, COOH + $H_2N—CH_2—COOH$ → OH, $CONH—CH_2—COOH$

Salicyclic acid Glycine Salicyluric acid

Fig. 18. *Conjugation of Salicylic acid with glycine.*

The amino acid conjugates are more water soluble than parent compound and hence are readily excreted in the urine as glycine conjugate salicyluric acid.

(*c*) Sulphate conjugation

This is also an important synthetic reaction and it is catalyzed by a family of enzymes known as sulpho-transferases which are present in the cell cytoplasm of liver and in many other organs. The essential co-factor or the activated transfering compound i.e., the sulphate donor is 3'-phosphoroadenosine - 5- Phosphosulphate (PAPS).

Sulphate conjugates are highly polar compounds and are readily excreted in the urine.

A typical reaction of ethereal sulphate synthesis is the reaction of sulphate (which is derived from sulphar containing amino acids such as cystine) with hydroxyl group and certain compounds containing an amino group. It can be shown by taking phenol as an example (shown in Figure 19).

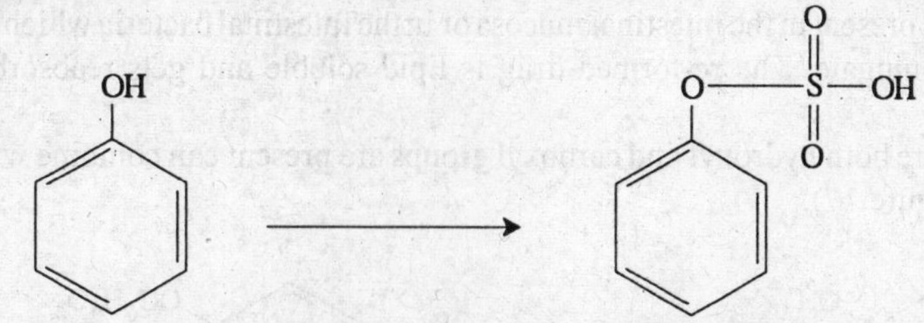

Phenol → Phenyl Sulphate

Fig. 19. *Conversion of phenol to phenyl sulphate as a result of sulphate conjugation.*

The ethereal sulphates are more water soluble than their parent compounds and are readily excreted in the urine.

(*d*) Methylation or methylated conjugates

Methyl transferases are found in the cytoplasm or in the endoplasmic reticulum of many organs such as liver, lung, kidney.

These enzymes generally use S-adenosyl methionine as a co-factor or act as a methyl donor. The sequnce of the reactions is shown below:

ATP + Methionine $\xrightarrow[\text{transferase}]{\text{Methionine adenosine}}$ S-Adenosylmethionine + Pyrophosphate + Phosphate

S-Adenosylmethionine + RZH $\xrightarrow[\text{transferase}]{\text{Methyl}}$ RZ—CH_3 + S-adenosyl homocysteine

where Z is: O, NH or S.

Methylation is a common biochemical reaction which is of a greater significance in the metabolism of endogenous compounds than the drugs. Methylation differs from conjugation processes as unlike the conjugation process the products formed by methylation have same or greater pharmacologic activity than the parent compound. *e.g.*, the conversion of norepinephrine to epinephrine (shown in Figure 20).

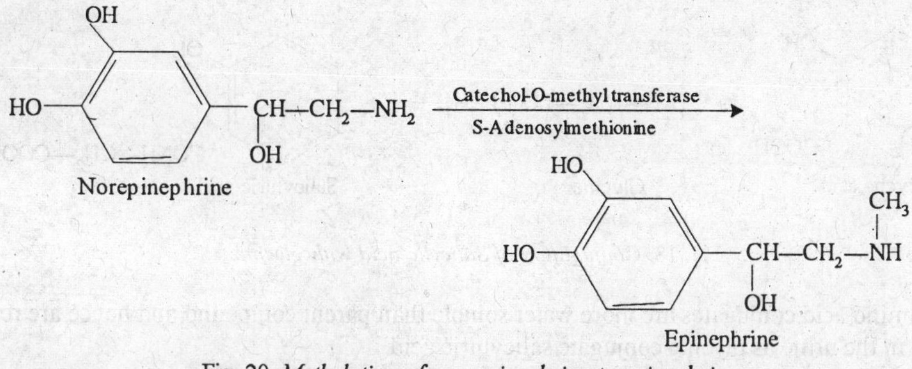

Norepinephrine $\xrightarrow[\text{S-Adenosylmethionine}]{\text{Catechol-O-methyl transferase}}$ Epinephrine

Fig. 20. *Methylation of non-epinephrine to epinephrine*

Other important endogenous compounds are also methylated like histamine, estradiol and thyroxine, etc.

(*e*) N-Acetylation of N-Acetylated conjugates

The acetylation reactions are called N-acetylation since a variety of primary amines undergo inactivation by N-acetylation. This reaction is catalysed by enzymes known as N-acetyl transferases which utilizes acetylcoenzyme (acetylco-A) as a co-factor which is found in liver, intestine, kidney, etc. Hepatic N-acetyltransferase catalyzes the acetylation of a number of commonly used drugs like isoniazid, P-aminosalicylic acid, hydralazine or shlphonamides (shown in Figure 21).

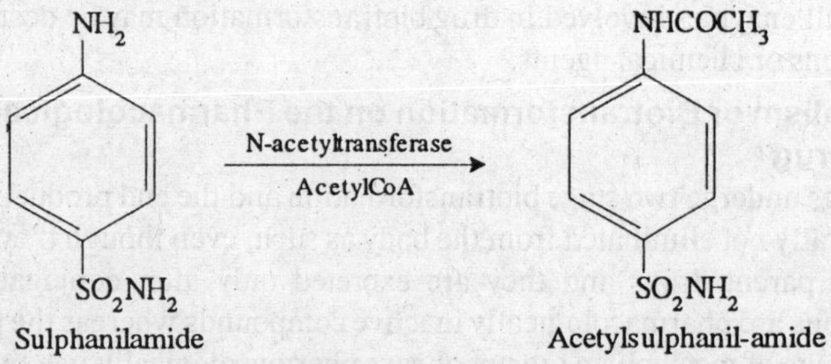

NH_2 — N-acetyltransferase / AcetylCoA → $NHCOCH_3$

SO_2NH_2 Sulphanilamide — SO_2NH_2 Acetylsulphanil-amide

Fig. 21. *An example of N-acetylation.*

Some aromatic compounds such as naphthalene are excreted in the urine as N-acetylcysteine derivation known as mercapturic acids. This type of conjugation is known as glutathione conjugation.

Major Pathways of Biotransformation

A single compound such as salicylic acid can be conjugated in different ways and it can also undergo several oxidative reactions and these metabolites also may get conjugated. Many drugs like salicylic acid which possess several functional groups can undergo chemical alteration by enzymic activity and can give rise to a variety of end products.

Usually one or two pathways account for the major metabolic alterations *e.g.*, for salicylic acid, the conjugates with glycine and glucuronic are the major metabolites. It is difficult to predict all the major pathways of biotransformation for all the functional groups, but the most probable major reactions of biotransformation of some important functional groups are predicted, for example, Amino (NH_2) group (aliphatic) undergo deamination glucuronide conjugation and aromatic amino groups undergoes acetylation, methylation and glucuronide conjugation.

The metabolic mechanisms, both in animals and humanS seem to be similar though there are some differences between different species and in each individual. The important factors which affect the drug metabolism are physiologic, genetic, pharmaceutical factors, drug interactions, stimulation and inhibition.

Summary of the Metabolism or Biotransformation of Drug

The action of drugs in the body are terminated by biotransformation and by excretion through kidneys and liver. One of the important characteristics of the drug is its lipid solubility, which enables the drug to cross the biologic barriers to reach sites where drugs can produce a pharmacologic effect. After some time the drugs are excreted through liver and kidneys from the body. This happens by the conversion of drugs into more water soluble material. However the lipid soluble drugs are not excreted to any significant degree by the kidneys or the liver. During biotransformation the drugs undergo chemical alteration and made more water soluble and less lipid soluble than their parent compounds. Biotransformation reactions are catalyzed by enzymes, it is significant to note that enzymes which catalyze majority of drugs are microsomal enzymes which act only on lipid soluble substrates.

Majority of drugs are metabolized by the enzymes resulting in their activation, inactivation or modification. The reactions which bring about these changes are: (*i*) oxidation-reduction, (*ii*) hydrolysis and (*iii*) conjugation or synthetic reactions. Oxidation—reduction and hydrolysis are collectively known as Phase I while conjugation is termed as Phase II reaction. Phase I is also known as Non-synthetic reactions and these reactions introduce polar groups like hydroxyl, amino, sulphhydryl, etc., into the drugs which makes the drugs more water soluble. The non-synthetic reactions can convert an inactive drug into an active agent or an active agent into another compound with less, equal, greater or different pharmacologic activity.

Phase II or conjugation reactions are a synthetic process by which a drug or its metabolite is combined with an endogenous substance resulting in various conjugates such as glucuronides, ethereal sulphates, methylated or acetylated and amino acid conjugates. Conjugation invariably results in inactivation of the compound.

The activity of all enzymes involved in drug biotransformation may be decreased or inhibited by pathologic conditions or chemical agents.

Effect of Metabolism or Biotransformation on the Pharmacological Activity of the Drug

Most of the drugs undergo two stage biotransformation and the end products of the non-synthetic reactions are generally not eliminated from the body as such, even though they are usually less lipid soluble than their parent drugs and they are excreted only after conjugation. The product of conjugation reactions are pharmacologically inactive compounds whereas the products of first stage or non-synthetic stage of metabolism are not always pharmacologically inactive compound. In fact, the Phase I or non-synthetic process may convert an inactive drug into an active agent or can change an active drug into another pharmacologically active compound. If there is a formation of another type of active compound or metabolite, it may have similar pharmacologic activity, or it may be less, equal, or more active, or this active, compound may possess different type of pharmacological activity or it might have greater toxicity. For example, codeine is partially biotransformed to morphine which has similar activity but it is more active than codeine. Another such example is that of aspirin whose active metabolite is salicylic acid which has similar and equal activity like that of its parent drug.

Table 1 shows some well-known examples of biotransformation and its effect on pharmacological activity of a drug.

Table 1

	Drug	*Activity of Drug*	*Metabolic Reaction*	*Activity of Metabolite*
1.	Methanol	Depressant	Oxidation	Shows different activity which is more toxic
2.	Aspirin	Analgesic	Hydrolysis	Equally potent analgesic
3.	Codeine	Analgesic	Oxidation	More potent analgesic
4.	Imipramine	Antidepressant	N-demethylation Desimipramine	Antidepressant (Same activity)
5.	Phenylbutazone	Antirheumatic and Uricosuric i.e. increasing the excretion of Uric acid in the Urine	Odidation	Metabolite I = antirheumatic Metabolite II Uricosuric
6.	Mercaptopurine	Anticancer drug	Non-microsomal oxidation (Mercapturic acid)	Loses anti cancer effect
7.	Phenobarbital	Sedative	Aromatic hydro-xylation	Loses sedative
8.	Procaine	Local anaesthetic	Hydrolysis (p-aminobenzoic acid)	Loses local anaesthetic effect
9.	Iproniazid	Antidepressant	N-dealkylation (Isoniazid)	Activity is changed and it shows antidepressive and antitubercular effect
10.	Parathion	Inactive	Oxidation	Insectide (toxic)

The rate of change of an active drug into another active drug determines the pharmacologic activity of the drug, *i.e.*, if the transformation is rapid the pharmacologic effect will be largely that of the metabolite and if the transformation is slow, the observed effect may be both of the parent drug and the metabolite.

Absorption of Drugs

The action of all the drugs except the drugs which are used topically (*i.e.*, at its site of application), depends on there ability to reach its site of action. The process and the rate at which a drug reaches its site of action and the movement of a drug from the site of application on in to the extracellular compartment of the body is known as absorption of drugs. The delivery of the drug to the tissue is dependent on an another process known as distribution of the drug.

A drug has to cross many biological barriers and membranes (most of them are lipid in nature) to reach at its site of action. A particular site of application or route makes a great deal of difference to the successful absorption of the drug. There are many factors other than the structure and composition of the membrane, determine the ease with which a drug is absorbed.

Most of the drugs are either weakly acidic or basic in nature. The acidic drugs are rapidly absorbed from the stomach. These drugs exist in the acidic medium of the stomach in a non-ionized from which favours their absorption. These drugs act rapidly when they are administered orally *e.g.* barbiturates, salicylates etc. The basic drugs are absorbed only when they reach the alkaline environment of the small intestine, where major component of the drug exist in an unionized form hence it facilitates their absorption. The action of these drugs are delayed when they are administered orally *e.g.* pethidine, ephedrine etc. If a drug can dissociate into its ions at given PH then it is more water soluble but if the dissociation of a drug is poor at a given PH, then it is insoluble in water and is more lipid soluble.

Routes of Adminstration

Drugs are adminstered by different routes which include oral, rectal, sublingual, parental, inhalation topical etc. The choice of route depends upon both convenience and necessity.

1. **Oral or Enternal Route :** The drugs are administered by mouth to the patients. It is the most convenient for access to the systemic circulation.

2. **Rectal Route :** The drugs are administered through the anal portal into the rectum or lower intestine. Rectal suppositories or retention enemas are administered in this way especially in pediatrics and geriatrics.

3. **Parenteral Route :** Parenteral route of administration other than the alimentary tract are called parental route. This route includes intravenous, intramuscular and subcutaneous routes. Parental route can be applied in cases of vomiting, diarrhoea, which might irritate stomach or when the patient is unconscious.

(*a*) **Intervenous route :** Drug is administered directly into the vein to give rapid action. This route is preferred when an emergency calls for an immediate response. Sometimes due to the rapidity with which drug enters the circulation, dangerous side effects may occur such as depression of cardio vascular function which is often called drug shock.

(*b*) **Intramuscular :** Soluble substances, mild irritants, suspensions and colloids can be injected through intramuscular injections. This route may be used where an immediate effect is not required but a prompt effect is desirable.

Sometimes there is irritation around the injection site. Depending upon the drug and other ingredients these injections should be administered under medical supervision and sterilization.

(*c*) **Subcutaneous :** In this route the drug is injected into the alveolar connective tissue just below the skin. Absorption is slower than the intramuscular route. This route is used for those drugs which are desired to spread the action slowly and for long duration. Examples of drugs given by this route are insulin and sodium heparin, neither of which is absorbed orally and these drugs should be absorbed slowly over many hours. The drugs used in the treatment of asthma, epinephrine are usually given subcutanceously to avoid the dangers of rapid absorption and other dangerous side effects like cardiovascular effect.

4. Inhalation route. Inhalation may be employed for delivering gaseous or volatile substances in to the systemic circulation, *e.g.* general anesthetics. In this case absorption is very rapid as the drugs can be delivered in to the alveali of the lengs, having permeable epithelial membrances, abundant blood flow and large surface area.

5. Sublingual Route (beneath the tongue). This route is employed when only small amounts of drugs are required to gain access to the blood. Patients with angina pectoris mah get quite prompt relief from an acute-attack by sublingual administration of nitroglycerin.

6. Topical Route. Topical administration is employed to deliver a drug to a particular area and it is effective only to the area it is applied. A large number of topical medicaments are applied to the skin although topical drugs are also applied to the eye, nose and throat. In man cutaneous absorption mainly occurs from the surface. It can also be absorbed through hair fallicles, sweat and sebaceous glands when the medicament is rubbed vigorously the absorption through hair follicles and glands increases. Since much of a drug that is absorbed through the epidermis diffuses in to the circulation without reaching a high concentration in some portions of the dermis hence systemic administration (affecting the body as a whole) of the drug should also be given along with topical administration.

Factors that Affect Absorption

In addition to the physiochemical properties of a drug molecules and biological membranes various factors affect the rate of absorption which are as follows:—

(*a*) Particle Size. The particle size of sparingly soluble drugs can affect their absorption. For example, the tablet which has large particle size of the active compound is not absorbed easily, whereas small particle size of the active compound is absorbed easily.

(*b*) Concentration. Higher concentration favours rapid absorption. Thus a 2% of solution of lidocaine will induce local anesthesia more rapidly than a 0.2% solution. After oral administration the concentration of drugs in the gut is the function of the dose. Drugs with low aqueous solubility quickly saturate the gastrointestinal fluids so that the rate of absorption tends to reach a limit as the dose is increased. The peplizing and solubilizing effect of bile and other constituents of the gastrointestinal contents assist in increasing the rate of absorption

Drugs that are administered subcutaneously or intramuscularly show a direct linear relationship between the rate of absorption and the concentration of drug in the applied solution.

When the drugs are administered in solid dosage forms such as capsules, tablets etc the rate at which the drug is released into solution may often be slower than other processes involved in absorption hence the dissolution becomes the rate limiting factor on absorption.

The rate of dissolution depends upon the surface area of the solid which in turn depends upon how finely the drug is subdivided. In oil solution or in oil-in-water (O/W) emulsions the rate of absorption may depend on O/W partition coefficients. The rate of absorption of a drug from a viscous solution may be determined by the show rate of diffusion of drug molecule through the solution.

(*c*) Area of Absorbing Surface. The area of absorbing surface is an important factor in determining the rate of absorption. Drugs can be absorbed better from small intestines than from the stomach because of larger surface area of the small intestine. Subcutaneous or intramuscular injection can be better absorbed if the site of application is massaged in order to spread the injected fluid from a compact mass to a larger absorbing surface.

The rate of absorption differs due to different routes of administration of drug for example the large alvealar surface of the lungs allows rapid absorption of gases, vapours etc.

(*d*) Physical State. Liquids are better absorbed than solids and similarly crystalloids are better absorbed than colloids.

(*e*) Ionization. The mucasal lining of the gastro-intestinal tract is impermeable to the ionized form of a weak acid or weak base. A weakly acidic and basic drug exists either in an unionized, lipid soluble component or as an ionized water soluble component. Unionized component is absorbed more rapidly than ionized component. The absorption of weakly acidic or basic drugs have already been discussed.

(*f*) Presence of other agents. Some times presence of other substances help in the absorption of the drugs far instance absorption of iron from g-astro-intestinal tract is enhanced by the presence of vitamin C.

Absorption of the drug is most rapid in the vascular tissues. Drugs are absorbed more rapidly from intramuscular sites than from less vascular subcutaneous sites. Despite the small absorbing surface for buccal or sublingual absorption the high vascularity of the buccal surfaces favours an unexpected high rate of absorption.

Vasoconstriction (narrowing of the luman of a blood vessel) also have a profound effect upon the rate of absorption. Absorption of the drug from the intestine occurs either by (*i*) simple diffusion *i.e.* a water soluble drug which has low molecular weight (alcohol, water, uera etc) can diffuse through the aqueous pores of the membrane, even highly lipid soluble drugs can be absorbed through simple diffusion process or by (*ii*) active transport-drugs of large molecular size are absorbed by active transport process which is a specialized process and is independent of the physical properties of the membrane. Drugs like steroids, glucose amino acids etc are absorbed by this method.

After absorption the drug is delivered to the tissues by the process of distribution.

Digestion and Absorption of Protein

Food is essential for the survival of the organism. Food provides necessary materials for tissue growth, repair and energy for doing work. Our food mainly consists of carbohydrates, proteins, fats, vitamins, minerals and water. Digestion of food stuff is accompanied by the variety of specific enzymatic activities are helped by the other secretions of the digestive glands. Digestion helps to break down compledx food substances into more simpler forms which are easily absorbed and assimilated by the cells. The digestion and absorption takes place in the alimentary canal which includes mouth, pharynx, oesophagus, stomach, small intestines, large intestines and the associated digestive glands. Mechanical activities like chewing and intestinal movements help in the process of mixing up of the digestive enzymes with the substrate.

The dietary proteins are mainly derived from meats and vegetables. The protein molecules consists of long chain of amino acids bound together by peptide linkages. The normal requirements of dietary proteins of an average man is 50 g/day. Proteins are broken down in to amino acids by series of enzymes which attack the peptide (–CONH–) linkages and it is generally known as peptidases. On hydrolysis the simple proteins yield only amino acid while the conjugated proteins yield amino acids along with non-amino group known as prosthetic group. The peptidases enzymes are extremely specific in their action. These enzymes can broadly grouped into two types known as

(*i*) Endopeptidases

(*ii*) Exopeptidases

Endopeptidases splits up the peptide linkage in the interior of the molecule and act on the long chain protein molecule and breaks it into shorter chain peptides. Whereas *Exopeptidases* linkage splits the terminal amino acids and act on short chain peptides and break it down in to amino acids.

Pepsin of gastric juice trypsin and chymotrypsion of pancreatic juice belong to the group of endopeptidases. Some examples of exopeptidases are amino peptidase, corboxy peptidase, tripeptidase,

Dipeptidase etc. The dietary proteins are acted upon by the enzyme pepsin in the stomach. The acid medium essential for activity is provided by hydrochloric acid of the gastric juice pepsin can digest all the different types of proteins in the diet including the collagen and convert it into smaller peptides. Upon entering the small intestine these partially broken down products are acted upon by the proteolytic enzymes of the pancreatic juice and intestinal juice. The digestion of proteins take place in the following way.

Gastric digestion

Pepsin is secreted by the central cells as an inactive precursor pepsinogen. This inactive pepsinogen is converted to active pepsin by the hydrochloric acid of the gastric juice or by pepsin itself. Pepsin acts optimally at PH 7.2 on all proteins (except keratin and protamines) and converts them into smaller peptides. Pepsin can coagulate caseinogen of milk into insoluble caesin.

Renin is another enzyme which can act upon the caseinogen to convert it into insoluble casein. The baby's stomach has a higher PH near about 5, which is helpful for renin action. It has least proteolytic activity so when milk is ingested it is clotted by renin and clot passes into small intestine for further action by intestinal proteolytic enzymes. As the baby grow the renin activity is dimished and proteolytic activity increases.

Intestinal digestion

The various secretions pouring into the lumen contain large number of enzymes. The shorter chain polypeptides, proteoses and peptones after gastric digestion pass into duodenum where these are acted upon by pancreatic enzymes like trypsin, chymotrypsin, carboxy peptidases etc. As the products of gastric digestion are in acid media it is neutralise by the alkalies of the pancreatic and bile juices.

Trypsin is secreted as inactive precursor trypsinogen which is activated to trypsin by the enzyme enterokinase secreted by the duodenum. It is destroyed in acid media.

Chymotrypsin is secreted by inactive chymotrypsinogen.

Carboxy peptidase and elastases are also secreted by pancreas which are converted into active enzymes by trypsin.

All these three enzymes produce peptides and a small portion of amino acids. They do not affect —S—S— linkage. The final products are amino acids and small peptides containing 2-6 amino acids.

$$\text{Proteins} \xrightarrow[\text{Trypsin Chymotrypsin and dastase}]{\text{Endopeptidases}} \text{Peptide}$$

$$\xrightarrow[\text{Carboxypeptidases}]{\text{Exopeptidases}} \text{Amino acids and small peptides}$$

Exopeptidases

It finalize the completion of digestion of proteins started by the endopeptidases. The tripeptidases and dipeptidases break down the respective tri and dipeptidases to monopeptides. Carboxypeptidases act on polypeptides and produce smaller tri and dipeptidases which are further broken down in to amino acids. Nucleoproteins are digested by specific enzymes like nuclease, nucleotidase etc.

Proteins $\xrightarrow{\text{Pepsin}}$ [(*i*) Proteoses (*ii*) Peptones (*iii*) Polypeptides]

$\xrightarrow[\text{Carboxy poly peptidase}]{\text{Trypsin, chymotrypsin}}$ (*i*) Polypeptides (*ii*) Amino acids $\xrightarrow{\text{Peptidases}}$ Amino acids

Schematic Representation of Digestion of Protein

The final digestion of proteins and their conversion in to individual amino acids occur in three locations *i.e.* in the intestinal lumen, in the brush bordered epithelial cells and in the cytoplasm of the muscosal cells. The digestion of protein is very efficient. Even if gastric and pancreatic proteolytic enzymes are absent as in gastrectomy protein digestion goes on efficiently.

Absorption of Protein

The ultimate purpose of digestion is absorption. The digested food is absorbed through the epithelial lining of the digestive tract into the blood. The mucus lining of the small intestine especially the villi and microvilli increases the surface area of the mucous membrane for absorption. Most of the absorption takes place through the villi of the upper part of the small intestine.

Most proteins are absorbed through the luminal membranes of the epithelial cells in the form of di and tripeptides and a few free amino acids. Approximately 225g of protein is absorbed per day from the small intestine. The amino acids are readily absorbed through out the entire length of the small intestine by active sodium dependent mechanism. Most peptide or amino acid molecules bind with a specific transport protein that also needs sodium binding before transport can occur. The sodium ion then moves down its electrochemical gradient to the interior of the cell and pulls the amino acid or peptide along with it. Therefore it is called co-transport or secondary active transport of amino acids and peptides. However a few amino acids do not require sodium co-transport or dependent mechanism and can be transported by special membrane transport proteins. Although all the amino acids are rapidly absorbed from the intestine, some are absorbed more rapidly than other. The order of rapidity is—glucine, alanine, cystine, glutamic acid, valine, methionine, leucine, tryptophane, isoleucine, norleucine, and isovaline. Laevorotatory and naturally-occuring forms are absorbed more rapidly than dextrorotatory forms. The absorption of l-aminoiacid is dependent on oxygen supply as any reduction will reduce absorption.

Amino acids compete with sugars for the absorptive mechanism. Sugars are absorbed more slowly from a sugar amino acid mixture than from a sugar solution. During the process of absorption amino acids accumulate in the mucosal cells and concentration gradient is developed for diffusion out of the cell into the portal blood. Most of the amino acids are not altered during absorption. However some of the amino acids like glutamic acid and aspartic acid undergo transamination with pyruvic acid and alanine is formed and released in the blood. This transamination is dependent on the rate of absorption.

Digestion of Fats

The fats present in our diet are called neutral fats or triglycerides, which are insoluble in water. A triglyceride is composed of glycerol nucleus and fatty acids. There are also some phospholipids, cholesterol, cholesterol esters in our diet. Neutral fats in our diet are of both animal and plant origin.

Fat splitting enzymes are absent in the saliva. The gastric lipase (fat digestive enzyme) is a weak enzyme and can act only on fats which are in the emulsified state such as milk fat or the fat in the yellow yolk of eggs.

The significant digestion of fats takes place in the small intestine by the action of the pancreatic lipase and intestinal juice lipase. Pancreatic lipase are of two types (*i*) esterases - acting on ester of monovalent alcohol or glyceryl esters of lower fatty acid (*ii*) the other type acts on the glyceryl esters of higher fatty acids and cholesterol esters. It is liberated in an inactive form and is activated by intestinal juice.

The hydrolysis of fat proceeds in stages. The preliminary step in fat digestion is emulsification *i.e.*, a process in which fat globules are broken down in small size so that water soluble digestive enzymes can act on the globule surface. Emulsification of fat is effected by the help of bile salts. Emulsification helps to increase the surface area of fat globules exposed to enzyme action. Pancreatic lipase and intestinal lipase acts on emulsified fats in presence of calipase and convert them into monoglycorides, free fatty acids and glycerol. The monoglycorides, free fatty acids form micelles with bile salts which are water soluble and get absorbed into the blood. These micelles are small spherical globules about 3 to 4 nanometers in diameter composed of 20 to 40 molecules of bile salt. The bile salt molecule is composed of sterol nucleus which is highly fat soluble and also has a polar group which is water soluble. These sterol nuclei of the bile salt aggregate together to form a small fat globule in the middle of the micelle. These polar groups in the molecules make them more water soluble and it is easily digested by the digestive fluids and can remain stable in solution form inspite of its large size.

The fatty acids containing more than 11 carbon atoms are re-esterified to triglycerides in the mucosal cells. The triglycerides in the mucosal cells. The triglyceride is coated with lipoprotein, cholesterol and phospholipids to form chylomicron. These *chylomicrons* are absorbed in to the lymphatics. Colipase is secreted in an inactive form by the pancreas along with lipase and is activated by trypsin. Pancreas also liberates a phospholipase A which converts lecethin into lysolecithin which is a detergent and emulsifier of food fat.

$$\text{Fat} \xrightarrow[\text{Agitation}]{\text{Bile}} \text{emulsified fat}$$

$$\downarrow \text{Pancreatic lipase}$$

Fatty acids and 2-monoglyecrides

Schematic Representation of Digestion of fat

Absorption of Fats

Fats are digested to form monoglycerides, free fatty acids and glycerols. Fat is not hydrolysed in the stomach and enters the small intestine as a coarse suspension. Fats are absorbed both in portal blood and lymphatic vessels. In fat absorption the bile salt plays an important role. Fat absorption takes place in three phases which are as fallows:

***(i)* Lumen Phase :** In this phase the fatty acids combine with bile salts to form polymolecular aggregates known as micelles in the lumen of the small intestine. These micelles are water soluble and in this form the monoglycerides and the fatty acids are carried to the surfaces of the microvilli.

***(ii)* Penetration Phase :** Micelles (40–20 Åin diameter) contain water soluble groups as well as fat soluble hydrocarbons of long chain fatty acids along with monoglycerides. The water soluble part of micelles face outwards and the fat soluble part inwards. The fats are presented in this form to the epithelial surface for absorption. The penetration of the particle is effected mainly by diffusion and is independent of temperature and does not require energy. The bile salts are turned back to the lumen to be used again.

***(iii)* Intracellular Phase (Phase of re-esterification) :** The micelles are taken up by the brush bordered epithelium of mucosa with the help of bile salts. The bile salts acts as a shuttle engine carrying the micelles to the brush border and enter into the cell and during this phase the fatty acids and monoglycerides are resynthesised to triglycerides intracellularly. Short or medium chained fatty acids are absorbed directly into the portal blood. Longer chain fatty acids are synthesised to triglycerides which are coated with lipoprotein, phospholipid and cholesterol. These are known as *Chylomicrons* and in this form it is absorbed in to the lymphatics. Synthesis of chylomicrons is an active process and requires energy which is derived from ATP or oxidative phosphorylation. In addition Magnesium ions and co-enzyme A are also necessary.

The fatty acids are activated first with the help of CoA which makes it activated fatty acid to form phosphatidic acid. This after losing phosphate becomes a diglyceride which later on becomes a triglyceride or converted into phospholipids. This process takes place in the endoplasmic reticulum.

The glycerol is absorbed into the blood, a small fraction of this may be used for synthesis of triglycerides. Short chain and medium chain fatty acids are absorbed into the blood. The long chain fatty acids monoglycerides are re-synthesised into triglycerides and absorbed in to lymphatics in the form of chylomicrons. Between 80 and 90 per cent of all fat absorbed from the gut is absorbed in this manner and is transported to the blood by way of the thoracic lymph in the form of chylomicrons.

Direct absorption of Fatty Acids into the Portal Blood

Small quantities of short chain fatty acids such as butterfat are absorbed directly into the portal blood rather than being converted into triglycerides and absorbed into the lymphatics. The cause of this difference between short and long chain fatty acid absorption is that the shorter chain fatty acids are more water soluble and mostly are not reconverted in to triglycerides by the endoplasmic reticulum. This allows direct diffusion of these fatty acids from epithelial cells in to the capillary blood of the villi.

Assay of Drugs

Assay is the estimation of potency of an active principle in a unit quantity of the preparation. The potency is the measure of the biological activity of a drug. It can be of three types:

1. Chemical assay
2. Biological assay
3. Immunological assay

1. In chemical assay the concentration of the active principle of the drug is estimated with the help of chemical methods. Depending upon the nature of the functional group present in the drug and its chemical properties, various chemical methods can be adopted. For example, the drug having carboxylic group can be estimated by direct titration with an alkali. The drugs having hydroxyl group or amino group can be estimated by acetylation. Similarly, aromatic amino group is estimated by diazotisation and so on.

Certain examples of the drug whose assay is determined by the help of chemical methods are salicylates, sulphonamides, narcotic analgesics and penicillin (the active principle of pencillin is estimated by iodometry) etc.

Recently instrumental methods of analysis are used to determine the assay of the drugs. The useful techniques are spectophotometry, fluorcmetry, mass spectrometry, high pressure—liquid chromatography, photometric methods etc. These methods are very useful as they are rapid, require smaller quantities of substance and can be readily compared with standard compounds.

2. Biological assay or bioassay. Biological assay is essentially a method of measurement in which a biological reaction is used as the indicator and it can be defined as the estimation of the amount of active principle present in a unit quantity of a drug in terms of biological response in the treated organism. Although biological responses are variable, bioassay can be devised to give almost any required degree of accuracy by repeating the tests and using statistical methods in the interpretation of the results.

Biological method is employed when the potency of an active principle of the drug cannot be estimated by chemical methods, *i.e.*, (*i*) when the chemical composition of the stabstance is not known though the substance has a specific action, (*ii*) when the chemical assay method is too complex, (*iii*) when drugs which differ in composition, but have the same pharmacological action or (*iv*) when the constant composition of a product cannot be determined by simple chemical means.

International Units of Biological Activity

International standards are necessary to compare the potency of the various biologically assayed compounds.

It is difficult to give a definite unit of a particular drug for the amount which causes a particular effect, as the biological effects of the drug vary from animal to animal, *i.e.* to express the potency of a drug in terms of rat or cat units, from time to time and from laboratory to laboratory.

If the substance to be bioassayed cannot be purified chemically then it has to be compared with a stable standard solution. In many cases international standard preparations are available. These reference standards can be obtained from pharmacopoeias or from WHO, for example, the standard preparation of insulin is a quantity of purified insulin, 0.04 mg, which contains 1 unit of activity, the standard preparation of digitalis consists of a mixture of dried and powdered digitalis leaves, 76 mg which contains 1 unit of activity.

A well planned assay should be so designed as to furnish evidence not only of the activity of the test preparation, but also of the limits of error of the test. The bioassay is based on the following principles which are as follows:

(*i*) A certain quantity of a drug produces the same degree of pharmacological effect in the same animal or animals of the same species, provided the other conditions are constant.

(*ii*) The drug which is to be tested should produce the same specific effect in all the animal species. However, a liver preparation used in the treatment of pernicious anaemia could not be experimented in animals as this disease occurs only in human beings.

(*iii*) The problems which arise due to individual variations must be kept minimum.

(*iv*) The preparation which is to be taken as standard should be highly pure and previously bioassayed.

(*v*) The reference standard should have the same active principle as the sample to be bioassayed.

(*vi*) The experiment conditions used for bioassay should be of high precision, reliability and reproductibility.

Bioassay is a measurement of concentration of the specific active component of a crude drug and this is determined by finding out the amount of the active ingredient or component required to produce a definite effect on a suitable test animal or organ which is to be tested under standard conditions. Bioassay can be performed on whole animals either singly or in groups or on isolated tissues. Some important methods for bioassay are as follows:

(*a*) *By direct comparison on same tissues*

In this method the sample is compared with the standard for its effect on the same tissue under identical conditions which can be done by the following methods:

(*i*) **Interpolation.** This method is based on drug response relationship, i.e. to study the effect produced by increasing doses of a drug which eventually reaches a steady level. Then, the log dose-response curve is plotted and the concentration of the unknown can be read from this graph.

(*ii*) **Matching.** In this method the doses of the standard and unknown are adjusted until the response matches with each other, *e.g.*, Acetylcholine released during nerve stimulation may thus be assayed by its effect in an isolated preparation against an acetylcholine standard. This method, however, is not very accurate, as it is difficult to estimate the margin of the error.

(*iii*) **Four Point assay.** This method is based on the principles of matching and interpolation and hence it is a more accurate method.

(*b*) *Parallel quantitative method*

In this method an unknown is tested against standard preparation in several different types of assay preparation. If they all show the same activity then the presumption that standard and unknown are identical is greatly strengthened.

(*c*) *Direct assay on several animals*

In this method a group of animals are treated with several doses of standard preparation in increasing amounts. Then, the minimum concentration, which is required to have positive response from 50 per cent of animals is determined. This value is represented as EDS (50) (effective dose for

50% with standard). In the same way the effective dose for positive response from 50 per cent for the unknown is also determined and is represented as EDu (50). (Effective dose for 508 unknown.) The concentration of the unknown solution can be determined from the following formula:

$$C = \frac{ED_s(50)}{ED_u(50)} \times \text{Conc. of standard}$$

Digitalis bioassay in guinea pigs are done by this method. The solution of the drug is infused slowly intravenously into an anaesthetized guinea pig until the heart stops. This procedure is repeated in several guinea pigs with solutions of the standard and unknown. The reciprocal of the ratio of the mean effective values of standard and unknown is then considered to be their activity ratio.

(*d*) ***Indirect Assay***

In this method the potency of a drug is calculated by comparing the log response curve of the unknown sample with that of the standard.

Bioassay of the crude ergot preparations are determined by the indirect method.

The procedures which are used for various bioassays should be of high sensitivity and accuracy. These procedures should also be highly specific in their action and their pharmacological response with the same dose should remain the same under identical conditions. Animals or tissues selected for bioassay must be easily available and must remain 'bioassay fit' for a sufficiently long time.

3. Immunological assay

This method is mainly used for estimating the potency of certain hormones. This method depends upon a chemical reaction between a hormone antigen (any substance which stimulates the production of an antibody) and its antibody. In this method the hormone is labelled with an radio-isotope *e.g.*, radio iodine and is taken along with unlabelled or natural hormone, both the hormones compete with each other to react with its specific antigen. It is observed that unlabelled hormone inhibits the reaction of the labelled hormone and hence the ratio of bound and free hormone known as B/F ratio gradually decreases and the concentration of unlabelled hormones increases. Now a number of B/F ratio to hormone concentration is plotted for the standard preparation. The unknown quantity of the unlabelled hormone is allowed to react with the mixture of labelled hormone and the antibody under the same set of conditions as the standards. The B/F ratio is obtained and the concentration of the unlabelled hormone is read from the graph.

Plasma concentrations of various hormones and other biological substances and drugs can be determined by this method.

The results of immunological assay are less reliable than the bioassy. Recently radio receptor assays have been developed which is based on the principles of both bioassay and radio immunoassay and this method is more specific and sensitive than radio immunoassay.

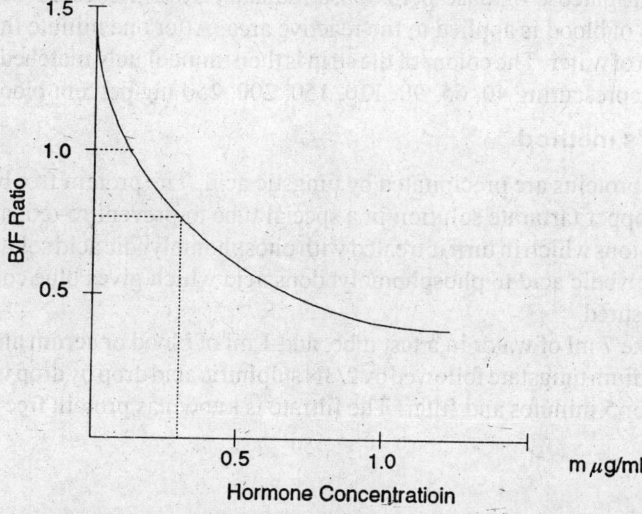

Chapter 4

Clinical Chemistry

Quantitative–Blood Serum

Determination of Sugar (Glucose) in Serum

A large number of methods have been developed to estimate blood glucose and most of the methods were based on the reducing property of glucose i.e. it can reduce certain metallic ions like cupric and ferric cyanide ions in hot alkaline solutions and the extent of glucose reduction can be determined by photometric, titrimetric or gasometric method. One of the method is of Folin and Wu. Besides glucose, there are many non-glucose reducing substances present in blood and therefore the values obtained by this method exceed the true glucose by 20-30 mg/100 ml. Later on, this method was improved by Nelson and Somogji and their values were more accurate as this method gave values exceeding the true glucose by 5–10 mg/ 100 ml only. Two recent methods glucose oxidase method and O-toluidine methods are quite specific and give true glucose levels, but these are rather expensive methods. Glucose can be measured in whole blood or plasma or serum, but the modern trend is to use plasma or serum. Serum is the clear fluid residue of blood from which the corpuscles and fibrin have been removed.

Glucose disappears rapidly from blood on standing due to its enzymatic break down to lactic acid. This can be prevented by adding 10 mg sodium fluoride per ml of blood.

Simple procedures for estimating approximate blood glucose levels are available which do not require special laboratory equipment. For example, the Dextrostix paper strip method. This is a strip test which estimates blood glucose using a single drop of whole blood. The tip of the strip is impregnated with a glucose oxidase-peroxidase indicator system covered with a semi-permeable membrane. A drop of blood is applied to the reactive area. After one minute the blood is washed off with a sharp stream of water. The colour of the strip is then immediately matched to a colour chart with the colour blocks representing 40, 65, 90, 130, 150, 200, 250 mg percent blood glucose.

1. Folin and Wu's method

Principle. The proteins are precipitated by tungstic acid. The protein free blood filtrate is boiled with an alkaline copper tartarate solution in a special tube to prevent re-oxidation. The cupric ions reduced to cuprous ions which in turn is treated with phosphomolybdic acid solution. The cuprous ions reduce phosphomolybdic acid to phosphomolybdous acid which gives blue colour. The intensity of blue colour is measured.

Procedure. Take 7 ml of water in a test tube, add 1 ml of blood or serum and mix gently. To this add 1 ml of 10% sodium tungstate followed by 2/3N sulphuric acid drop by drop with constant shaking Allow it to stand for 5 minutes and filter. The filtrate is known as protein free filtrate.

Take three 25 ml graduated Folin-Wu tubes (Figure 1). Add 2 ml of protein free filtrate (or 1 ml of filtirate plus one ml of water, if very high blood sugar values are expected), 2 ml of working standard glucose solution and 2 ml of water into the tubes marked "unknown", "standard" and "blank" respectively. Pipette out 2 ml of alkaline copper tartarate in each tube and mix. Keep the tubes in a boiling water bath for 8 minutes. Remove the tubes and cool them by immersing in cold water.

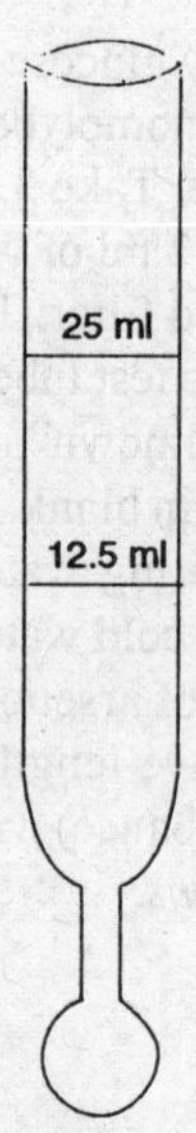

Fig. 1.

Add 2 ml of phosphomolybdic acid solution to each tube and allow to stand for a few minutes. Fill the tubes upto the 25 ml mark with water and mix thoroughly by repeated inversion. Then adjust the wavelength of the photometer to 420nm or put in a blue filter. Set the instrument to 100% transmittance (or zero absorbance) with blank and read the transmittance of unknown and standard.

Calculations:

$$\text{Blood glucose (mg/ml)} = \frac{Au}{As} \times \text{mg glucose standard} \times \frac{100}{0.2}$$

$$= \frac{Au}{As} \times 0.2 \times \frac{100}{0.2}$$

$$= \frac{Au}{As} \times 100$$

Au: Absorbance of unknown

As: Absorbance of standard

Reagents:

(*i*) *Sodium tungstate:* (10%)- 10 g of $Na_2WO_4.2H_2O$ is dissolved in water and diluted to 100 ml.

(*ii*) *Sulphuric acid (2/3 N):* 10 ml of concentrated sulphuric acid is diluted to 500 ml with water and standardised against 2/3 N sodium hydroxide.

(*iii*) *Alkaline copper tartarates:* 40g of anhydrous sodium carbonate is dissolved in about 400 ml of water, 7.5 g tartaric acid followed by 4.5 g crystalline copper sulphate are added and dissolved. Then the volume is made upto 1000 ml with water.

(*iv*) *Phosphmolybdic acid:* 200 ml of 10% Sodium hydroxide and 200 ml of water are added to 35g of molybdic acid and 5g of sodium tungstate and the solution is boiled for half an hour. After cooling the volume is made upto 350 ml with water. 125 ml of 85% phosphoric acid is added and the volume is made upto 500 ml with water.

(*v*) *Stock standard glucose solution:* 100 mg of glucose is dissolved in saturated benzoic acid solution and made upto 100 ml with saturated benzoic acid solution (benzoic acid is used as a preservative).

(*vi*) *Working standard glucose solution:* 10 ml of stock standard glucose solution is diluted to 100 ml with saturated benzoic acid solution. 1 ml of this solution contains 0.1 mg of glucose.

2. Nelson-Somogyi Method

Somogyi suggested a modified method of determination of glucose in the blood.

Principle. Blood is deproteinized by zinc hydroxide and barium sulphate solution which gives protein free filtrate containing practically no reducing substance other than glucose. This filtrate is heated with alkaline copper tartarate and then treated with arsenomolybolate reagent. The cupric ions are reduced by glucose into cuprous ions which in turn reduce the colourless arsenomolybdic acid to coloured arsenomolybdous acid. The intensity of the colour is measured.

Procedure. Take 3.5 ml of water in a test tube and add 0.1 ml of blood or serum and mix gently. To this add 0.2 ml of 0.3N barium hydroxide. Mix and add 0.2 ml of 0.3 N zinc sulphate and shake vigorously and filter. The filtrate thus obtained is protein free filtrate of the serum or blood.

Lable three test tubes as "unknown", "standard" and "blank". Pipette out 1 ml of the protein free filtrate in "unknown" marked test tube, 1 ml of working standard glucose solution in "standard" and 1 ml of water in blank. Add 1 ml of water and 2 ml of alkaline copper tartarate to each tube and keep them in a boiling water bath for 10 minutes. Remove the tubes from the bath and cool them by immersing in cold water.

Add 1 ml of arsenomolybdic acid to each tube and dilute to 10 ml with water. Mix thoroughly. Adjust the wave-length of the photometer to 680 nm or put a red filter. Set it to 100% transmittance (or zero absorbance) with blank and read the absorbance or (transmittance) of unknown and standard.

Calculations:

$$\text{Blood glucose (mg/ml)} = \frac{Au}{As} \times \frac{100}{0.025} \times 0.05$$

$$= \frac{Au}{As} \times 200$$

Au = Absorbance of unknown

As = Absorbance of standard

Reagents:

(*i*) *Zinc sulphate (0.3 N):* 5 g of $ZNSO_4.7H_2O$ is dissolved in water and made upto 100 ml.

(*ii*) *Barium hydroxide (0.3 N):* 5.6g of $Ba(OH)_2.8H_2O$ is dissolved in hot water and made upto 100 ml. This should be standardised against 0.3 N Zinc Sulphate using phenolphthalein as indicator.

(*iii*) *Alkaline copper tartarate:* It is prepared in two parts — (Solution A & B).

Solution A: 25 g of anhydrous sodium carbonate, 20 mg of sodium bicarbonate, 25 g of sodium potassium tartarate and 200 g of anhydrous sodium sulphate are dissolved in about 700 ml water and diluted to 1000 ml.

Solution B: 15 g of crystalline copper sulphate is dissolved in sufficient water and 0.05 ml of concentrated sulphuric acid is added. Then the volume is made upto 100 ml.

96 ml of solution A and 4 ml of solution B are mixed just before use.

(*iv*) *Arsenomolybdic acid:* 25 g of ammonium molybdate is dissolved in 450 ml of water. 21 ml of concentrated sulphuric acid is added slowly. 3 g of disodium hydrogen arsenate ($Na_2HAsO_47H_2O$) is dissolved in 25 ml of water and added to the acid molybdate mixture. The solution is stirred and kept in an incubator at 37°C for 48 hrs. It is stored in a glass stoppered brown bottle.

(*v*) *Stock standard glucose solution:* 100 mg of glucose is dissolved in saturated benzoic acid solution and is made upto 100 ml with saturated benzoic acid solution.

(*vi*) *Working standard glucose solution:* 5 ml of stock standard glucose solution is diluted to 100 ml with benzoic acid solution. 1 ml of this solution contains 0.05 mg of glucose.

3. O-toluidine Method

This is one of the most widely used methods for determination of glucose. The method is simple and specific. One of the advantages of this method is that it can be directly applied to plasma or serum without precipitation of protein.

Principle. Glucose reacts with O-toluidine in glacial acetic acid at 100°C to form blue green coloured N-glucosylamine. The intensity of the colour is measured colorimetrically.

Procedure. There are two procedures: (*i*) direct and (*ii*) indirect.

The indirect method is used for haemolysed plasma or serum (The blood or serum from which red blood cells are removed). The direct method is used for clear un-haemolysed plasma or serum.

(*i*) *Direct procedure:* Take 4ml of O-toluidine reagent in three tubes labelled "Unknown", "Standard" and "Blank". Add 0.02ml of serum, standard glucose solution and water to the respective tubes. Mix and keep the tubes in a boiling waterbath. Remove after 15 minutes and cool under running tap.

Adjust the wavelength of the photometer to 630 nm. Set it to 100% transmittance (or zero absorbance) with blank and read the transmittance or absorbance of unknown and standard.

(*ii*) *Indirect method:* Take 0.9 ml of 3% trichloroacetic acid in 3 tubes labelled "unknown", "standard" and "blank". Add 0.1 ml of blood or serum, standard glucose solution and water to the respective tubes and mix. Centrifuge "unknown" until clear. Transfer 0.2 ml from each tube into correspondingly labelled tubes containing 4 ml of O-toluidine reagent. Remaining procedure is proceeded as in the direct procedure.

Calculation:

$$\text{Blood or serum glucose (mg/100ml)} = \frac{Au}{As} \times 100$$

Au–Absorbance of the unknown

As–Absorbance of the standard.

Reagents:

(*i*) *Trichloroacetic acid (3%):* 3g of trichloroacetic acid is dissolved in water and diluted to 100ml.

(*ii*) *O-toluidine reagent:* About 300ml glacial acetic acid is taken in a 1 litre volumetric flask. 1.5g of thiourea and 9.5g of sodium borate are added successively and dissolved. 80ml of O-toluidine is added. The volume is made upto 1.0 litre with glacial acetic acid.

(*iii*) *Standard glucose solution:* The stock standard glucose solution is made by dissolving 100mg of glucose solution in saturated benzoic acid solution.

Interpretation: The normal range of fasting blood glucose concentration depends upon the method of determination. The concentration of true glucose determined by O-toluidine method varies from 60-95mg/100 ml blood. Nelson Somogyi gives a range of 65-110 mg/100 ml of blood. A range of 80-120mg/100ml of blood is obtained by Folin and Wu's method. Plasma or serum glucose is about 10% higher as compared to blood glucose. The commonest cause of raised blood glucose level (fasting hyperglycaemia) is diabetes mellitus in which the level may rise upto 500mg/100ml of blood or above. Low blood sugar values are found after administration of insulin.

Determination of Serum Cholesterol

Cholesterol present in the blood occurs both free and as esters. The anlysis of serum samples for cholesterol involves the isolation of cholesterol from other substances, separation of the unesterified lipid bound cholesterol from the more soluble esters and subsequent conversions to a substance which absorbs or emits light.

For many years simple methods using the Libermann-Burchard reaction, the green colour resulting from the reaction of cholesterol with acetic anhydride and sulphuric acid were used quite often for detection of free cholesterol and the ester cholesterol may react a little differently. Hence for really accurate determination, the ester should be hydrolysed before estimating the total cholesterol but this cannot be done in the simple methods. Recently total cholesterol in serum is determined by using the reaction of cholesterol with ferric chloride and sulphuric acid.

Sackett's method for total cholesterol is discussed here which is a modification of Bloor's method.

Sackett's Method for Total Cholestrol

Principle. Proteins are precipitated and cholesterol is extracted by an alcohol-ether mixture. The extract is evaporated and residue is dissolved in chloroform. A green colour is developed by Libermann-Burchard reaction and it is measured colorimetrically.

Procedure. Measure 12 ml of alcohol-ether mixture and transfer it to a dry centrifuge tube and add 0.2 ml of serum into it slowly. Cork the tube and shake vigorously for one minute. Keep the tube in a horizontal position for half an hour, so that the precipitate is evenly distributed along the tube. Centrifuge it at 1500 r.p.m. for 5 minutes. Decant the supernated fluid completely in a small beaker and it is evaporated to dryness on a water bath. Dissolve the residue after evaporation in 5 ml of chloroform, lable the tube "Unknown". Label two other test tubes "Standard" and "Blank" and pipette out 5 ml of working standard cholesterol solution (containing 0.4 mg of cholesterol) and 5 ml of chloroform into the tubes respectively. Add 2 ml of acetic anhydride-sulphuric acid mixture to each of the three tubes. Mix and keep in the dark for 15 minutes.

Adjust the wave-length of the photometer to 680 nm. Set the photometer to 100% transmittance (or zero absorbance) with blank and read unknown and standard.

Calculations:

$$\text{Serum Cholesterol (mg/100ml)} = \frac{Au}{As} \times \frac{100}{0.2} \times 0.4$$

$$= \frac{Au}{As} \times 200$$

Au: Absorbance of the unknown

As: Absorbance of the standard

Reagents:

(*i*) *Alcohol-ether mixture:* Mix 3 volume of ethanol and 1 volume of ether.

(*ii*) *Chloroform:* absolutely anhydrous and pure.

(*iii*) *Accetic anhydride and sulphuric acid mixture:* Mix 20 ml of acetic anhydride with 1 ml of concentrated sulphuric acid. Make freshly just before use.

(*iv*) *Stock standard cholesterol solution:* Mix 200 mg of pure cholesterol in chloroform and make upto 100 ml with chloroform. Keep well stoppered.

(*v*) *Working standard cholesterol solution:* 1 ml of stock standard cholesterol solution is diluted to 25 ml with chloroform. 5 ml of this solution contains 0.4 mg of cholesterol.

Precaution: All the glass-wares used in the procedure must be absolutely dry.

Interpretation: Serum cholesterol varies from 150-250 mg/ 100 ml in healthy young adults of this about 1/3 to 1/2 is present as free cholesterol and the remaining is esterified. The level rises with age that may go upto 300 mg/100 ml in elderly.

An increase in serum cholesterol known as hyper-cholesterolaemia is found in diseases like obstructive jaundice, hypothyroidism, diabetes mellitus.

A decrease in serum cholesterol (hypocholesterolaemia) is found in hepatocellular damage, anaemia, acute infections etc.

Some Important Tests for Cholesterol

1. **Salkowski test or sulphuric acid test.** Dissolve a few crystals of cholesterol in a little chloroform and add an equal volume of concentrated sulphuric acid. Colour changes from bluish red to cherry red and purple in the chloroform layer whereas the acid assumes a green fluorescence.

2. **Libermann-Burchard (Acetic anhydride-Sulphuric acid test).** Dissolve a few crystals of cholesterol in 2 ml of chloroform in a dry test tube. Add 10 drops of acetic anhydride and 1 to 3 drops of concentrated sulphuric acid. The solution becomes first red, then blue and finally bluish green in colour.

Quantitative—Urine

Estimation of Glucose in Urine

Urine consists of water and water soluble waste products which are removed from the blood stream by the kidneys and excreted as urine. Normal urine consists of approximately 95% water—the rest being made up of organic constituents (like urea, uric acid, creatinine, ethereal sulphate, acetone body etc.) and inorganic constituents (like ammonia, chlorides, phosphates, sulphates, carbonates, sodium, potassium, calcium etc).

Chemical examination of urine may yield useful information in many abnormal conditions e.g., infections of kidney, ureters, bladder and urethra is indicated by the presence of pus, red blood cells and organisms in the urine. Reducing substances in the urine indicate diabetes mellitus. Similarly the presence of bile may indicate liver diseases.

Diagnostic test for sugar in urine

There are two tests which are commonly used to show the presence of sugar in urine: (*i*) Benedict test (*ii*) Fehling's test. Among these two tests Benedict's test is more satisfactory and is used more frequently. Benedict's test is discussed here which is as follows:

Benedict's test. Some substances are capable of reducing blue cupric sulphate to red cuprous oxide when boiled in alkaline solution. If reduction occurs, a yellow to brick red precipitate of cuprous oxide is formed.

Procedure. Take 5 ml of Benedict's reagent in a test tube and add 5 ml of urine and mix. Heat the test tube for 5 minutes in water bath and cool spontaneously. A greenish to yellowish brown to reddish brown precipitate is formed which shows the presence of sugar. A white precipitate may be produced due to the presence of phosphate, but this is not recorded.

If the solution appears green due to the suspension of a yellow precipitate, then the presence of reducing substance or sugar is indicated as "trace". If the solution shows a yellow tinge, result is reported as +, an organge precipitate as ++ and a brick red precipitate as +++.

Reagents: Benedict's reagnet prepared by heating a mixture of 173 g of sodium citrate, 100 mg of anhydrous sodium carbonate in about 100 ml of water in a beaker. To this, a solution of copper sulphate is added by constant stirring, cool and transfer to a one litre volumetric flask and make upto the mark with water.

Benedict's copper reduction method has been adopted for use by the diabetic patients in their own home using a standard dropper. 5 drops of urine, 10 drops of water and a tablet containing copper sulphate and sodium hydroxide are added to a test tube. The heat evolved by the solution of sodium hydroxide boils the mixture, and if there is presence of sugar in the urine, it forms the precipitate. A small colour chart is included, so that a rough assessment of the amount of sugar present in the urine can be made.

Another convenient test for detection of glucose in urine is "Clinistix- strip test". In this test, strip paper impregnated with glucose oxidase, O-toluidine and a vegetable peroxidase is available. The test-end of the strip paper is dipped in the urine and removed. If the paper turns blue within a minute, it confirms the presence of glucose in the urine. The glucose-oxidase of the strip oxidises the glucose to gluconic acid and hydrogen peroxide in the presence of atmospheric oxygen.

The gluconic acid so formed gives blue colour with O-toluidine in presence of peroxidase, positive results can be obtained with concentrations of glucose above 10-100 mg / 100 ml.

For most purposes a sufficiently good idea of the amount of glucose can be obtained from the qualitative tests. For example, urine containing 0.2% sugar or more, the supernatant fluid is colourless when the precipitate is allowed to settle down. A greenish precipitate is obtained with sugar in the region of 0.25%, a greenish yellow at about 0.5% and a yellowish orange with about 1%. Sometimes an accurate determination of the concentration of glucose in urine is required, in such case the amount of sugar has to be estimated by quantitative analysis. A common and convenient method is that of Benedict's.

This is a titrimetric method based on the reducing property of glucose. The method is not specific for glucose and measures all the reducing carbohydrates.

Procedure. Perform Benedict's qualitative test with the urine and note the colour of the precipitate. The urine should be diluted according to the colour of the precipitate.

Green precipitate	—	No dilution
Yellow precipitate	—	Dilute 1 in 2
Orange precipitate	—	Dilute 1 in 5
Red precipitate	—	Dilute 1 in 10

The dilution should be done in a volumetric flask.

Fill the diluted urine in the burette after prior rinsing. Note the initial reading. Pipette out 10 ml of Benedict's quantitative reagent in a clean porcelain dish. Add approximately 5 g of solid sodium carbonate and a few pieces of porcelain to reduce the risk of bumping. Keep the dish over a burner and the contents of the dish should be allowed to boil. Urine is added from the burette to the boiling solution in the porcelain. If the solution in the porcelain gets concentrated, a little water may be added into the dish. Continue the addition of urine from the burette until the solution in the dish becomes colourless. This is the end point. Repeat the titration for concordant values.

Calculation: It is known that 10 ml of Benedict's reagent is reduced by 20 mg of glucose. If 'a' ml of diluted urine is required to reduce 10 ml of Benedict's quantitative reagent:

$$\text{Glucose in } a \text{ ml of diluted urine} = 20 \text{ mg}$$

$$\text{Glucose in 100 ml of diluted urine} = \frac{20 \times 100 \text{ mg}}{a}$$

$$\text{Glucose in 100 ml of undiluted urine} = \frac{20 \times 100 \times D}{1000}$$

$$= \frac{20 \times 100}{a \times 1000} \times D_g = \frac{2D}{a} \text{ g}$$

D = Urine was diluted to D times.

Results: Urine glucose concentration $= \frac{2D}{a}$ g/100ml

Reagents: Benedict's quantitative reagent consists of:

Crystalline copper sulphate	—	18 g
Anhydrous sodium carbonate	—	100 g
Sodium citrate	—	200 g
Potassium thiocyanate	—	125 g
Potassium ferrocyanide (5% soln)	—	5 ml
Water to make up	—	1000 ml

Sodium carbonate, sodium citrate and potassium thiocyanate are dissolved in warm water to make about 800 ml of solution and cooled. Copper sulphate is dissolved separately in about 100 ml of water and added to the first solution. Then 5 ml of 5% potassium ferrocyanide solution is added and the volume is made upto 1000 ml with water. 10 ml of this reagent is reduced by 20 mg of glucose.

Diagnostic Test for Salts in Serum and Urine

Chloride, sodium and potassium ions play an important part in maintaining normal water distribution between cells, plasma and interstitial fluid due to osmotic pressure. Together with bicarbonate they form the most important part of cation-anion balance.

Test for salt in serum

The sodium and potassium of whole blood appear to be present entirely in the ionized state. About 90 per cent of the total blood sodium is found in plasma whereas potassium is primarily present in the cells. Whole blood sodium or potassium determination have little clinical significance, hence plasma or serum is ordinarily used for analysis.

The concentration of serum sodium and potassium is determined either by flame photometry or by chemical analysis based on the precipitation of a metal complex.

Sodium is chemically determined by precipitation in the form of triple salt, Uranyl zinc sodium acetate which dissolves in water but is insoluble in ethanol, from protein free serum sample. The triple salt thus precipitated should be determined by gravimetric or titrimetric or photometric method.

The chemical method used mostly for serum potassium is precipitation of potassium, as the insoluble cobaltinitrite followed by titrimetric or photometric estimation.

The determination of the concentration of the potassium and sodium by flame photometry is based on the following principle:

The sample in solution form is introduced in the form of a fine continuous spray into a non-luminous gas using either natural acetylene, propane or butane gas. By the use of colour filter or diffraction grating the emitted light of a wavelength characteristic for the ion being analyzed, is isolated and focussed on a photoelectric cell. The electrical response to the photoelectric cell is measured on a suitable meter previoulsy calibrated to convert the electrical impulse to ion concentration either by direct reading or by reference to a calibration curve.

Normally the range of serum sodium is 300 to 355 mg /100 ml or 130 to 154 m Eq per litre. The potassium concentration of serum is 14 to 22 mg /100 ml or 3.6 to 5.6 m Eq per litre.

An increase in serum potassium is found in severe Addison's disease in acute bronchial asthma, in fever etc. The highest values seen in disease are 9 and 10 m Eq/litre.

A reduced serum potassium is reported due to severe vomiting or diarrhoea during the attacks of familial periodic paralysis.

The sodium ions have immense importance in the maintenance of osmotic pressure and acid-base balance. Significant decrease in sodium ion is noted in pregnancy, pneumonia, Addison's disease etc.

Tests for chlorides in serum

Chloride content in blood serum is commonly expressed in terms of milli-equivalents (mEq) of chloride per litre of the sample, since the major functions of chloride in the body are concerned with osmotic pressure regulation and acid base balance. One mEq of chloride ion corresponds to 35.5 mg or 58.5 mg of sodium chloride. The general relationship between two of expressing chloride content is as follows:

$$\text{mEq per litre} \times 5.85 = \text{mg chloride as NaCl per 100 ml}$$

The normal value of chloride varies from 98 to 108 mEq per litre of serum or plasma which is equivalent to about 570 to 620 mg of sodium chloride per 100 ml. Chloride ion concentration indicates the state of carbon-dioxide balance since chloride shifts to compensate for the carbon-dioxide changes in serum. For example, high chloride values are often accompanied by low bicarbonate values as in acidosis or the reverse is true in alkalosis. Chloride changes are also seen in gastro-intestinal disturbance involving loss of gastric juice (*e.g.* in emesis or dehydration) and in pathological disturbances in which the water balance or acid base balance of the blood is altered.

Chloride ion in serum can be determined by various methods, *e.g.* Schales and Schales method, Vanslyke and Hiller and coulometric method.

The method of Vanslyke and Hiller is simple and quick though the end point is not so clear, this method is based on the principle that protein free serum is shaken with an excess of solid silver iodate and filtered. Chloride present reacts with the insoluble silver iodate to form insoluble silver chloride and soluble iodate, which passes into the filtrate. On the addition of iodide to the filtrate, the iodate reacts to produce free iodine which is then titrated with standard thiosulphate.

In Schales and Schales method, the sample is titrated with standard mercuric nitrate solution using diphenylcarbazone as indicator. Chlorides present react with the added mercuric ions to form soluble, but undissociated mercuric chloride. When an excess of mercuric ion is added, the indicator turns purple. The end point is sharp.

The change in normal concentration of the salt, *i.e.*, sodium and chloride indicates certain diseases, e.g. there is loss of sodium and chloride from alimentary tract by vomiting, diarrhoea, urine and sweat. Severe diarrhoea leads to loss of sufficient sodium chloride which reduces the plasma chloride considerably. Low serum sodium and chloride levels are observed in a number of acute infections like meningitis, lobar pneumonia etc.

An increase in plasma sodium and chloride is found in acute nephritis, diseases of the urinary tract, enlarged prostate, water depletion etc.

Continuous over consumption of sodium chloride leads to hypertension or high blood pressure.

Test of Salts in Urine

Detection of Chloride in Urine: Take 5 ml of urine in a test tube and add concentrated nitric acid and a few drops of silver nitrate. White precipitate of silver chloride shows the presence of chloride ions which darkens on standing and is soluble in ammonium hydroxide. Depending upon the amount of chloride precipitate, it is reported as normal, diminished or absent.

Approximately 170 to 250 mEq of chloride is excreted per day in an average adult. This amount is closely dependent upon the chloride content of the food ingested. In Pneumonia and certain other acute infectious diseases the chloride ions are markedly diminished. A decrease is also seen in nephritis associated with edema.

Determination of Sodium and Potassium: Sodium and potassium in urine can be estimated by flame photometer. The intensity of the emitted colour at the wavelengths specific for each of the ions is compared to a standard curve and the concentration is determined.

Generally on a normal diet 130 to 220 mEq (3 to 5 g) of sodium (an average of 120 mEq per litre) and 25 to 100 mEq (1 to 4g) (an average of 40 mEq per litre) of potassium is excreted daily in the urine by a healthy normal adult. The sodium excretion is greatly influenced by variations in intake of sodium and in disease varies between very wide limits. An excessive loss of sodium is found in renal diseases. The excretion of sodium and potassium increases in aldosteronism, alkalosis and in fever. The determination of ability to retain sodium or potassium is determined by deprivation test and the excretion of these ions are determined by a loading test. These tests are of excellent diagnostic value for diseases due to derangement of production of adrenocortical or pituitary hormones.

Salt-deprivation test: The rate of excretion of sodium and chloride ions are controlled by endocrine hormones, derangement of production of these hormones occur in Addison's disease, hypo or hyper aldosteronism, Cushing's syndrome etc. The derangement results in impaired ability to conserve sodium chloride on a low salt diet or to excrete it during overloading. Both the salt deprivation test and salt loading test have been used but the former is more extensively used.

Test: A low salt diet is ingested for three days. Fluid intake is also controlled during the test period and urine and blood are collected at selected intervals. All samples are analyzed for sodium and chloride.

Interpretation: A four hour collection during the third day should reveal a chloride output less than 125 mg per 100 ml of urine. Excretion of 125 to 225 mg per ml indicates hypoadrenalism or Addison's disease while values more than 225 mg indicate low adrenal cortical output. The volumes of urine output, the sodium concentration and the potassium concentration can also be used as parameters for this study in place of chloride ion.

Detection of Cholesterol in Urine

Examination of urinary deposit sometimes yield valuable information either of a positive or negative character. The constituents of such urinary deposit are classified into three main groups e.g (*a*) chemical substance —cystalline or amorphous (*b*) cells from the blood or various parts of the renal tract (*c*) casts.

Examples of chemical substances which may be present both normally or pathologically *i.e.* showing the presence of a disease are phosphates, oxalates, calcium carbonate, calcium sulphate, uric acid, urates, amino acids and organic substances like hippuric acid, indigo, cholesterol etc.

Different type of cell present in the urine are red blood cells, pus cells, epithelial cells etc.

Different types of casts present in the urine are Hyaline, epithelial, granular casts etc.

Cholesterol crystals occur very rarely in urine and are found in some diseases of kidneys and renal tract. The crystals are typical as rectangular or rhomboid plates with notched corners (Figure 2). These crystals are slightly soluble in cold water and ethanol but readily soluble in them when hot. They are soluble in acetic acid as well.

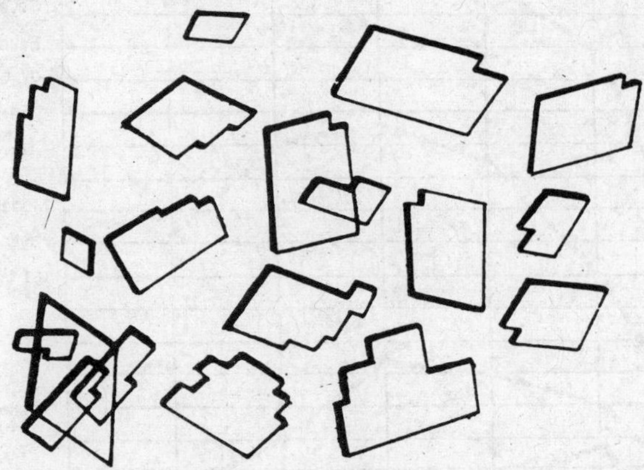

Fig. 2. *Cholesterol crystals.*

Detection of Diabetes

Blood glucose analysis is of great value in detection of diabetes. Due to hormonal control blood sugar level is maintained constant at a value of 70-120mg of glucose per 100ml. Though several hormones are involved in this, the most important ones being insulin and glucagon. Diabetes is caused by hormonal imbalance of these two hormones. This leads to decreased production of insulin, which results in acumulation of sugar in the blood. Diabetetes is detected by testing a urine sample for the presence of sugar in the urine. By diet control the presence of sugar in the urine may come down to normal range. But this may not be so in the blood where the sugar level may be beyond the normal range. Thus the severity of the diabetic condition, that is, when the blood sugar level is 160-180 mg/dl, can be known only by blood sugar tests.

One of the important blood sugar tests for detecting the diabetic condition is the sugar tolerance test or glucose tolerance test known as GTT.

Method: This test is usually carried out in the morning after a night's fast but if necessary it may be done some four or five hours after the last food was taken. Blood samples are taken for the determination of fasting blood sugar. After withdrawing the blood sample the patient is orally given 50g or 1.75 g of glucose/kg body weight and blood samples are taken subsequently at half, one and two hours intervals and the blood sugar level is determined. In the normal case the blood sugar level rises to about 150 after an hour, then starts falling and at the end of two hours the normal level is restored. In diabetic condition the sugar level rises higher at 200-300mg and stays high for a longer time i.e. 3-6 hours. Further the peak value is reached in longer time (2 hours). In non-diabetics the peak of the curve is usually between half and one hour. Typical sugar tolerance values in normal diabetic and other conditions are shown in Fig. 3.

For estimation of glucose, sodium fluoride (100mg/ml) has to be added to the blood immediately after collection. This is done to inhibit blood enzymes which alter the amount of glucose in blood during storage.

Detection of Anaemia

Anaemia can be defined as the reduction in the circulating red cell mass. Since this information is difficult to obtain, it is more practical to define anaemia in terms of haemoglobin (Hb) concentration and red cell counts. Different standards are adopted in different institutions. If the haemoglobin is less than 10g percent or the RBC count is less than 3.8 million/c mm, it is considered as an anaemic case and a complete haemogram should be done, to find out the type and cause of anaemia.

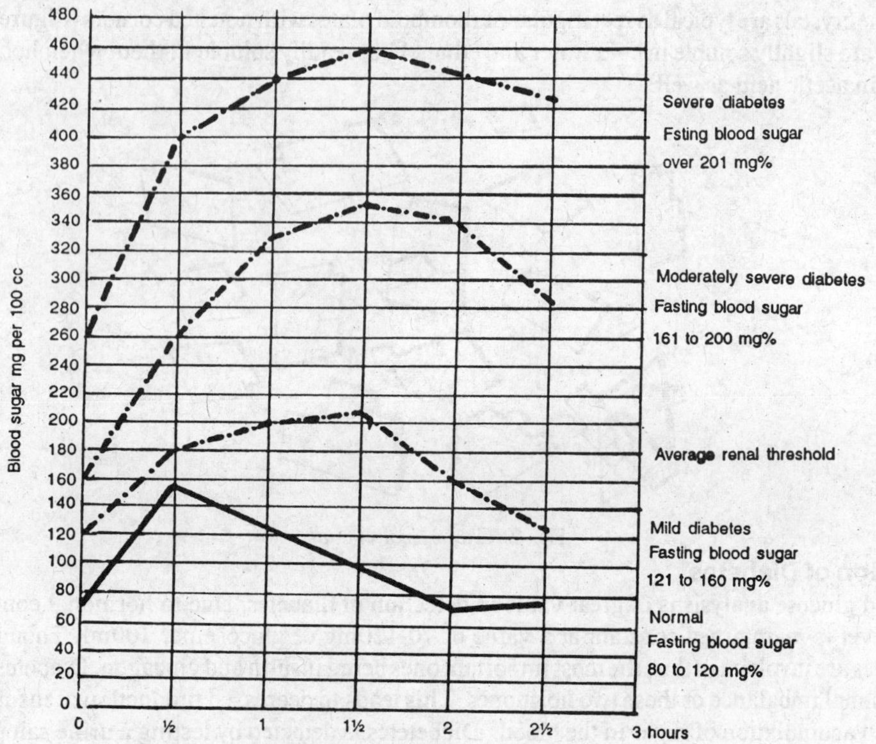

Fig. 3. *Glucose tolerance curves..*

Anaemia can be caused due to several factors like loss of blood, decreased formation of red cells and haemoglobin, iron deficiency, Vitamin B_{12} or Folic acid deficiency, malnutrition etc.

Anaemia can be diagnosed by estimating haemoglobin concentration and the red çell count.

Estimation of Haemoglobin (Hb Concentration)

Principle. Haemoglobin in the red blood corpuscles is converted into acid haematin by oxidation of haem. The brown colour so developed is matched against standard brown tinted glass in a comparator by direct vision. Reading is taken directly as gHb/100ml blood. Capillary or venous blood may be used for the estimation.

Apparatus and Reagent

Apparatus needed for the estimation is known as Haemoglobinometer box (Sahli's) which contains the following apparatus:

(*i*) **Comparator.** A slot accommodates the Sahli tube. Non-fading standard brown tinted glass pieces are provided on either side of the slot for colour matching. An opaque white glass is fitted at the back to provide uniform illumination.

(*ii*) **Haemoglobin tube (Sahli Adam tube).** Graduated in g percent of Hb (2-24) as well as percentage (20-140) as in Figure 4 (a).

(*iii*) **Haemoglobin pipette.** It has only one mark indicating 20 cmm (0.02ml) blood. (Figure 4 (b)).

(*iv*) A thin glass rod stirrer ordinary glass tube dropper with a teat.

It bears graduations in g percent on one side and in percentage on the other 100 percent being equivalent to 17.3g Hb per 100ml of blood.

Reagent: N/10 hydrochloric acid (0.1N HCl) solution, distilled water, pricking needle, cotton swab, alcohol.

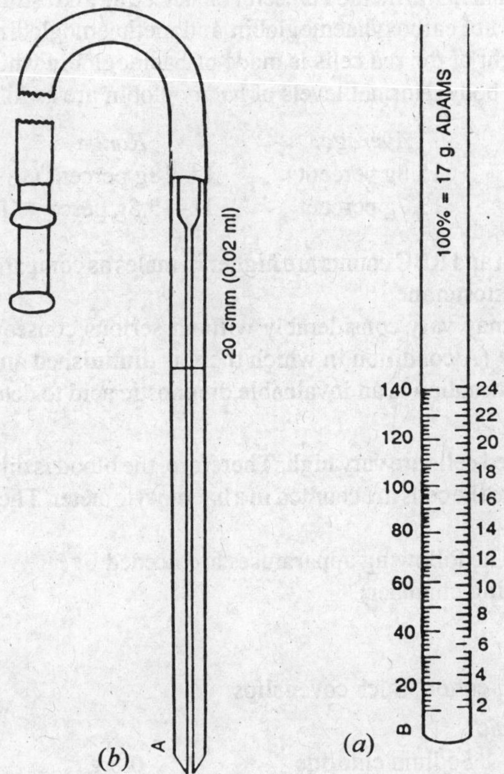

Fig. 4. *Sahli-Adam haemoglobin tube.*

Procedure. Take haemoglobin tube and note the graduations. Clean and dry the tube, add 3g or 20% N/10 hydrochloric acid and put it aside. Get a finger prick with a sterile needle and wipe away the first one or two drops of blood with a cotton swab.

When a good sized drop forms at the puncture site, clip the tip of the haemoglobin pipette in it and draw blood upto 20c mm mark, by gentle sucking on the plastic mouth piece fitted to the rubber tube, the tip should not press against the skin nor move out of the drop of blood, otherwise air will enter the pipette and the whole procedure will have to be repeated again. Transfer the blood from the pipette into the Sahli tube, suck a small amount of acid into the pipette and expel it again into the tube, repeat this twice to make sure that no solution remains in the pipette. Mix the contents quickly and gently by shaking the tube before any clotting of blood occurs. Put the tube back in the comparator and wait for 10 minutes. During this time the red cells rupture liberating the haemoglobin into the acid solution; the acid reacts with haemoglobin and converts it into acid haematin which is brown in colour. Take out the tube from the comparator, add a few drops of distilled water and stir the contents with the glass rod gently. Continue to add water drop by drop, stirring the contents each time till the colour of the solution matches the standard. Read the upper meniscus since it is a coloured solution and note the reading.

Results. The extent of error with this method is about 8-10% which can be reduced to about 5% by taking three readings; first, when the colour is slightly darker than the standard; second, when the colour exactly matches the standard and the third, when the colour is slightly lighter than the standard.

The average of the three readings gives a more accurate result and it is expressed as.

Haemoglobin concentration = g per 100ml blood.

Discussion. Haemoglobin may be present in the blood in many forms, *e.g.* oxyhaemoglobin, carboxyhaemoglobin, methaemoglobin etc., hence it is important to know the limitations of any particular method. Sahli's and haematin method is useful far detecting and estimating oxyhaemoglobin and reduced haemoglobin but not carboxyhaemoglobin and methaemoglobin.

About 90% of the dry weight of the red cells is made of haemoglobin which plays a vital role in the normal functioning of the body. Normal levels of haemoglobin are as follows:

		Average	*Range*
Males	=	15.8g percent	14–18g percent (16± 2)
Females	=	13.7g percent	12–15.5g percent (14 ± 2)

Haemoglobin concentration and RBC counts are higher in males as compared to females due to the action of the male hormone testosterane.

Although the Hb content may vary considerably without serious consequences but any severe anaemia causes tissue hypoxia (A condition in which there is diminished amount of oxygen in the tissues). The acid haematin estimation is an invaluable diagnostic acid to detect anaemia.

2. The Red Cell Count

Principle. The number of red cells are very high. Therefore, the blood is diluted 200 times with an appropriate diluting fluid before the cells are counted in a haemocytometer. Their number in undiluted blood can then be calculated.

Apparatus and Reagents: The following apparatuses are needed:

1. Haemocytometer (counting chamber).
2. RBC Pipette.
3. Hayem's fluid.
4. Pricking needle, alcohol, cotton, thick cover slips.

Composition of Hayem's Fluid:

Sodium chloride	0.5 g
Sodium sulphate	2.50 g
Mercuric chloride	0.25 g
Distilled water	100 ml

The mercuric chloride is antifungal and antibacterial. The diluting fluid is isotonic with blood.

Haemocytometer or counting chamber: It is a single solid heavy glass slide. There are three narrow parallel platforms in its middle third portion and which extend across it. The central platform or floor piece is 1/10mm lower than the other two and is divided into two parts by a shoot transverse gutter or trench as shown in the Figure 5.

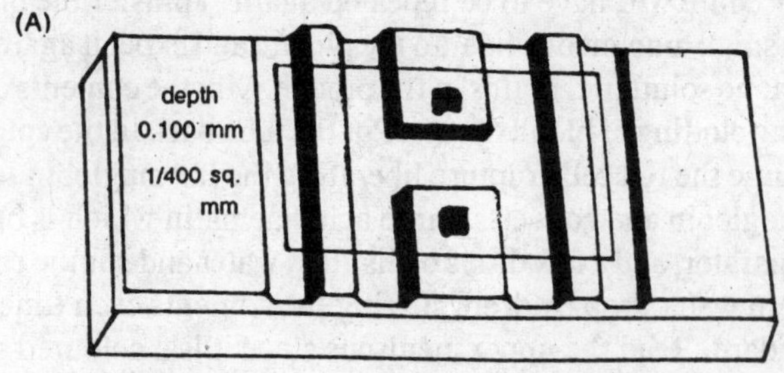

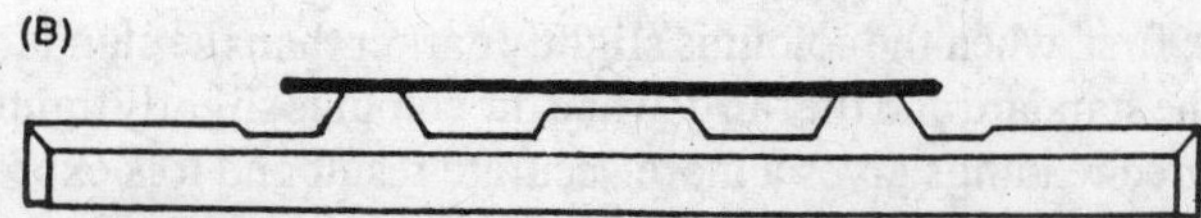

Fig. 5. *Haemocytometer (counting chamber).*
(A) Surface view with the cover slip in position (B) Side view with the cover slip in position

Each platform is ruled with a counting grid for red and white blood cells. When the cover slip is placed in position, a space of 0.1mm deep is left between the under surface of the cover slip and the counting platform.

The counting chamber in common use has improved Neubauer's ruling. There are 400 small squares grouped into 25 groups of 16 squares each by two lines called trans lines. Squares, for counting white blood corpuscles, four groups of 16 square each are also etched on counting platform as shown in Figure 5. In every make the dimensions and volumes of the squares are always specified on the chamber surface. Each chamber can be examined under low and high power objectives as in Figure 6.

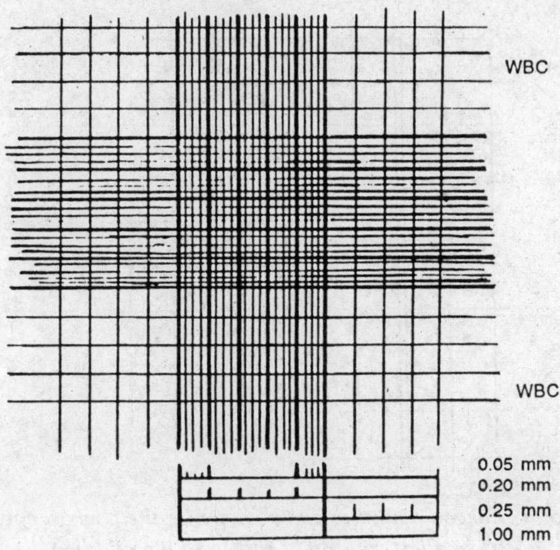

Fig. 6. *Improved Neubauer's chamber. The red blood cells are counted in five groups of 16 smallest squares each. (size 0.05mm)*

RBC Pipette. It has a narrow glass stem graduated in tenths, with three markings 0.5, 1.0 and 101. The glass capillary tube widens into a small bulb containing a red glass bead. The bulb narrows down again into a glass capillary tube and at this point it is marked 101 as shown in Figure 7.

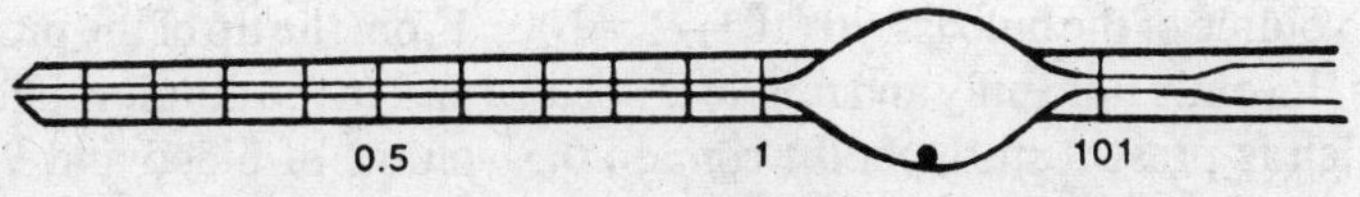

Fig. 7. *RBC pipette.*

The end of the tube is attached to a thin rubber tube with a plastic mouth piece. The red bead helps in mixing the contents of the bulb and for a quick identification of the RBC pipette.

The markings 0.5, 1.0 and 101 indicate the volumes of blood and diluent and do not indicate cmm of blood and diluent.

Procedure. Place the counting chamber on a flat surface with a blank white paper under it. Then put the cover slip on the surface of the chamber near the microscope. Draw the blood from finger tip through the RBC Pipette upto 0.5 mark to be followed by Hayem's fluid to the mark 101. Mix the contents of the bulb thoroughly for one minute, with the precaution that the red cells towards the end of the pipette should not be centrifuged. Discard first few drops from the pipette and charge the counting chamber with the dilute blood by placing the tip of the pipette on the surface of the chamber touching the edge of the cover slip at an angle of 45° to the horizontal. Allow the diluted blood to flow evenly and slowly under the coverslip by capillary action. Remove the pipette quickly as soon as the counting platform is covered. There should be no overflow into the central or side trenches as the excess fluid will lift up the cover slip and give false high counts. If any diluted blood spills over into the trenches, the chamber and the cover slip must be cleaned and chamber recharged.

Place the recharged haemocytometer on the stage of the microscope and locate the central 1mm sq. area for red cell counting under low power. Allow the cells to settle for one minute before starting the cell count. Focus the lines and cells under low power to get a general impression of the distribution of the RBC over the counting grid as shown in Figure 8.

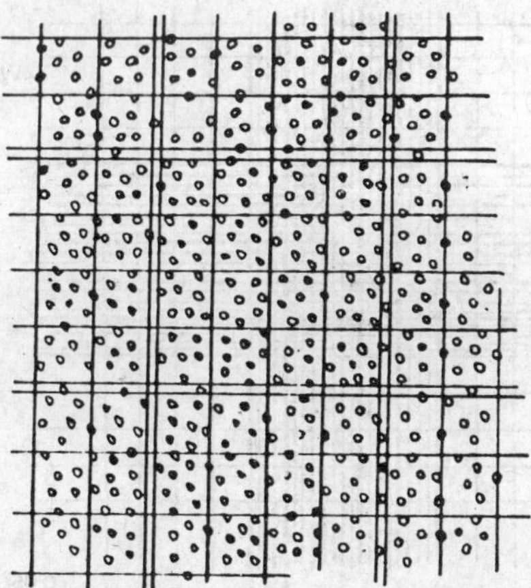

Fig. 8. *View of a charged counting chamber under the microscope. A group of 16 smallest squares is shown in the middle.*

If large discrepancies are seen in different parts then the spread is evidently uneven and the counts are not reliable. In such case, chamber should be cleaned and recharged.

The cells are counted in the four corner and one middle square each with 16 smallest squares i.e. in 80 squares. Switch to high power objective and adjust the position of the haemocytometer so that a group of 16 smallest squares comes into view. Count the RBC in each square and those lying on its left and lower lines, ignore those on its right and upper lines so that one can avoid counting the cells twice. Write the count for each small square as it is made.

Observation: The volume of the bulb is 100 (101 – 1 = 100). From the tip of the pipette to the mark 1.0, the stem contains Hayem's fluid only and this does not take part in the dilution. Thus 100 volumes of diluted blood (which is present in the bulb) contain 0.5 volume of blood and 99.5 volumes of Hayem's fluid, giving a dilution of half in hundred. Therefore, dilution obtained is one in two hundred *i.e.* 1 : 200 (200 times).

Calculation: The area of the smallest square is 1/400 sq mm and volume is 1/4000 cmm (side = 1/20mm, depth=1/10mm, volume = $1/20 \times 1/20 \times 1/10$ = 1/4000 cmm or mm^3.

Example: Suppose number of cells is 80 squares = 480.

Number of cells in one square = 480/80 = 6.

Volume of one square = 1/4000 cmm.

Thus there are 6 RBC in 1/400 cmm of diluted blood. The number of cells in 1 cmm will be $4000 \times 6 = 24000$ as the dilution employed is 1 : 200 the number of cells in 1 cmm of undiluted blood is $24000 \times 200 = 4800000$.

Result is expressed as: RBC Count = 4.8 millions/cmm.

Multiplying the number of cells in 80 squares (480 in this example) by 10,000 will give the RBC count in millions per cmm blood *i.e.* by putting four zeros in front of 480.

Normal RBC count: Normal adult levels are as follows:

	Mean	*Range*
Male (16-17 Yrs)	5.0 millions/cmm (m^3)	4.75-5.5 million/cmm
Female (16-70 Yrs)	4.5 million/cmm	4.0-5 million/cmm.

The higher count in males is due to testosterone. With the method of RBC counting the error is said to be ± 15%.

RBC count less than 3.8 millions/cmm indicates anaemic conditions in a person.

Chapter 5

First Aid for Accidents

First aid is the immediate treatment given to the victim of an accident or sudden illness before the proper medical aid is reached. Everyone must know what first aid should be given in such cases. The main aims of first-aid is to preserve life, to prevent worsening of the casualty's condition and to promote recovery.

Certain Important Rules of First Aid

1. In case of any accident if the cause of the accident is still there, then it should be immediately removed, e.g. live electric wire, pillars or logs on body, a room with poisonous fires etc.
2. Should attend to the casualty immediately, quietly and without fuss. A golden rule of first aid is to "haste, slowly"
3. People should not crowd around as fresh air is essential.
4. Bleeding should be stopped immediately.
5. Casualty must be treated for shock by moving the patient as little as possible and very gently.
6. Minimum and precise first aid should be given which is essential to save life and prevent the condition from worsening.
7. If necessary, artificial respiration should be given immediately.
8. Arrangement should be made at the earliest to provide proper medical care by a doctor or a hospital.

1. Cuts, Abrasions and Bruises

Minor cuts and abrasions: These wounds are caused by any sharp instruments like knife, razor etc. If it is a clean wound, then apply little tincture of iodine or spirit and cover the wound with a clean piece of gauze or surgical dressing.

Extensive and deep cuts: In such cases, wound should be covered with a sterilised dressing and immediate medical help should be arranged.

Dirty wounds: In such wounds cleanliness is essential for the prevention of infection and everything that comes in contact with a wound should be scrupulously clean. In such cases wash the wound with boiled water and soap and then apply tincture of iodine and cover the wound by a sterile piece of gauze or bandage. In case of dirty wounds or deep cuts due to rusted objects there is danger of titanus and hence a doctor should be approached for giving antititanus injection.

In case of wounds which are accompanied by haemorrhage the casualty should be placed in a suitable position, because bleeding is less when the patient sits and it is still less when the patient lies down. Bleeding part should be elevated except in the case of a fractured limb. Any visible foreign body should be removed with a piece of clean cotton and gauze. Apply and maintain pressure directly or indirectly. Wound should be bandaged and injured part should be immobilised.

Bruise: A person may get a bruised finger as a result of being caught in a door or get a black eye after a blow. In such cases cold water compression is given to the injured part.

Precaution should be taken that ointment should be an antiseptic ointment or wet dressing over the wound should be sterilised properly.

Doctor should be immediately contacted if the wound is deep.

2. Bleeding

Bleeding can be from capillaries, from an artery or a vein.

In capillary bleeding, blood is bright red and oozes out from the wound. In such cases wound should be covered with a sterile piece of gauze and bandaged tightly.

In bleeding from an artery, blood is bright red and comes out in spurts, and in bleeding from a vein, the blood is blackish red and flows in a steady stream.

To check bleeding firm pressure with thumb can be applied at the bleeding spot or at the upper part of the artery or over the part of the vein distant to the wound to stop bleeding. Injured part can be raised above the level of the heart to check bleeding.

In case of bleeding from nose, head should be thrown backwards and an icebag or a pad dipped in cold water should be kept on the patient's nose, forehead and at the back of the neck.

In the case of internal bleeding which occurs with in the cavities of the body such as those of chest or the abdomen, actual bleeding is not visible in this case and it has to be detected from the signs and symptoms, such as paleness of face, rapid and weak pulse, sweating, rapid and shallow breathing etc. If there is any indication of internal bleeding immediate arrangements should be made to remove the patient to a hospital. Patient should be kept in a supine position and his head should be lowered. Any tight clothing should be loosened round the chest, abdomen and waist, to allow the patient to breathe more freely.

3. Fractures

Bone sometimes breaks as a result of an accident, there is intense pain and swelling of the injured part at the site of the fracture. In such cases the patient should not be moved without immobilising the injured limb with a splint. The injured leg should be tied to the other leg at several levels. Fractured arm can be tied with a splint and should be supported by a broad arm sling. (Fig. 1).

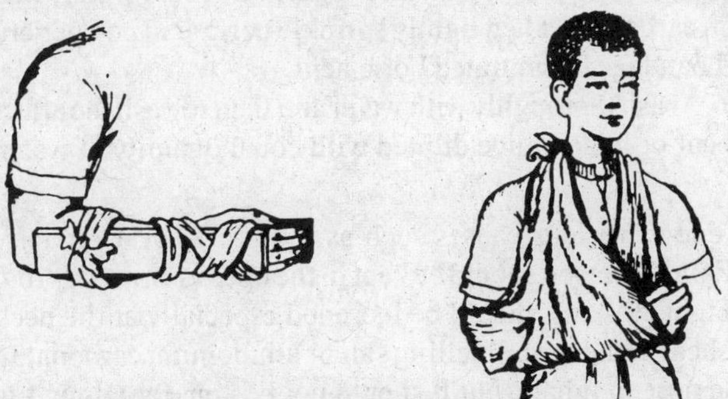

Fig. 1. *Application of splint and arm sling for fracture of forearm*

Broken necks and backs: If a person's back or neck has broken, be very careful when shifting the casualty. Try not to change the patient's position, but if the casualty has to be shifted, it should be done very carefully. The casualty should be put on a stretcher without bending.

Broken ribs: These are very painful but almost heal on their own. It is better not to splint or bind the chest. The best treatment is to give aspirin and rest.

Broken bones that tear through the skin: These types of fractures have great risk of getting infected. It is always better to get help from a doctor. Clean the wound and exposed bone with boiled water and never put the bone back into the wound until the wound and the bone are absolutely clean.

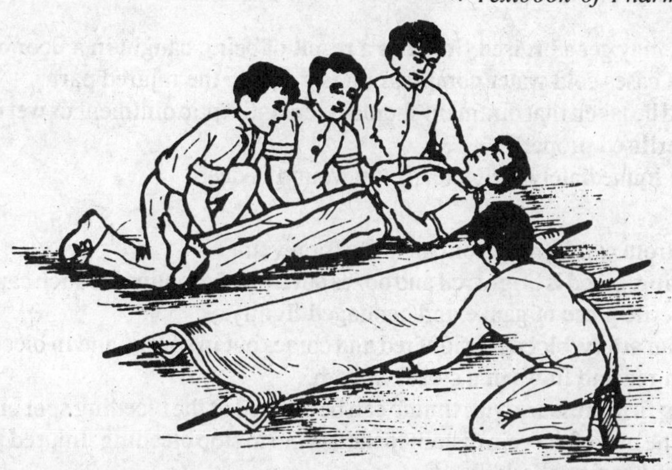

Fig. 2.

4. Burns

Minor burns: These are usually caused as a result of contact with a hot vessel or steam or by spilling of very hot water or oil etc. The burned part may become red or it may develop blisters.

In such cases some antiseptic cream like Burnol or Savlon should be applied at the burnt area after washing the area gently with water. The burnt area can be submerged in clean cold water for some time. Warm drinks may be given to the patient.

In case of extensive burns, *i.e.* if a person's clothes catch fire, smother the flame by wrapping the victim in a thick blanket or a rug or coat. The casualty should be covered up with warm clothing. Blisters should not be opened or punctured. Plenty of fluid should be given, stimulants like tea or coffee can also be given. Medical help should be arranged immediately.

Chemical burns: These can be caused by strong chemicals like acid, alkalis, bromine, phenol etc. Any soiled clothing should be immediately removed and affected area should be thoroughly washed by clean water.

Burns due to strong acid: After washing thoroughly with cold water, wash the affected area with 5% solution of sodium carbonate. Then lightly sprinkle boric acid powder on the wound and bandage with wet compressed bandage of saturated boric acid.

Burns due to alkali: Wash thoroughly with water and then rewash the affected part with a weak acid solution such as vinegar or lemon juice diluted with equal quantity of water.

5. Fainting

Fainting may be due to various causes such as overwork, over-anxiety, excessive heat, severe injury, bleeding etc. Fainted person should be kept in the open or in an airy room to ensure an abundant supply of fresh air. Tight clothing should be loosened especially at the neck, chest and waist. Cold water should be splashed on the face, smelling salt or ammonium carbonate or crushed onion should be kept in front of the nose to inhale, but it should not be kept for a long time.

Precaution should be taken that nothing is given by mouth till the person can swallow. When the patient recovers, some hot stimulating drink like tea or coffee should be given. If fainting is prolonged, proper medical aid should be arranged.

In case of fits or convulsions which are common in children and may occur at the onset of fever or due to other causes like epilepsy etc., people should not be allowed to crowd around the patient. A piece of thick cloth should be kept between the teeth to avoid the tongue from being bitten. Nothing should be given by the mouth till the patient regains consciousness and is able to swallow.

6. Poisonous bites

1. *Bite by a rabid dog:* Bite by a rabid dog causes a disease known as rabies or hydrophobia, which is fatal. In such cases some quantity of blood should be allowed to flow out of the wound and then the

wound should be washed thoroughly with clean water and soap. The dog should be kept under observation and casualty should be taken to the doctor for advice on anti-rabic injections.

2. *Snake Bite:* All the snakes are not poisonous. A bite by poisonous snake usually show two punctured wound about 2.5 cms apart, from which reddish fluid oozes out and there is pain, swelling and discolouration at the site of the wound.

In case of a bite by a very poisonous snake, send for a doctor at once. Meanwhile, make the patient lie down and complete rest should be given. Casualty should not be allowed to walk. A piece of cord or a handkerchief should be tied as close to the wound as possible, but this bandage should be tied between the wound and the heart to stop the flow of the venom to all parts of the body.

Bleeding should be encouraged from the wound so as to wash out the poison from the body. For this purpose, keep the part low and also keep the affected part in the ice water.

In case of very poisonous snake bite like rattle snake bite, incision of 1/4 inch, deep and 1/4 inch long should be made through each fang mark and mechanical suction like sucking by mouth should be applied.

A First Aid Box

In every house a small first aid box containing the following articles should be kept:

1. Absorbent cotton wool pads.
2. Sterile gauze pieces
3. Roller bandages
4. A roll of adhesive tape
5. Tincture iodine
6. Rectified spirit
7. A pair of scissors and forceps
8. An antiseptic cream like Savlon, Burnol etc.
9. Antiseptic lotion like Dettol
10. A bottle of smelling salt
11. Toilet soap
12. Baking soda or Sodium-bicarbonate
13. Electric torch

Detection of Hallucinogens and Poisons—Antidotes for Poisoning

Some substances when taken into the body in fairly large quantities can be dangerous to health and can cause death; such substances are known as poisons. They get into the body either by swallowing, by breathing poisonous gases or by injection.

Majority of poisons are non-specific and diagonosis are usually made from the history obtained from the patient's relatives or on circumstantial grounds such as finding tablets beside the patient. Certain clinical symptoms can suggest the cause of poisoning, for example:

(*a*) Pinpoint pupil, vomiting and depressed respiration suggest over doses of morphine and related alkaloids.

(*b*) Widely dilated pupil, bladder distension, arrhythmias and pyramidal tract involvement suggests, tricyclic anti depressant toxicity.

(*c*) Sweating, tinnitus (a ringing or roaring sound in the ears), deafness, strongly suggest acute salicylate poisoning.

(*d*) Delirium, hallucinations suggest alcohol, atropine, antihistamine pesticides, methyl bromide poisoning

Treatment should be given immediately and should not be delayed by spending excessive time in attempting to identify the poison. Basic therapeutic principles for the treatment are:

1. Maintenance of Respiration

A clear airway is necessary. This can be ensured by removal of dentures, vomitus, foreign bodies etc. Patient should be turned to one side to prevent tongue falling backwards causing obstruction to the air passage and to avoid aspiration of vomitus and mucus. Oxygen should be given.

2. Inactivation or Removal of Poison

When a liquid or solid poison is in contact with the patient's body or clothes, these must be removed and washed off to avoid any absorption.

3. Vomiting, Gastric Aspiration and Lavage

When the poison has been swallowed, conscious patient should be given an emetic i.e. two table spoon of salt to a tumbler of water, to induce vomiting or they can also be given activated charcoal.

If the patient is unconcious or if the poison is corrosive like acids or alkalis, vomiting should not be induced. In such cases the poison should be neutralised by giving an antidote.

In few cases of poisoning gastric aspiration and lavage is done which is a process of draining the fluid from stomach, by suction and washing the stomach or rectum by tipid water. Lavage (washing of stomach) should be done till clear fluid comes out. In corrosive poisoning 100 ml of milk may be left in stomach or in case of salicylate poisoning a solution of sodium-bicarbonate is left behind.

4. Dilution of the Poison

The poison must be diluted by giving large quantities of cold water which will dilute the irritant and lessen the irritating effect. Some soothing drinks like milk, barley water, raw egg or flour mixed with water can also be given.

Some Common Poisons and Their Antidotes

The Corrosive Group

1. Acid poisoning. It is caused by strong acids like hydrochloric acid, nitric acid and sulphuric acid.

Symptoms: In such cases patient may complain of burning throat, pain in mouth, abdomen and tenesmus (meaning a painful ineffectual straining to empty the bowel or bladder). Patient may have acidic, brown or black vomiting containing blood, mucous membrane. Difficulty in swallowing and breathing is also noted.

Antidote: In such cases vomiting should not be induced and gastric lavage should not be performed. Plenty of water should be given to the casualty to dilute the acid. Acid should be neutralised by giving two table spoons of chalk, milk of magnesia, tender coconut water, beaten egg etc.

2. Alkali poisoning. It is caused by Caustic soda, Caustic Potash, Ammonia etc.

Symptoms : Mouth and throat show corrosion, vomit is alkaline, thick, slimy, and dark, containing stringy mucous and blood. There is burning pain in gastrointestinal tract, difficulty in swallowing and breathing, nausea and vomiting.

Antidote : Stomach wash, and emetics should not be used to induce any vomiting. Alkali should be neutralised by acids like dilute vinegar (acetic acid) lemon or orange juice. Plenty of water or coconut water should be given. Milk, ghee, olive oil and white of an egg may be given. Prescribed dose of Morphine can be given for pain.

3. Poisoning by disinfectants. Disinfectants like Carbolic acid, lysol, phenyl etc. cause poisoning.

Symptoms: There is burning pain sensation in gastrointestinal tract, nausea, thirst, convulsions, shock, respiratory failure etc.

Antidote: Vomiting should not be induced. Four spoons of magnesium sulphate (Epsom salt) in a litre of water should be given by mixing it with a tea cupful of medicinal paraffin. Castor oil can also be given.

4. Hallucinogens. Hallucinogens are psychotomimetic or psychotogenic drugs. These drugs produce depersonalization, changes in mood and a variety of effects on memory and learned behaviour.

These drugs produce hallucinations hence these are known as hallucinogens. Some examples of these drugs are LSD (lysergic acid diethyl amide) mescaline, hashish, ganja, bhang etc.

5. Atropine (alkaloid). Bella donna, alkaloid (the deadly night shade) Dhatura poisoning. *Symptoms :* In such cases, face is flushed along with dialation of pupil. dysphagia, impaired vision, hallucinations, giddiness, delirium, excitement and convulsions may be noted.

Antidote: Stomach wash should be given with potasium permanganate solution. Respiratory stimulant along with oxygen should be given. The specific antidotes of morphine poisoning are Naloxona and Nalorphine. These drugs produce dramatic reversal of morphine induced respiratory depressions and are generally administered intravenously.

6. Alcohol. Smell of alcohol in breath and vomiting is noted. Tongue becomes dry and furred. Patient may develop abnormal behaviour, reaction of pupil may vary from extreme dilation to extreme contraction. It may result into tremors, shallow breathing, incoordination, coma, and even may lead to death.

Treatment: Stomach wash, and emetics are used. Oxygen inhalation is also helpful.

7. Mercury poisoning. It is a protoplasmic poison.

Symptoms : These are salivation, thirst, burning of throat, abdominal pain, vomiting, bloody diarrhoea, shock etc.

Antidote: White of egg in water should be given, followed by milk. Any emetic can be used to make the casualty vomit. It should be removed by lavage with water or 6 table spoons of activated charcoal per litre. Magnesium sulphate can be given by mixing with water - Dimercaprol injections can be given.

8. Salicylate poisoning. This poisoning is most commonly caused by aspirin injection. *Symptoms:* are flushed face tinnitus, abdominal pain, vomiting, convulsions, pulmonary edema, dehydration and coma.

Antidote: Airway should be cleared, gastric lavage, or stomach wash should be done by warm water or water containing one table spoon activated charcoal per 200 ml water. Dehydration should be treated by giving saline water.

Chapter 6

Causes of Common Diseases and Their Treatment by Drugs

The word 'disease' literally means departure from normal health. We can characterise health as a state of complete physical, mental and social well-being or a feeling of complete harmony with the environment. But the environment in which the human being or any other organism lives is a hostile environment. There are many outside influences which disrupt the delicate internal balance of the body and upset the harmony and thus leads to a disease. Diseases could be due to many causes like extremes of temperatures, lack of proper nourishment, parasite, micro and macrobiological life which invade living organisms etc.

Depending upon the various causes the diseases could be broadly classified as follows:

(*a*) **Infective diseases.** All the infectious diseases are caused by germs. These germs can spread through air, water, insects etc., for example typhoid, small pox etc.

(*b*) Diseases due to physical and chemical burns and poisons.

(*c*) Diseases due to injuries.

(*d*) Diseases due to nutritional deficiencies such as rickets, scurvy etc.

(*e*) Diseases due to new growth such as cancer.

(*f*) Diseases due to degenerative origin such as thickening of arteries and liver cirrhosis, Oto-sclerosis (thickening of ear ossicles leading to deafness).

(*g*) Miscellaneous diseases like diseases of allergic origin.

(*h*) **Hereditary diseases.** A few diseases are hereditary i.e. diseases which are transferred from one generation to the other. Example : diabetes, asthma, mental disorders etc.

Some Common Insect Borne Diseases and Their Treatment

1. Malaria

Malaria is caused by a protozoa known as malarial parasite. This infection is transmitted by female anopheline mosquitoes from one person to the other. The malarial parasite can be detected in a blood smear under a microscope. Malaria is characterised by a high fever with shivering. Sometimes along with high fever there is vomiting, headache and severe body ache.

Treatment: Anti-malarial drugs like chloroquin or camoaquin is given. Any analgesic like Novalgin etc., is given for body or headache. Any paracetamol, (example crocin) etc., may be given to reduce high temperature.

Prevention: In control of malaria the most important measurement is to check the growth of mosquitoes, which can be done by spraying DDT and keeping the surroundings clean, water should not be allowed to get accumulated anywhere as stagnant water is the breeding place of mosquitos.

Two tablets of camaquin can be taken weekly to prevent malaria.

2. Filariasis

This disease is characterised by the swelling of the legs or some other parts of the body. The disease is commonly known as elephantiasis due to its resemblance to a leg of an elephant.

In India Filariasis is caused by a worm known as Wuchereria bancroft and W. malayi. The infection is transmitted by culex mosquito from one person to the others.

At the beginning of the disease one gets repeated attacks of fever off and on and there is inflammation of the lymphatic system where the adult worm lives.

Prevention and treatment: Once the swelling appears there is no other treatment except surgical operation to reconstruct the affected part.

The control of culex mosquitoes is very difficult as common insecticide like DDT is ineffective to kill this type of mosquito. A drug diethyl carbamazine has been shown to kill the microfilaria though it has no action in curing this disease. Weekly intake of diethyl carbomazine tablet is a good preventive measure against filaria.

3. Plague

Plague is caused by a germ known as Pasturella pestis, a deadly bacterium. It goes from rat to rat through the rat fleas. When the rats die of plague, the fleas leave the dead rat and then these fleas can bite a man and can inject plague germs into his blood stream.

Now effective treatment is available against plague. Tetracycline and streptomycin are effective in the treatment of plague. But one should immediately report to the doctor as this disease is deadly. The control measures include the control of rat population, removal of garbage and use of insecticides to kill the fleas. Innoculation of antiplague vaccine is very useful for protection against plague.

Common Air-borne Diseases, Their Control and Treatment

1. Diphtheria

It is an infectious disease which affects tonsil, phrynx and larynx and is caused by specific bacteria. It usually affects children up to five years of age. It may start as sore throat, mild fever and sometimes vomiting and headache. In diphtheria, throat or tonsils show a gray membrane which may spread down and cause difficulty in breathing.

Prevention and treatment: Prompt treatment with diphtheria anti-toxin should be given. Patient should be kept in well-ventilated room and children should not be permitted in that room.

Every child should be innoculated with triple vaccine which prevents an infant from diptheria, whooping cough and tetanus.

2. Whooping Cough

It is a highly infectious disease causing inflammation of the respiratory passages with severe attacks of cough.

Prevention and treatment: There is no specific treatment, but it is necessary to keep the child under nedical supervision for the relief of cough.

The disease can be prevented by giving triple vaccine to the infants.

3. Influenza

It is commonly known as "Flu" and is highly infectious. It causes fever and pain all over the body. The disease is caused by various types of influenza viruses.

Prevention and treatment: There is no specific curative drug for this disease. An attack of influenza is followed by severe weakness which can be overcome by taking nourishing food and rest.

4. Measles

It is a highly infectious disease causing fever, inflmmation of the air passages and a rash all over the body. It attacks especially children below five years.

Prevention and treatment: Any person adult or a child affected by measles should be kept in isolation and in a well-ventilated room. There is no specific treatment for this measles. Analgesics and antipyretics may be given for body pain and fever. Sulpha or Penicillin drug can be given to prevent any secondary infection.

5. Mumps

It is an infectious disease, causing fever, difficulty in opening the mouth and painful swelling of parotid glands which lie just below the lobe of the ears.

It is caused by the virus of mumps. Patient should take bed rest to avoid complication. There is no other specific treatment or preventive vaccine.

6. Common cold

It is not a disease but a symptom and is caused by a virus and starts with irritation of the nose, feeling of discomfort, headache, slight sore throat, sneezing etc. All the symptoms disappear within four to seven days, if there is no other complication.

Person suffering from cold should take rest in a well-ventilated room. Aspirin, hot saline gargles and hot drinks help in reducing the discomfort. It the patient gets high fever or loss of voice or ear-ache, a doctor must be consulted.

7. Tuberculosis (T.B.)

It is an infectious disease and spreads through air, contaminated food, cough of the infected person, dust or through direct contact. It is caused by a microbacterium known as Tubercle bacillus. Tuberculosis, especially, pulmonary T.B is a major health problem in India. The main causes of spreading of this disease is due to lack of diagnostic and treatment facilities, poverty, unhygenic conditions, over-crowding and ignorance about the disease, irregular treatment, insanitary habits, indiscriminate coughing etc. spreads the disease.

The symptoms of this disease are low fever, raised pulse rate, tiredness, loss of appetite, loss of weight, breathlessness and one of the most important symptoms is cough with sputum which may sometimes be blood tinged. An X-ray or the chest can detect early stages of the disease and screening of the chest with X-ray can detect only very advanced disease. The early diagnosis, proper and continuous treatment, proper diet, and hygienic conditions can cure this disease completely.

Prevention and treatment: The discovery of potent anti-tuberculosis drugs like streptomycin, isoniazide, PAS (paramino salicylic acid) have helped in complete cure of this disease. These drugs can be administered orally and should be given in adequate doses and for a prolonged period. Along with drugs it is necessary to give the general line of treatment like complete rest, keeping the patient in an airy room with good ventilation, proper nourishing food like plenty of milk, egg etc. should be given.

Some preventive measures are—use of handkerchief, while coughing or sneezing. Sputum must be collected and disposed carefully. Patient should be kept under medical supervision. BCG vaccination (anti-tubercular vaccine) should be given to other members. The new-born babies should be given a dose of BCG vaccine.

Some Common Water-borne Diseases

1. Cholera

This is an infectious disease which is caused by a germ known as Vibrio cholera. The cholera germ can infect a healthy person when contaminated food or drink is taken by mouth. The patient starts passing frequent stools and gets repeated vomiting. The disease can be diagnosed by the microscopic examination of the stool or vomit.

Prevention and treatment: The cholera patient should be immediately shifted to hospital and should be kept separately in a room or in isolation ward. As there is rapid loss of fluid and salt through stool and vomit, the patient should be given saline water intravenously. In order to relieve thirst, sips of water or strong black coffee may be taken in small sips.

For cholera great emphasis is laid on its prevention and early treatment. The drinking water must be disinfected and the sanitation must be maintained properly, so that the flies do not breed anywhere and do not spread the disease by contaminated food. The food should be kept well covered. Cholera vaccine should be given to people. It is very effective and gives protection against cholera.

2. Typhoid

The germs of this disease are carried through water and is known as salmonella typhi. The patient suffering from typhoid gets fever for three or four weeks.

Prevention and Treatment: The typhoid fever is treated with a specific antibiotic known as Chloromycetin. These modern drugs control the temperature within a few days but proper rest and food is essential for some days, otherwise the disease may relapse.

Since this disease spreads through contaminated water and food, great precaution should be taken to disinfect the water and contamination of food by flies and dust should be prevented.

Prophylactic typhoid innoculation is available against this disease.

3. Dysentery

Dysentery of two types (*i*) bacillary dysentery (*ii*) amoebic dysentery. These can be differentiated by the examination of stools. Bacillary dysentery infection may be caused by micro-organisms known as shigella bacillus or Flexneri bacillus or Sonnei bacillus. Amoebic dysentery is caused by an micro-organism known as Entamoeba hystolytica.

Bacillary dysentery spreads mainly by flies whereas amoebic dysentery spreads through water. It is an acute intestinal disease with frequent loose motions along with blood and mucus, fever and abdominal pain. Headache and vomiting may be there in a few cases. It is likely to become a chronic disease if it is not properly treated.

Treatment: Bacillary dysentery can be treated by giving sulphaguanidine tablets, mexaform tablets or chlorostrep tablets in prescribed doses. Plenty of water and fruit juices should be given. Amoebic dysentery can be cured with Enterovioform, Flagyl etc. Tinilazole tablets should be given in prescribed doses. In both the cases diet should be restricted. Fatty, fried and spicy food should be avoided.

Some Disorders of Digestive System

1. Jaundice

Jaundice is yellow discolouration of the skin, mucus membrane, conjunctiva etc. It is important to note the colour of the stool and urine as it helps in the diagnosis. Jaundice may be due to unusual destruction of red blood corpuscles and it is a symptom which indicates that the bile pigment produced in the liver are more than what the liver can dispose off. The excess thus gets retained in the blood and that is why the eye and the skin becomes yellow. The colour of the bile or bile pigment is derived from the breakdown of the haemoglobin present in the red blood cells. Therefore, when for any reason there is an excessive destruction of red cells, there is an excess of bile pigment. Jaundice may also occur when the liver is affected and is unable to deal with the products of red blood cell destruction that takes place normally. This happens in an acute condition called infectious hepatitis and is caused by a virus. Hepatitis is the common cause of jaundice especially in small children. The hepatitis virus passes from the stool of one person to the other through contaminated water or food.

The symptoms are greyish colour stool, yellow coloured urine and yellow coloured eyes, nausea, vomiting, loss of appetite, constipation, tenderness in the region of the liver.

Prevention and treatment: Antibiotics do not work against hepatitis. In fact, some medicines cause damage to the sick liver. Patient suffering from jaundice should take rest and drink plenty of liquid and fruit juice like sugar-cane juice etc. Fatty food should be avoided.

Some important preventive measures are that sick persons stool should be buried or burned carefully as the germs spread through it. The patient should be kept very clean and the persons attending the patient should wash their hands whenever they attend the patient. Hepatitis can also be transmitted by giving injection, hence only sterile or disposable needles should be used.

Common Disease of the Respiratory System

1. Asthma

It is characterised by periodic attack of breathlessness. The difficulty in breathing is felt mainly in expiration. The syptoms are mainly due to sudden narrowing of the bronchial tubes due to spasm of the bronchial muscles and swelling of the mucous membrane. During asthmatic attack breathing becomes very difficult and one has to put all muscles of respiration into action. Patient may have cough and wheezing noises in the chest. The asthmatic attack can last for a few hours or days. The attack may reoccur at intervals of a few days or weeks or months. It is not a contagious disease, but is more common in children with relatives who have asthma. It is generally worse during certain months of the year or at night. Sometimes serious and prolong asthma may develop into emphysema, which is a very serious and incurable condition of the lungs.

Prevention and treatment: Asthma may be caused by eating or breathing, certain things to which the person is allergic. Common causes of allergic reactions are due to dust, certain medicines, hair from cats and other animals, specific food, feather pillows etc.

The patient having an asthmatic attack should be kept in a comfortable position and the room should have clean fresh air. Patient should be given lot of liquid. For mild attack rephedrine tablets may be given. If the attack is very severe adrenaline injection or aminophylline injection can be given subcutaneously. If the person is wearing tight clothing it should be loosened. In rare cases, round worms cause asthma; in such cases deworming medicines should be given. If the person has fever or attack lasts more than three days tetracycline or erythromycin capsules can be given. A person with asthma should avoid the things he or she is allergic to. Persons with asthma may improve when they move to a different area or to a different climate.

Asthma patients should protect themselves from cold and should take early treatment if they suffer from infection.

A Common Disorder of Nervous System

1. Epilepsy

The affected person gets attack of fits very often. Patients suffering from epilepsy might get a peculiar feeling just before the fit, which may cause convulsions, loss of consciousness and violent movements, the eyes often roll back and there may be some frothy material at the mouth. Such attacks may come either from internal disease of the brain such as tumour, abscess or by some damage of brain at birth or by high fever in infancy or by tape worm cysts in the brain.

Prevention and treatment: Epilepsy is not infectious. People who have the tendency to get fits regularly should not be allowed to swim, climb a tree or work with a machinery, as they might injure themselves. During the fit, proper care should be taken such as people should not be allowed to crowd near the patient, to ensure free passage of air, tight clothes should be loosened near the waist and neck, a piece of thick cloth should be kept between the teeth to avoid the tongue from being bitten. Nothing should be given by mouth till the patient is conscious. The medicines do not cure epilepsy, but they help to prevent fits and these medicines should be taken life long, for example, Phenobarbital, diphenyl-hydantoine etc.

Some Other Common Diseases

1. Piles

Pile is the enlarged and lengthened varicose veins just under the mucous membrane and the skin of the anal canal. The exact causes of piles in a majority of the cases are unknown although it has been found that things like sedentary habits, constipation or raised pressure in major venous channels may cause piles. The veins of the anus are covered with thin mucosa which tears easily by a constipated hard faecal mass and thus causes bleeding. Sometimes the veins may elongate so much that they may protrude outside the anal canal and this condition is known as prolapse. The prolapse may reach a severe degree, when the blood inside these veins get clotted and the condition is known as prolapse. The prolapse may reach a severe degree, when the blood inside these veins get clotted and the condition is known as thrombosed piles. These piles may even ulcerate and get severe inflammation and all these conditions are extremely painful. Severe piles may cause anaemia due to loss of blood.

Prevention and treatment: In the earlier stages piles can be controlled by some specific exercises, taking plenty of fluids, taking sufficient fats like ghee, oil etc. in the diet, non-vegetarian food should be avoided, laxative should be taken regularly. To get relief from the pain the patient can sit in a tub filled with warm water, in which little potassium permanganate is added. Certain bitter plant juices like witch hazel etc., when applied on the piles, help to shrink them. If these simple measures fail or if the piles are complicated, then surgical treatment is required.

2. Leprosy

It is a communicable disease and there are lots of misconceptions about it in the society, hence every individual should be made aware about the cause, prevention and proper treatment of this disease. Leprosy is caused by a micro-organism known as Leprosy bacillus or M. Leprae.

The clinical features or signs of leprosy differ greatly depending partly on the stage of the disease and partly on the type of the reaction that occur in the body in response to the invasion of the leprosy bacilli. If the body shows resistance to these invading bacillus then it is known as non-lepromatous and is generally non-infectious, but if the body does not offer much resistance to the invading organisms then it multiplies and spreads all over the body. This type of leprosy is known as lepromatous and is very infectious from the early stages.

The signs and symptoms of the non-lepromatous type are appearance of one or more patches on the skin and there is loss of sensation or feeling in some parts of the body such as in hands and feet. These patches do not have any itching, burning or pain and are never ivory white like the patches of leucoderma.

The signs and symptoms of lepromatous type are slight change in colour and texture of the skin, swollen nerves that form thick cords or lumps under the skin, large chronic sores that do not hurt or itch, in advanced stages deformities occur viz., atrophy of fingers, toes, ulcers on hands, feet etc.

Leprosy is not transmitted through air or water and it is not hereditary and a casual contact is not enough to acquire the disease. It spreads only through close and prolonged contact with only the infectious cases. It has earlier been mentioned that all the cases are not infectious. The contact is either to skin to skin or through clothes or the utensils.

Prevention and treatment: Adequate treatment of leprosy results in cure in all cases. An early diagnosis and treatment can cure leprosy before developing deformities and the infectious cases on proper treatment become non-infectious and are harmless from the public health point of view. The best medicines for treating leprosy are Sulfones like Dapsone tablets. Patient should be kept under medical supervision.

Unfortunately 85% of leprosy cases do not get any proper treatment due to ignorance and fear of being excluded from the society.

The large open sores and gradual loss of hands and feet are not caused by the disease itself and can be prevented. These symptoms result because of lack of feeling, a person no longer protects himself against the injury. For example, when a person with leprosy gets a blister on his or her feet, there is no sensation of any pain and if they do not take any precaution, then the blister turns into an open sore and gets infected and the infection slowly spreads into the bones and destroys them. It can be prevented by protecting the body from cuts, bruises, blisters etc., person should not walk bare-footed, if there is an open sore it should be kept very clean and further injury to the sore should be avoided. Patients should wear gloves if they are doing any work to protect them from injuries.

As for preventive measures every one should get examined once in a year by the family physician for any sign of leprosy and if there is any symptom, person should start the treatment immediately. It is a moral duty of every individual to spread an awareness about leprosy so that every leprosy patient gets proper treatment and should not be rejected by the society.

Indian Medicinal Plants and Trees

The use of plants or plant extracts for medicinal purposes has been in use for thousands of years, herbalism and folk medicines both ancient and modern have been source of much useful therapy. The drugs of plant origin have been used as diuretics, carminatives, expectorants, analgesics, antidepressant drugs, anaesthetics, insecticides, etc. Diseases like rheumatic fever, diabetes, hypertension, arthritis etc., have been successfully treated by these drugs.

Some important medicinal plants are discussed here.

1. Adathoda Vasica

This plant belongs to Acanthaceae family. In Hindi it is known as "Adulasa, in Tamil Adadodai, in Sanskrit as Vasaka.

It is found throughout the plains of India and in the sub-Himalayan tracts upto 4000 ft.

The plant is of immense medicinal value. Its leaves and roots are used in cough as expectorant, asthama, bronchitis, whooping cough etc. Its leaves are also used in rheumatism and has insecticidal properties.

Flowers, leaves and roots also have antiseptic properties.

2. Ocimum Sanctum

It belongs to family Labiatae. In Hindi it is known as Tulsi, in Tamil, Telugu and Sanskrit as Tulasi. This plant is found throughout India.

Medicinal properties: Its leaves are used as expectorant, diaphoretic, antiperiodic etc. Leaves are also useful in the treatment of bronchitis, gastric disorders. An infusion of the fresh leaves in boiling water is a remedy for nausea, vomiting, earache etc. Dried leaves are used as snuff in ozaena which are used as cure for headache and bad colds.

Seeds are used in disorders of the genito-urinary system. They are also taken internally as an antidote to poison.

Roots are used as diaphoretic in malarial fevers.

Fresh roots, leaves, stems are applied to the bites of mosquitoes. Whole plant is used in snake bite and scorpion sting.

3. Hibiscus Rosa – Sinensis

It belongs to family Malvaceae. It is also known as shoe flower (English), in Hindi it is called Jasum. in Sanskrit as Joba, in Tamil Sembarithi. It is cultivated in gardens throughout India.

Medicinal properties: Flowers are used as emollient, aphrodisiac and refrigerant, decoction of flowers are used in bronchial catarrh. Leaves are used as emollient, aperient and laxative, a decoction of leaves are used as a lotion in fevers. Roots are used for coughs, in venereal diseases, fever, gonorrhoea and menorrhagia.

4. Mangifera Indica (Mango-tree)

It belongs to family Anacardiaceae. In Hindi it is known as Aam, in Tamil Mamaram, in Telegu Amramu. It is generally cultivated in the tropics.

Medicinal properties: Ripe fruits are used as laxative, astringent, diuretic. It is also useful in haemorrhage in uterus, lungs or intestines.

Unripe fruit. These are useful in ophthalmia and eruptions. It exhibits antibacterial, antifungal activities.

Rind of fruits also show astringent, and stimulant properties.

Seeds are used in asthma. Seed kernel has astringent taste and is used as anthelmintic and also in bleeding piles.

Bark has astringent taste and is used in diptheria and rheumatism. Gum is used as a substitute for gum arabic. It is used in dressings for cracked feet and for scabies.

5. Azadirachta Indica

It belongs to Meliaceae family. In Hindi it is known as Neem tree, in Tamil as Vepa, Sanskrit Nimba. It is planted all over India.

Medicinal properties: Bark has astringent, antiperiodia properties and is used in syphilis.

Gum has proteolytic activity and is used in Catarrhal.

Leaves have antiviral, antineo-plastic properties. It is used as an antiseptic in many skin diseases like boils, eczema, ring worm, scabies etc. Flowers—They possess anthelmintic properties. Dry flowers—tonic, stomachic etc.

Oil is stimulant, antiseptic, used in rheumatism and skin rheumatism. (an acute fever associated with streptococcal infattion).

Neem plant, in general, is used in inflammation of gums, sores, fever, spleen complaint, tumours, measles, small pox, cholera, nausea, fever, various parts of the plant are used in cosmetics as the extracts of bark, leaf, stem, root etc., are beneficial for skin.

6. Ficus

Commonly known as fig family. It belongs to Moraceae family. Figs in their natural and dried forms are highly nutritive food recommended for growing children. They are great health promoter, being rich in all the necessary nutritious elements. In dry figs the sugar content is high. Figs are recommended for digestive problems.

F. Earica: In Hindi, it is known as anjira, in Tamil Simaiyathi, Telugu - Anjuru.

Medicinal properties : It is an anticancer compound. Fruits are used as demulcent, emollient and as nutritious fruit, laxative, useful in anaemia.

Milky juice from the fresh green fruit is acrid in taste which is used to destroy warts.

F. glomerata: The fruits, roots, bark, and leaves of this plant are useful in the cure of menstrual irregularities, urinal diseases, mouth wounds and diabetes. The bark decoction helps to protect fertilised ovum in womb.

7. Solanum trilobatum

It belongs to family Solanaceae. In Sanskrit it is called Alarka, in Tamil Tuduvalai, Telugu Lavaste. It is mainly found in Tamil Nadu, Gujarat, Maharashtra etc.

Medicinal properties: Root and leaves - These are bitter in taste and are prescribed in consumptive cases in form of decoction.

Berries and flowers are given in cough; Decoction of the plant is useful in chronic bronchitis.

8. Phyllanthus niruri

It belongs to family Euphorbiaceae. In Hindi it is known as Jaramla, in Tamil Kilkkaynelli, Malayalam Kizhkkynelli. This plant is found throughout the hotter parts of India from Punjab to Assam and southwards to Travancore. This plant is used as diuretic in gonorrhoea and other troubles of genito urinary tract.

Fresh shoots are given in dysentery. Fresh roots are an excellent remedy for jaundice. Powdered leaves and roots along with rice water is used to lessen swellings and ulcers. A milky secretion of the plant is applied on offensive sores. Leaves contain a bitter substance, called phyllanthin which is a popular remedy against fever.

9. Grasses

(i) Eragrostis cyanosuroides (Dharbe grssx)

It belongs to family Gramineae. In Hindi it is known as Durva, Tamil as Attudharbe, Sanskrit—darbha. It is found throughout in hot and dry places of India. It finds use as diuretic and stimulant in dysentery and in menorrhagia.

(ii) Cymbopogon citratus (Lemon grass)

It belongs to family Gramineae. In Hindi it is known as Gandhatrina, Tamil- vasana pillu. It is grown in northern parts of India and in south India also it grows in Mysore.

The oil distilled from lemon grass is of immense medicinal importance. Citral is the principal constituent of essential oil. It gives immediate relief in obstinate vomiting, flatulent colic. It is used in emulsion form, for the treatment of cholera. It is mixed with sweet oil and is very useful remedy for sprains, chronic rheumatism, muscular pains, lumbago etc. (lumbago-a painful rhceumatic cendition).

(iii) Cymbopogan juarancusa

It also belong to family Graminaeae. In Hindi it is known as Lamjak, Sanskrit Lamaggaka. It is found from Kashmir to Assam and in north Western plains of India.

The grass is used to purify blood. It is also an useful remedy for the treatment of coughs, chronic rheumatism, cholera, gout, fever (gout-excess of uric acid, type of metabolic disease).

Essential oil obtained from this grass is also used.

(iv) Cynodon dactylon

It belongs to family Gramineae. In Hindi it is known as Dubha, Tamil Arugampullu, Sanskrit Durva. It is found throughout India ascending to 8000 ft. in the Himalayas.

Juice of the grass is used in the treatment of fresh cuts, wounds, in hysteria, epilepsy, insanity. It is useful in catarrhal ophthalmia.

10. Greens

(i) Tinospora cordifolia

It belongs to Menispermaceae family. In Hindi it is known as Gulancha. In Tamil Sindal, Telugu Somida.

Medicinal value: Stem is bitter and is used as stomachic, antiperiodic. It is also used as aphrodisiac.

The juice of the plant is good for liver complaints. It is also useful in cough. The juice of the plant is also helpful in the treatment of paralysis. Starch obtained from roots is nutritious and is used in chronic diarrhoea and dysentery.

(ii) Basella alba

It belongs to family Basellaceae. In Hindi it is known as Poi, in Tamil Vasalakkirai, Telugu- Batsala. It is found throughout India, wild or cultivated. It is a climbing herb and is used as wholesome vegetable.

Medicinal value: Its leaves are demulcent and diuretic and are useful in gonorrhoea. It is particularly useful in piles. Leaves have digestive power and are used for diabetes. It cures burns and are used for prickly heat.

Juice of leaves is used in constipation particularly in children and pregnant women.

(iii) Portulaca oleracea

It belongs to Portulacaceae family. In Hindi it is known as Khursa, Tamil Pulikkirai, Telugu Pappukura, Malayalam - Pasalai. It is found throughout India upto 5000 ft. in the Himalayas. It is a herb and mainly its leaves are used as a vegetable.

Medicinal values: Its leaves are useful in scurvy and liver diseases. The juice of the stem is applied to prickly heat. Seeds are vermifuge.

(*iv*) *Portulaca quadrifida*

It also belongs to Portulacaceae family. In Hindi it is known as Chhota uniya, in Tamil—Passal-aikkirai, in Telugu—Goddupavili in Malayalam—Sirupasalai. It is found throughout the warmer parts of India.

Medicinal value: Fresh leaves are used as external application in erysipelas and as a diuretic in dysuria. Seeds are vermifuge. [erysipelas—An acute inflammation of skin and subcutaneous tissues caused by streptococcus pyogenos and giving rise to acute feverish illness].

(*v*) *Spinach*

It is considered the broom of the stomach and is wonderful cleanser. It is rich in iron. It is blood cleanser.

Chapter 7

Some Medicinally Important Inorganic Compounds

Compounds of Aluminium

1. Alum (KAl $(SO_4)_2$ 12.H_2O)

Alum is potassiuma aluminium sulphate.

Preparation: It is prepared by adding a concentrated solution of potassium sulphate to a hot solution of aluminium sulphate, the solution thus obtained is concentrated and cooled when characteristic crystals of alum separate out.

$$\underset{\text{Potasium sulphate}}{K_2SO_4} + \underset{\text{Aluminium Sulphate}}{Al_2(SO_4)_3} + 24H_2O \longrightarrow \underset{\text{Alum}}{2KAl(SO_4)_2 \cdot 12H_2O}$$

Properties: It is a colourless, transparent crystalline substance, soluble in water, insoluble in alcohol. On heating it looses its water of crystallisation at 200°C and becomes anhydrous. It gives reactions characteristic of aluminium, potassium and sulphate.

Uses: It is used as pharmaceutical aid and astringent. It can be used to prepare styptic pencil for stopping the bleeding in small cuts.

2. Aluminium Hydroxide Gel

It is an aqueous white colour suspension of hydrated aluminium oxide along with varying amounts of basic aluminium carbonate. In these preparations the percentage of aluminium oxide (Al_2O_3) should not be more than 4.4% or less than 3.5% by weight. Sometimes preservatives like sodium benzoate (not more than 0.5%) are added. Glycerine, sucrose, menthol oil, pepermint oil may be added as a sweetening and flavouring agents.

Preparation: It is prepared by adding an aluminium salt solution to sodium or ammonium carbonate. When unstable aluminium carbonate is formed as one of the product, the unstable aluminium carbonate hydrolyses to give aluminium hydroxide.

$$Al_2(SO_4)_3 + 3Na_2CO_3 \longrightarrow \underset{\text{Aluminium carbonate}}{Al_2(CO_3)_3} + 3Na_2SO_4$$

$$Al_2(CO_3)_3 + 3H_2O \longrightarrow \underset{\text{Aluminium hydroxide}}{3Al(OH)_3} + 3CO_2 \uparrow$$

It is a white viscous suspension. When it is kept standing for sometime, a clear liquid gets separated. It should be stored in well closed containers and should not be allowed to freeze.

Uses: It is a very effective slow acting antacid. Since it is slow acting, it can control gastric acidity for longer period. Another advantage is that it does not produce carbon-dioxide and it does not get absorbed in alimentary canal. It is also used as an astringent and antiseptic. It is also used as mordant in dyeing fabrics.

3. Dried aluminium hydroxide gel

It is white, light amorphous powder which comprises not having less than 47% of Al_2O_3 when ignited to constant weight. It is odourless and tasteless. It is insoluble in water and alcohol, but it is soluble in mineral acids and excess of alkali solutions. It should be stored in an air tight container in a cool place.

Uses: It is used externally as a mild astringent and internally as an antacid and protective.

4. Bentonite

It is a colloidal hydrated aluminium silicate which occurs naturally. It is a fine pale buff or cream coloured powder. It has no smell and is insoluble in water and swells to about twelve times its volume in water.

Uses: It is good pharmaceutical aid. Used as a base for many pharmaceutical preparations including plasters and ointments. It is an ingredient of calamine lotion.

5. Aluminium acetate or burrows solution

Preparation: It is prepared by mixing aqueous solution of lead or barium acetate and aluminium sulphate.

$$3(CH_3COO)_2\,Ba + Al_2(SO_4)_3 \longrightarrow \underset{\text{Aluminium acetate}}{2(CH_3COO)_3Al} + 3BaSO_4 \downarrow$$

Uses: It is used in the treatment of superficial fungal infections.

6. Aluminium monosterate [$C_{17}H_{35}COOAl(OH)_2$]

Uses: It is used as a gelling agent in the preparation of sterile procaine Penicillin G with aluminium sterate suspension.

Some aluminium compounds are used as Gastric antacids. These are agents that lower the acidity of the gastric contents.

7. Dihydroxy aluminium amino acetate ($NH_2CH_2COOAl(OH)_2(H_2O)–x$)

This is a salt of aluminium hydroxide and glycine. It has a more rapid and more prolonged action.

Neutralising capacity of dihydroxy aluminium sodium carbonate $(HO)_2$. $AlO.CO_2Na$ and hydrated magnesium aluminate (a chemical combination of hydroxides of aluminium and magnesium) are enhanced by their absorptive and buffering properties. They are given in the liquid form in large amounts.

(Ag $Al_2O_6H_4$, Mg AlO_4H_3 and Mg_2 $2AlO_6H_5$)

(Magnesium aluminium and aluminium oxyhydroxides) and high capacity gastric antacids. They are very reactive and strongly anti-proteolytic.

Compounds of Phosphorus

1. Phosphoric Acid (H_3PO_4)

Preparation: (*i*) It is prepared on a small scale by oxidising phosphorous to phosphorus pentoxide in a current of hot air and treating phosphorous pentoxide with water to form phosphoric acid.

$$P_2O_5 + 3H_2O \longrightarrow 2H_3PO_4$$

(*ii*) It can be prepared on large scale by digesting crushed rock phosphate somberite ($Ca_3(PO_4)_2$) with calculated quantity of dilute sulphuric acid, filtering and evaporating to a syrup.

$$Ca_3(PO_4)_2 + 3H_2SO_4 \longrightarrow 3CaSO_4 + 2H_3PO_4$$

It is a colourless, odourless liquid. It is miscible in water and alcohol. When heated, it loses water and finally forms metaphosphoric acid.

Uses: It is used as a pharmaceutical aid in the preparation of dilute phosphoric acid I.P. which is used in the treatment of lead poisoning. It is used in beverages, dental cements etc.

2. Dilute Phosphoric Acid

It contains 10% W/W of phosphoric acid. It is prepared by dilution of phosphoric acid by pure distilled water.

It is a clear odourless liquid and is miscible in water and alcohol.

Uses: Its uses are the same of phosphoric acid.

3. Hypophosphorus Acid (H_3PO_2)

Preparation: It is prepared by treating calcium hypophosphate with oxalic acid. The insoluble calcium salt is filtered off and the filtrate is collected and concentrated.

$$\underset{\text{Calcium hypophosphite}}{Ca(H_2PO_2)_2} + \underset{\text{Oxalic acid}}{H_2C_2O_4} \longrightarrow \underset{\text{Calcium oxalate}}{CaC_2O_4} + \underset{\text{Hypophosphorous acid}}{2H_3PO_2}$$

It is a clear yellowish liquid with slight acidic odour and taste. It is miscible with water and alcohol.

Uses: It is used as an antioxidant in pharmaceutical prepations. These are regarded as brain tonics. However it is believed to be pass-out of the body unchanged.

Compounds of Arsenic

1. Arsenous Anhydride (As_2O_3)

It is poisonous in nature, the highest tolerated single dose for adults is 0.005g.

Preparation: It is prepared by roasting sulphur compounds of arsenic in a stream of air.

$$2As_2S_3 + 9O_2 \longrightarrow \underset{\text{Arsenous anhydride}}{2As_2O_3} + 6SO_2$$

It is a white powder and dissolves very slowly in water, but readily dissolves in hydrochloric acid and alkalis. On heating it sublimes and settles on the walls of the test tubes.

Uses: It is used externally and orally. It is used in certain skin diseases. It is prescribed in pills for treating anaemia.

2. Sodium Arsenate ($Na_2HAsO_4.7H_2O$)

Preparation: It is prepared by oxidising Arsenic (III) oxide to arsenic (v) oxide which is then treated with sodium carbonate.

$$As_2O_3 + 2HNO_3 \longrightarrow As_2O_5 + H_2O + NO_2 + NO$$

$$As_2O_5 + 2Na_2CO_3 + H_2O \longrightarrow 2Na_2HAsO_4 + 2CO_2$$

It is a colourless, odourless crystalline substance and is soluble in water.

Uses: The drug is used in the form of aqueous solution for subcutaneous injections. It is also used as general tonic.

3. Aromatic Arsenicals.

These are valuable therapeutic agents. They are: (*i*) compounds in which arsenic is trivalent and (*ii*) compounds in which arsenic is pentavalent. Trivalent As is present as either As=O (or) - As= As- group. The former is called 'arsenoxide' form which has the greatest therapeutic activity. All arsenic compounds are toxic. Pentavalent As is reduced to trivalent As (active) in the body, the arsenoxide compounds react with thiol groups in the organism. The thiol groups of parasites are destroyed thus.

$$C_6H_5As\begin{matrix} \diagup OH \\ \diagdown OH \end{matrix} + 2RSH \longrightarrow C_6H_5As\begin{matrix} \diagup S{-}R \\ \diagdown S{-}R \end{matrix} + 2H_2O$$

(*a*) *Carbasone [N—carbamoyl arsanilic acid or P—ureido benzyen arsonic acid]:* It is used orally in the chemotherapy of intestinal ameobiasis and intravaginally as suppositories in the treatment of Vaginitis. It is absorbed in the gastrointestinal tract (when taken orally) and excreted in the urine. It is a safe arsenical in clinical use.

(*b*) *Tryparasomide [monosodium-N-(Carbomyl-methyl-arsenitate)]:* It was used in the treatment of syphilis of the C^{NS} and in trypanosomiasis. The drug was administered intravenously.

(*c*) *Acetarsonide (3-acetoamids-4-hydroxy-benzene-arsonic acid):* It is used in the treatment of amoebiasis, trichomonas vaginitis and vincent's angina.

(*d*) *Glycobiarsol:* It is also a low toxic antiamoebic.

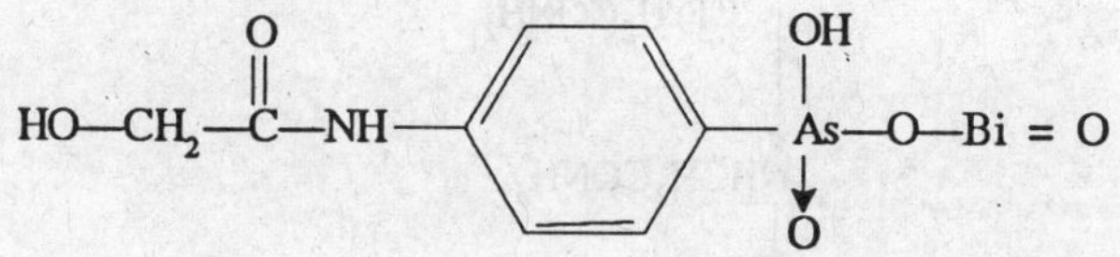

$$HO{-}CH_2{-}\overset{\overset{\displaystyle O}{\|}}{C}{-}NH{-}C_6H_4{-}\underset{\underset{\displaystyle O}{\downarrow}}{\overset{\overset{\displaystyle OH}{|}}{As}}{-}O{-}Bi = O$$

(*e*) *Oxophenarsine hydrochloride:* 2-amino 4 arsenophenol hydrochloride was used in the treatment of syphilis.

(*f*) *Neoarsphenamine* is sodium 3,3 - diamino 4-4 'dihydroxy arseno benzene —N— Methanol sulphosylate or Neodiarsenol.

Preparation of carbosone, trypersamide, acetorsol-oxophenergene and Neorsphenemine from aniline. (See page 82)

(*g*) Melarsoprol is used as a trypanocide. (See top of page 82)

Other important organic compounds used as pharmaceutics are:

4. Sodium Cocodylate $(CH_3)_2AsOONa$

It is soluble in water and is used in chronic skin diseases.

5. Copper Cocodylate $Cu[(CH_3)_2AsO_2]_2$

It is present with iron cocodylate Fe $[(CH_2) AsO_2]_3$ in the preparation of Ferroamprin which is used as a hematinic.

Compounds of Iron

1. Ferrous fumarate $(C_4H_2Fe\ O_4)$

Preparation: It is obtained by double decomposition when a solution of hot ferrous sulphate is added slowly to a solution of sodium fumarate with constant stirring when ferrous fumarate gets separated. It is filtered and dried.

It occurs as reddish brown powder. It has astringent taste and is sparingly soluble in water and alcohol.

Uses: It finds use as a haematinic, like other iron salts. When other iron preparations do not get tolerated, this salt gives better results. Ferrous fumarate is given in the form of tablets.

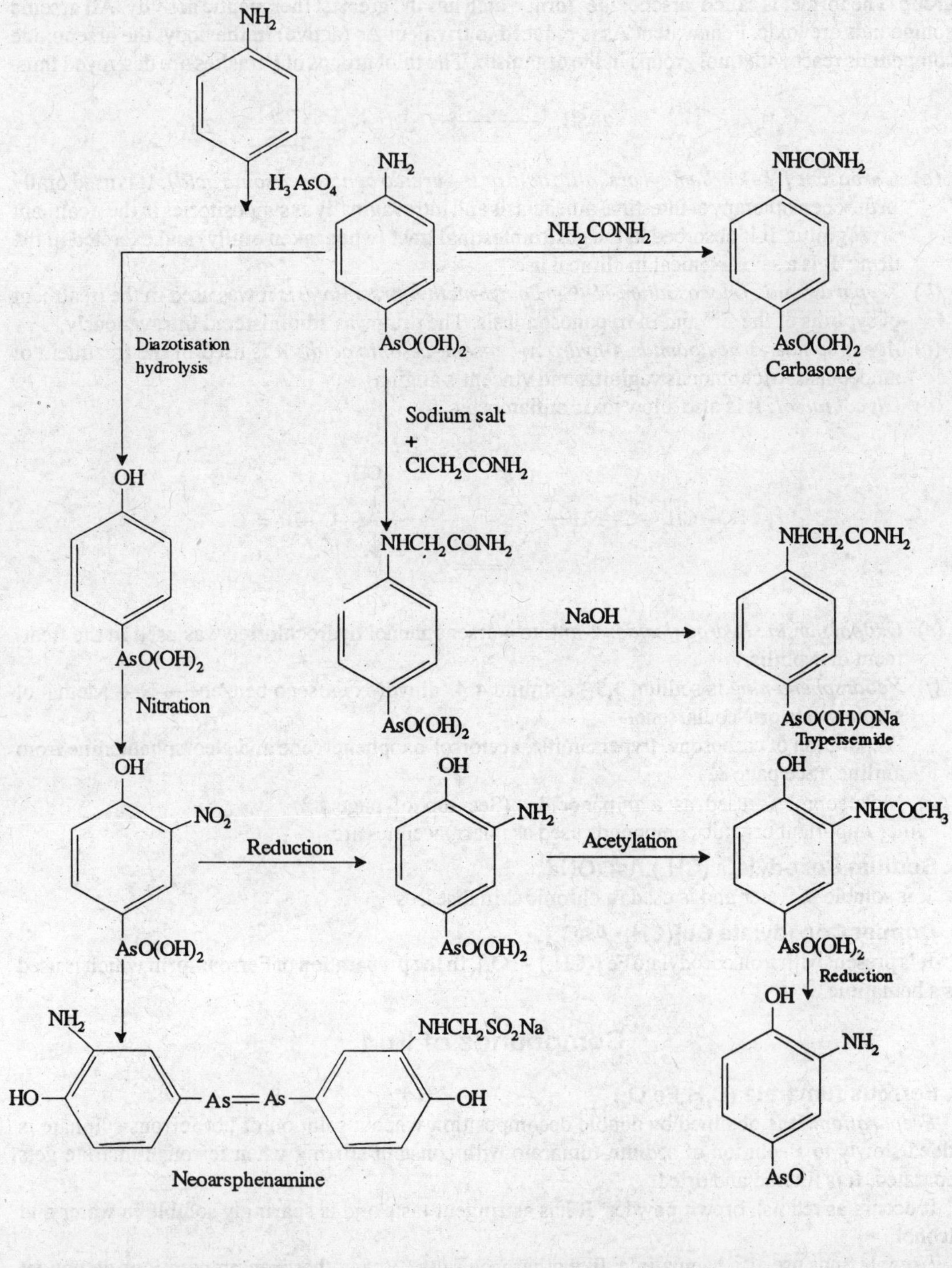

NH_2 (aniline) $\xrightarrow{H_3AsO_4}$ NH_2–C_6H_4–$AsO(OH)_2$

NH_2–C_6H_4–$AsO(OH)_2$ $\xrightarrow{NH_2CONH_2}$ $NHCONH_2$–C_6H_4–$AsO(OH)_2$

Carbasone

NH_2–C_6H_4–$AsO(OH)_2$ → Diazotisation hydrolysis → OH–C_6H_4–$AsO(OH)_2$

NH_2–C_6H_4–$AsO(OH)_2$ → Sodium salt + $ClCH_2CONH_2$ → $NHCH_2CONH_2$–C_6H_4–$AsO(OH)_2$

$NHCH_2CONH_2$–C_6H_4–$AsO(OH)_2$ $\xrightarrow{NaOH}$ $NHCH_2CONH_2$–C_6H_4–$AsO(OH)ONa$

Trypersemide

OH–C_6H_4–$AsO(OH)_2$ → Nitration → OH, NO_2–C_6H_3–$AsO(OH)_2$

OH, NO_2–C_6H_3–$AsO(OH)_2$ → Reduction → OH, NH_2–C_6H_3–$AsO(OH)_2$

OH, NH_2–C_6H_3–$AsO(OH)_2$ → Acetylation → OH, $NHCOCH_3$–C_6H_3–$AsO(OH)_2$

OH, $NHCOCH_3$–C_6H_3–$AsO(OH)_2$ → Reduction → OH, NH_2–C_6H_3–AsO

OH, NO_2–C_6H_3–$AsO(OH)_2$ → HO–$C_6H_3(NH_2)$–As=As–$C_6H_3(NHCH_2SO_2Na)$–OH

Neoarsphenamine

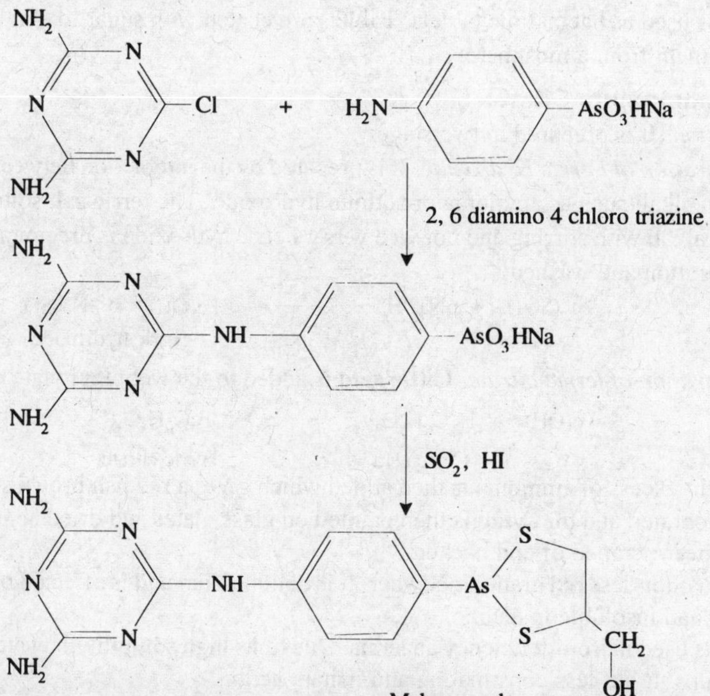

$$\text{2,6-diamino-4-chloro triazine} + H_2N\text{—}C_6H_4\text{—}AsO_3HNa \longrightarrow \text{triazinyl—NH—}C_6H_4\text{—}AsO_3HNa \xrightarrow{SO_2,\ HI} \text{Melarsoprol}$$

NH2, N, N, Cl, +, H2N, AsO3HNa, NH2, 2, 6 diamino 4 chloro triazine, NH2, N, N, NH, AsO3HNa, N, NH2, SO2, HI, NH2, N, N, NH, As, S, S, N, CH2, NH2, OH

Melarsoprol

2. Ferrous gluconate—$[CH_2(CH)(CHOH)_4CO_2]_2Fe.2H_2O$

Preparation: It is prepared from Gluconic acid which is obtained from D-glucose by oxidation. Gluconic acid is treated with ferrous carbonate to form soluble ferrous gluconate.

$$\underset{\text{Gluconic acid}}{HOH_2C\text{—}(CHOH)_4\text{—}COOH} + \underset{\text{Ferrous carbonate}}{FeCO_3} + H_2O \longrightarrow \underset{\text{Ferrous gluconate}}{[HOH_2C\text{—}(CHOH)_4\text{—}COO]_2 Fe.2H_2O} + CO_2\uparrow$$

It is a pale yellowish green powder. Aqueous soln is acidic to litmus. It is affected by light and ferrous is oxidised to ferric.

Uses: It is used in the treatment of anaemia.

3. Ferrous Sulphate

Preparation: It is obtained by dissolving iron in excess of dilute H_2SO_4. Ferrous sulphate is obtained as green crystals on filtration and concentration and cooling.

$$Fe + H_2SO_4 \rightarrow FeSO_4 + H_2$$

In all these operations, exposure to air is prevented because ferrous sulphate is oxidised readily on exposure to moist air and the crystals get coated with brownish yellow ferric sulphate.

It occurs as greensih crystals having astringent taste. It is soluble in water; on heating it is decomposed to ferric oxide and sulphur dioxide and sulphuric acid.

Uses: It is used as haematinic. It is also used in dyeing fabrics, tanning leather, in photography and also as a disinfectant. It is the most commonly used iron preparation used in anaemia due to iron deficiency.

4. Dried Ferrous Sulphate or Ferrous Sulphate Tablets

These are prepared from dried ferrous sulphate by compression. Dried ferrous sulphate is having not less than 80% and not more than 90% of $FeSo_4$.

Uses: It is used as haematinic tablets. Tablets are coated with sugar to eliminate gastric irritation and protect them from atmosphere.

5. Ferric Ammonium Citrate

Preparation: It is prepared in two stages.

(*i*) *Preparation of Ferric Hydroxide:* It is prepared by the interaction between a ferric salt solution and an alkali such as ammonia or sodium hydroxide. The ferric salt solution should be added to the alkali with stirring and not vice versa. Ferric hydroxide is precipitated and it is collected by filtration and washed.

$$Fe_2(SO_4)_3 + 6NaOH \longrightarrow \underset{\text{Ferric hydroxide}}{2Fe(OH)_3} + 3Na_2SO_4$$

(*ii*) *Preparation of ferric citrate:* Citric acid is added to the wet precipitate of ferric hydroxide.

$$Fe(OH)_3 + \underset{\text{Citric acid}}{H_3C_6H_5O_7} \longrightarrow \underset{\text{Ferric citrate}}{FeC_6H_5O_7} + 3H_2O$$

A slight excess of ammonia is then added which gives a reddish brown solution. This solution is evaporated and the syrup is then painted on glass plates and dried at 40°C. The dried scales have been scraped of and backed.

It is an odourless red granular powder. It is deliquescent and is affected by light. It is soluble in water and insoluble in alcohol.

Uses: It is used in iron-deficiency anaemia. Due to its high solubility in water it can be used in the form of syrups. It has less constipating and irritant action.

6. Iron Sobitex Injections

Iron Sobitex is a sterile aqueous solution of a complex of iron, sorbitol and citric acid, stabilised with dextrin. It is used as a source of iron.

7. Ferrocholinate

It is a chelate prepared by reacting equimolecular quantities of freshly precipitated ferric hydroxide with choline hydrogen citrate. It is less toxic and less irritant than ferrous gluconate. It is administred orally.

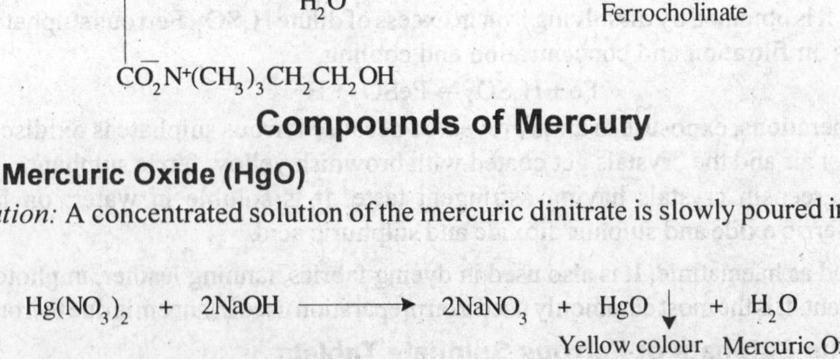

H_2O
CH_2CO_2H
O
C — O Fe O — $C.CH_2CO_2^-N^+(CH_3)_3CH_2CH_2OH$
HO_2CH_2C — O O — C=O
CH_2
H_2O
$CO_2^-N^+(CH_3)_3CH_2CH_2OH$

Ferrocholinate

Compounds of Mercury

1. Yellow Mercuric Oxide (HgO)

Preparation: A concentrated solution of the mercuric dinitrate is slowly poured into a dilute alkali solution.

$$Hg(NO_3)_2 + 2NaOH \longrightarrow 2NaNO_3 + \underset{\text{Yellow colour Mercuric Oxide}}{HgO\downarrow} + H_2O$$

All the operations of mixing should be performed in the dark, otherwise black colour Hg_2O is obtained. It is an orange yellow heavy amorphous powder. It is odourless and gets discoloured on exposure to light.

It is insoluble in water and ethanol though it is readily soluble in dilute HCl. It is stored in a tightly closed container protected from light.

Uses: It has mild antiseptic action. It is used as an antibacterial ointment. In mild inflammation of eyes, 0.1% ointment is used to treat conjunctivitis.

2. Oleated Mercury

Oleated mercury has the equivalent of 20% of Yellow mercuric oxide. (limits 19 to 21).

Ingredients:	Yellow Mercuric Oxide	200g
	Liquid Paraffin	50g
	Oleic acid	750g

It is obtained by triturating mercuric oxide thoroughly with liquid paraffin until it gets thoroughly subdivided. Oleic acid is then added and contents are mixed thoroughly. The mixture is heated at 50^0 triturating occasionally until combination gets effected. On cooling a yellowish preparation is obtained.

It has to be stored in an air tight container because it can develop rancidity due to oxidation of Oleic acid. It must also be protected from light. If mercury globules get seperated, this should not be used.

Uses: It finds use as an anti-ineffective and as a pharmaceutical necessity.

3. Mercurous Chloride HgCl (Calomel)

Preparation: It is obtained as a sublimate by heating a mixture of mercury and mercuric chloride.

$$Hg + HgCl_2 \longrightarrow \underset{\text{Mercurous chloride}}{2HgCl}$$

It is a dull white heavy powder. It is insoluble in water, alcohol and acid.

It is kept in a tightly closed container and should be protected from light.

Uses: Calomel has been used for centuries as a cathartic but in recent years it has been replaced by other drugs, due to cases of mercury poisoning. It is also used as ointment for the treatment of eczema and as a prophylactic syptulls. prophylactic = measure taken to prevent a disease)

4. Mercuric Oxide (eye ointment)

It is prepared from yellow oxide of mercury. It is used as an ophthalmic oinment for the treatment of conjunctivitis. It is also used as oinment for some skin infections like eczema etc.

5. Ammoniated Mercury or Mercury Amidochloride NH_2HgCl

Preparation: It is prepared by pouring a cold solution of mercuric chloride into cold ammonia solution.

$$HgCl_2 + 2NH_3 \longrightarrow \underset{\text{Ammoniated Mercury}}{HgNH_2Cl} + NH_4Cl$$

The precipitate is washed with very dilute ammonia solution and dried in the dark below 30°C. Washing is necessary to remove ammonium chloride but prolong washing should be avoided.

Properties: It is snow white amorphous powder, soluble in water, alcohol and ether.

Uses: It is used as dermal ointment in 3-10% strength and also as cosmetic ointment.

6. Mercury with Chalk (Grew Powder)

It consists of 31-35% W/W of mercury and 62-70% W/W of chalk *i.e.* $CaCO_3$.

Ingredients:	Mercury	33g
	Dextrose	1g
	Chalk	66g

Preparation: First all the ingredients are mixed and then triturated in a porcelian mixture until the mixture attains a uniform pale grey colour, and the mixture is ready only when no metallic globules are visible under a magnifying lens.

It is kept in tightly closed container in a cool place.

Uses: It is used as purgative.

Chapter 8

Biological Role of Some Inorganic Compounds

Sodium and Its Compounds

The metallic element sodium was discovered by Sir M. Davy in 1807, when he electrolysed sodium hydroxide. Sodium belongs to group I of the Periodic table. The metals in this group are mono-valent, strongly electropositive and are soluble in water. They form strong alkaline hydroxide on combining with water. Sodium and potassium are the two most important metals in this group from the point of view of pharmaceutical interest.

Sodium is found abundant in nature, but always in combination with other elements such as Sodium chloride (NaC! 2 to 3% in sea water and as rock salt), as chile salt petre ($NaNO_3$), as sodium carbonate, as nardite (Na_2SO_4 anhydrous) in chile, as sodium carbonate Na_2CO_3 in certain springs etc.

Sources

Common salt or sodium chloride is found in sea water, in salt, wells, inland—lakes and deposits of rock-salt.

Other sources of sodium are in most cereals, milk, fruits and vegetables. During cooking sufficient amount of sodium chloride is added.

Biological Role of Sodium and Its Compounds

Biological role of sodium is as follows:

1. It is the main base of the body.
2. It is one of the most important cation responsible for maintaining the internal environment by maintaining the osmotic pressure, volume and composition of the extra cellular fluid.
3. Chloride of the gastric juice comes from sodium chloride of blood. Thus it supplies the raw material for formation of hydrochloric acid.
4. It forms the alkali reserve of the body and acts as a buffer of blood and tissue.
5. It maintains neuromuscular conductivity and irritability.
6. It maintains cardiac rhythmicity and contractivity.
7. It helps in maintaining the peripheral resistance of blood vessels.
8. In diet it adds to the tastefulness of food.

The salts of sodium such as its chloride, carbonates, bicorbonates etc. are important salts which are present in the extra cellular fluid of the body. The extracellular fluid plays an important role as a transport medium for various substances moving into and from the cells. Some of the substances present in this body fluid exist as ionized particles and when these are placed into an electrical field they migrate to their respective electrodes, hence these are also known as electrolytes, examples : Sodium chloride which finds use as an electrolyte plays an important role in maintaining osmotic pressure.

Human requirment of salt is 1-2 grams daily. Sodium deficiency or excess in body cause marked physiological disturbances. Mild sodium decrease causes loss of appetite, lethargy and cramps. Massive sodium loss is usually accompanied by a shock, a fall in the blood pressure, limbs become cold, pulse rate becomes high and cardiac out-put is low. Symptoms like vomiting, an orexia, muscle cramps etc., may also be there. Sodium deficiency is found in Addison's disease when there is increase in the renal excretion of sodium leading to a fall in blood sodium level.

Mild sodium deficiency can be treated by giving sodium chloride solution orally. (0.9% sodium chloride solution). For severe sodium deficiency isotonic saline solution can be given intravenously. In Addison's disease where there is continuous loss of sodium the patient is advised to have additional sodium chloride in the diet.

Increased sodium retention is associated with retention of water causing oedema. The treatment is directed towards the reduction of the sodium level. 5% dextrose in water is given till serum sodium level decreases.

Less sodium intake is advisable for the patient suffering from high blood pressure as more sodium in the body causes more fluid, i.e. more blood volume which will increase the blood pressure.

Sodium carbonates and bicarbonates are generally used as an antacid and as a pharmaceutical aid. The reactions of the body fluids are regulated by their hydrogen ion or hydrion (H^+). The concentration of hydrions are maintained by the ratio of carbonic acid ($H.HCO_3$) and bicarbonate ($B.HCO_3$). Thus a fall in carbonic acid or a rise in bicarbonate level will decrease the ratio and the hydrion concentration will decrease leading to alkalosis. Similarly a rise in carbonic acid or a fall in bicarbonate will increase the ratio leading to the rise in hydrion, causing acidosis. It can be corrected by slow injection of Sodium bicarbonate and alkalosis can be treated by administering ammonium chloride intravenously.

In case of severe dehydration (loss of fluid from the body) due to chlolera, acute diarrhoea etc. an isotonic saline (sodium chloride solution) along with isotonic sodium bicarbonate should be given intravenously in the ratio of 2 : 1.

In case of mild dehydration a fluid consisting of 1 teaspoon sodium chloride, Sodium bicarbonate, 1 teaspoon of Potassium citrate, 2 teaspoons of Glucose, boiled water (1 litre) can be given orally.

Sodium iodide is used in cases of hypothyroidism as a source of Iodine. It is also used as an expectorant.

Sodium nitrites can relax smooth muscle especially of the smaller blood vessels, because of this property. Sodium nitrite is used in angina pectoris as a vasodilator. It is also used as an antidote to cyanide poisoning.

Sodium citrates and salicylates. These salts are used as analgesics and antipyretics.

Sodium citrate is oxidized by the tissues to an equivalent amount of sodium bicarbonate. Therefore it is useful as a systemic antacids. It also prevents clotting of blood, hence are useful as anticoagulants. It is available both in the form of tablets and as an injection.

Sodium metabisulphite finds use as an anti-oxidant.

Potassium and Its Compounds

Metallic potassium was first isolated by Sir Humphrey Davy by the electrolysis of fused hydroxide.

Sources

Potassium compounds occur in the earth's crust especially as silicates. Most of the land plants contain considerable amounts of potassium compounds which they obtain from soil. Potassium does not occur free in nature and is found only in combined state.

An important source of Potassium are carbanalite. Other sources of potassium are cereals, fruits like oranges, banana, green leafy vegetables, tomatoes. Water also contains considerable amount of potassiun and sodium. Coconut water contains about 70m. equilitre of potassium.

Some important compounds of potassium are:

Potassium acetate	(CH_3COOK),
Potassium bicarbonate	($KHCO_3$),
Potassium acid tartarte	CH(OH)COOK | CH(OH)COOK
Potassium bromide	(KBr)
Potassium chloride	(KCl)
Potassium citrate	CH_2COOK | C(OH)COOK || CH_2COOK.
Potassium hydroxide	(KOH)
Potassium iodide	(KI),
Potassium permanganate	($KMnO_4$)

Biological Role of Potassium and Its Salts

Potassium is present in the intracellular fluid of the body. The concentration of potassium is important for the regulation of volumes and tonicity of extracellular and intracellular body fluid. Potassium is one of the major intracellular cation which is essentially a cation of the cells and hence its distribution is related to the cell mass. About 70% of the total potassium is in the muscles. About 20% in the brain and 10% in the skin and subcutaneous tissues. Human requirment: Average daily requirment of potassium in an adult is 4g.

Biological role of potassium is as follows:

1. With sodium, it forms the main base of the body and is the chief intracellular cation.
2. It is responsible for maintaining the bioelectric potential of the cell.
3. It is responsible for maintaining cardiac-rhythmicity.
4. Excess or lack of potassium depresses neuromuscular conduction.
5. It has a balancing effect along with sodium on the main base content of the body.
6. In experimental animals lack of potassium retards growth and may prove fatal also.
7. Application of high potassium containing solution to excitable tissues causes stimulation due to its depolarising action.

Deficiency: Potassium deficiency is known as hypokalemia. The causes of potassium deficiency are: (*i*) poor intake (*ii*) poor absorption (*iii*) loss due to diuresis, alkalosis, diarrhoea etc. (*iv*) adrenocortical hormone excess (*v*) familial periodic paralysis etc.

Symptoms are weakness of muscles, anorexia, thirst, lethargy etc. Potassium deficiency leads to retention of sodium. The normal activity of both skeletal and smooth muscles is impaired. Hypokalemia is associated with arrhythmias and tachycardia.

Treatment: Potassium is usually given by mouth. The drugs used for hypokalemia are potassium chloride, potassium bicarbonate, potassium citrate.

In emergency cases potassium can be given intravenously by drip method. Potassium should never be given directly into a vein as this may cause sudden death from cardiac arrest. Generally the solution which is given intravenously contains sodium chloride, sodium lactate, potassium chloride.

Hyperkalemia or Excess of Potassium

Causes are: (*i*) failure of the kidney to excrete potassium (*ii*) overdosage with potassium salts (*iii*) unusual release of intracellular potassium in burns, severe infection etc. (*iv*) in patients with Addison's disease due to lack of aldosterone.

Some of the symptoms are muscle weakness, paralysis, diarrhoea etc. It may cause various arrhythmias leading to cardiac arrest, skeletal muscles may show paralysis.

Treatment: Extra potassium should be avoided in the diet. Potassium excretion can be enhanced by using glucocorticoids. Dextrose and sodium bicarbonate is also given to reduce potassium concentration.

Therapeutic uses of some salts of potassium are as follows:

Potassium acetate is used as diuretic alkalizer and diaphoretic.

Potassium acid tartarate finds its use as saline purgative.

Potassium bi-carbonate is a useful antacid and electrolyte replenisher.

Potassium bromide is a useful sedative.

Potassium chloride is used as a diuretic and electrolyte replenisher. It is an ingredient of sodium chloride injection (Ringer's solution) and sodium lactate injection (Ringer lactate solution).

Potassium citrate. This salt of potassium is a useful osmotic diuretic, diaphoretic and expectorant.

Potassium hydroxide. It finds use as a pharmaceutical aid. Its aqueous solution is used as a laboratory reagent and 5% W/V of total alkali calculated as potassium hydroxide. It should be stored in lead free green coloured bottle with a rubber stopper. Potassium iodide is used as an expectorant and as a source of iodine. It is also an important laboratory reagent.

Potassium permanganate is a powerful oxidising agent and is used as an anti-infective or as a disinfectant. Its action is due to the liberation of oxygen which oxidizes the protein of the bacteria and kills them. It is used as mouth wash and gargle.

Calcium and Its Compounds

Plants and animals need calcium salts for body building. Protective shells of marine animals are mainly made of calcium carbonate whereas bones and teeth are made of calcium phosphate.

Sources

Calcium occurs in nature only in the combined state e.g. as

(*i*) Carbonates—lime stone, chalk, dolomite, $CaCO_3$, $MgCo_3$ etc.

(*ii*) Fluoride—fluors-par CaF_2, fluorapatite etc.

(*iii*) Silicates—$CaSiO_3$ (constituents of many rocks)

(*iv*) Phosphate—phosphorite $Ca_3(PO_4)_2$

(*v*) Sulphate—Gypsum $CaSO_4 . 2H_2O$

(*vi*) Chloride—chlorapatite.

Food is an important source of calcium e.g. milk and milk products, green leafy vegetables, cereals like ragi, potato, cheese etc.

Human requirment. An average adult requires 0.5 g of calcium per day. Pregnant women and lactating mother require about 1g daily and children about 0.7g.

Various compounds of calcium of pharmaceutical importance are as follows:

Calcium amino salicylate: $\left[\left(NH_2\text{–}C_6H_3(OH)\text{–}COO^-\right)_2 Ca\right]$

Calcium carbonate	($CaCO_3$)
Calcium chloride hydrated	($CaCl_2 \cdot 6H_2O$)
Calcium gluconate	(HO CH_2 $(CHOH)_4$ $COO)_2$ Ca. H_2O
Calcium hydroxide	$Ca(OH)_2$
Calcium lactate	($CH_3CH(OH)$ $COO)_2$. Ca. $5H_2O$
Soda lime	
Calcium mandelate	$C_6H_{11}O_6Ca$
Calcium phosphate	$Ca_3(PO_4)_2$
Calcium dibasic phosphate	$CaHPO_4 \cdot 2H_2O$
Plaster of Paris	$CaSO_4 \cdot 1/2H_2O$

The biological role of calcium is as follows:

1. Calcium helps in the formation of bones and teeth.
2. It helps in clotting of blood and curdling of milk.
3. It is essential for maintaining neuro-muscular excitability.
4. It helps to maintain cardiac contractility. Excess calcium will make the heart to stop in systole.
5. It is also responsible for excitation contraction coupling of neuro muscular apparatus after a stimulus.
6. It is actively involved in the membrane transport mechanism.
7. Lack of ionic calcium leads to tetany which can be prevented by administration of calcium salt.
8. It may help ciliary movement.

Calcium is an essential constituent of bones and plays an important role in body hameostasis. Any disturbance in calcium metabolism effects various cellular functions.

About 99% of the body calcium is present in bones and the bone minerals exist in the form of a complex salt containing calcium, phosphate and carbonate. Calcium is distributed extracellularly and most of the calcium present in blood is present in the plasma.

Calcium is an important component of the cell membrane and controls its permeability and electrical properties. Calcium is an essential factor between excitation and contraction in all forms of muscle. It is essential for the formation of bone and teeth. The ionic calcium present in the blood plasma is essential for the coagulation of blood as it helps in the conversion of prothrombin into thrombin.

Deficiency. Deficiency of calcium or hypocalcemia results in vitamin D deficiency, hypoparathyroidism renal failure. The symptoms are convulsions, laryngeal spasm, skin and nails show atrophy i.e. decrease in size of a normally developed organ or tissue. Apart from vitamin D deficiency calcium and phosphorous deficiency also causes rickets in children.

Excess. Hypercalcemia or excess of calcium occurs in hyper-parathyroidism, hypervitaminosis. Symptoms are lethargy, anorexia, weakness, nausea, vomiting, severe constipation etc. Excess calcium gets deposited in various tissues, arteries, muscles etc. Severe hypercalcemia may develop some mental changes, which is a medical emergency.

Plaster of Paris

Formula: $CaSO_4 \frac{1}{2}H_2O$ (calcium sulphate hemi-hydrate)

Calcium sulphate occurs in nature having two molecules of water ($CaSO_4.2H_2O$) as monoclinic prisms in mineral gypsum and in granular masses called alabaster. It also occurs as ($CaSO_4$) in a mineral known as Anhydrite.

Preparation: When gypsum is heated to about 125°C, it loses 3/4th of its water and forms, the hemihydrate, which is known as plaster of paris.

$$\underset{\text{Gypsum}}{CaSO_4.2H_2O} \xrightarrow[125°C]{\text{heated}} \underset{\text{Plaster of pari}}{CaSO_4.½H_2O}$$

It should be stored in a cool dry place and should also be protected from moisture and heat.

Properties: It occurs as a white hygroscopic powder. It is odourless and tasteless and is sparingly soluble in water. Its solubility decreases sharply with the rise in temperature. It is almost insoluble in alcohol. When Plaster of Paris is mixed with water to form a paste, it turns hard in a few minutes.

The process of setting involves two stages. In the first stage the hemihydrate of the Plaster of Paris changes into Orthorhombic dihydrate and in second stage it changes to gypsum and then hardens. The first stage is known as setting stage and the second as hardening stage.

$$\underset{\substack{\text{Hemithydrate} \\ \text{Plaster of Paris}}}{CaSO_4.1/2\,H_2O + 3/2\,H_2O} \xrightarrow[\text{Setting Stage}]{\text{I Stage}} \underset{\substack{\text{(Orthorhombic,)} \\ \text{dihydrate}}}{CaSO_4.2H_2O} \xrightarrow[\text{hardening stage}]{\text{II Stage}} \underset{\substack{\text{(hard cast)} \\ \text{Gvpsum} \\ \text{(Monoclinic)}}}{\text{Gypsum } CaSO_4.2H_2O}$$

Uses:

(1) It is used as a surgical aid for making casts, moulds and bandages, which are used in fractures or dislocated bones to prevent the movement of broken limbs.
(2) It finds use in dentistry for taking impressions.
(3) It is also used as fire proofing material.
(4) It is also used for making black board chalk.

Therapeutic Uses of Calcium Salt

Oral supplement of calcium salts are used to prevent calcium deficiency. They are also necessary for growing children, pregnant women and lactating mothers, in patients on long term corticosteriod therapy, rickets, vitamin D deficiency etc.

Calcium gluconate and chloride are used in hypocalcemia.

Calcium chloride is used as electrolyte replenisher, it is used as an ingredient in sodium chloride and sodium lactate injection. Anhydrous calcium chloride is used as drying agent.

Calcium salts are used for treating ureticaria and non-specific intestinal colic pain.

Calcium amino salicylates are used as antitubercular drug.

Calcium carbonate is used as a base for tooth powder and as an antacid.

Calcium mandelate finds use as an antibacterial.

Calcium phosphate is used as pharmaceutical aid and as a gastric antacid.

Calcium sulphate hemihydrate is used as Plaster of Paris which is used in orthopaedic surgery, for making plaster casts. It is used in dentistry for taking impressions.

Hypercalcemia is treated by reduction of dietary intake and by gluco-corticoids.

Iodine and Its Compounds

Sources

Iodine is mostly obtained from sea weeds, chile salt petre and various brines, sea fish and cod liver oil are also valuable sources of iodine.

Various pharmaceutical preparations of iodine are as follows: There are three types of iodine solutions: (*i*) weak iodine solution (*ii*) strong solution (*iii*) aqueous solution.

Weak iodine solution is also known as tincture of iodine. It contains 2% W/V of iodine and 2.5% W/V of potassium iodide.

Strong iodine solution. It contains 10% W/V of iodine and 6% W/V of potassium iodide.

Aqueous Iodine solution (lugal's solution). It contains 5.0% W/V of iodine and 10% W/V of potassium iodide.

(*ii*) Iodoform (organic iodide).

(*iii*) Iodophors (complex compound of Iodine with surfactants like non-ionic detergents, quarternaries and macro-molecules).

(*iv*) *Iodised oil injection.* It is a sterile preparation of single dose containers. It is prepared by ethyl esters of the fatty acid with hydroiodic acid.

(*v*) *Non-staining iodine ointment.* It contains 5% iodine with arachis oil and yellow soft paraffin. Sometimes 5% methyl salicylate is also added in the ointment, then it is known as methyl salicylate ointment.

(*vi*) *Povidone.* It contains 5% of iodine with polyvinyl pyrrolidone.

Biological Role of Iodine and Its Preparations

Biological role of Iodine

(1) Iodine is probably the oldest and the most widely used antiseptic.

(2) Iodine is also one of the most important substances required for the synthesis of thyroid hormones.

(3) It is also a powerful bactericidal, sporicidal, fungicidal, amoebicidal, viricidal and germicidal.

(4) Elemental iodine, if ingested, can cause severe gastrointestinal irritation and shock.

(5) Iodine tincture is usually employed for disinfection of skin prior to surgery.

(6) Iodine glycerin preparation is applied in the treatment of pharyngitis and tonsilitis.

(7) Iodine ointments are used as fungicides in the treatment of ring worm.

(8) Iodine is also used in several radio-opaque preparations used for diagnostic procedures.

Iodides are used as expectorants and are probably the oldest remedy for thyroid disorders like enlargement of thyroid gland known as goitre.

Iodoforms is used as antiseptic agent. It has an unpleasant odour.

Iodophors are non-irritating, non-staining germicidals. They are also used as local antiseptics in ring worm infections and in vaginal infections. Iodophors also find use as disinfectant for disinfections of operating rooms, wards, nurseries, equipments etc.

Iodine is present in food in inorganic form and is absorbed from gastrointestinal tract.

Human requirment. Daily dietary requirments of a healthy adult is 100-200 g.

Iodine deficiency. Lack of Iodine leads to goitrous condition i.e., swelling of thyroid gland.

If mother suffers from lack of iodine, there is a chance of having dwarf babies or the new born may have other deformities.

Iodine deficiencies may occur in areas where the element is lacking in the soil which may give rise to the iodine deficiency diseases like goitre. The use of iodized or sea salt has helped in prevention of these diseases.

Excess. It iodine intake is very high then most of the extra iodine is excreted by the kidney.

Copper and Its Compounds

Sources

Copper is present in all living cells of both plants and animals. It is associated with formation of chlorophyll.

It occurs native in nature or in combined form. It is obtained from copper pyrites ($CuFeS_2$), Cuprite (Cu_2O), Copper glance (Cu_2S), Malachite ($Cu(OH)_2 . CuCO_3$).

Food is also an important source of copper. It is present in leafy vegetables, fruits, nuts, liver, legumes, cereals, poultry product etc.

Important compound of copper used in pharmaceutical preparations is copper sulphate : $CuSO_4.5H_2O$

Biological Role of Copper and Its Compounds

Copper is present in the body as copper protein complex and its important biological roles are as follows:

1. It catalyses certain reactions specially haemoglobin synthesis in mammals.
2. Some important copper containing enzymes are cytochromes, tyrosinase, uricase, ascorbic acid oxidase etc.
3. Haemocyanin, a copper-protein complex present in red blood corpuscles of crab, spider and snail, is responsible for carriage of oxygen like haemoglobin in man.
4. Some mammals contain copper as copper-protein complex known as haemocuprien in red blood corpuscles.
5. Abnormal copper metabolism leads to a disease known as Wilson's disease (characterized by pathological changes in the liver).
6. Brain and liver of infant and foetus contain more copper than adult.
7. It plays an important role in the utilisation of iron for the formation of haemoglobin.
8. Copper is also present in some coenzymes as a cofactor as in acyl-coenzyme A dehydrogenase.
9. Liver is the chief store house of the copper.

Copper sulphate is used as an emetic which evokes vomiting by irritation of gastro-intestinal tract. It also possesses astringent action. It is also used as fungicide and algacide.

Human requirment. Daily requirment of copper is 2mg.

Deficiency of Copper. It has certain role in animal nutrition as new born lambs suffer from "Sway back" in absence of copper.

Copper deficiency has not been described in adults although children long on a milk diet may show an anaemia responsive to traces of copper with iron.

When a congenital deficiency in the haemostatic mechanism for copper exists, then it gets accumulated in liver, brain, the cornea of the eye and in other tissues causing Wilson's disease.

Copper deficiency produced in animals causes fatal hypochromic microcytic anaemia. Copper favours absorption of iron from intestines and also its transport from and to tissues.

Copper deficiency also causes abnormalities in bone formation and greying of hair is experienced in animals.

Copper deficiency may lead to Menke's disease which shows the symptoms like unstable body temperature, abnormal bone formation, steely hair etc.

Zinc

Sources

Zinc occurs widely in animal tissues and it is on average 20-30 microgram per 1g of fresh tissue. Biochemically, zinc occurs in association with some metalloenzymes, such as carbonic anhydrase, aldolase, carboxy peptidase etc. It also occurs bound to R.N.A.

Zinc is not found in the native state. It is obtained from Zinc blende ZnS, Calamine ($ZnCO_3$), Zincite (Zno) and Franklinits—($ZnO.Fe_2O_3$, the first two being more abundant.

An average mixed diet contains 10-15mg Zn, which is more than adequate. Meat, liver, eggs, fish, cereals are good sources of Zinc. It is readily absorbed from intestines.

Various compounds of Zinc of Pharmaceutical importance are as follows:

(*i*) Zinc oxide
(*ii*) Zinc oxide compound paste
(*iii*) Zinc oxide ointment
(*iv*) Hydrous zinc oxide ointment
(*v*) Zinc sulphate $ZnSO_4 . 7H_2O$
(*vi*) Zinc stearate $(C_{17}H_{35}. COO)_2.Zn$
(*vii*) Zinc undecylenate $(CH_2. CH (CH_2)_8 COO)_2Zn$
(*viii*) Zinc gelatin
(*ix*) Zinc undecylenate ointment.

Biological Role of Zinc

(1) It is an element essential for normal growth, reproduction and longevity of animals.
(2) It is a component of several enzymes e.g. alcohol dehydrogenase, alkaline phosphate, carbonic anhydrase etc.
(3) Zinc is necessary for maintaining the plasma concentration of vitamin A as Zn is an important constituent of Retinene reductase which helps in vitamin A metabolism.
(4) It is an important micro-nutrient which is a co-factor in number of enzyme systems especially carbonic anhydrase which is concerned in the excretion and transport of carbon-di-oxide.
(5) Its one of the most vital function may be concerned with the synthesis of R.N.A. (ribonucleic acid).

Zinc deficiency. A deficiency of Zinc in soil can lead to its deficiency in the usual food sources and can result in a deficiency in the human. Delayed wound healing and impairment of taste are earlier symptoms. A deficiency can also occur in alcoholic cirrhosis. The cirrhotic liver contains less zinc than normal livers.

Zinc deficiency causes impaired growth and also causes delayed sexual development.

Chapter 9

Blood and Haematological Agents

Composition of Blood

It is a connective tissue whose matrix is a liquid known as plasma in which the formed elements such as blood corpuscles (RBC), white blood corpuscles (WBC) and platelets remain suspended. It is a red, thick and opaque fluid which is alkaline in reaction. Its specific gravity is 1.055-1.065, pH 7.3-7.4 and average amount is 5 litres which varies with body weight (about 7.7% of body weight)

Blood is a complex fluid which is scarlet red when taken from an artery and darkish when taken from a vein. It has a tendency to solidify when it stands, known as clotting. When clotting is prevented by anticoagulants such as heparin, oxalate etc. it is seen to consist of two parts—(I) Plasma, (II) Solid cellular portion.

I. Liquid part is known as plasma—which is 55% of the volume of blood, and consists of 91% to 92% of water and 8%-9% solids. Solid portion consists of the following substances.

1. *Organic substances (7.1%-8.1%):* There are mainly three types of organic substances
   - (*a*) Proteins (7%) —there are four types of proteins namely Albumin, Globulin, Fibrinogen, Prothrombin. These plasma proteins exert an osmotic pressure of 25-30mm and help in regulating blood volume and blood's viscosity. Fibrinogen helps in blood clotting.
   - (*b*) (*i*) Internal secretion or hormones—these are the chemical messengers from endocrine glands.
     - (*ii*) Antibodies—they are important in immunity reactions and contribute to blood' viscosity.
     - (*iii*) Enzymes e.g. amylases, carbonic anhydrases etc.
   - (*c*) (*i*) Non-protein nitrogenous substances—Urea, creatinine, ammonia, amino acids etc. Mainly these are the products of tissue activity which are transported from tissues to kidney or skin for excretion.
     - (*ii*) Neutral fats—e.g. cholestetrol, phospholipids, pentoses etc. These are food stuffs absorbed from intestine to tissue for utilization and storage.
2. *Respiratory gases:* e.g. Oxygen and Carbon-di-oxide.
3. *Inorganic substances:* e.g. Sodium chloride, Calcium, Potassium, Bicarbonate, Iodine, Iron etc.

II. Solid cellular portion (which is approximately 45% of the total volume of blood) in which free cells known as formed elements are suspended in the plasma. The cells are known as corpuscles or blood cells. The relative volume of plasma and corpuscles can be determined by haemotocrit. There are three types of corpuscles:

1. Red blood cells (RBC) or Erythrocytes (4.2-6.4 million/C.m.m)

RBC are produced in the bone marrow of flat bones vertebrae, ribs and heads of femur and humerus etc. These are biconcave discs having a diameter of 7.2 microns, these cells do not have nucleus, their life is 120 days. It contains red pigment haemoglobin. RBC helps in the transport of oxygen and carbon-di-oxide.

2. White Blood Cells (WBC) or Leucocytes

There is usually one WBC for every 500 RBC. In normal adult their number is about 5000-10,000/cmm of blood. These cells do not contain any pigment. There are five types of leucocytes: (*i*) Neutrophil polymorph (*ii*) Eosinophil (*iii*) Basophil (*iv*) Lymphocyte (*v*) Monocyte. The first three are together known as granulocytes since their cytoplasm contains granules. The other two *i.e.* lymphocyte and monocyte are known as non-granular leucocytes.

The main function of W.B.C. is body defence against infection by the phagocytes and production of antibodies by lymphocytes.

3. Platelets or Thrombocytes (Thrombo = clotting, cyte = cell)

These are non nulceated fragments of the cytoplasm of the very large cells. They are shaped like small flattened discs of about 2.4 micrometres in diameter. They contain many small granules. Their number is approximately 250,000-500,000/cmm. Platelets develop to some extent in spleen and bone marrow. Thrombocytes main function is to assist in haemostasis *i.e.* prevention of bleeding and stopping of bleeding (coagulation of blood) when injuries occur to the blood vessels.

The composition of the blood can be illustrated in brief in the chart given below:

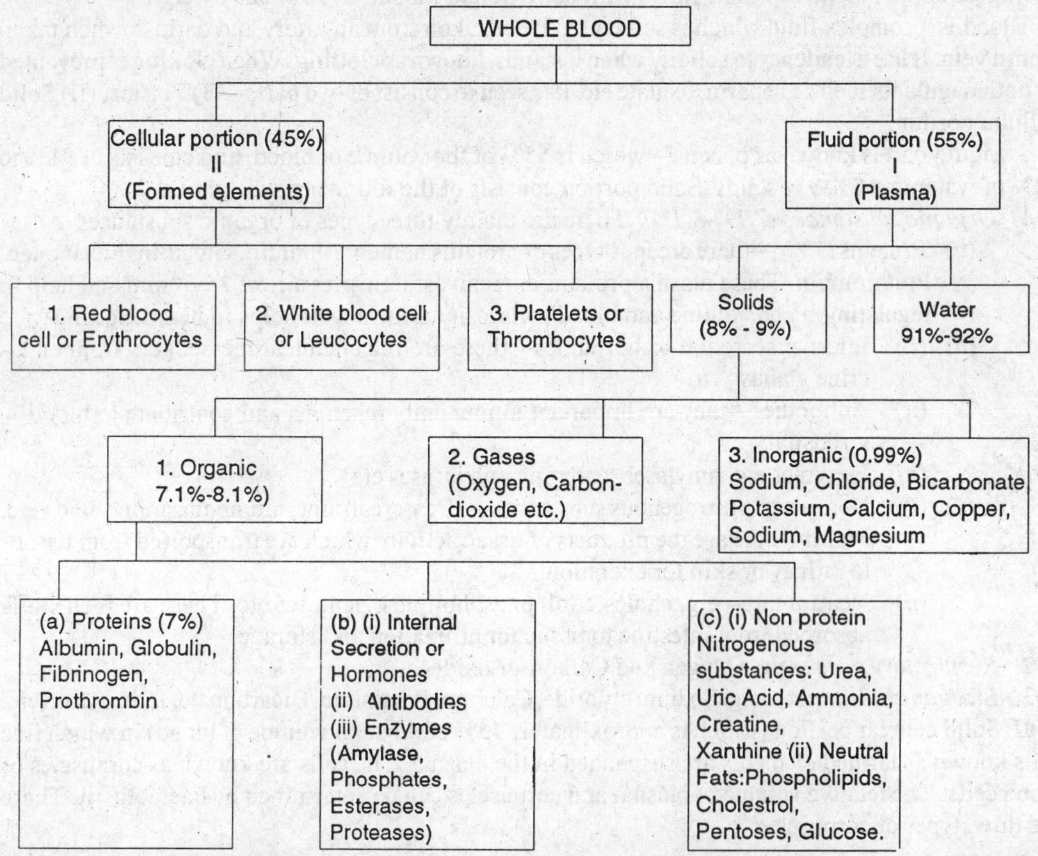

Along with the lymph and tissue fluids blood serves as the connecting link between the individual cells of distant organ and tissues. It carries the nutrients derived from digested food materials to tissue cells for growth and supplying energy. Blood carries oxygen from air through lungs to tissues and

removes carbon dioxide from tissues through lungs to air. It helps in maintenance of water content of tissues and regulating the body temperature etc.

Blood Grouping and Matching

The branch of haematology which includes the study of blood groups, their types and the preparation of blood for use in transfusion is usually referred to as Immunohaematology. There are four common blood groups namely A, B, O and AB. The blood group of an individual can be determined by the following methods:

Blood groups can be classified depending on the presence or absence of certain antigens known as agglutinogens in the red blood cells. The red cells have one, both or neither of the two agglutinogens called A and B correspondingly the plasma contains one, both or neither of two antibodies known as agglutinins called Anti A or α and anti B or β. Agglutinin antibodies cause cells coated with non-self antigens to clump together. The clumping is called agglutination and this process i.e. (agglutination) of the cell indicates the agglutinogens they possess, which in turn indicates the group to which they belong.

Procedure: Two antisera one Anti-A and another Anti-B are prepared. Blood from the subject to be tested for blood group is drawn in a syringe containing 0.4ml of 3.8% sodium citrate.

Then on a white porcelain slab or on a slide one drop each of known serum Anti A or α and anti B or β is placed. Separate pipette should be used for the each type of serum. The group of serum is marked with a pencil. Just near the drops of antisera, one or two drops of the corpuscular suspension from the bottom of the citrated blood are placed on the slide as shown in Fig. 1. The cells are mixed well by rotating or tilting slab or slide. It is allowed to stand undisturbed for a few minutes. The slides are then examined for clumping or agglutination under a microscope—Agglutination appears as red clumps separated from clear serum. Absence of agglutination appears as homogenous suspension of red blood cells. Sometimes difficulties may arise in blood grouping procedure when infected blood cells may string together in rouleaux due to an excess of globulin proteins in the antiserum. This rouleaux formation should not be mistaken with agglutination. The observation can be written as follows:

Sl. No.	*Anti A*	*Anti B*	*Conclusion*	*Interpretation*
1.	Agglutination	No Agglutination	A— antigen on cell	A Group
2.	No Agglutination	Agglutination	B—antigen on cell	B Group
3.	Agglutination	Agglutination	Both A and B Antigens	AB Group
4.	No Agglutination	No Agglutination	Neither A nor B Antigens	O Group

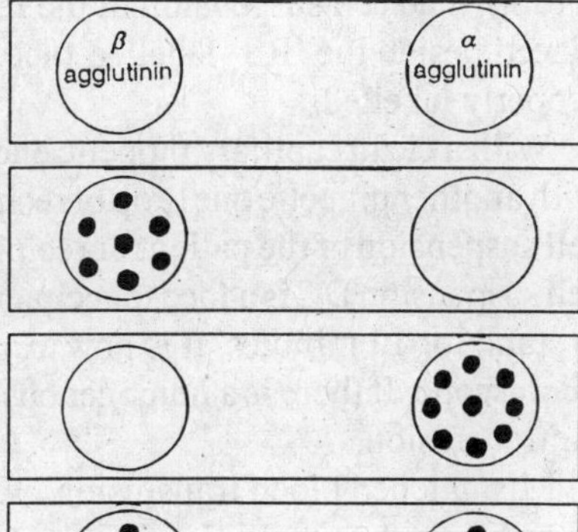

Group O blood gives no agglutination since no agglutinogens are present in these cells to be clumped by testsera

Group B (β agglutinogens give agglutination with group A serum)

Group A (α agglutinogens give agglutination with group B serum)

Group AB (gives agglutination with both testsera)

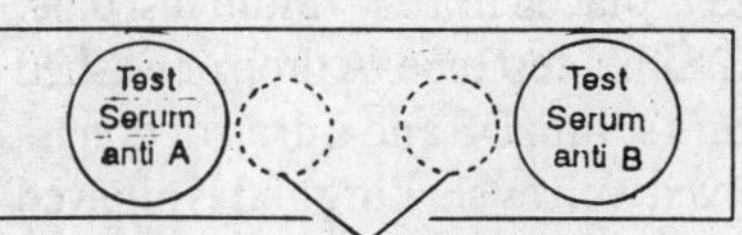

Patients carpuscular suspension

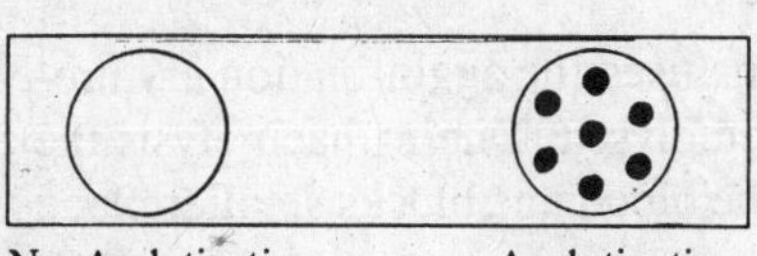

No Agglutination Agglutination

Fig. 1.

In human blood these natural agglutinogens have reciprocal relationship with natural antibodies known as agglutinins present in the serum. In other words the A group blood contains anti B or β-agglutinins and B group blood contains anti A or α-agglutinins in the serum. This relationship can be illustrated as follows:

S. No.	*Blood Group*	*Erythrocyte contains (agglutinogen)*	*Plasma contains (agglutinin)*
1.	A	A	Anti B or β
2.	B	B	Anti A or α
3.	AB	A & B	Nil
4.	O	Nil	Anti A or α Anti B or β

The knowledge of blood group is of immense importance in blood transfusion. In conditions like severe bleeding, shock, burns, severe anaemia or patients undergoing surgical operation, diseases in which there is severe loss of blood then blood from a donor has to be given to the patient and this process is known as blood transfusion. Before a transfusion, precaution must be taken to ensure that the blood of donor and recipient are compatible. If the wrong blood is given to a patient, clumps of red blood cells may block small blood vessels in vital organs e.g. lung or brain. The subsequent haemolysis of agglutinated cells may lead to haemoglobin in the urine and eventually to kidney failure and death.

The following chart indicate the compatibility of the blood groups:

S. No.	*Group*	*Can give blood to*	*Can receive blood from*
1.	Group AB	Only AB	From any group
2.	Group A	AB and A	A and O
3.	Group B	AB and B	B and O
4.	Group O	Any group	Only O

It is evident from the above chart that blood group AB is universal recipient and blood group O is universal donor.

It is essential to use the same blood group from donor to the recipient. The compatibility of the blood group can be tested by cross agglutination test known as Direct-Cross-Matching.

Cross matching should be done before every blood transfusion.

Procedure

1. Preparation of red cells and serum. Blood is drawn from the recipient or patients. It is allowed to clot and centrifuged. The clear serum is transferred in a clean test tube, labelled as RS (recipient's serum). The cell suspension of the recipient blood should be kept in a clean tube, labelled as RC, is placed beside the 'RS' labelled tube. Similarly donor's serum DS and donor's cells DC should be properly labelled.

With a clean capillary pippette one drop of the donor's serum DS is placed in 10 × 75mm test tube, with another pippette one drop of recipient's serum RS is placed in another test tube. A drop of washed cell suspension of the patient or recipient's RC is added to the donor's serum DS and a drop of donors cell suspension DC is added to recipients serum RS. The cells are mixed well by shaking and is allowed to stand for 10 minutes. It is now examined for agglutination with a hand lens and finally under the microscope. If there is a homogenous suspension of red cells on both the sides of the donor is suitable for transfusion.

Mismatched blood transfusion *i.e.* if incompatible blood is transfused the agglutination erythrocytes block the small blood vessels and eventually they are haemolysed. Due to haemolysis the haemoglobin is released in the blood and it gets accumulated in the blood and blocks smaller blood vessels in the coronary, pulmonary and renal circulation.

The incompatible blood transfusion has severe effect on the patient and it can be fatal. Whole blood is transfused when treating for haemorrhage when the replacement of blood volume is important, other blood components may be separated for transfusion in special circumstances. They include platelets and white cells as lymphocytes.

Blood for transfusion has to be checked for any unwanted contaminants before it is used. Viral hepatitis, syphilis and AIDS are among the diseases which have been contracted following transfusion.

Another important factor in blood transfusion is **Rhesus or Rh blood groups** named after the Rhesus monkey used in the original investigations.

The Rh groups are of outstanding clinical importance. The original discovery was made by injecting red cells of rhesus monkey into rabbits. The rabbit responded to the presence of an antigen in these cells by forming an antibody which agglutinates rhesus red cells. It was found that if the immunized rabbit's serum is tested against human red cells, agglutination occurs in 85 per cent, of these people are called Rh+ (positive) and no agglutination occured in 15 per cent and these people are called Rh–(negative).

There are several varieties Rh antigen and Rh antibody; the most common Rh antigen is called D and its antibody is called anti D. The inheritance of Rh antigens are controlled by genes. The cells which possess D are known as Rh+ ive and those without it are known as Rh–ive.

If the blood of a person who is Rh+ ive is transfused into a person who is Rh -ive the recipient's body slowly produces an antibody to agglutinogens. The antibody is termed as the anti-Rh factor. There may not be any indication of this in compatibility following the first transfusion but the second transfusion is often fatal as the anti Rh factor causes severe damage to the transfused erythrocytes.

Complications arise sometimes when a Rh -ive woman bears a Rh +ive foetus. Normally there is no mixing of the foetal and maternal blood cells although the two circulations run close together in the placenta. Occasionally foetal blood can pass through placenta into mother's blood which results in the production of anti Rh factor in mother's blood. This anti Rh factor is capable of crossing the placental membranes and enter into the blood of the foetus and it haemolyses and destroys the foetal red blood cells, it is known as haemolytic disease of the new born known as 'HDNB'. This disease can be treated by giving the foetus several transfusions while it is developing inside the uterus.

Rhesus grouping. Anti D is an immune antibody. Anti D is usually incapable of causing the agglutination of D-cells (Rh +ive) on its own. For this reason it is called an incomplete antibody. The cells become coated by anti-D and agglutination may occur when another protein is added, usually albumin.

In a simple test, a drop of 20% bovine albumin is added to a drop of antiserum containing anti-D. To this a drop of patient's cells in saline is added. After 3 to 5 minutes incubation at room temperature agglutination indicates Rh +ive cells.

Physiological Function of Plasma Protein

Plasma comprises approximately 55% of blood and it is about 5% of the total body weight. Plasma consists of 9% of solids and 91% of water of the solids about 7-8% is protein and the rest are salts, ions, organic molecules. Plasma proteins consist of four major fractions namely Albumin 3.5-4.58%, Globulins 1.6-3.28%, Fibrinogen 0.2-0.48%, Prothrombin 0.02-0.048%.

Functions of Plasma Proteins

The main functions of plasma proteins are as follows.

1. **Maintenance of Colloidal Osmotic Pressure.** All the fractions of plasma proteins maintain colloidal osmotic pressure. Albumins contribute 80% of the osmotic pressure whereas globulin and fibrinogen contribute about 20%. The total osmotic pressure exerted by plasma proteins is of the value of 25mm Hg in the living body.

2. Fibrinogen and prothrombin are the principal factors in blood clotting.
3. **Maintenance of Viscosity.** Plasma proteins maintain the viscosity of blood which maintains peripheral resistance of arterioles, thereby help maintaining blood pressure.
4. **Maintenance of acid-base balance of blood.** The proteins are good buffers and they have the power to accept H^+ ions.
5. **Carriage of Other Substances.** Polysaccharides, lipids, and lipid soluble substances phospholipids, cholesterols, steroids, hormones, bilirubin, metal like Iron, Zinc etc. are carried by plasma proteins.
6. **Production of Immune bodies.** Many antibodies against varieties of antigens such as organisms of diphtheria, mumps, chickenpox, typhoid, streptococci, staphylococci, hepatitis etc. are closely associated with globulin fraction of the plasma proteins. Purified-globulin are utilized for immunizing against these infections.
7. **Carriage of Carbon dioxide.** The amono acid content of the proteins helps formation of carbamino compound and thus helps carriage of carbondioxide.
8. Fibrinogen and globulin fractions help in rouleaux formation and thus help in increasing sedimentation rate of the corpuscles.
9. **Regulation of Fluid interchange between blood and tissue.** Fluids are regularly interchanged between tissues and blood due to the presence of normal concentration of proteins. This exchange maintains homeostasis of internal environment.
10. **Reserve Proteins.** During prolonged starvation plasma poteins contribute to supply the protein need through amino acid pool, though there is no direct evidence that plasma proteins are directly interchangeable with tissue proteins.
11. **Repair of Damaged tissue.** When plasma proteins exude into tissue spaces as in inflammation or internal haemorrhage, it forms a good medium for growth of various tissues essential for repair.

Role of Blood as Oxygen Carrier

Atmosphere is the source of oxygen and lungs act as a local depot from which it is supplied to the tissues. This delivery system consists of lungs and cardiovascular system which is the transport by which oxygen is supplied to the tissues. The efficiency of oxygen depends on the quantity of oxygen available in lungs, efficiency of pulmonary gas exchange, efficiency of pulmonary blood flow and the capacity of blood to carry oxygen to the tissues.

Transport of Oxygen by Blood

Oxygen diffuses through the respiratory membrane from the alveolar air and into the blood. The oxygen carrying capacity of blood is about 20 ml gaseous oxygen per 100ml of whole blood. The presence of respiratory pigment, haemoglobin in red blood cells accounts for about 98% of the oxygen transported to the tissues. It is carried as oxyhaemoglobin.

The venous blood carries about 14 to 15 ml of oxygen per 100 ml of blood and is oxygenated in the lungs and the blood leaves the lungs as arterial blood carrying about 19 to 20ml of oxygen per 100ml of blood. Oxygen is carried in the blood as follows:

(*i*) **In Physical solution.** The amount of oxygen carried in physical solution depends on the pressure gradient, temperature and solubility coefficient 1.0ml of blood absorbs 0.024ml of oxygen at 37°C at a pressure of 760mm Hg. The partial pressure of oxygen in the alveoli is 100mm Hg, so at this pressure of oxygen 1ml of blood will be $\frac{100 \times 0.024}{760} = 0.003$ ml. Thus 100 ml of arterial blood will carry 0.3 ml of oxygen whereas at this alveolar partial pressure 100 ml of venous blood will carry 0.12ml of oxygen in physical solution.

(*ii*) **In chemical combination with haemoglobin.** Arterial blood oxygen is about 19 to 20ml per 100ml and the oxygen in physical solution is about 0.3ml. This difference is due to combination in the haemoglobin. Haemoglobin has a complex molecular structure consisting of four sub-units. Each sub-unit contains a group called haem attached to a protein called globin. Each sub-unit may combine with one molecule of oxygen by a process called oxygenation, the derivative formed is called oxyhaemoglobin, oxygen is released from the haemoglobin by deoxygenation. Since there are four sub-units, upto four molecules of oxygen can load to each molecule of haemoglobin.

$$\text{Haemoglobin} + O_2 \xrightleftharpoons{\text{Oxygenation}} \text{Oxyhaemoglobin}$$

$$Hb_4 + 4O_2 \rightleftharpoons Hb_4O_8$$

The combination of oxygen with haemoglobin takes place stepwise as follows:

$$Hb_4 + O_2 \rightleftharpoons Hb_4O_2$$

$$Hb_4O_2 + O_2 \rightleftharpoons Hb_4O_4$$

$$Hb_4O_4 + O_2 \rightleftharpoons Hb_4O_6$$

$$Hb_4O_6 + O_2 \rightleftharpoons Hb_4O_8$$

In standard conditions of temperature at 37°C and 760mm Hg pressure each gram of haemoglobin may combine with 1.36ml oxygen. In an adult male there are usually 13.0g to 18.0g haemoglobin in every 100ml of blood and in women it is between 11.5g to 16.5g.

The process of oxygenation and deoxygenation takes place very rapidly i.e. it takes place in seconds, and this process depends on three main factors, oxygen tension, carbon dioxide tension and temperature. Sometimes the presence of poisons such as carbon monoxide disrupts oxygen transport by haemoglobin.

Under resting conditions 100 ml of arterial blood transports about 19.3ml of oxygen, carrying 0.3ml in physical solution and 19.0ml in chemical combination with haemoglobin at a tension of 96.0mm Hg, in the blood of the capillaries, for supply to the tissues. The oxygen tension of the tissues is about 35.0mm. Due to this pressure gradient and absorption of the carbon dioxide by the blood oxygen passes from the blood to the tissues within fraction of a second and thus the oxygen tension of blood is lowered to 40mm Hg in resting conditions. Usually 5.0 to 6.0ml of oxygen per 100ml of blood is extracted by the tissues under resting conditions. The oxygen difference between arterial and venous blood is 19.0 – 14.0 = 5.0 ml per 100 ml.

The coefficient of oxygen utilization

$$\frac{\text{oxygen taken up by the tissues}}{\text{oxygen content of arterial blood}} = \frac{5}{19} \times 100 = 26\%$$

The coefficient of oxygen utilization is taken as an index of the activity of the tissues as it depends on the degree of tissue activity *e.g.* during exercise it may be raised to 80%. The normal coefficient of oxygen utilization by the heart is about 63%.

Blood Pressure

The pressure exerted by the flow of blood on the walls of the blood vessels is called blood pressure. Blood pressure is dependent on many factors such as (*i*) strength of contraction of cardiac muscle (*ii*) resistance of walls of arterioles (*iii*) total blood volume (*iv*) blood viscosity etc. out of these, power of contraction of cardiac muscle is the main factor.

When the heart muscle is at its maximum contraction, the pressure is known as Systolic Pressure. It fluctuates with the physiological condition of exertion, mental state, sleep or meals.

Minimum pressure is exerted by the blood when the heart is relaxing and it is called Diastolic Pressure.

The difference between two pressures systolic and diastolic is known as Pulse Pressure. The elastic walls of blood vessels help to maintain the blood pressure at normal level. The peripheral resistance offered by arterioles is the important factor in regulation of blood pressure. Similarly blood volume in body and condition of blood viscosity influence blood pressure.

The normal blood pressure in a young adult is 120mm (Hg) systolic and 80mm (Hg) diastolic. High systolic pressure of 150 and above and diastolic pressure more than 100 is known as high blood pressure or hypertension. Blood pressure below 90/60mm Hg is known as low blood pressure or Hypotension.

Hypertension

When walls of arteries and arterioles harden, there is more resistance to blood flow, and heart has to contract more forcefully to cope with the peripheral resistance which enhances both systolic and diastolic pressure. This is called hypertension or high blood pressure. Gradually to compensate this, more power is exerted by heart muscle which results in dilation of heart especially left ventricle. In course of time the heart cannot cope up with this extra hard work and the heart fails causing dependent edema which leads to congestive cardiac failure.

Hypertension is of two kinds:

(*a*) **Primary or essential hypertension.** Primary hypertension is when the real cause is not known. Primary hypertension may be due to age, anxiety, heredity, peripheral resistance due to narrowing of arterioles etc. It is classified into two groups:

  (*i*) *Benign hypertension:* It is not so serious and is chronic.

  (*ii*) *Malignant hypertension:* In this case the blood pressure rises to dangerous levels and complication like renal failure may occur and requires prompt treatment.

(*b*) **Secondary hypertension.** This type of hypertension is due to definite causes like nephritis, adrenal tumours, etc. Symptoms depend upon the severity of hypertension. Some common symptoms are headache, sleeplessness, palpitation, retinal haemorrhage etc.

Treatment. The secondary hypertension is controlled by treating the cause.

Primary Hypertension

In case of primary hypertension when the real cause is not known, the patient has to be on constant medication and controlled diet.

Diet: Less sodium intake is an important step in treatment. If there is more sodium in the body there will be more fluid i.e. more blood volume and higher blood pressure.

Body weight should be on the lower side. Patient should take more rest. Tranquillizers and sedatives are given to reduce anxiety and tension. Systolic B.P. is mainly due to contractile power of heart and diastolic due to vascular resistance.

Normal B.P. in young adults is systolic 120 diastolic 80 Mild hypertension systolic 140 diastolic 90-100.

Severe hypertension is when systolic 180 and diastolic 100-120.

Treatment of Mild Hypertension

Salt restriction and thiazide diuretics are given to control the pressure. Moderate hypertension can be treated by reserpine, diuretic and hydrallazine or methyldopa (Aldomet) tablets. In case of hypertensive crisis Aldomet, hydrallazine or diazoxide can be given intravenously which reduces blood pressure in short time.

Systolic pressure is highly variable depending upon the factors of emotion, exercise etc. It indicates the force of contraction of heart; the more is the blood volume, the more is the systolic pressure. Diastolic pressure is the index of the peripheral resistance and it is the constant pressure in the arteries due to tonicity of the arterioles and capillaries. Diastolic pressure is the level at which the heart is pumping blood and if the diastolic pressure is raised heart is doing more work. Therefore diastolic pressure is more important than systolic pressure.

Hypotension: Blood pressure is much below normal values *i.e.* about 100/60 in adults. Hypotension should be treated otherwise the patient may suffer from occasional Syncape i.e. a sudden suspension of consciousness due to diminution of the blood supply to the brain. In case of hypotension, patient should be given high protein diet like meat, egg etc. Some important drugs are haemotinies and anabolic steroids, nandrolone, methan dienone etc.

Measurement. The instruments for measuring blood pressure is known as sphygmomanometer. This instrument consists of rubber bag covered with a cloth envelope which is wrapped round the upper arm over the Brachial artery. One tube connects the inside of the bag with a manometer containing mercury. Another tube connects the inside of the bag to a hand operated pump with a release valve.

Method. The cuff is wrapped around the arm just above the elbow and air is pumped into the rubber bag and the pressure of the air inside the bag is raised to about 200mm Hg. which is enough to flatten the brachial artery in the arm and blood flow in the artery is thus stopped. The microphone of a stethoscope is placed over the brachial artery, the pressure inside the cuff is slowly lowered by means of a valve. A pressure is reached at which the blood can just squeeze through the artery each time. The left ventricle contracts, the artery makes tapping sounds named after Korotkov who first described this pressure corresponds to systolic blood pressure. The cuff pressure is now lowered still further until the blood flows through the artery even between ventricular contractions. The Korotkov sounds suddenly become muffled and disappear. The cuff present at this point corresponds to the diastolic blood pressure.

Coagulation or Clotting of Blood

Blood is a fluid when it is inside the blood vessels of the living body but when blood is taken out and kept in a test tube at room temperature without addition of anticoagulants it become jellified within 5-6 minutes. This transformation of blood from liquid to jelly like mass is known as coagulation or clotting of blood. Coagulation of blood is a major chemical defense against blood loss.

Mechanism of Blood Coagulation

Coagulation of blood is the result of a chain reactions and the mechanism of blood coagulation can be divided into three stages:

The *I stage* in this process is the production of a lipoprotein called thromboplastin at the site of injury.

The *II stage* is the conversion of prothrombin to thrombin.

The *III stage* is the conversion of fibrinogen to fibrin. A series of cogulating factors has been found to be essential for clotting process some important coagulation factors are as follows:

Factor No.	*Common Name*
I	Fibrinogen
II	Prothrombin
III	Thromboplastin
IV	Calcium ions
V	Hereditary labile factors like activator globulin, proaccelerin.

VI	Active form of factor V known as Accelerin.
VII	Proconvertin—Serum Prothrombin Conversion Accelerator (SPCA)
VIII	Antihemophilic Factor (AHF)
IX	Plasma Thromboplastin Component (PTC)
X	Stuart power factor.
XI	Plasma Thromboplastin Antecedent (PTA)
XII	Hageman factor.
XIII	Fibrin stabilizing factor, Fibrinose.

In the I stage the production of thromboplastin may originate from undamaged tissues by two processes known as (*i*) *extrinsic mechanism*—when tissue extract with the help of factors V, VII, X produces thromboplastin outside the blood vessels and (*ii*) *intrinsic mechanism*—when there is production of thromboplastin inside the blood vessels with the help of surface contact factors (X, XII) factors V, VII, IX, X and Platelet factor.

In the II stage the thromboplastin acts on another inactive plasma protein known as prothrombin in presence of calcium ions (factor IV) and converts it into active thrombin.

In the third stage thrombin, in turn, acts on another plasma protein called fibrinogen and converts it into fibrin, which is insoluble and when blood cells get attached to the fibrin strands a, blood clot is formed. The mechanism of blood clotting can be represented as follows:

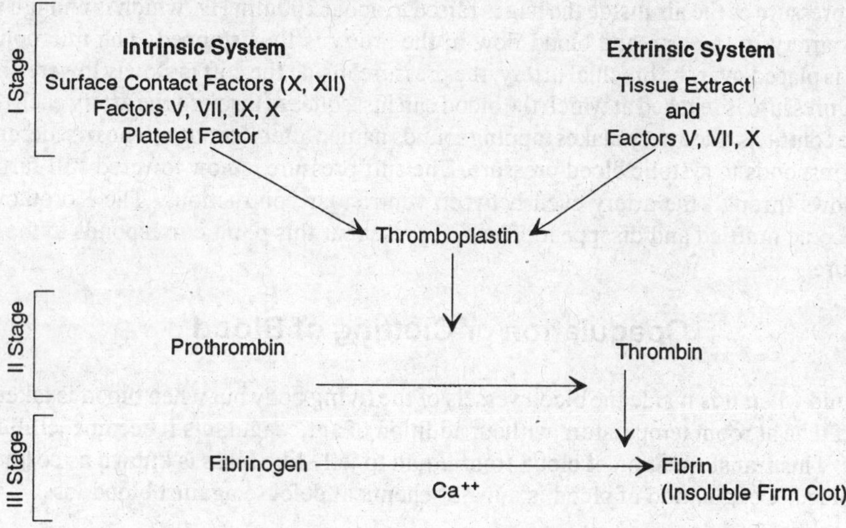

As it is important for blood to cogulate at the site of injury, it is equally important that blood does not coagulate in a healthy vessel. Blood coagulation is prevented in healthy vessels by substances secreted from the vessel wall, most important among these substance is *Prostacyclin*. If a vessel is injured, the damaged part does not produce prostacyclin and the platelets coagulate.

Haemotological Agents

Coagulants

These drugs help in blood coagulation and hence are useful in treatment of severe haemorrhage. These drugs help to stop oozing of blood from small blood vessels and capillaries. These drugs can be classified into two groups:

(*i*) Local
(*ii*) Systemic

(*i*) *Local:* These agents control oozing of blood from minute vessel and are generally applied locally, e.g. gelatin sponge used for controlling oozing of blood from surface wounds, Thrombin, Oxycel is surgical gauze etc.

(*ii*) *Systemic:* These drugs are given orally or parenterally and stop bleeding by enhancing coagulation of local blood vessels, e.g., Fibrinogen, Vitamin K, Rutin, Plasma etc.

Some important cogulants are described below:

1. Vitamin K. It is a fat soluble vitamin and exists in two forms Vitamin K_1, and Vitamin K_2; both are derivatives of naphthaquinone. Another vitamin known as Vitamin K_3 of menadione is a lipid soluble, synthetic naphthaquinone compound and is as active as Vitamin K. Vitamin K occurs as a bright yellow crystalline powder, while menadione is a light yellow oil. Vitamin K is essential for the biosynthesis of prothrombin and factors VII, IX, X. Vitamin K is produced by the flora of the human intestine. The fat soluble Vitamins K_1 and K_2 are absorbed in the presence of bile salts. It is an useful antihaemorrhagic agent as it is able to maintain a normal concentration of prothrombin and other clotting factors in blood plasma by increasing their biosynthesis in liver. Vitamin K can be given by mouth or intravenously. Vitamin K_1 tablets of 5mg and ampoules 10 or 50mg per ml of Vitamin K_1, oxide is available. Menadione sodium bisulfite injection is also available which can be used in emergencies by injecting intravenously.

2. Protamine Sulphate. Protamines are proteins having low molecular weight. It is a white or grey-yellow hygroscopic amorphous or crystalline powder. It has slightly astringent taste. Protamines are found in the sperm or mature testes of certain fish. Protamines are basic due to presence of basic amino acid arginine.

Protamine sulphate is available as 1 per cent solution. It is mainly used as heparin antagonist i.e. it promptly arrests the anticoagulant effects of heaprin. 1.0mg of protamine sulphate neutralizes the anticoagulant effect of 100 units of heparin. These are used as an injection which is sterilised by filtration and is stored at 2°–8°C. It is used as intravenous injection but sometimes it may cause a sudden fall in blood pressure dyspnoea or, transitory flushing etc.

3. Dry-Thrombin. It is an enzyme which can convert human fibrinogen into fibrin. Dried thrombin can be prepared from pooled liquid plasma obtained from human subjects. By a suitable fractional technique the prothrombin fraction can be separated. Prothrombin can be converted into thrombin in presence of calcium ions and human thromboplastin. This solution can be sterilised by filtration and it can be freezed-dried.

It is used topically for its clotting effect on blood. Thrombin can be applied as a powder or in solution form. Dried thrombin should be stored in sterile containers in an atmosphere of nitrogen and at low temperature.

Anticoagulant Drugs

Anticoagulant are substances which interfere with blood coagulation and thus prolong the coagulation time. These drugs can be classified into:

1. Those used for preventing clotting of blood inside the vascular system of the body it can be either (*a*) rapidly acting drug e.g. heparin etc. (*b*) slow acting drug e.g. coumarin derivatives, indandione derivatives etc.
2. Those drugs which prevent clotting of blood in vitro.

Certain drugs can be used both in vivo as well as in vitro.

Some important anticoagulants are as follows:

1. Coumarine

These are slow acting oral anticoagulants i.e. these drugs are effective when administered by mouth. Coumarin ring consists of a benzene ring fused with an α-pyrone ring. The first Coumarin compound which was isolated from sweet clover plant was bishydroxy coumarin or Dicoumarol; it is a derivative of 4-hydroxy coumarin. Main chemically related compounds have been studied but only a few are clinically useful. Examples of such drugs are nicoumalone and warfarin.

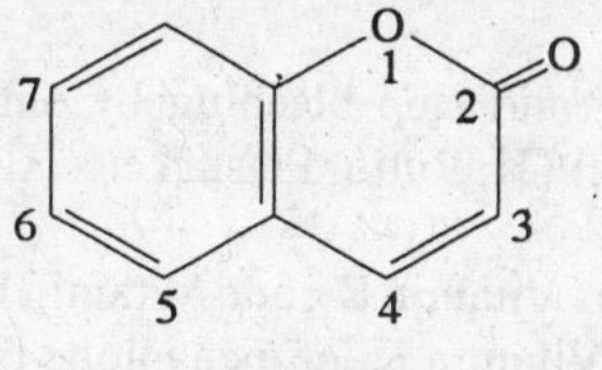

Coumarin

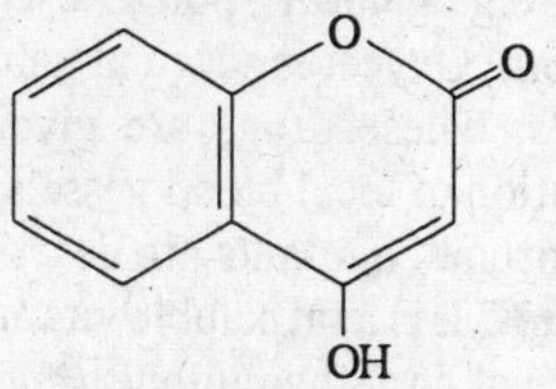

4-Hydroxy Coumarin

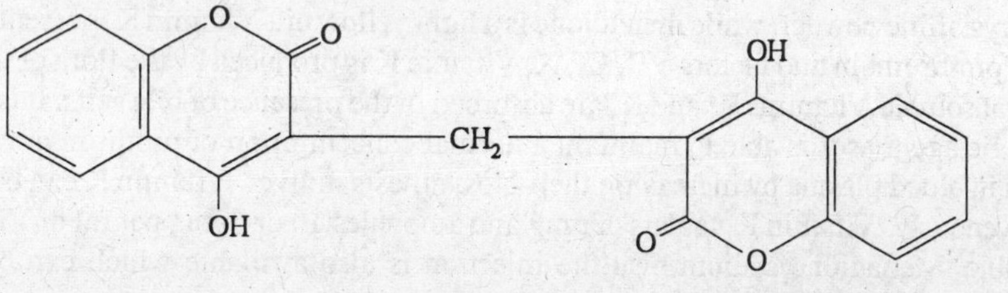

Dicoumarol

The therapeutic action of coumarin derivative is due to their ability to prolong the synthesis of prothrombin and factors VII, IX, X by the liver. Coumarin therapy is found to increase the plasma antithrombin levels. Bishydroxy coumarin has no anticoagulent action in vitro. Bishydroxy coumarin has prolonged anticoagulant therapy and is quite effective when given orally. It is used in the treatment of pulmonary embolism (obstruction due to blood clot), traumatic injuries to blood vessels, vascular surgery etc.

2. Indandione Derivatives

These are also slow acting anticoagulants and includes compounds like phenindione, diphenindione, anisindione etc. Among these the compound used initially was phenindione and others differ from it in the duration of action and other toxic effects. Phenindione is a derivative of indane which consists of a benzene ring fused with cyclopentane ring is named as 2-phenyl-indane-1,3 dione. It has the following structure:

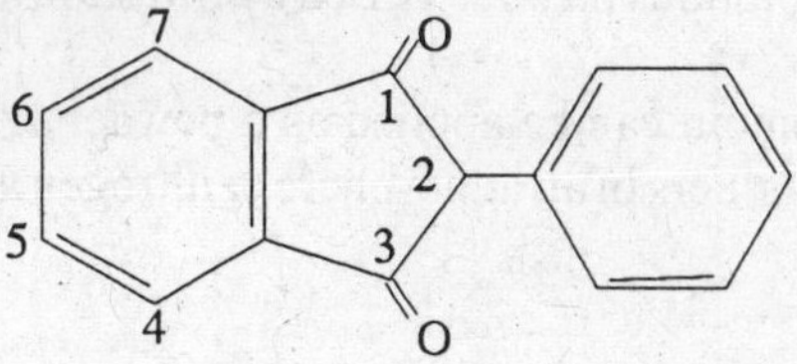

Phenindione

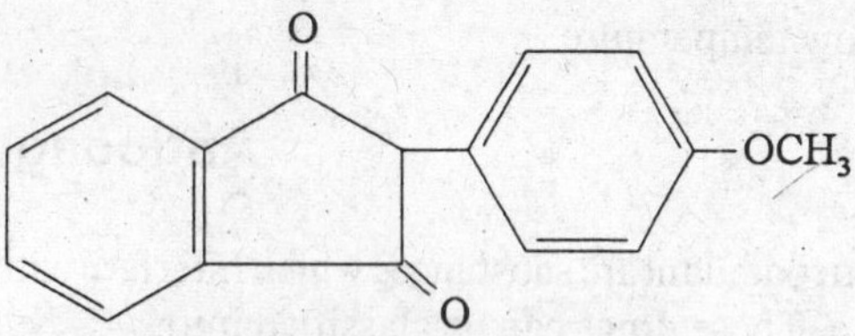

Anisindione

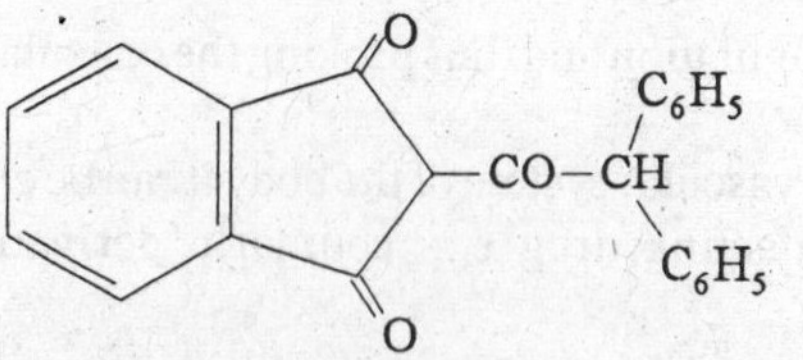

Diphenindione

The anisindione is a creamy white crystalline powder, diphenindione and phenindione is creamy white coloured crystalline material and its anticoagulant activity is similar to Coumarin compounds. It has rapid onset and short duration of action. With the initial therapeutic dose, significant action is seen within 24 hours. Diphedione and Anisindione is a prothrombopenic anticogulant with actions and uses similar to Bishydroxycoumaru.

Sometimes it shows some adverse reactions like skin rash, extensive oedema, jaundice etc. Some of these compounds turn the urine redish orange. Diphendione may cause nausea, vomiting etc.

3. Heparin

It is fast acting anticoagulant and is obtained from most cells of liver. Commercial heparin is obtained from the lung and the intestinal mucosa of pigs and cattle. Heparin is a mucopolysaccharide. It is white or pale coloured amorphous powder. It is hygros copic in nature and resistant to all kinds of chemical agent.

Heparin prevents the clotting of blood both in vivo and vitro. It probably acts on all the three stages of coagulation. Heparin is not effective orally but it is well absorbed after subcutaneous injection. It is mainly metabolized by a liver enzyme known as heparinase. Its calcium and sodium salts are also used. Heparin is a strong electro negative compound and its anticoagulant activity is attributed to its strong electronegative charge. It forms complexes with postively charged proteins including several proteins concerned with blood coagulation and inhibits its action. When Heparin is administered in therapeutic doses, it prolongs the clotting time of blood. Heparin is used in acute conditions in which a rapid reduction in coaguiability of the blood is desired. It is also used for prevention of clotting of blood sample.

4. Citric acid

Citric acid is obtained from Citrus fruits like oranges, lemons etc. It is also produced synthetically from glycerol. Citric acid is 2-hydroxy propane 1, 2, 3, tricarboxylic acid. It forms salts with potassium and sodium.

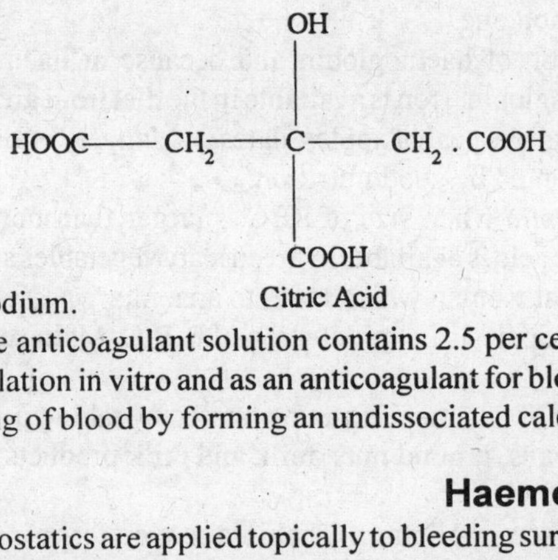

$$HOOC-CH_2-C(OH)(COOH)-CH_2 \cdot COOH$$

Citric Acid

$$HOOC-CH_2-C(OH)(COOH)-CH_2 \cdot COONa$$

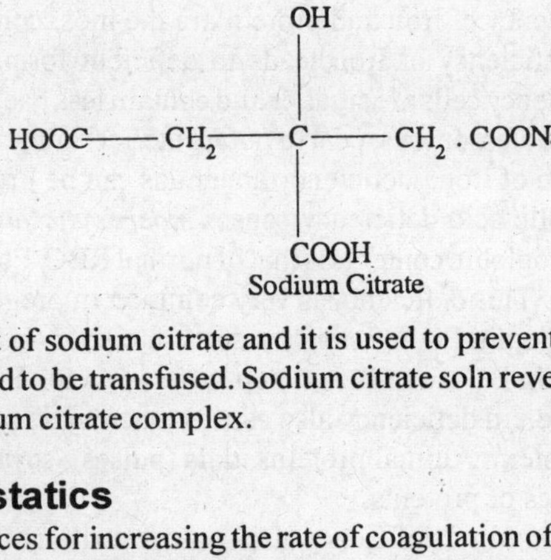

Sodium Citrate

The anticoagulant solution contains 2.5 per cent of sodium citrate and it is used to prevent blood coagulation in vitro and as an anticoagulant for blood to be transfused. Sodium citrate soln revents the clotting of blood by forming an undissociated calcium citrate complex.

Haemostatics

Haemostatics are applied topically to bleeding surfaces for increasing the rate of coagulation of blood. These agents help to arrest haemorrhage by controling capillary oozing and venous bleeding. Important example of haemostatic drugs are transexamic and amino caproic acid.

1. Amino Caproic Acid

It is white crystalline odourless and tasteless substance which is soluble in water. Aminocaproic acid is $H_2N(CH_2)_5COOH$ and it is structurally related to lysine. It reduces fibrinolytic activity by blocking plasminogen activation by competitive activity.

It is used in the treatment of severe haemorrhage. It can be administered either by mouth or by slow intravenous injection which is diluted with sodium chloride injection. It is rapidly absorbed when it is given orally. This drug is mainly excreted by the kidney. A dilute solution of this drug is used for rinsing after dental extraction. It is also used in the treatment of disorders of fibrinolyses is enhanced such as cardiac by pass, major thoracic surgery etc. Adverse effect—skin rash, hausea, diuresis etc.

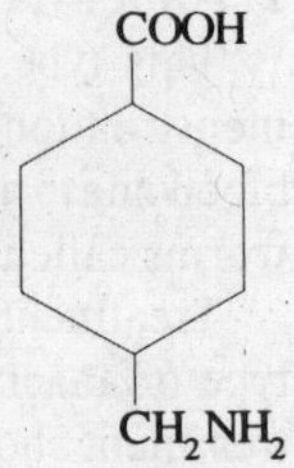

COOH

CH_2NH_2

Tranexamic Aci

2. Tranexamic Acid

It is a white crystalline powder, soluble in water. Its action and uses are similar to those of aminocaproic acid although it is more potent than aminocaproic acid.

Anaemia—Causes and Control

Anaemia is a state in which R.B.C. as well as amount of haemoglobin is less in blood. Blood performs several important functions like carriage of oxygen to different parts of the body and carriage of carbon dioxide to the lungs from the tissues. Anaemia is also defined as "decrease in oxygen carrying capacity of the blood. It also carries nutrition, defence materials, hormones, enzymes and various salts. It has several mechanisms to keep acidity, fluidity, pressure etc. constant, hence deficiency of blood leads to a deficiency of functions of all organs of the body.

There are several causes of anaemia. The main one are: (*i*) deficiency of raw materials required to produce blood in the bone marrow (*ii*) excessive loss of blood from the body and (*iii*) excessive destruction of blood in the body itself, a process known as haemolysis.

Sign, Symptoms and Types of Anaemia

Anaemic patient looks pale, this paleness is observed in the nails, lips, tongue and the conjunctiva of the eyes. There is loss of appetite, lack of energy, lack of interest, breathlessness, palpitation of heart, general weakness etc.

Anaemia can be mild, moderate or severe. When the haemoglobin level is over 70% it is mild anaemia, between 45% and 70% is moderate anaemia and below 45% is severe anaemic condition.

1. Deficiency or nutritional anaemia

This anaemia is due to deficiency of iron, folic acid, Vitamin B_{12} and proteins. Among these, deficiency of iron and protein are the most common one.

Deficiency of iron leads to deficient formation of haemoglobin and because of haemoglobin deficiency cells are smaller and contain less haemoglobin. Iron is available in the diet from green leafy vegetables, onions, peal of potato etc., certain fruits like peal of apples, dates etc. Jaggery is also a rich source of iron. Deficiency anaemias can be prevented by good nutrition.

Folic acid deficiency causes *macrocytic anaemia* when size of RBC is larger than normal but haemoglobin contents is that of normal RBC. Folic acid is available in green leafy vegetables and fresh fruits. This deficiency is very common in pregnant women which leads to anaemia.

Vitamin B_{12} deficiency results into poor erythropoiesis i.e. production of R.B.C. Main sources of vitamin B_{12} are milk and non-vegetarian food.

Protein deficiency also causes macrocytic type of anaemia. It can be treated by taking a diet rich in protein. Animal proteins, dals (pulses), soyabeans, ground nuts, milk and milk products are rich sources of proteins.

Vitamin C deficiency also causes nutritional anaemia. Vitamin C helps to convert ferric ion (Fe^{+++}) into ferrous (Fe^{++}), in which form iron is absorbed and thus it helps in normal haemoglobin formation. All the citrus foods are rich sources of Vitamin C.

This type of nutritional anaemia can be treated by giving proper nutritional diet and also by proper medicines either in tablet form or by injection. Sometimes in case of severe anaemia blood transfusion is also required.

2. Anaemia due to loss of blood

This type of anaemia is due to heavy blood loss due to accidents or it can be chronic e.g. menstruation. Any internal haemorrhage also causes anaemia which can be detected by vomiting of blood due to an ulcer in the stomach or duodenum cancer of stomach or cirrohosis of liver. Sometimes worms called hookworms also cause anaemia by sucking small quantities of blood from duodenum.

Treatment should be done by treating the causes or factors (described as above) responsible for this type of anaemia. Patient should be kept under medical supervision and proper diet and complete treatment should be given.

3. Anaemia due to excessive destruction of blood or Haemolysis

Haemolysis is a process by which the red blood cells are destroyed in the circulation and this causes anaemia. This condition is not common. Examples of this type of anaemia are snake venoms, which cause anaemia due to haemolysis, bacterial infections like septicaemias, lead poisoning, incompatible blood transfusions etc.

In certain diseases *e.g.* malaria, there is excessive destruction of red blood cells which cause anaemia.

Detection of anaemia is discussed earlier in Chapter 4 of this book.

Antinaemic Drugs

Some of the important drugs which are effective in treating different types of anaemias are as follows:

1. Drugs effective in iron deficiency anaemias

Iron is specific in the treatment of iron deficiency anaemias. Iron is available in meat, liver, bajra, whole wheat flour, rice bran, tamarin pulp, bengal gram, dates etc. The approximate daily dietary requirement of iron is 8-18mg for children, 15-20mg for women, 10-15mg for men. Iron from food resources are absorbed in ionisable ferrous form. Absorption of iron is maximum from duodenum.

Some of important iron containing drugs, used orally are:

(*i*) *Hydrated ferrous sulphate I.P.* ($FeSO_4 7H_2O$): These are transparant green crystals with a metallic and an astringent taste. In presence of moist air the crystals become brownish yellow due to oxidation to ferric sulphate and it should not be used. 300mg hydrated ferrous sulphate tablets are available.

(*ii*) *Dried ferrous sulphate 200mg tablets are available:*

(*iii*) *Ferrous sulphate syrup:* It contains 40mg of ferrous sulphate in each millilitre.

(*iv*) *Ferrous gluconate:* It is a yellowish grey powder or granules and are available as 300mg tablets.

(*v*) Ferrous fumarate—It is a reddish brown salt and relatively resistant to oxidation. It is available as 200mg and 300mg tablet form.

(*vi*) *Ferric ammonium Citrate:* It is given in a liquid form but it stains teeth and tongue.

(*vii*) *Ferrocholine citriate:* It is a iron chelate complex in which ferric iron is bound in a protein molecule. It is a well tolerated drug and is given to children and also given to patients who can not tolerate ferrous iron salts dose—300mg 3 times a day after meals.

Some drugs are used parenterally i.e. in injection form examples are:

(*i*) *Iron dextran:* It is available as 5ml ampoules containing about 50mg of elemental iron per ml. It can also be given intravenously.

(*ii*) *Saccharated iron oxide (Ferri-venin):* It is a complex of ferric hydroxide and sucrose in 5ml ampoules containing 20mg of iron per ml. It is distinctly alkaline.

(*iii*) *Iron sorbitol citric acid complex:* It contains iron, dextran, sorbitol, and citric acid and stabilized with gelatin. It is given intramuscularly.

(*iv*) *Iron dextrin:* It is complex of ferric hydroxide and is available in 5ml ampoules containing 29mg of elemental iron per ml of solution.

Pyridoxime HCl is also used in the treatment of iron deficiency anaemia. Pyridoxine HCl takes part in haemoglobin formation and utilization of iron. Pyridoxine deficiency may cause a sort of iron deficiency anaemia which is resistant to oral or parenteral iron therapy but responds to pyridoxine therapy. Dose is generally 10 to 50mg orally daily in divided dose or intramuscularly.

2. Drugs effective in megaloblastic anaemias

Megaloblastic anaemia is characterised by the presence of abnormally large RBC but the haemoglobin content is normal. Megaloblastic anaemia is due to Vitamin B_{12} and folic acid deficiency.

Some important drugs used in megaloblastic anaemia are as follows:

(*i*) *Vitamin B_{12}:* Vitamin B_{12} can be given parenterally or orally. Vitamin B_{12} is synthesised by micro-organism in soil, water and animal intestine. Non-vegetarian foods like fish, egg yolk, liver etc. are rich in vitamin B_{12}. Commercially vitamin B_{12} is obtained as a side product of streptomycin industry.

(*ii*) *Folic acid (Pteroylmono glutamic acid):* Chemically it is pteridine, para aminobenzoic acid and glutamic acid. It is commonly present in green leaves or vegetables, liver, yeast etc. Folic acid is an orange yellow micro crystalline powder which is odourless, tasteless and slightly soluble in boiling water. Folic acid tablets contain 5mg of the drug. Dose 5 to 20mg can be given orally. It can also be given parenterally.

(*iii*) *Folinic acid (citrovorum factor):* Chemically is N^5 formyl tetra hydrofolinic acid. It was originally found in the liver. It is available in 1ml ampoules. It is only used in the treatment of toxicity due to folic acid antagonists like methotrexate.

Other drugs used in megaloblastic anaemias are cyanocobalamin injection, hydroxycobalamin injection etc.

List of some haematinic products (the drugs which increase haemoglobin of the blood) available:

Trade Name	*Composition*
Autrin (capsules)	Vit. B_{12}, Folic acid, ferrous fumarate, Vit. C.
Calron (liquid)	Vit. B_{12}, Vit. A, Vit. B_1, B_2, B_6, D_3, Niacinamide, Sodium and Calcium glycerophos
Dexorange (liquid)	Haemoglobin, Vit. B_{12}
Ferroredoxon (capsule)	Ferrous fumarate, Vit. B_{12}, Folic acid, Vit. C.
Anaemidox (capsule)	Ferrous fumarate, Folic acid, Vit. B_{12}, Vit. C etc.
Rubraton (liquid)	Ferric ammoneit, B_{12}, Folic acid
Iberol (tablet and liquid)	Ferrous sulphate, B_{12}, Folic acid, B Complex.
Hapatoglobin (liquid)	Prot liver, Vit. B_{12}, Oxyhaemoglobin, Ferric ammoneit, Nicotinic acid.

Structure of Haemoglobin

Haemoglobin is the red pigment of the blood and comprises a chromoprotein. It consists of a non-specific simple protein globin and a specific prosthetic group haem or heme which is an iron containing pigment. The proportion of globin and heam is 96% and 4% respectively.

The empirical formula of haemoglobin is $(C_{738}\ H_{1166}\ FCN_{203}\ O_{208}\ S_2)^4$ and its molecular weight is 64,500. One molecule of haemoglobin contains 4 atoms of iron and 8 atoms of sulphur.

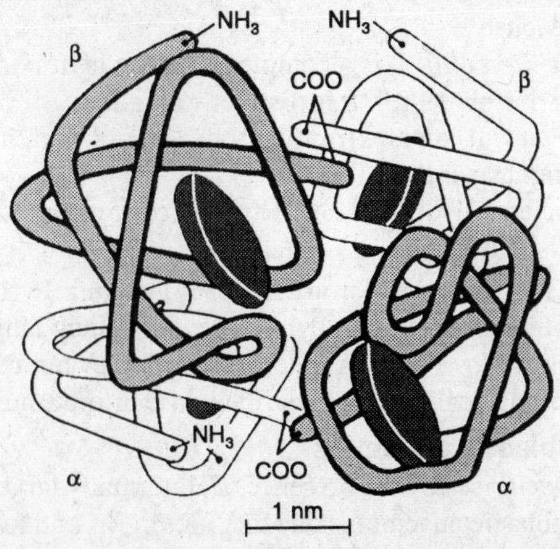

Diagrammatic representation of a molecule of haemoglobin showing 4 sub-units. There are 2 α and 2 β polypeptide chains, each containing a heme moiety. These moieties are represented by the disks.

Structure of Heme

Heme is a compound of protoporphyrin (consisting of 4 pyrrole groups joined together) and a metallic iron, hence it is a metallo-porphyrin in which metal is the iron.

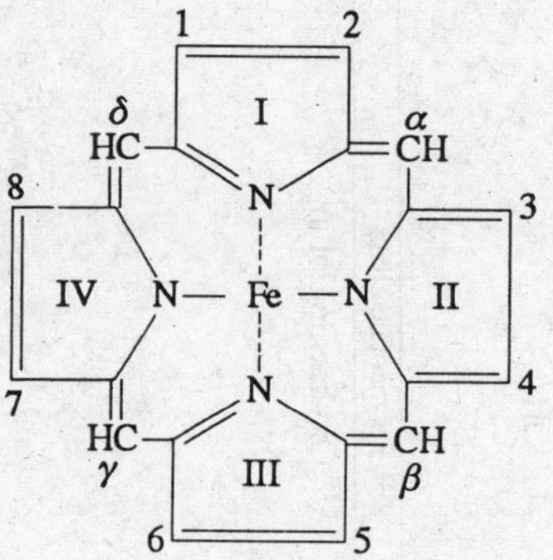

Porphyrin unit

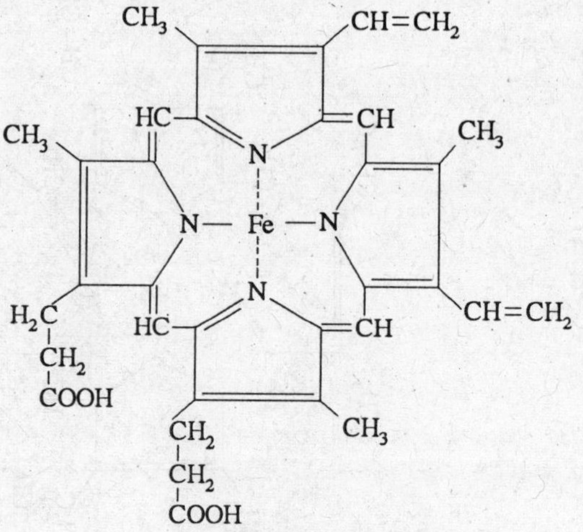

Protoporphyrin
(parent compound of haem)

Porphyrin unit consists of heterocyclic compounds containing a ring structure which is formed by linking four pyrrole rings by methylidyne (—CH==) bridges to form a porphrin. The carbons in the four pyrrole rings (which are not linked with the methylidyne bridges) are numbered as 1 to 8. The nethylidene bridges are referred to as α, β, γ and δ. The two hydrogen atoms in the —NH groups of pyrrole I and III are replaced by ferrous iron (Fe^{++}) which will occupy the centre of the compound ring structure but if the iron is in (Fe^{+++}) state the surplus +ive charge may be balanced by combining with an anion as Cl^-. In protoporphyrin which is the parent compound of heme, the position 1 to 8 are substituted as—position C-1 = CH_3, C-2 = Vinyl, C-3 = Methyl, C-4 = Vinyl, C-5 = Methyl, C-6 = Propionic acid, C-7 = Propionic acid, C-8 = Methyl group. The porphyrin ring can be readily synthesized from simple substances like glycine etc. Porphyrin are found in plants and animals. It also forms the colouring matter of the plant pigment, chlorophyll and brown pigment of egg shell.

Human heamoglobin has 0.34% of iron and has a minimum M.Wt. of 16,400 which represents a single iron and hence a single heme unit per molecule but the M.Wt. of haemoglobin is 64,500 which means four units (each containing one heme) form an aggregate molecule of haemoglobin.

Structure of Globin

The globin of haemoglobin molecule consists of four polypeptide chains on either side. In each side the polypeptide chain is divided into two chains: one an α-chain containing 141 amino acid and a β-chain containing 146 amino-acids hence here are 287 amino acids in each side arranged in α and β chains. Each chain is attached with the iron of the haem—moiety which is already linked with N of the pyrrole rings in haem is now linked with the N of the imidazole group of the polypeptide. A bond of iron remains free which combines with oxygen. X-ray crystallography studies have shown that haemoglobin is elliptical in shape which is formed by the molecular chains.

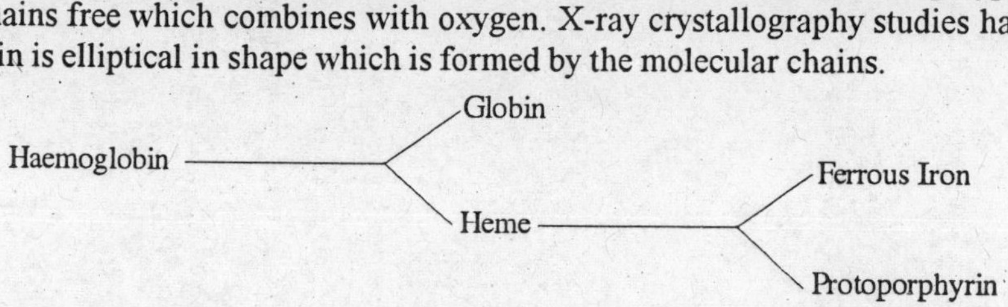

Haemoglobin → Globin; Heme
Heme → Ferrous Iron; Protoporphyrin

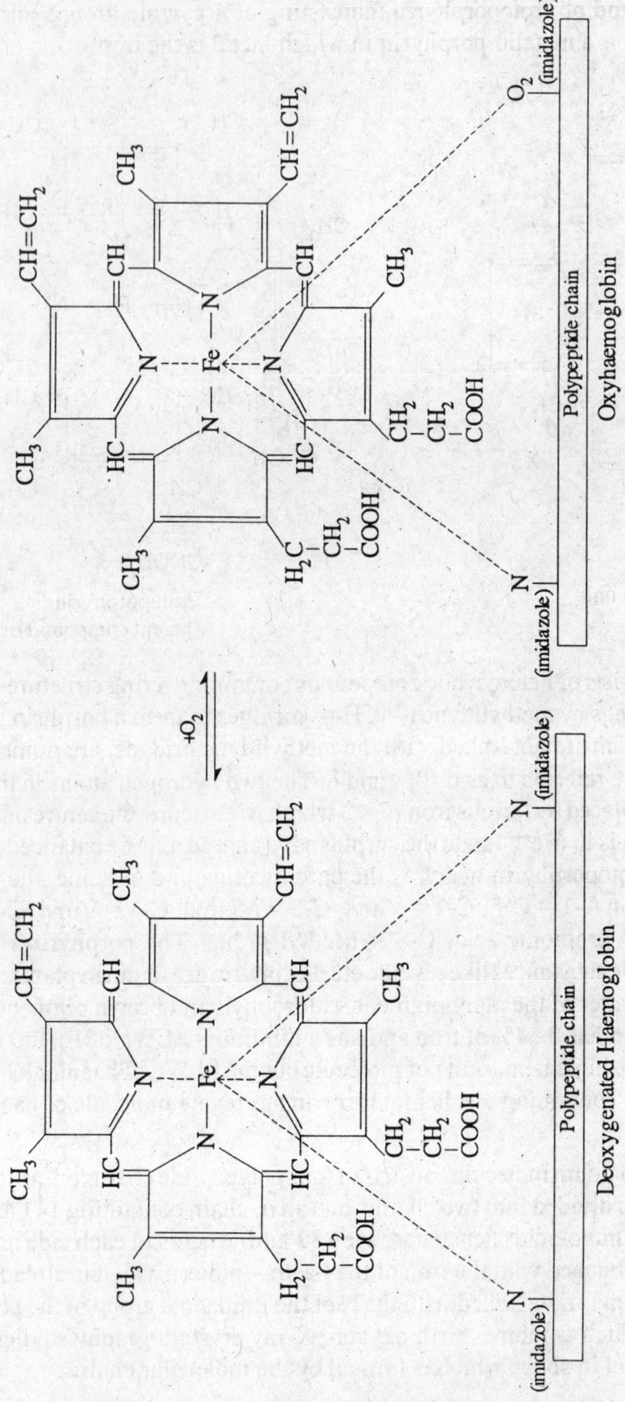

Polypeptide chain
(imidazole)
Deoxygenated Haemoglobin
+O2
Polypeptide chain
(imidazole)
Oxyhaemoglobin

The iron content of haemoglobin is about 0.34% and the iron content of the total amount of blood in an adult is about 3g. In haemoglobin iron remains in ferrous form and when it combines with oxygen it is not converted into ferric form, in other words, oxyhaemoglobin is oxygenated haemoglobin and not oxidized haemoglobin. The combination of haem with globin helps in keeping the iron in ferrous state and at the same time it also helps to combine molecular oxygen loosely with reversible potentiality i.e. the molecular oxygen can either combine or separate out at ease.

Analysis of Blood

Determination of Blood Urea

Urea is one of the end products of protein metabolism. Blood always contain some urea, the level varying between 15 and 40mg per 100 ml. In general, a blood urea of over 50mg per 100ml indicates renal failure, diarrhoea, vomiting, urinary tract obstruction, diabetes, or bleeding from gastrointestinal tract. In childhood or in pregnancy, low values are often found.

A decreased urea value is found in pronounced kidney, liver disorders, congestive heart failure, etc.

The determination of urine urea and blood urea is of major clinical value as index of kidney function.

The estimation of urea is usually performed on whole blood. Any of the routine anticoagulants can be added to the blood sample to prevent clotting. Among anticoagulants, sodium fluoride or ammonium oxalate should not be used as anticoagulants as sodium fluoride is an enzyme poison and is unsuitable for urease method, ammonium oxalate is not used as all the routine techniques depend on the measurement of ammonia.

Principles of Blood Urea Determination

Most methods of estimating urea are based on the action of the enzyme urease which decomposes urea to form Ammonium carbonate.

$$(H_2N)_2C{=}O \xrightarrow[\text{(Enzyme + Water)}]{\text{Hyrolysis}} 2NH_3 + CO_2 \xrightarrow{\text{Water}} \underset{\text{Ammonium Carbonate}}{(NH_4)_2CO_3}$$

The ammonia produced may be measured directly by colorimetric methods. Alternatively it may be separated by aeration or distillation and then determined colorimetrically or titrimetrically.

The Urease Method

The urea is converted into ammonia by incubation with enzyme urease. The proteins are then removed by precipitation with zinc hydroxide. The ammonia liberated by the urease is measured by adding Nessler's solution, the resulting yellow colour is proportional to the amount of ammonia present.

Procedure: In two centrifuge tubes marked as Test, and Blank, are taken and the following solution is pipetted out into the centrifuge tubes and the contents are mixed well.

	Test	*Blank*
Isotonic sodium sulphate	3.0 ml	3.2 ml
Whole blood, serum or plasma	0.2 mg	—
Urease suspension	0.2 ml	0.2 ml

The tubes are incubated in a water bath at 37°C for 20 minutes.

0.3ml of zinc sulphate solution and sodium hydroxide solutions are added to the above tubes to precipitate protein and tubes are centrifuged.

Now set up another three tubes marked as Test, Standard, and Blank and pipette the following solution:

	Test	*Standard*	*Blank*
Distilled water	5 ml	5 ml	5 ml
Supernatant fluid (Equivalent to 0.1 ml of blood in the 'Test' fluid)	2 ml	—	2 ml
Standard solution (Equivalent to 0.02 mg of nitrogen)	—	2 ml	—

A pinch of sodium citrate is added to each tube to prevent clouding on addition of the Nessler's solution.

Now 1ml of Nessler's reagent is added to each tube and mixed well. The clear solution is read immediately against the blank solution in colorimeter at 480mμ.

Calculations: Molecular weight of urea CO $(NH_2)_2$ = 60

One molecule contains 2N i.e. 2 × 14 = 28 parts by weight of 1mg of Nitrogen therefore 1 mg of N is contained in 2.14mg urea.

The blood is diluted 1 in 20, therefore, 2ml of supernatant = 0.1ml blood 2ml of standard = 0.02mg Nitrogen, 2.14mg urea = 1mg Nitrogen

$$\text{mg urea}/100\text{ml} = \frac{T-B}{S-B} \times 0.02 \times \frac{100}{0.1} \times -$$

$$= \frac{T-B}{S-B} \times 42.8$$

T = Test reading

S = Standard reading

B = Blank reading

Interpretation: The level of urea is influenced by the protein content of the diet. A high protein diet raises the serum urea level and low protein diet lowers it.

Serum urea is elevated in a number of renal diseases, acute and chronic glomerulo nephritis, nephrosclerosis, hydronephrosis, etc. Dehydration and shock may also raise serum urea.

Serum urea is decreased in acute hepatic insufficiency and also in pregnancy.

Estimation of blood urea is one of the oldest test of renal function to assess the functional efficiency of the kidneys.

The blood urea clearance test is a measure of the efficiency with which the kidney excretes urea. The necessary data includes the concentration of urea in blood and urine and the volume of urine excreted in unit time. This method is based on the view that with abundant urine, the urea excretion per minute equals the urea contained in a constant volume of blood.

The recent technique of estimating blood urea is as follows:

A 10 μl drop of patient sample is deposited on the clinical chemistry slide and is evenly distributed by the spreading layer (clinical slide is a dry multilayered analytical element coated on a clear polyester support). Water and non-proteinaceous components then travel to the underlying reagent layer where the urease reaction generates ammonia. The ammonia then diffuses through the semipermeable membrane to react with ammonia indicator.

The semipermeable membrane allows only ammonia to pass through to the color forming layer where it reacts with the indicator.

After a fixed incubation period, the reflection density of the dye is measures spectrophotometrically through the ployester support using the white background of the spreading layer as a diffuse reflector.

Slide Ingredients: Urease and N-propyl-4 (2,6 dinitro-4-chlorobenzyl) quinolinium ethane sulphonate. Other ingredients include pigments, binders, buffers, surfactants, and a polymer cross linking agent.

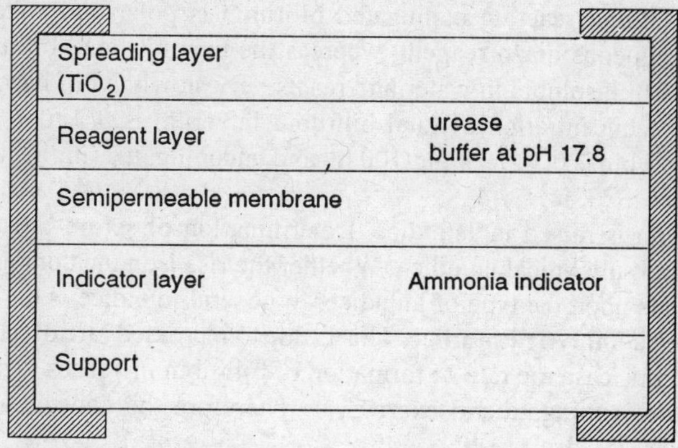

Wave length 670 mm
Assay time/ temperature-Approx. 5 minutes at 37°C.

Estimation of Bile Pigment in Serum

The bile pigment, Bilirubin, in serum is estimated by converting it in a purple compound known as azobilirubin, with the help of diazotided sulphanilic acid also known as Ehrlich's reagent. The resulting purple colour is measured colorimetrically. Unconjugate d bilirubin is measured by adding methanol in a concentration that does not precipitate the proteins.

Procedure: Three tubes are labelled as "unknown", "blank" and "standard". Pipette 0.2 ml of serum and 1.8 ml of water into unknown and 2 ml of water in blank. Pipette 2 ml of working standard bilirubin solution into "standard" and 0.5 ml of diazo reagent to unknown and standard and 0.5 ml of diazo blank is added to "blank". 2.5 ml of absolute methanol is added to each tube and read against water after 30 minutes at 540 nm.

Calculations: Serum total bilirubin (mg / 100ml) $= \frac{Au - Ab}{As} \times \frac{0.02}{0.2} \times 100$

$$= \frac{Au - Ab}{As} \times 10$$

Au = Absorbance of unknown
Ab = Absorbance of blank
As = Absorbance of standard

Reagents: (*i*) Diazo reagent—It is prepared in two parts, Solution A—1.0 of sulphanilic acid is dissolved in 15ml of conc. hydrochloric acid and made up to 1 litre with water. Solution B—0.5 g of Sodium nitrite is dissolved in water and diluted to 1 litre. Just before use the diazo reagent is prepared by mixing 0.3 ml of solution A and 10 ml of solution B.

(*ii*) Diazo blank—15 ml of conc. hydrochloric acid is added slowly to about 900 ml water in a one litre volumetric flask and diluted to the mark.

(*iii*) Absolute methanol.

(*iv*) Stock standard bilirubin solution—10 mg of bilirubin is dissolved in and diluted to 100 ml with chloroform.

(*v*) Working standard bilirubin solution—10 ml of stock standard bilirubin solution is diluted to 100 ml with chloroform. This solution contains 0.02 mg of bilirubin/2 ml.

The conjugated or direct reacting bilirubin is measured as above by substituting absolute methanol with water unconjugated or indirect reacting bilirubin is obtained by difference. Serum bilirubin shows two distinct types of reaction, one is a fast or direct reaction, and the other one is indirect or delayed reaction. The direct reacting conjugated bilirubin is polar and water soluble and reacts immediately with the aqueous diazo reagent, whereas the unconjugated bilirubin is non-polar and soluble in chloroform but insoluble in water and reacts very slowly with diazo reagent.

Interpretation: The concentration of total bilirubin in serum is 0.2 to 1.1 mg/100 ml, that of conjugated (direct) bilirubin is 0.1 to 0.4 mg/100 ml and unconjugated (indirect) bilirubin is from 0.1 to 0.7 mg/100 ml.

Total serum bilirubin is raised in Jaundice. Determination of serum bilirubin is helpful in the diagnosis of clinical and subclinical jaundice. Whether the rise is in the conjugated or unconjugated fraction or both depends upon the type of jaundice. In general joundice is divided into :

(*a*) Pre-hepatic or haemolytic jaundice. This is due to increased rate of haemolysis arising from a variety of causes. In this case the rate of formation of bilirubin increases to such an extent that the liver cells are unable to conjugate and excrete it. Therefore the concentration of unconjugated bilirubin in serum is increased.

(*b*) Mepatic jaundice. This is usually caused by viral hepatitis, toxin and certain drugs. The damaged livér cells are unable to conjugate bilirubin and therefore unconjugated bilirubin in serum is raised. In this case conjugated bilirubin in serum is also frequently raised.

(*c*) Obstructive jaundice. This is caused by obstruction of biliary channels due to the presence of concentration of conjugated bilirubin in serum is increased.

The recent technique of estimating bilirubin concentration in serum is based on a modification of the classical diazo reaction.

The test is performed on the same clinical chemistry slide (dry multi-layered analytical element coated on a clear polyester support). A 10μl drop of patient sample is deposited on the slide and is evenly distributed by the spreading layer. This layer provides a reflective background for measuring the diazo products of bilirubin and contains all reagents necessary to determine total bilirubin.

The method uses dyphylline and surfactant, to dissociate unconjugated bilirubin from albumin. Unconjugated bilirubin, conjugated bilirubin and albumin linked bilirubin subsequently react with the diazonium salt 4-(N-carboxy methyl sufamyl) benzinediazonium hexafluorophosphate to produce azobilrubin chromophores that have similar molar absorptivities and absorbance maxima about 520nm. The buffered control layer which is below the spreading/reaction layer, contains a ploymeric quaternary amine (a mordant) coated on a transparent support. The mordant is used to stabilize the azo dye generated in the reaction layer and to increase the sensitivity of the assay.

The concentration of total bilirubin is determined by measuring the azo bilirubin chromophores at two wave lengths through the transparent support. Reflectance measurements at 540nm and 460nm are taken after 5 minutes incubation at 37°C. The measurement at 460nm is used to correct for spectral interferences.

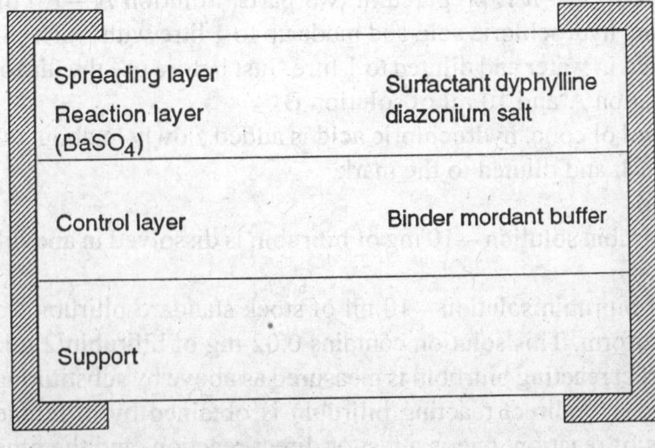

Wave length 540 and 460 nm
Assay time/ temperature: Approx. 5 minutes at 37°C.

Slide Ingredients: Dyphylline, 4-(N-carboxymethyl sulfamyl) bezene diazonium hexafluro-phosphate, copoly (styrene-co-N-vinyl benzyl quaternary salt-co-divinyl benzene). Other ingredients include pigment, binders, buffer, surfactants, ploymer cross linking agent and a preservative.

Estimation of Total Protein in Serum

Serum proteins transport drugs and metabolites and maintain plasma osmotic pressure. Most serum proteins are synthesized in the liver with the exception of gamma globulins. One of the most important serum proteins produced in the liver is albumin. The total serum protein can be used for evaluation of the nutritional status.

The heat coagulable proteins such as albumin and globulin present in serum can be detected by Heat Coagulation Test.

Procedure: 10ml of clotted blood is centrifuged to separate serum and the cells serum is decanted into a test tube. Equal volume of water is added to it and mixed well. PH is adjusted to 5.4 with the help of chlorphenol red indicator and 1% acetic acid. The solution is boiled. A dense coagulum is formed showing the presence of albumin and globulin in the serum.

Total proteins in serum are determined by micro-kjeldahl method. This method is based on direct nesslerization, making the appropriate correction for non-protein nitrogen. Fibrinogen in plasma is determined by isolation as fibrin, followed by digestion and direct nesselerization. Albumin is determined by analysis of the fluid, remaining after precipating the globulin fraction with 23% sodium sulphate solution. Globulin is serum estimated by subtracting the albumin from the total protein content.

Normal values for plasma proteins in g per 100ml of plasma are as follows :

Albumin—4.6-6.7

Globulin—1.2-2.3

Fibrinogen—0.3-0.6

Increased plasma protein levels are noted in dehydration due to diarrhoea, excessive burn, Addison's disease, multiple mycloma, tropical diseases, etc. Increase globulin content is found in malignancy, liver cirrhosis, etc.

Decreased plasma protein is found in impaired protein synthesis due to malnutrition, vitamin deficiencies etc. Fibrinogen values are increased in pneumonia and in other infections, low total serum protein is also found in pregnancy, excessive intravenous fluid administration, cirrhosis of liver, chronic alcoholism, heart failure, etc.

Different serum proteins *i.e.* Fibrinogen, Globulin, Albumin can be separated from oxalated plasma and the precipitated proteins can be tested by Biuret, Millons of Hopkins-cale Tests.

Preparation of oxalatted plasma: 9.0ml blood is collected and mixed with 1ml of 0.1M sodium oxalate in a test tube centrifuged for 5 minutes and plasma can be separated.

Separation of fibrinogen, globulin, albumin: 0.5ml of half saturated ammonium sulphate when fibrinogen is precipitated.

To 0.5ml of serum (separated from blood), 0.5ml of saturated ammonium sulphate is added and mixed when globulin is precipitated. Precipitated globulin is removed through centrifugation to the clear filtrate, powdered ammonium sulphate is added when albumin is precipitated.

The samples of the precipitated fibrinogen, globulin and albumin can be tested by Biuret or Millon's or Hopkins-cale Tests.

The Biuret Test

Procedure: 2.0ml of the protein solution in a test tube. Then 2.0ml of sodium hydroxide is added to it. Copper sulphate solution is added dropwise and mixed well.

A purplish violet or pinkish colour indicates the presence of protein.

Reagents

(*i*) Sodium hydroxide 10% solution in water.
(*ii*) Copper sulphate 0.5% solution in water.

Estimation of Total Proteins and Albumin Based on Biuret Method and BCG Method

The peptide bonds present in polypeptide molecular complex, bind cupric ions in an alkaline medium to give a violet colour.

The Bromocresol Greem (BCG) dye in a buffered acid medium selectively binds albumin and forms a complex. During the reaction, the initial yellow colour quantitatively changes to green.

Reagents Used

1. Biuret Reagent
2. BCG Reagent
3. Standard Solution
4. Serum samples are collected for determination of total proteins of Albumin.

Procedure: (Biuret Reagent) Take three test tubes marked as Blank 1, Standard 1 and Test 1 and follow the following procedure:

Total Proteins	*Blank* (B_1) *ml*	*Standard* (S_1) *ml*	*Test* (T_1)*ml*
Reagent 1 Biuret reagent	1.0	1.0	1.0
Sample	—	—	0.02
Standard	—	0.02	—
Distilled water	0.02	—	—

Mix the test tubes and allow to stand at room temperature for 20 minutes. Read the absorbance of Test (AT_1), Standard (AS_1) and Blank (AB_1) at 546nm.

BCG method for Albumin, take three test tubes and mark as Blank (B_2), Standard (S_2) and Test (T_2) respectively and follow the following procedure:

Albumin	*Blank* B_2 *ml*	*Standard* S_2 *ml*	*Test* T_2 *ml*
Reagent 2 BCG reagent	1.0	1.0	1.0
Sample	—	—	0.01
Standard	—	0.01	—
Distilled water	0.01	—	—

Calculations:

$$\text{Total protein concentration (g/dl)} = \frac{AT_1 - AB_1}{AS_1 - AB_1} \times \text{T.Pr.conc.}$$

$$\text{Albumin concentration (g/dl)} = \frac{AT_2 - AB_2}{AS_2 - AB_2} \times \text{Albumin concentration}$$

$$\text{Albumin/Globulin ration} = \frac{Al_b(g/dl)}{T.Pr.(g/dl) - Al_b(g/dl)}$$

Normal values:

Total proteins	= 6.0-8.5g/dl
Albumin	= 3.2-5.5g/dl
Albumin/Globulin ratio range	= 1.1-2.2

The recent technique of determination of total protein in serum or plasma is based on the Biuret reactions, which produces a violet complex when protein reacts with cupric ion in an alkaline medium.

A 10 μl drop of sample is deposited on the clinical chemistry slide and is evenly distributed by the spreading layer into the underlying reagent layer.

When the fluid penetrates the reagent layer, the reagent diffuses upto the spreading layer and reacts with protein. The reaction between protein and copper tartrate takes place largely in the spreading layer where the protein is confined because of its high molecular weight.

The amount of coloured complex formed is proportional to the amount of total protein in sample and is measured by reflectance spectrophotometry.

Slide ingredients: Copper sulphate, tartaric acid, lithium hydroxide. Other ingredients include binders, surfactants and a polymer cross linking agent.

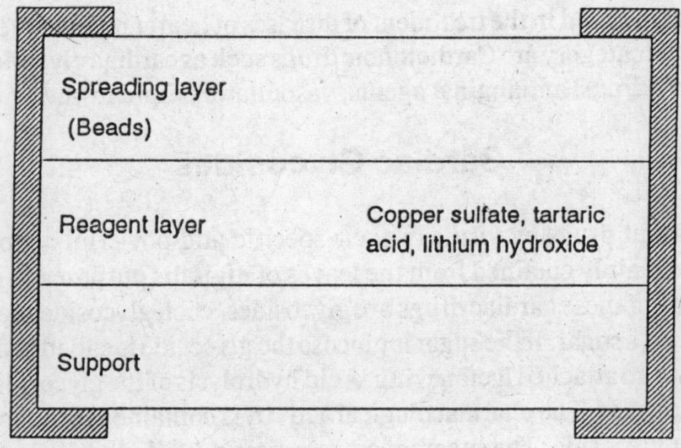

Wave length 540 and 460 mm
Assay time/ temperature-Approx. 5 minutes at 37°C.

Ketone Bodies: Three substances are collectively known as ketone bodies namely acetoacetate, β-hydroxybutyrate, and acetone. The ratio [3 hydroxybutyrate/(acetoacetate)] in blood varies 1 : 1 and 10 : 1. Higher than normal values present in the blood or urine constitute Ketonemia or hyperdetonemia. The overall condition is called ketosis. Acetoacetic and 3-hydroxybutyric acids are both moderately strong acids and are buffered when prenol in the blood. However, their continual excretion in equantity entails some loss of buffer cation, that progressively depletes the alkali reserve, causing keto acidosis. This may be fatal in uncontrolled diabetes mellitus, ketosis is always associated with some abnormalities of carbohydrate metabolism. Ketosis occurs in fasting of during carbohydrate deprivation. Pathologically it is most severe in diabetes mellitus.

Chapter 10

Cardio Vascular Drugs

Cardio Vascular drugs are used in the treatment of diseases of heart and blood vessels. Different kinds of drugs included in this category are Cardiotonoic drugs such as cardiac glycosides, antihypertension drugs, antiarrhythemic drugs, antianginal agents, vasodilators, lipid lowering agents etc.

Cardiac Glycosides

These are very important drugs by virtue of their specific and powerful actions on the heart. The cardiac glycosides are mainly obtained from the leaves of digitalis purpurea or digitalis lanta plants. The active constituents of these cardiac drugs are glycosides, each glycoside represents the combinations of an aglycone with a sugar. If the sugar is glucose the glycoside is called as a glucoside. Aglycone is a steroid nucleus with an attached lactone ring. Acid hydrolysis of the glycoside results in separation of the aglycone and the sugar. The pharmacological activity is contained in the aglycone which is more potent than the parent glycoside. The sugars increase water solubility, cell penetrability and potency of the aglycones.

Digitalis has several cardiac glycosides and it is the most important drug for the treatment of congestive cardiac failure. It is also used for oedema, diuretics etc.

By the action of digitalis, systolic contraction and myocardial efficiency is improved and cardiac output is also increased. In case of cardiac failure the heart beat is faster, digitalis slows the heart beat and thereby improves its functions. In patients suffering from oedema, the digitalis acts as a diuretic due to better blood supply to kidneys.

Some common preparations of digitalis are Digoxin (crystalline, glycoside obtained from Digitalis

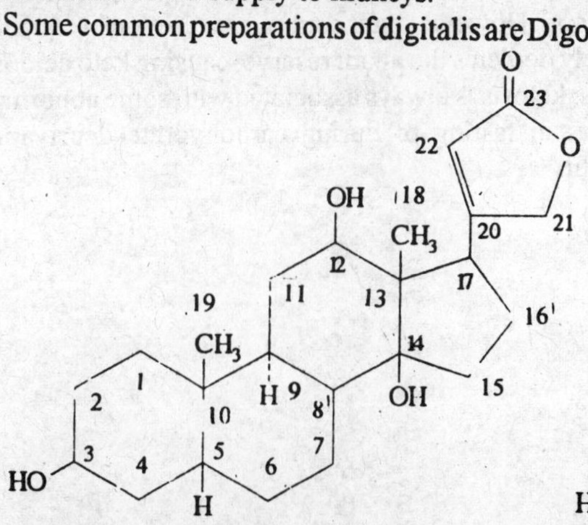

Digoxigenin (aglycone)

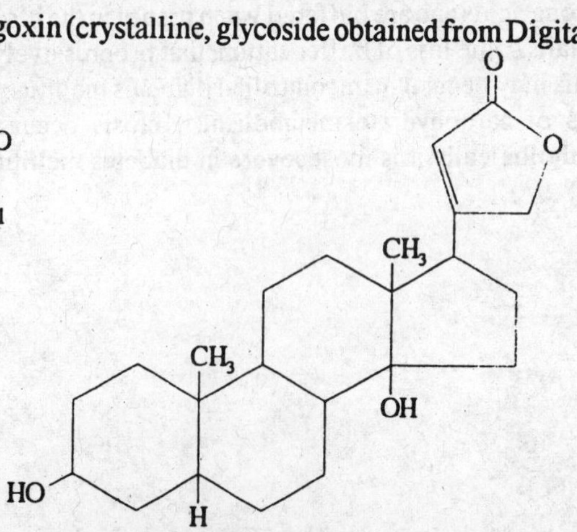

Digitoxigenin (aglycone)

lanta, containing digoxigenin, aglycone and Digitoxose sugar) digitoxin (crystalline glycoside obtained from Digitalis purpurea, it contains Digitoxose sugar and digitoxigenin glycone), etc.

Digoxin is the choicest drug and is used extensively due to its moderately fast onset of action as it is quickly absorbed. It can be given by mouth or parenterally. Digoxin contains Digoxigenin aglycone and digitoxose.

Digitoxin is having the sugar part similar to digoxin but aglycone differs by not having a hydroxyl group at position-12. They both contain the same sugar known as digitoxose.

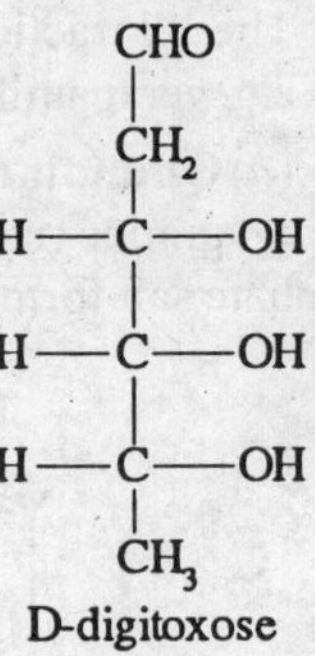

D-digitoxose

Digitalis shows some adverse reactions such as headache, vomiting, visual disturbances etc. The digitalis tends to increase the contractility of the heart muscle and tends to decrease the size of heart in case of heart failure. It may have direct action on contractile protein of heart muscles.

Another drug is Quabin which is a glycoside of strophanthus gratus, its aglycone is quabagenin and the sugar is L-rhamnose. The solutions are sterlised by autoclaving. It can be adminstered orally.

Digitalis produces cardiovascular and extracardiac actions.

Cardio Vascular Action

The major pharmacologic effect of digitalis is its direct action on the myocardium. It increases the force of syostolic contraction of the heart rate. Large doses of digitalis depress the conduction velocity of heart. Digitalis helps in retention of sodium into the cells during diastolic depolarization and helps in faster dispersion of calcium. Excessive calcium ion concentration leads to cardiac arrest in systole. These actions of digitalis enhance the contractility, automaticity and pacemaker activity in the heart. Digitalis also prevents the re-entry of potassium into cell after repolarization. These alterations of fluxes of sodium and potassium are responsible for electrophysiological effects of digitalis.

Antiarrhythmic Drugs

The drug therapy used for disturbances of cardiac rhythum or cardiac arrhythmias, are known as antiarrhythemic drugs. If the heart beats are more than 180 beats per minute it may lead to death. Anti-arrhythemias drugs are used to correct cardiac arrhythmias.

The actions of antiarrhythmic drugs are as follows:

(*a*) It slows the diastolic depolarization in the myocardiac cells.
(*b*) It elevates the threshold potential of myocardiac cells.
(*c*) It may slow the potassium efflux from the myocardiac cells during repolarization.
(*d*) It may prolong the spontaneous diastolic depolarization.

An ideal antiarrhythmatic drug should improve metabolic efficiency in the contractile tissue of the heart without increasing metabolic demands. The compound should produce minimum adverse effect and it should reduce intrinsic rhythmicity of the conducting pathways.

The various antiarrhythmic drugs can be categorised as follows :

(*i*) Drugs that depress the myocardium—These drugs control cardiac rhythm by depressing diastolic depolarization and conduction velocity, e.g. quinidine lignocaine, phenytoin, procainamide, etc.

(*ii*) Drugs action through the autonomic nervous system—These drugs include (*a*) β-adrenergic blocking drugs e.g. propranolol and (*b*) adrengic neurone blocker e.g. bretylium (*c*) cholinergics–edrophonium.
These drugs act by reducing cardiac activity by decreasing or preventing beta adreno receptor stimulation, their action causes decreased automaticity and conduction velocity. The refractory period is prolonged.

(*iii*) Digitalis (discussed earlier).

(*iv*) Antagonists of calcium—These drugs inhibit the migration of calcium across the cell membrane. The contraction of myocardium is highly dependent on the extracellular calcium concentration e.g. verapamil, prenylamine etc.

(*i*) (*a*) **Quinidine.** This drug depresses the myocardium. Chemically it is a dextrorotatory isomer of the drug quinine (used as antimalarial drug) and is a cinchona alkaloid, quinidine is genarally used as bisulphate salt form. It has the following structure.

CHOH—CH—N—CH_2

CH_3O CH_2

CH_2

N CH_2—CH—CH−CH=CH_2

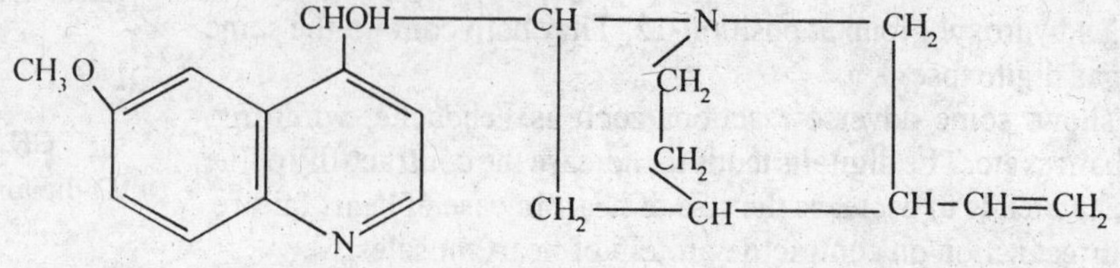

Quinidine

Quinidine sulphate is a white crystalline substances. It is sparingly soluble in water and is stored in an airtight and light resistant container. Quinine bisulphate is soluble in water.

The pharmacological actions of quinidine can be considered as cardiac action and as an extra cardiac actions.

The cardiac actions are mainly due to its myocardiac depressant property. Its main functions are :

(*a*) It depresses the entry of sodium into the cell during depolarization and thus depresses the automaticity and the depressed automaticity helps to suppress arrhythmias.

(*b*) Quinidine helps to reduce the excitability of the cardiac tissue.

(*c*) It helps to prolong refractory period of all cardiac tissues by depressing the potassium efflux during repolarization. Prolongation of the refractory period prevents the heart from responding to premature or rapid stimulation.

(*d*) It shows the conduction velocity in all cardiac tissues, this property along with the increased refractory period and decreased excitability help to reduce cardiac rate in arrhythmias.

The extra cardiac actions are :

(*a*) It reduces blood pressure and thus helps to reduce left ventricular load.

(*b*) In addition to its depressant action on the myocardium it also depresses skeletal muscle.

Quinidine is completely absorbed from the intestine. It is given orally. It can be toxic if it is adminstered intravenously.

Preparation and dosage

Quinidine gluconate injection 80 mg/ml. It is used for intramuscular administration.

Quinidine sulphate I.P. tablets or capsules dose 200-400 mg.

Therapeutic uses

1. It is used for conversion of atrial fibrillation (rapid twitching of the muscular wall of auricle of the heart) to normal sinus rhythm.
2. It can successfully treat arrhythmia in patients with hypertension.
3. It is useful for the treatment of paroxysmal ventricular tachycardia (disease where there is excessive rapidity of heart's action).
4. The development of premature beats during acute myocardial infraction, can be treated with quinidine.

Adverse effects : Nausea, vomiting and diarrhoea, headache, tinnitus etc.

(*i*) **Procainamide.** It *p*-amino-N(2-diethylaminoethyl) benzamide. It is mainly used as its hydrochloride salts. Its salts are soluble in water. It has the following structure.

Its cardiac actions are similar to those of quinidine. Procainamide is effective by oral, intramuscular and intravenous routes. This drug induces a moderate fall in blood pressure. It prevents arrhythmias under cyclopropane anesthesia and is useful in the treatment of junctional and ventricular arrhythmias.

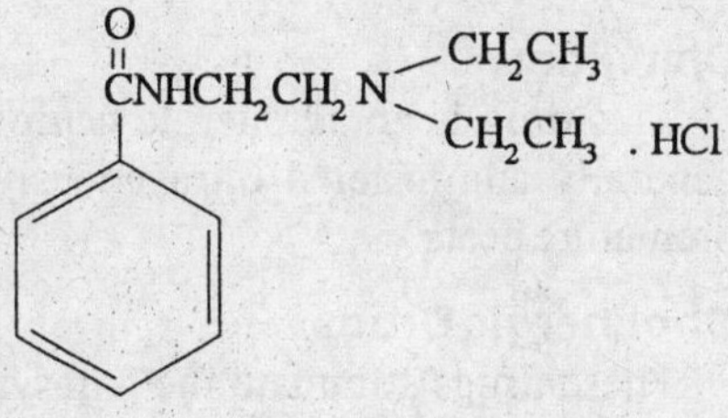

O
$CNHCH_2CH_2N(CH_2CH_3)_2$. HCl

(*ii*) (*a*) The β-adrenergic agents reduce cardiac activity by decreasing beta adrenergic receptor stimulation. These drugs usually reduce myo-cardial oxygen requirements and improve exercise tolerance in patients with angina. These drugs also reduce blood pressure by their action on the heart and reduction in cardiac output. Propanolol and some of their beta blocks have a direct depressant effect on the heart similar to that of quinidine. These drugs are also useful in the treatment of angina pectoris.

Some important β-Adrenergic Drugs are :

Propanolol Hydrochloride

It is a water soluble compound and its solutions are sterilised by autoclave method. Chemically it is named as 1-isopropylamino-3-(1naphththyloxy) propane-2-ol hydrochloride, it has the adjoining structure.

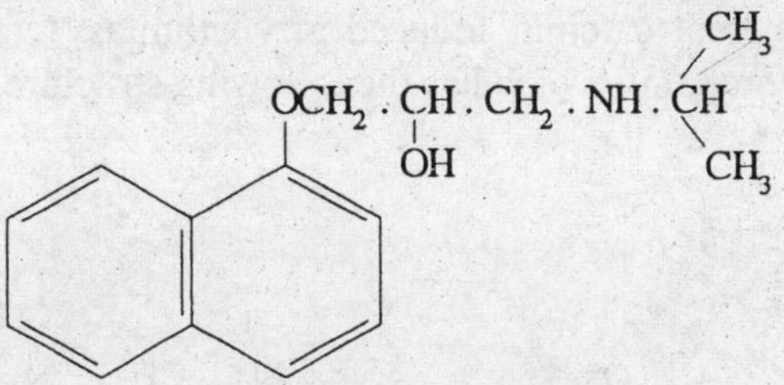

Propranolol have similar depressant effection on the heart, to that of quinidine, recently it has been found that propranolol alters mood and is used in psychiatric disorders, prolonged use of propanolol can cause fatigue, muscle crams, lethargy and hallucinations.

$OCH_2 . CH(OH) . CH_2 . NH . CH(CH_3)_2$

Propranolol hydrochloride

Uses. It is used as antiarrhythmic drug and is a useful drug for angina pectoris, hypertension etc.

Other related compounds are alprenolol hydrochloride, oxprenolo hydrochloride, pindolol, practolol. All the four compounds have the common side chain i.e. 2-hydroxy-3-isopropyl amino propoxy $(CH_3)_2CH.NH.CH_2\ CH\ (OH)\ CH_2O$ but it differs in the aromatic system.

Pindolol and Practolol have partial agonistic activity. Practolol also has intrinsic sympathomimetic

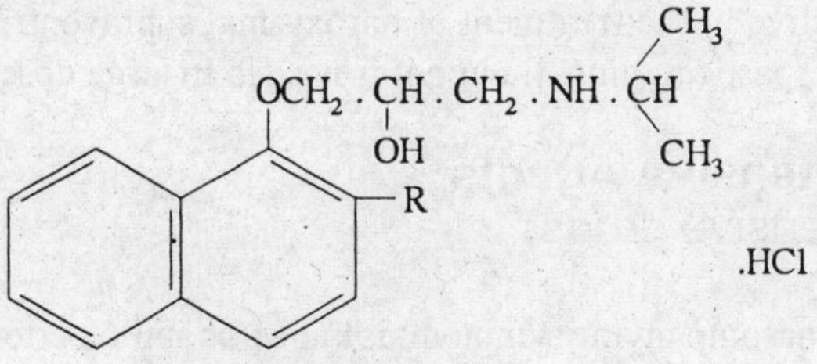

$OCH_2 . CH(OH) . CH_2 . NH . CH(CH_3)_2$ .HCl

Alprenolol hydrochloride
$R = CH_2{=}CH . CH_2—$

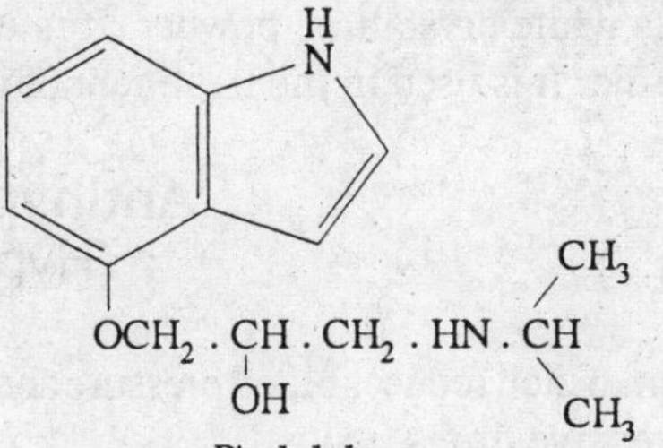

H
N
$OCH_2 . CH(OH) . CH_2 . HN . CH(CH_3)_2$

Pindolol

$OCH_2 . CH(OH) . CH_2 . NH . CH(CH_3)_2$
$NHCOCH_3$

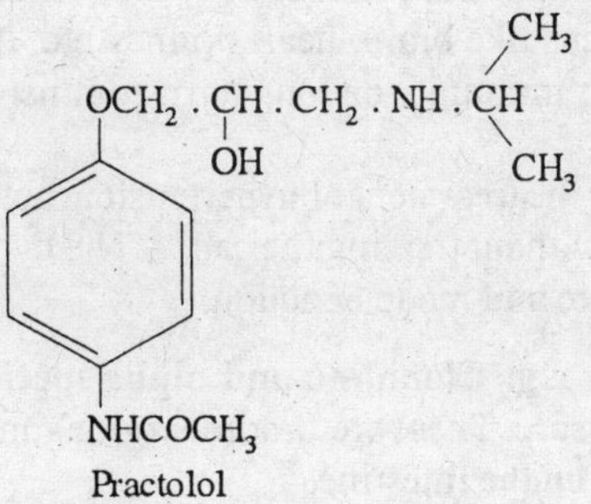

Practolol

activity. These drugs are rapidly absorbed orally and are metabolized by the liver.

Bretylium

This drug is an adrenergic neurone blocker. This drug has direct action on myocardium. It is generally adminstered intravenously. The major use of this drug is suppression of ventricular premature beats.

Cholinergic Drugs

These drugs stimulate the parasympathetic nerves such as acetyl choline, which decreases the automaticity and increases the resting membrane potential and also the refractory period. It gets hydrolysed by an enzyme acetyl cholinesterase to choline and acetic acid. The drug Edrophonium is a short acting cholinesterase inhibitor. It is useful in the treatment of arrhythmia by preventing the destruction of acétyl choline.

Antagonists of Calcium

The calcium antagonist drugs selectively block the transmembrane calcium influx into the excited heart muscles without affecting the sodium influx. These drugs produce negative inotropic effect i.e. effecting the force of cardiac contractions and thus reduces conduction velocity. Calcium blockers can prevent calcium induced arryhythmias. One of the most important calcium antagonist drug is Verapamil which has the following structure.

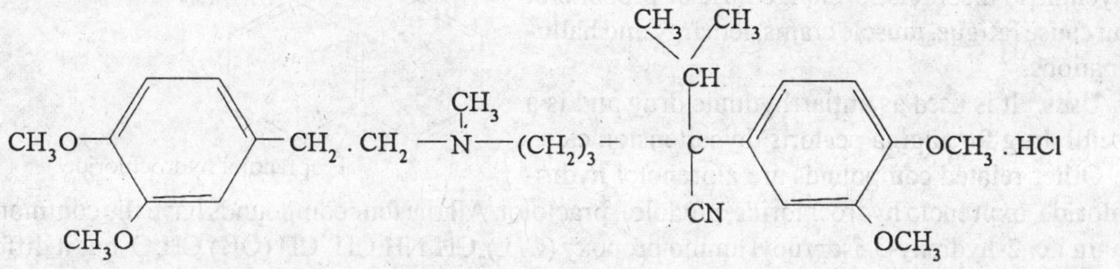

CH_3O CH_3O — $CH_2 . CH_2$ — $N(CH_3)$ — $(CH_2)_3$ — $C(CN)(CH(CH_3)_2)$ — OCH_3 OCH_3 . HCl

Verapamil hydrochloride

Verapamil hydrochloride is a synthetic pavaverine derivative and chemically it is named as 5- [N-(3,4 dimethoxy phenethyl) methyl aminol]-2 (3,4 dimethoxy phenyl) -2-isopropyl valeronitrile. It occurs as white crystalline powder. It is effective in the treatment of paroxysmal supraventricular tachycardia. It is used in the treatment of vasospastic angina, though it is needed in large doses.

Antihypertensive Agents
(Hypotensive drugs)

Any agent which reduces blood pressure and its accompanying symptoms is known as antihypertension or hypotensive drugs.

An ideal antihypertensive drug should reduce both systolic and diastolic pressure, it should not reduce circulation to vital organs like brain, heart, kidney etc. It should not have toxic effects and should not produce tolerance or immunity on long term adminstration and it should have sufficient duration of action.

The drugs which are used in the treatment of hypertension, act by reducing the cardiac output and the total peripheral resistance without treating the cause. The hypertensive drugs could be classified as follows according to their site and mode of action.

(*a*) **Drugs acting centrally e.g. Clonidine and alpha methyldopa**. These drugs reduce both systolic and diastalic blood pressure. These are β-adrenergic agonist in action. These are lipid soluble drugs and are well absorbed from the intestine.

Clonidine

Chemically it is named as 2-(2,6 dichloro anilino) -2-imidazoline. Its hydrochloride salt occurs as a white crystalline powder. It is kept in well closed, light resistant containers. It is adminstered either orally or by slow intravenous respose.

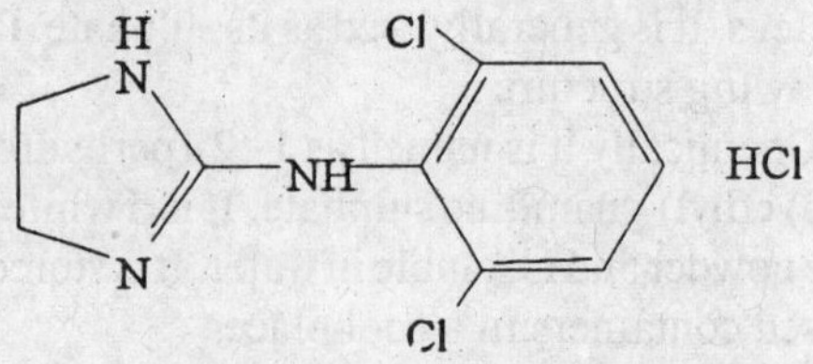

It produces short hypertensive response by prolonged reduction of systolic and diastolic blood pressure. It also inhibit peripheral sympathetic activity. It is a lipid soluble drug. It finds use in hypertension and migraine. This drug stimulates the alpha adrenergic receptors in the brain.

Clonidine hydrochloride

Alpha Methyldopa (Aldomet)

It is a white powder and is available in 250 mg tablets or as injection. Chemically it is named as (-)-3-(3,4 dihydroxyphenyl-2-methyl-1-alanine sesquihydrate.

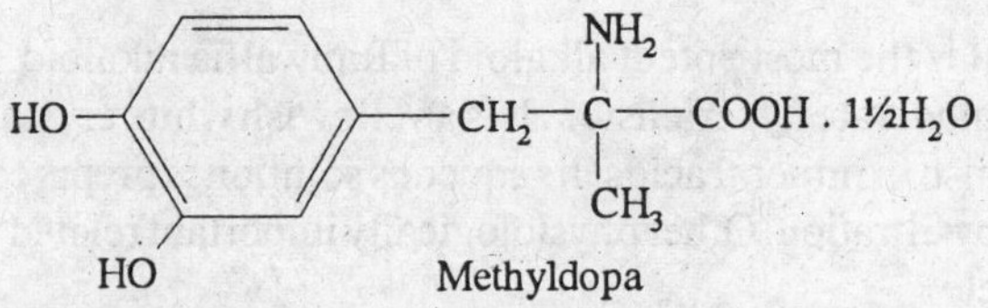

Methyldopa acts by inhibiting the enzyme dopa decarboxylase, which helps in synthesis of noradrenaline. It stimulates the alpha adrenergic receptors in the vasomator centre and exerts hypotensive effect within the central nervous systems by its conversion to methyl noradrenaline which is a patent adrenergic agonist.

Methyldopa

It is used in the treatment of moderate to severe hypertension.

(*b*) **Ganglion blockers** *e.g.* Mecamylamine, pempidine, hexamethonium, pentoliniumtartrate, Chlorisondamine etc.

These drugs act on autonomic ganglia by blocking the transmission of impulses in the sympathetic and para-sympathetic autonomic ganglion.

Pentolinium Tartrate

It is a quaternary ammonium compound and is chemically named as 1,1-dimethyl-N,N-penta methylene di-(pyrrolidinium)-di-(hydrogen tartrate). It is white powder and is soluble in water. It is generally adminstered in the injection form.

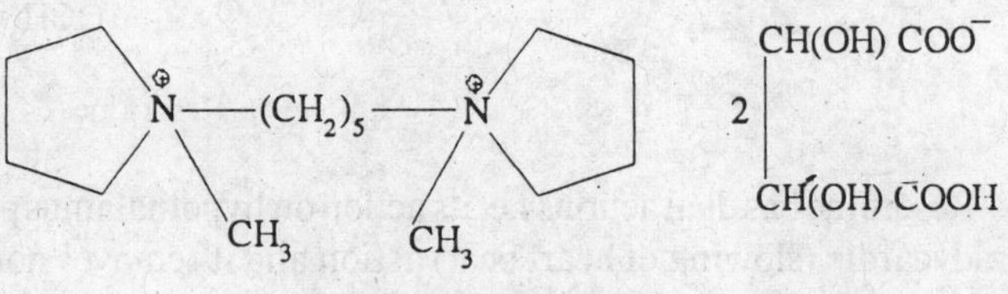

Pentolinium Tartrate

Adverse effect—Faintness, weakness, nausea and vomiting.

(*c*) **Drugs acting on postganglionic sympathetic nerve endings.** These drugs are of two types i.e. Adrenergic blockers—e.g. Guanethidine, Bretylium, Bethanidine etc., and Catecholamine depletors e.g. Reserpine.

The adrenergic blockers act on the post-ganglionic sympathetic neurones without depletion of the neurotransmitter.

Guanethidine Drugs

It was synthesized in 1961 by Maxwell et.al. It is usually adminstered by mouth.

Guanethidine has a perhydroazocine hetero cycle nucleus. It is generally used as its sulphate. It has the following structure :

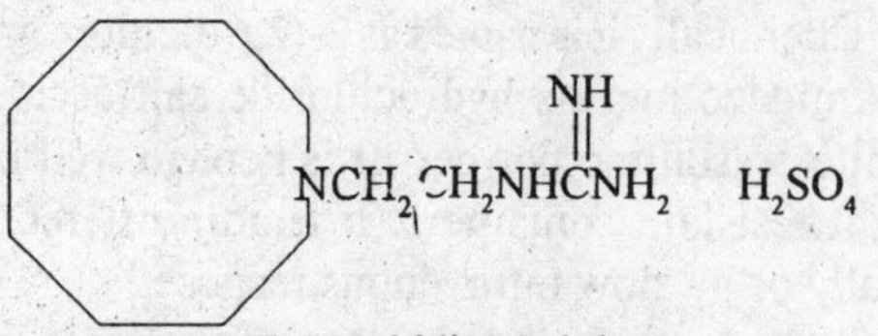

$NCH_2CH_2NHC(=NH)NH_2 \quad H_2SO_4$

Guanethidine Sulphate

Chemically it is named as 1-(2-(perhydroazocin-1-yl) ethyl) guanidine sulphate. It is a white crystalline powder and is soluble in water. It is stored in well closed containers in a cool place.

It is employed in the treatment of moderate or severe hypertension.

Guanethidine acts by inhibiting the release of the transmitter noradrenaline at the sympathetic nerve terminals. It also helps in depletion of noradrenaline stores at the sympathetic nerve endings. It is completely absorbed from the gastrointestinal tract.

Adverse effects—diarrhoea, muscle aching etc.

Reserpine

It is the most potent alkaloid of Rauwalfia alkaloid. It ia used extensively for the treatment of mild to moderate hypertension. It is a yellowish white crystalline powder. It is a sparingly soluble in water and also in mineral acids. Its aqueous solution is prepared by dissolving it in a suitable acid and is sterilised by filtration. Other physiologically important related rauwalfia alkaloids are deserpidine, rescinnamine etc.

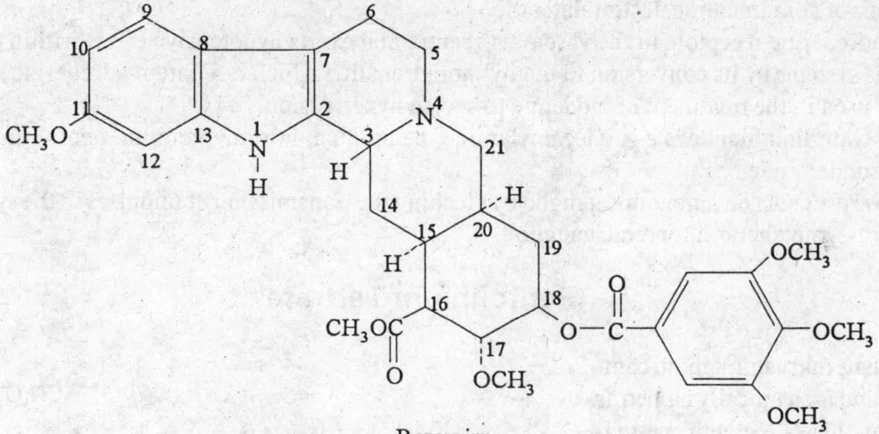

Reserpine

Reserpine has dual actions i.e. its action on hypothalamus producing hypotensive, tranquilising and bradycardic (slowing of heart beat) action and it removes nordrenaline stores from adrenergic nerve endings and dilates blood vessels and reduces blood pressure. Its sedative action helps in removing anxiety and tension. Reserpine controls blood pressure very effectively and has minimum side effect. Reserpine is often given along with thiazide diureties. Reserpine is used in treating hypertensive emergencies and is generally adminstered intravenously.

Adverse effects : In large doses reserpine may produce sedation and mental depression, nasal stuffness, gastric hyperacidity etc.

(*d*) **Vasodillators** *e.g.* Hydrallazine, Prazosin, Sodium nitroprusside etc. These drugs are discussed later in this chapter.

(*e*) **Beta blockers.** These drugs are either alpha adrenergic blocking agents or Beta adrenergic blocking agents. The alpha adrenergic blocking agents are of relatively little value in the treatment of essential hypertension. β-adrenergic block drugs reduce both systolic and diastolic blood pressure e.g. phenoxy benzamine, chlorpromazine, isopropylartennol, sotalol, dihydrogenated ergot alkaloids etc.

Phenoxy Benzamine

It is an alpha adrenergic blocking agent and has a long duration of action. It can be given orally or intravenously. Chemically it is N-(2-chloroethyl)-N-(1-methyl-2-phenoxyethyl) benzylamine hydrochloride.

(*f*) **Drugs which act refluxly by stimulating bare receptors e.g. Veratrum.** The primary site of their action is the afferent receptors in the heart and in the carolid sinus area. These have many adverse effects like nausea, vomiting, burning sensation etc.

(*g*) **Diureties—e.g. Thaiazides, chlorthalidone etc.** These drugs reduce systolic and diastolic pressure. These drugs reduce sodium in plasma and produce cardiac output and thus treat hypertension.

(*h*) **Drugs which block renin-angio tension-aldosterone axis e.g. captopril, spironolactone, saralasin etc**. These drugs can be used orally as well as parenterally.

(*i*) **Miscellaneous drugs.** *e.g.* MAO inhibitor-pargyline etc. These drugs inhibit the action of monoamine oxidase (MAO) located inside the mitochondria.

Anti Anginal Agents

Anti anginal agents are used in angina pectoris, which is not a disease but a symptom of myocardial ischemia produced in many ways by narrowing of the lumen of the coronary arteries due to atherosclerosis when heart does not get proper nutrition and oxygen. It is generally produced by exertion, tension, anxiety, excessive smoking or obesity etc.

Beside the coronary dilators, following drugs are used in the treatment of angina pectoris.

1. Nitrites.
2. Beta-adrenergic blocking agents.
3. Calcium blockers.
4. Miscellaneous drugs.

1. Nitrites

Both inorganic and organic nitrites dilate the coronaries. Inorganic nitrates are inactive while organic nitrates are very important antianginal agents because of the qualitative similarities between the actions of nitrites and organic nitrates. The whole group is generally referred to as "Nitrites".

Nitrites have direct relaxant action on smooth muscles and it increases the exercise tolerance in the patients suffering from angina by reducing the oxygen consumption of the heart.

Inorganic Nitrites—e.g. sodium nitrite, bismuth subnitrate etc.

Organic Nitrites—e.g. amyl nitrite, glyceryl trinitrate (nitroglycerin), Isosorbide dinitrite etc. Among this nitroglycerine is the most commonly used drug.

Glyceryl Trinitrate OR Nitroglycerin. It is a nitrate of glycerol. It occurs as a colourless, odourless, oily liquid. It is sparingly soluble in water and is stored in air tight containers, in a cool place as it explodes on heating.

It is unstable when stored in plastic containers. Hence, it is stored in glass bottles.

Nitroglycerin gives immediate relief from paraoxysmal dysponea of attack, generally takes place at night. These are effective vasodilators and are used for the treatment of congestive heart failure due to acute myocardial infarction. This drug reduces myocardial oxygen demands, relaxes the muscle and have direct vasodilating action of caronary blood vessels and thereby act quickly in relieving anginal attacks. It is also used for renal and gallbladder colicpain.

Nitroglycerine solution is given in tablet form of 0.3, 0.4 or 0.6 mg. potency. These are readily absorbed from sublingual mucosa (i.e. beneath the tongue) and are most effective when given in this way. It has the following structure.

CH_2ONO_2
|
$CHONO_2$
|
CH_2ONO_2

Glyceral Trinitrate

Amyl Nitrite. It is a volatile liquid and supplied in glass capsule. It is a potent coronary vasodilator. Amyl nitrite is rapidly absorbed from lungs, hence it is administered by inhalation. Amyl nitrite inhalation and intravenous adminstration of sodium nitrite are used for the treatment of cyanide poisoning. Amyl nitrite causes characteristic changes in pressure gradients and heart murmurs and helps in diagnosis of congenital and rheumatic valvular disease.

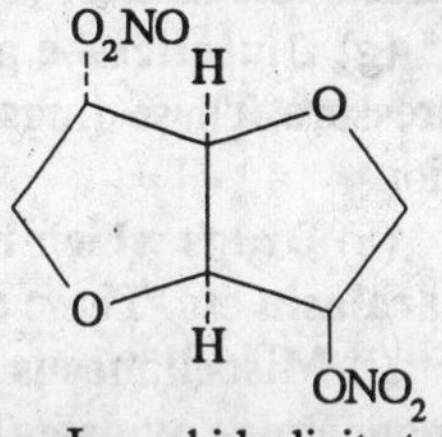

Isosorbide Dinitrate. It is a white crystalline powder and is sparingly soluble in water. In presence of excessive heat, it is stored in tightly closed containers. It has the adjoining structure.

O_2NO H O O H ONO_2

Isosorbide dinitrate

It is adminstered orally or sublingually. It is used in the treatment of angina pectoris and in congestive heart failure.

Dipyridamole

It is a synthetic heterocyclic compound and occurs as yellow crystalline powder. It acts as vasodilator and is used in the prophylaxis of angina pectoris. It finds use as antithrombotic drug.

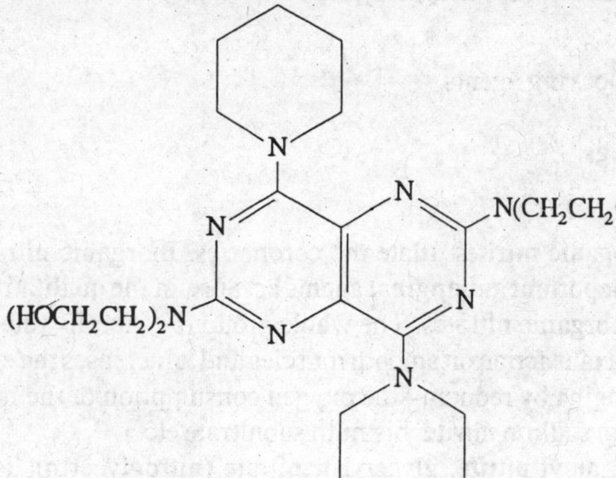

N N N $N(CH_2CH_2OH)_2$ $(HOCH_2CH_2)_2N$ N N N

Dipyridamole

Vasodilator

Vasodilators are those drugs which cause dilation of blood vessels. These drugs are used for the treatment of peripheral vascular diseases. Precautionary measures are very important for the successful treatment of these diseases, such as to avoid factors which cause vasospasm (*e.g.* smoking), postural exercises to increase local circulation, proper precaution against cold, treatment of underlying disorders like diabetes, anaemia etc.

Some vasodilator drugs are :

1. *Alpha-adrenergic blocking agents.* These drugs mimic the responses obtained as a result of stimulation of adrenergic nerves e.g. Phenoxybenzamine, azapetine, tolazaline etc.

2. ***Beta-adrenergic stimulant*** **e.g. Nylidrin, isoxsuprine, salbutanol etc.**
3. ***Anticoagulants*** **e.g. Heparin, coumarin derivatives etc.**
4. ***Miscellaneous agents*** **e.g. Nicotinic acid, papaverine, cyclandelate.**

Tolazoline Hydrochloride

Chemically it is 2-benzyl-2-imidazoline hydrochloride and has the adjoining structure.

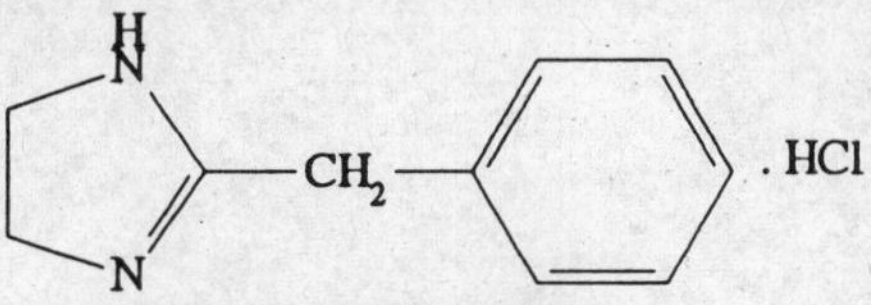

It is a white crystalline powder and is soluble in water. It produces an increase in the force of contraction of myocardium and tachycardia. Tolazoline dilates the peripheral blood vessels. It is available as 25 mg tablets and injections. It finds use in the treatment of peripheral vascular disorders.

Tolazoline Hydrochloride

Isoxsuprine Hydrochloride

It is the beta adrenergic stimulant. It is used for the treatment of central and peripheral vascular diseases.

It is chemically named as 1-(4-hydroxy phenyl) -2-(1-methyl-2-phenoxy ethylamino) propanoil -1-hydrochloride. It occurs as a white crystalline powder and is soluble in water.

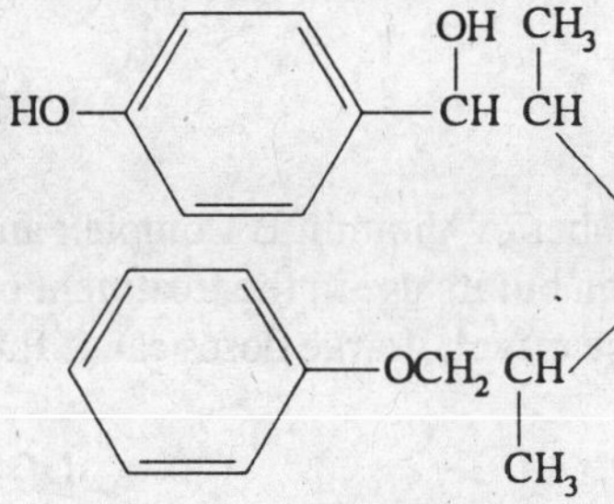

It increases the blood flow of resting uterine muscles, hence it is used in the treatment of threatened abortion or premature labour and is also useful in the treatment of peripheral vascular diseases. This drug can cause palpitation, nervousness etc. It is available as tablets and injection form.

Isoxsuprine

Sodium Nitroprusside

It is an arteriolar vasodilator and acts as a short acting hypotensive. It has specific vascular smooth muscle relaxant action. Its chemical formula as follows:

$$Na_2\ Fe\ (CN)_5.2H_2\ NO.2H_2O$$

and it is named as sodium nitrosyl penta cyano ferrate III dihydrate. Sodium Nitroprusside occurs as reddish brown crystalline powder and is soluble in water. Its aqueous solution slowly decomposes, hence it should be freshly prepared for injections. It is useful in the treatment of hypertension and in acute myocardiac infarction.

Hydrallazine Hydrochloride

It is an antihypertensive drug and is an arteriolar vasodilator.

It is chemically named as phthalazin -1-yl hydrazine. It is a white crystalline compound and is soluble in water.

This drug causes peripheral vasodilatation. It finds use as vasodilator and also for the treatment of moderate to severe hypertension.

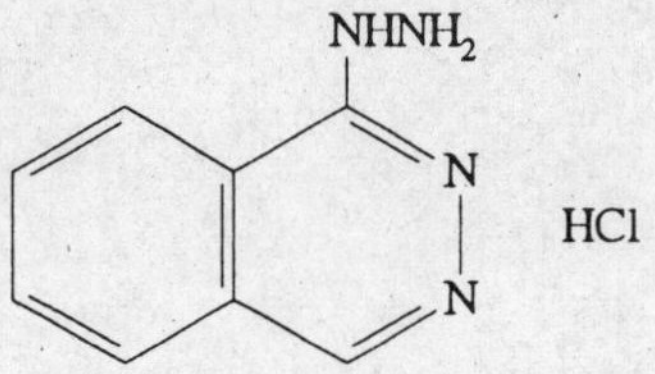

Adverse effects : headache, palpitation

Hydrallazine hydrochloride

Papaverine

It is an opium alkaloid. Chemically it is named as 6.7-dimethoxy-1-(3,4 dimethoxy-benzyl) isoquinoline.

Papaverine has shown coronary dilation in animal expriments but its value in the treatment of human angina pectoris is doubtful. It is used in Arterial embolism which is generally a complication of heart disease.

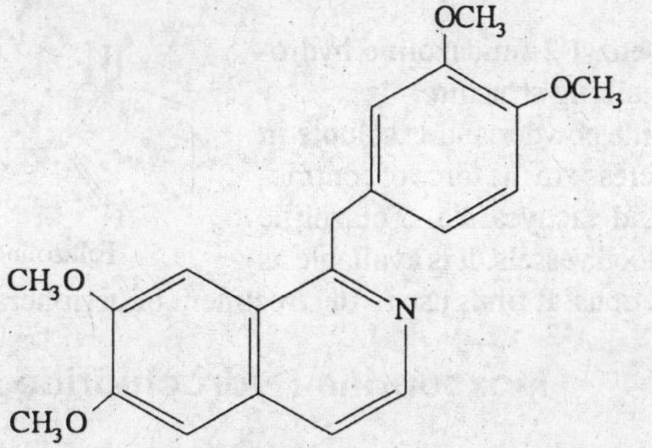

OCH_3 OCH_3 CH_3O CH_3O N

Papaverine

Nicotinic Acid

It is a member of Vitamin B Complex and is a weak vasodilator. This drug on large doses induces hypotension but its use in the treatment of angina pectoris is doubtful.

Adverse effect. Large doses cause flushing and burning of skin and dizziness.

Inositol Nicotinate

Chemically it is named as hexa nicotinate of hexahydric. It is insoluble in water. It is hydrolysed to nicotinic acid. It is used as vasodilator and is used for the treatment of peripheral vascular disorders and cerebral vascular disease. It is also used to reduce hypercholesterolaemia.

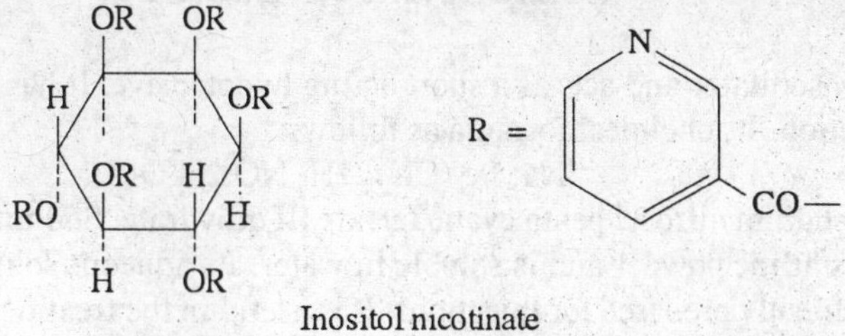

OR OR H OR OR H RO H H OR R = N CO—

Inositol nicotinate

Methyl Nicotinate

It is a vasodilator which is used in creams and ointments for relieving pain in muscular rheumatism, lumbago etc.

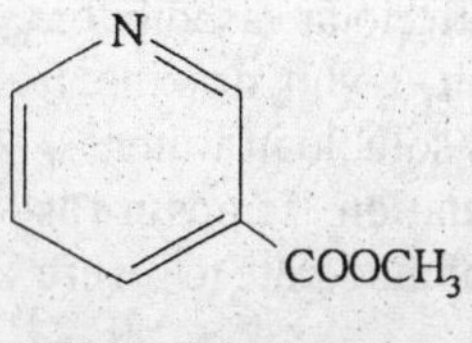

N $COOCH_3$

Methyl Nicotinate

It is a white crystalline powder and has a characteristic odour. It is soluble in water.

Chapter 11

Sulphonamides

Sulphonamides are antimicrobial compounds containing sulphonamido (SO_2NH_2) group. These are synthetic chemotherapeutic agents. These were the first of antibacterial drugs to be used successfully for the treatment of bacterial infections in human. These are active against certain gram positive and gram negative organisms. These are mainly bacteriostatic in action and are used in the treatment of many diseases like gonorrhoea, blood poisoning, scarlet fever, tonsillitis, sinus infections, urinary tract infections etc.

The history of discovery and development of sulphonamides is very interesting. In 1908 Glemo successfully prepared sulphanilamides (p-amino benzene sulphonamide), but not much interest was shown in this compound. Later on, almost after thirty years, Domagk in 1930, while working on azo dyes demonstrated the efficiency of prontosil (a dye with sulphonamides side chain) against the growth of streptococci both in vitro and in vivo. Once it so happened that Domagk's daughter got a deadly infection in her finger and all attempts failed either to cure the disease or to stop the spread of this deadly infection. Fortunately Domagk gave prontosil azo dye orally to her daughter and the girl recovered completely. Later on this dye was found to be active against inhibiting the growth of streptococci both in vivo and vitro. This dye is prepared by diazotising sulphanilamide and then coupling with m-phenylene diamine.

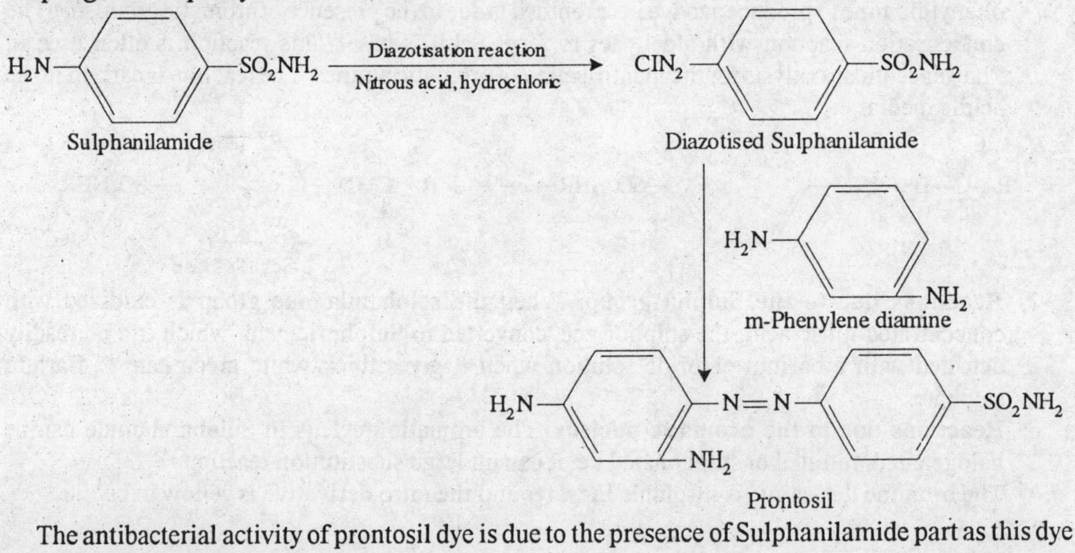

H_2N–C_6H_4–SO_2NH_2 → (Diazotisation reaction; Nitrous acid, hydrochloric) → ClN_2–C_6H_4–SO_2NH_2

Sulphanilamide — Diazotised Sulphanilamide

+ m-Phenylene diamine (H_2N–C_6H_4–NH_2) ↓

H_2N–$C_6H_3(NH_2)$–N=N–C_6H_4–SO_2NH_2

Prontosil

The antibacterial activity of prontosil dye is due to the presence of Sulphanilamide part as this dye is reduced to Sulphanilamide in vivo.

Properties of Sulphanilamides

The Sulphanilamide has the following structure.

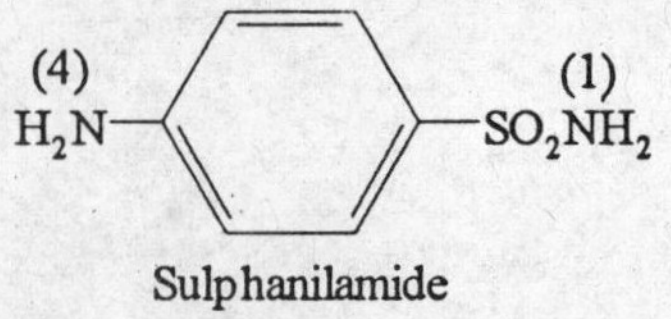

(4) H_2N—C6H4—SO_2NH_2 (1)

Sulphanilamide

These are crystalline, odourless, yellowish white powdered substances. Most sulphanilamides are amphoteric in nature having both acidic and basic properties. They show basic character due to the presence of an amino group in the aromatic ring and like a base they can dissolve in acids with the formation of salts.

$HCl.H_2N$—C6H4—SO_2NH_2

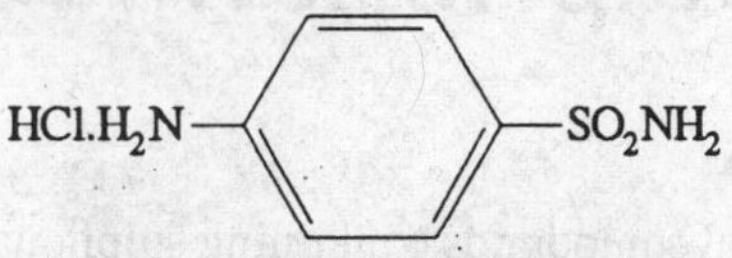

Sulphanilamide Hydrochloride

The acidic properties of the Sulphanilamides are due to the presence of imide group hydrogen which can be substituted by metals with the formation of salts.

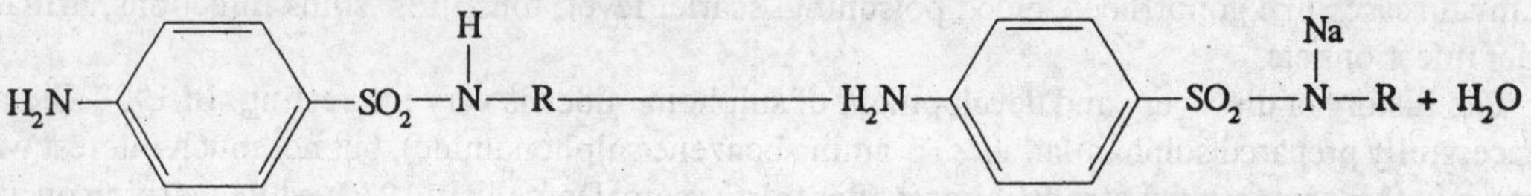

H_2N—C6H4—SO_2—N(H)—R ⟶ H_2N—C6H4—SO_2—N(Na)—R + H_2O

Substituted Sulphanilamide Sodium Salt of Substituted Sulphanilamide

The acidic forms dissolve poorly in water but are readily soluble in organic solvents, viz., alcohol, acetone etc. and have specific melting point. The acidic properties of these drugs are more pronounced than the basic properties.

The chemical properties of the Sulphanilamide are due to the presence of three functional groups namely the aromatic amino group ($C_6H_5NH_2$), Sulphogroup (SO_2) and imide group (—NH).

1. **Reactions due to the aromatic group.** (*a*) The diazotisation reaction and combination with phenyldiamine to produce azodye i.e. prontosil is due to the presence of aromatic group. (*b*) The condensation reaction with aldehydes to form Schiff's base. This reaction is often used in Pharmaceutical analysis for the identification of aromatic amines. The reaction is carried in the acidic media.

R—C(=O)—H + H_2N—C6H4—SO_2NHR ⟶ R—C(H)=N—C6H4—SO_2NHR

Schiff's Base

2. **Reactions due to the Sulpho group.** When the sulphanilamide group is oxidized with concentrated nitric acid, the sulphur gets converted to Sulphuric acid, which can be readily detected with a barium chloride solution when it gives thick white precipitate of Barium sulphate.

3. **Reactions due to the aromatic nucleus.** The aromatic nucleus in Sulphanilamide can be halogenated, nitrated or sulphonated i.e. it can undergo substitution reaction.

The bromine derivative is insoluble in water and the nitro derivative is yellow in colour.

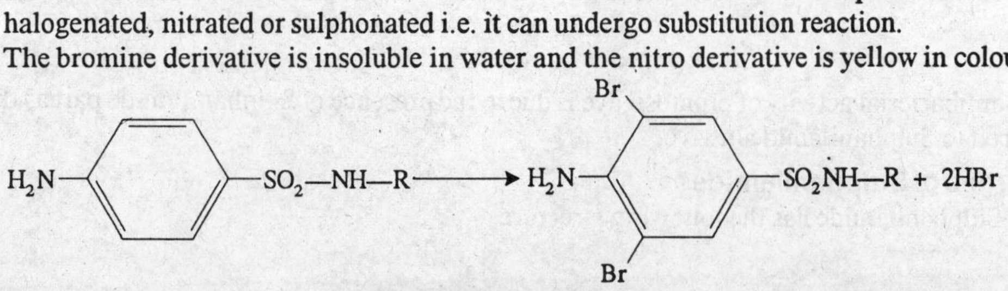

H_2N—C6H4—SO_2—NH—R ⟶ H_2N—C6H2(Br)(Br)—SO_2NH—R + 2HBr

4. All the sulphanilamides can be thermally decomposed.
5. **Reactions due to imide group (—NH).** The hydrogen atom of this group reacts with salts of the heavy metals like $CuSO_4$, $COCl_2$ etc. and forms coloured chelate compounds. The colour is specific for each sulphanilamide drug.

The methods for the quantitative determination of the Sulphanilamide drugs are based on their chemical properties and the nature of the functional group.

The content of the substance in the drug can be estimated by photocolorimetric method which is based on the ability of the sulphanilamides to yield coloured products in reactions with aldehydes, salts of heavy metals, and azo dye formation. The colour intensity could be compared with that of a reference solution.

Many substituted sulphanilamide were prepared and their antibacterial activity was tested. Some of their *structure activity relationship* are as follows :

1. The physiological activity of the sulphanilamide due to the sulphanil radical in the molecule.

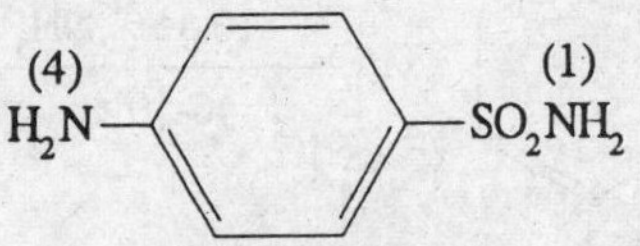

(4) H_2N—[benzene ring]—SO_2NH_2 (1)

Sulphanilamide

H_2N—[benzene ring]—SO_2^-

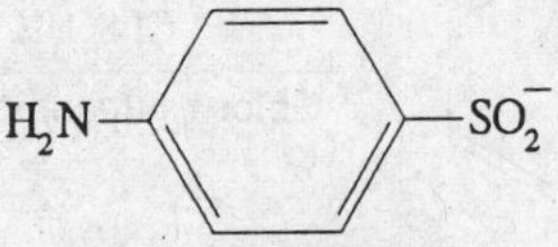

Sulphanil Radical

2. The removal of the amino group from position 4 to some other position of the aromatic ring results into the complete loss of physiological activity.
3. The presence of *p*-amino group is important for its antibacterial activity and if this group is substituted it should be done only by such radicals or groups which can be converted to free amino group in the body.
4. The introduction of additional substituents into the aromatic ring reduces physiological activity.
5. The substitution of the hydrogen of the SO_2NH_2 group with a heterocyclic aromatic nuclei gives highly potent compounds.
6. When various radicals are introduced into the sulphamide group, the physiological activity can either increase or diminish depending on the nature of the substituted radical.
7. The presence of methoxy group in a mole is important for prolonged action of the drug. The methoxy group ensures the higher binding of the sulphanilamide with plasma proteins which helps in prolonged action of a drug.
8. The substitution of the NH_2 group have also produced many potent compounds.
9. When amino group (designated as N^2) is in the meta or artho position, the compound becomes physiologically inactive.

Mechanism of Action of Sulpha Drugs

These are bacteriostatic in action. The bacteriostatic nature of sulphonamides is due to their structural similarity to para amino benzoic acid (PABA), which is a compound for the normal functioning of some of the vital process in bacteria.

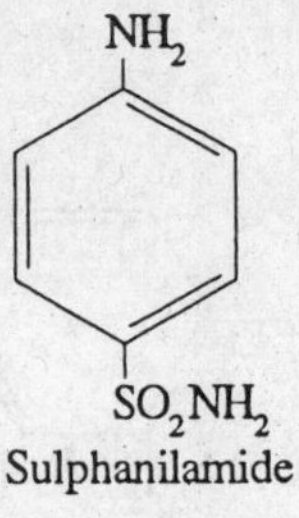

NH_2 — [benzene ring] — SO_2NH_2

Sulphanilamide

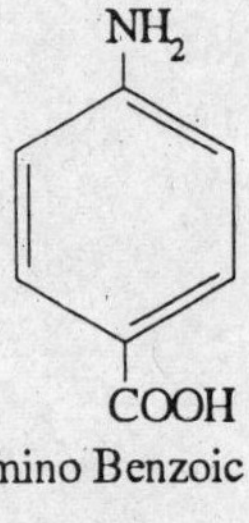

NH_2 — [benzene ring] — COOH

Para Amino Benzoic Acid (PABA)

Folic acid derived from PABA is important for bacterial metabolism. Sulphonamides are antimetabolites of PABA, as due to structural similarity the bacteria takes up sulphonamide which inhibit the enzyme folic acid synthetase, that is involved in the conversion of PABA to folic acid. Thus the metabolism is blocked and the growth of bacteria stops.

The major pathway in biotransformation of Sulphonamides is the acetylation of the Nitrogen of paraamino group (N_4). The free and acetylated sulphonamides are mainly eliminated in urine.

Some important sulphonamide drugs are as follows :

1. Sulphanilamide

It is the parent compound of all the Sulpha drugs. It is prepared from acetanilide in the following manner:

NHCOCH₃ (acetanilide) —[$ClSO_3H$, Chloro sulphonic acid]→ pN-acetyl Sulphanilyl Chloride (ASC) ($NHCOCH_3$ / SO_2Cl) —[10-15% NH_3, 40-50°C]→ p-Nacetylsulphanilyl amide ($NHCOCH_3$ / SO_2NH_2) —[15-20% HCl, 100°C]→ (NH_2HCl / SO_2NH_2) —[NaOH]→ Sulphanilamide (NH_2 / SO_2NH_2)

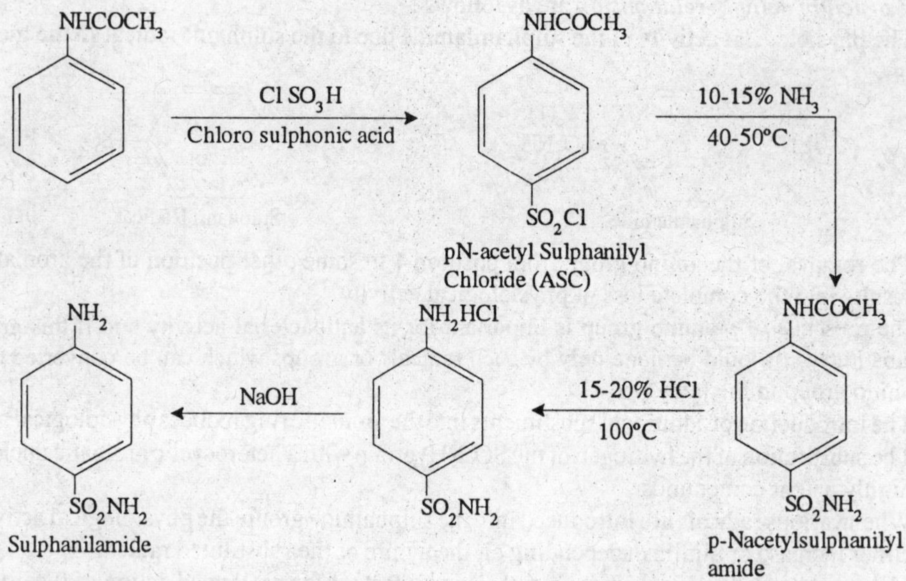

Uses : Sulphanilamide is important in the control of 'cocci' infections such as pneumococci, streptococci, meningococci etc. These are used in urinary tract infections.

Adverse effect. Systemic side effects such as nausea, vomiting, exfoliate dermatitis, urinary tract disturbances etc.

2. Sulphadiazine

Chemically it is N[1] (pyrimidin-2-yl) Sulphanilamide. It is prepared by condensing 2-amino pyrimidine with ASC in the following manner.

2-Aminopyrimidine + ClO_2S–(ASC)–$NHCOCH_3$ —[(i) pyridine, (ii) alkali]→ (pyrimidin-2-yl)NH–SO_2–(C₆H₄)–NH_2

Sulphadiazine

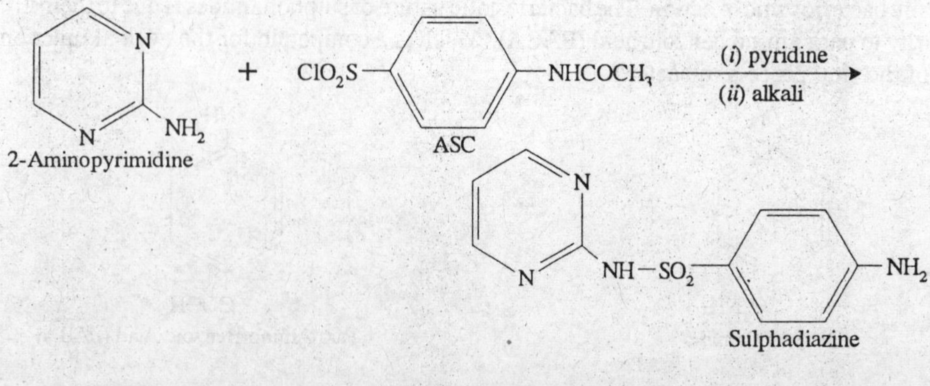

Therapeutic Uses. It is used for general purposes as it is not toxic and is easily absorbed and excreted. It is a short acting sulpha drug and is used for systemic infections and is usually administered orally. It is used in the treatment of meningococcal meningitis.

3. Sulphapyridine

Chemically it is N^1-2 pyridyl-sulphanilamide. It was the first N^1 heterocyclic sulphanilamide to be used as a chemotherapeutic agent. It is obtained by the condensation of ASC with 2 aminopyridine in the following manner.

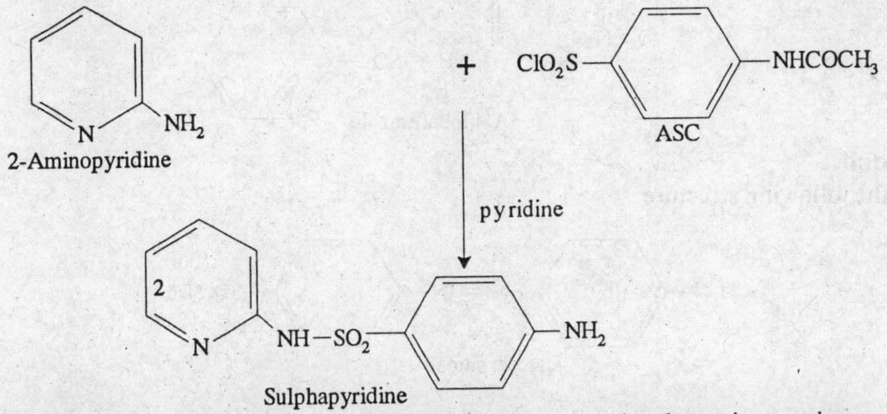

2-Aminopyridine + $ClO_2S-C_6H_4-NHCOCH_3$ (ASC) $\xrightarrow{\text{pyridine}}$ $C_5H_4N-NH-SO_2-C_6H_4-NH_2$

Sulphapyridine

Therapeutic Uses: It is very active against pneumonia causing bacteria, meningococcal and streptococcal infections. It has several toxic side effects like nausea, vomiting, urinary tract disturbance etc. and more over it is readily acetylated in the body and the acetyl sulphapyridine crystals are deposited in the kidneys which may damage the kidneys.

4. Sulphathiazole or Cibazole

Chemically it is N^1-2-thiazolyl sulphanilamide. It is prepared by condensing ASC with 2-aminothiazole and hydrolysing the product in alkaline solution.

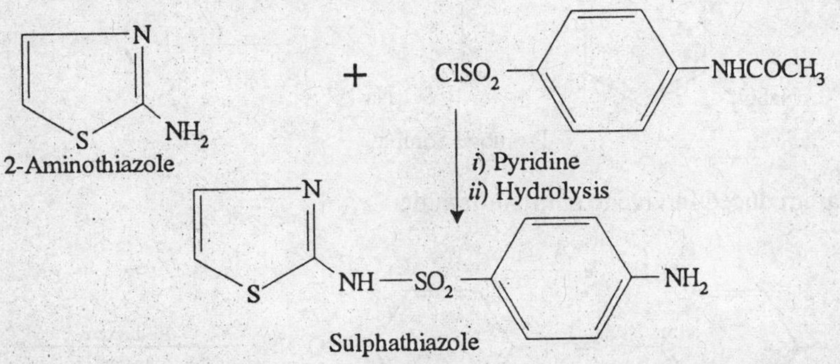

2-Aminothiazole + $ClSO_2-C_6H_4-NHCOCH_3$ $\xrightarrow{\text{i) Pyridine, ii) Hydrolysis}}$ $C_3H_2NS-NH-SO_2-C_6H_4-NH_2$

Sulphathiazole

Uses : It is used for general purposes and it is a highly bacteriostatic sulphonamide drug. It is mainly used in staphylococcal infections and in bubonic plague. It is not a very toxic drug. (bubonic plague is inflammation of the glands).

5. Sulphafurazole—(Sulphisoxazole)

Chemically it is N^1 (3,4 dimethyl isoxazol-5-yl)—Sulphanilamide. It is prepared by condensing 3,4 dimethyl-5-aminoisoxazole with ASC.

Uses : It is a short acting sulphonamide. It is rapidly excreted and is in high concentration in plasma. It is less acetylated and both free and acelytated forms are soluble over a wide range of pH and are used in urinary tract infection.

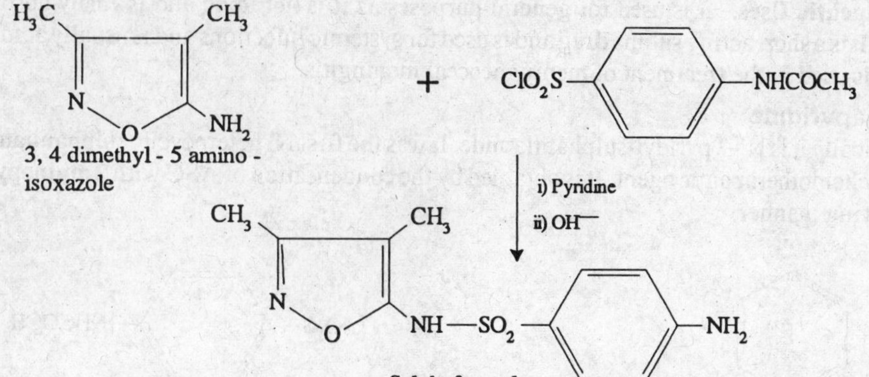

H_3C, CH_3, N, O, NH_2 — 3, 4 dimethyl - 5 amino - isoxazole + ClO_2S–C6H4–$NHCOCH_3$ → i) Pyridine ii) OH^- → CH_3, CH_3, N, O, NH—SO_2–C6H4–NH_2

Sulphafurazole

6. Prontosil

It has the following structure.

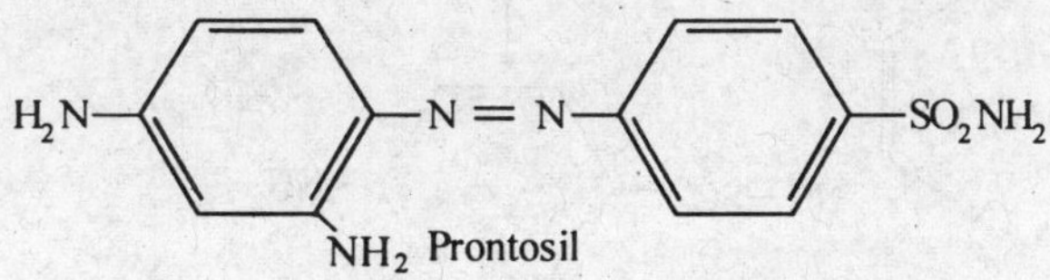

H_2N–C6H3(NH_2)–N = N–C6H4–SO_2NH_2

Prontosil

Preparation of prontosil is described earlier in this chapter. It was used in the treatment of streptococcal sore throat, streptococcal fever, acute endocarditis etc. Due to its low solubility it was generally used as its hydrochloride derivative as it is more soluble. Later on prontosil dye was replaced by prontosil soluble which is superior drug and has greater solubility and greater activity. Prontosil soluble has following structure.

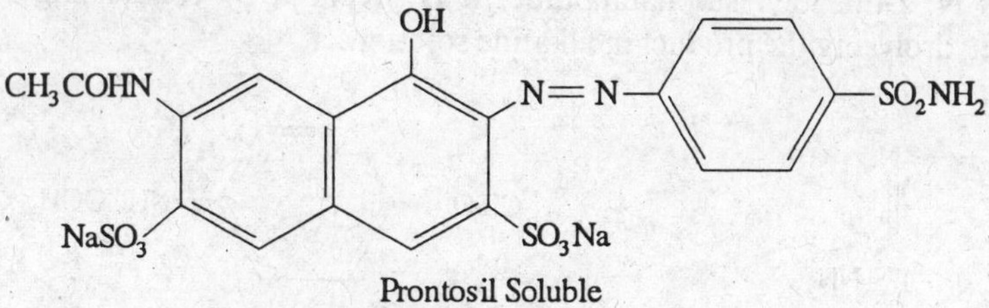

OH, CH_3COHN, $NaSO_3$, SO_3Na, N=N–C6H4–SO_2NH_2

Prontosil Soluble

Prontosils are reduced in vivo to sulphanilamide

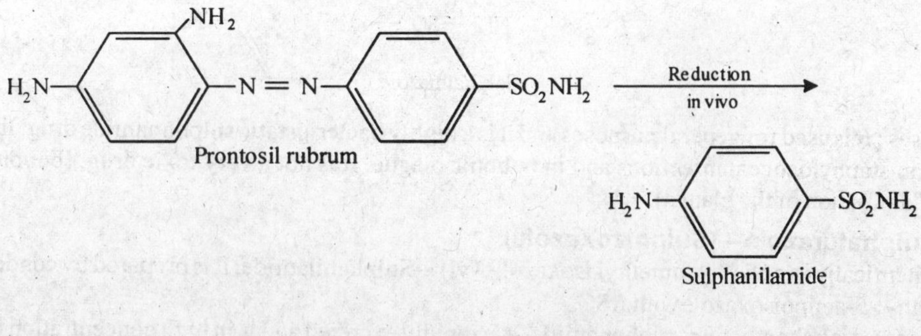

H_2N–C6H3(NH_2)–N = N–C6H4–SO_2NH_2 —Reduction in vivo→ H_2N–C6H4–SO_2NH_2

Prontosil rubrum

Sulphanilamide

Chapter 12

Vitamins

Vitamins are biologically active organic compounds of a diverse chemical nature. Vitamins are essential for maintenance and normal growth of the life in humans, animals and some bacteria and micro-organism. These are supplied to the organisms through food. In general fresh and natural foods like vegetables, fruits etc., contain all the necessary vitamins. When there is a lack of vitamins in an organism, it is known as hypovitaminoses for example deficiency of vitamin C causes a disease known as Scurvy, Vitamin D deficiency causes rickets. Diseases associated with the absence of vitamin in an organism are called avitaminoses. An excess of vitamins in an organism also result in certain diseases known as hypervitaminoses. For example, excess of vitamin K result in a severe disease known as thrombophlebitis (when there is blood clot formation in a vein with inflammation of the vessel lining).

Vitamins can be synthesised and are employed as drugs for therapeutic uses.

Vitamins were discovered by polish biochemist K. Funk who described vitamin as compounds containing nitrogen and called it is as vitamine due to their vital function and basic nature. But later on it was found that some vitamins do not contain nitrogen and therefore the term vitamine was modified to vitamin.

Classification of Vitamins

They are broadly classified into two groups namely :

1. **Fat soluble vitamins.** Fat soluble vitamins includes vitamins A, D, E and K.
2. **Water soluble vitamins.** This group includes vitamins of B group, vitamin C etc.

Generally the vitamins belong to either of these two groups, though some exceptions are also there like vitamin H which is insoluble both in fat and in water. The fat soluble vitamins can be stored in the body and excessive administration can be toxic where as water soluble vitamins are excreted easily and excessive administration have low toxicity. There are some chemical substances which are chemically similar to a vitamin and these can be converted easily into vitamins in vivo, such compounds are known as provitamins or precursons e.g. B. carotene, Ergosterol etc. As the chemical structure of vitamins were established it was possible to classify vitamins according to their chemical structure which is as follows :

The chemical classification divides all the vitamins into four main group.

1. Vitamins of alliphatic series: e.g. vitamin C or Ascorbic acid etc.
2. Vitamins of the alicyclic series e.g. vitamin of group A or Retinol etc.
3. Vitamins of the aromatic series e.g. vitamins K etc.
4. Vitamins of the heterocyclic series: e.g. vitamin B complex etc.

A normal individual gets adequate supply of vitamins from a well balanced diet, vitamin deficiency, however, can occur in various ways, such as poor dietary intake, inadequate absorption from the gastro-intestinal tract, interference with utilisation of increased demand. Deficiency of vitamins results in various diseases, when vitamins are used as drugs for therapuetic purposes.

Physiological Function of the Vitamins

The biological or physiological function of the vitamins are due to their co-enzymes which are active form of the vitamins e.g. Thiamine has thiamine pyrophosphate co-enzyme. Pantothenic acid in co-enzyme A vit. B-12 in Cobamide coenzymes etc.

Some important vitamins, their sources, deficiency, diseases and treatment is discussed here.

1. Vitamin A has two groups of vitamins namely :

(*a*) **Vitamin A_1, A-xerophythol.** It occurs generally in animals either in free state or in combination with fatty acids.

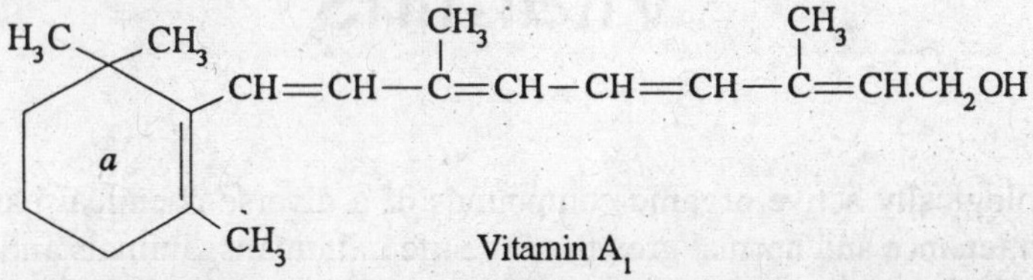

H_3C, CH_3 — $CH{=}CH{-}\overset{CH_3}{\overset{|}{C}}{=}CH{-}CH{=}CH{-}\overset{CH_3}{\overset{|}{C}}{=}CH.CH_2OH$; ring *a* with CH_3

Vitamin A_1

a = β-ionone ring

It is a yellow crystalline vitamin which is insoluble in water but it is soluble in alcohol and fixed oils.

(*b*) **Vitamin A_2 or Retinol.** It has a structure similar to Vitamin A, and differs only in β-Ionone ring, which has another double bond hence it is called dehydroretinol.

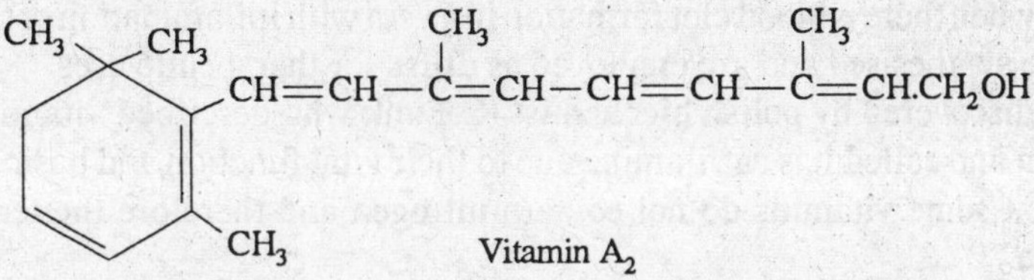

CH_3, CH_3 — $CH{=}CH{-}\overset{CH_3}{\overset{|}{C}}{=}CH{-}CH{=}CH{-}\overset{CH_3}{\overset{|}{C}}{=}CH.CH_2OH$; ring with CH_3

Vitamin A_2

Vitamin A_2, Vitamin A_1 and A_2 are similar in their properties and action.

Vitamin A has important provitamin known as carotenes which exists as α, β, γ out of which β is the most important one.

Source : Vitamin A is found in fish liver oil, cod-liver oil, milk, egg yolk, sweet potatoes, tomatoes, notably in carrots. An adult person needs 5000 units daily.

Deficiency diseases: It causes night blindness, in later stages true kerato malacia (ulceration and softening of the cornea) may occur leading to blindness. It may cause retardation of growth, dryness of skin and hair etc.

Therapeutic uses. Vitamin A is produced synthetically for medical purposes.

Vitamin A is available in concentrated solution form or in tablet form. Cod-liver oil capsules are also available. In the treatment of night blindness, vitamin A is administered in the dose of 50,000 to 75,000 units usually in the form of halibur liver oil.

An excess of vitamin A causes nausea, weakness deramatitis etc.

2. Vitamin B complex. This complex includes all the water soluble vitamins found in yeast, liver, rice polisings etc. It comprises of the following groups.

(*i*) (*a*) **Vitamin B_1 or thiamine :** Thiamine is basic in nature and forms salts with acid. Its hydrochloride salts are the most commonly used ones. It's structure is given on the following page.

These are found as colourless crystals and are soluble in water. Its biological activity is due to thiamine pyrophosphate co-enzyme.

Source : It is found in rice, wheat, yeast, egg, meat, beans etc.

Deficiency diseases : Causes loss of apetite, gastrointestinal disturbances, hypotension and a disease known as "Beriberi" which is a form of neuritis with pain, paralysis and oedema of the extremities.

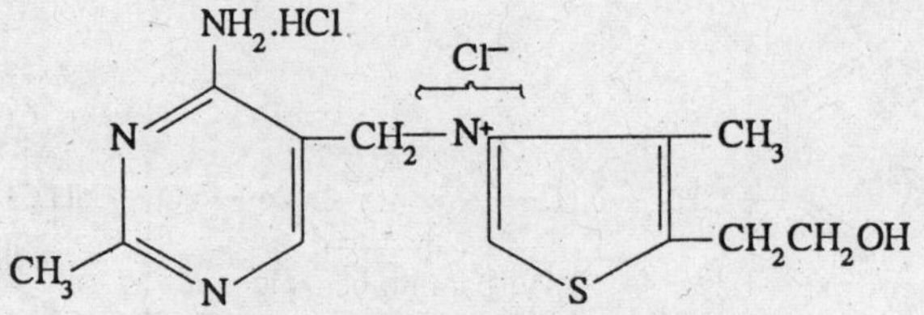

$$\text{Thiamine: } NH_2.HCl,\ Cl^-,\ CH_2-N^+,\ CH_3,\ CH_2CH_2OH,\ CH_3$$

Thiamine Chloride Hydrochloride

Therapeutic uses : Thiamine hydrochloride tablets containing 5 to 50 mg of the drug is given for therapeutic use. Thiamine hydrochloride injections are also available containing 25 mg of the vitamin per ml. Vit. B, deficiency can be prevented by taking proper diet like cereals particularly rice, beans, peas etc.

3. Vitamin B$_2$ or Riboflavin. It is a yellow crystalline water soluble substance. Its active coenzymes are Flavin mononucleotide and flavin adenine dinuclotide. Its aqueous solution gives yellowish green fluorescence.

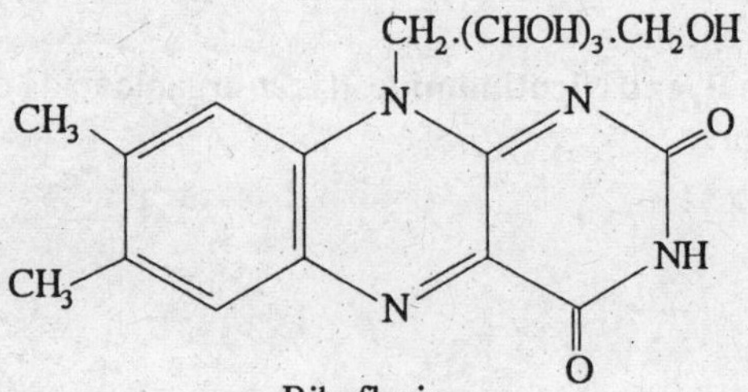

$$CH_2.(CHOH)_3.CH_2OH$$

Riboflavin

Occurrence: It is found in yeast, liver, wheat, egg yolk, milk, fish etc.

Deficiency diseases. The average healthy adult requires about 2-3 mg of Riboflavin daily.

The deficiency of this vitamin leads to inflammation of the tongue, cracking of lips and corners of the mouth, known as cheilosis etc.

Therpeutic uses. Riboflavin is administered either orally or parenterally. Riboflavin tablets containing 2 mg of the vitamin are administered 5 to 10 mg daily in therapeutic doses. Riboflavin injection contains 10 mg of riboflavin per ml. It is usually given with other members of vitamin B group.

Vitamin B$_3$ or pantothenic acid. It is an organic acid and has the following structure.

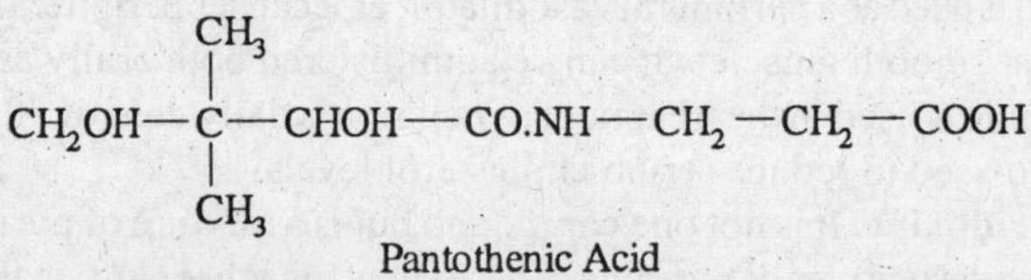

$$CH_2OH-C(CH_3)_2-CHOH-CO.NH-CH_2-CH_2-COOH$$

Pantothenic Acid

It is the constituent of coenzyme A. All acetylation reactions are influenced by this enzyme. The daily human requirement of pantothenic acid is about 5 mg.

Occurrence. The richest sources of this vitamin are liver, kidney, wheat, peanuts, cereals, milk etc.

Deficiency diseases. The symptoms of pantothenic acid deficiency is not well-known in man. In animals it results in dermatitis, arrest of growth, neuromuscular degeneration etc.

Therapeutic uses. Calcium pantothenate is used in Streptomycin toxicity. Rheumatoid arthritis and as a nutritional supplement along with other members of B-complex group. A low vitamin diet containing pantothenic acid antagonist causes fatigue, headache, nausea etc.

4. Folic acid or pleroyl glutamic acid. It has the following structure.

N N COOH
N CH$_2$ NH CO NH.CH.CH$_2$.CH$_2$.COOH
N glutamic acid
OH pamino benzoic acid
2 amino 4 hydeory pleridine
Pteroicacid

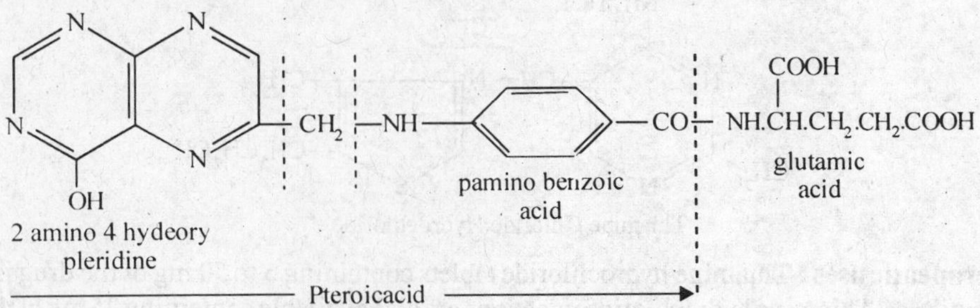

Pteroye pleroyl glutamic glutamic acid (Folic acid)

Its active coenzyme is tetrahydrofolic acid.

Occurrence. In animals it is found in liver, kidney, yeast. It is also found in spinach, banana, lemons, strawberries etc.

Deficiency diseases. The deficiency of folic acid cause macrocyclic anaemia.

Uses : For the treatment of macrocyclic anaemia. Folic acid tablets are administered orally. Its sodium salts can be administered parentally.

5. Nicotinic acid, Vitamin B$_5$ and Nicotinamide. It is an organic acid i.e. Pyridine-3-carboxylic acid and has the following structure.

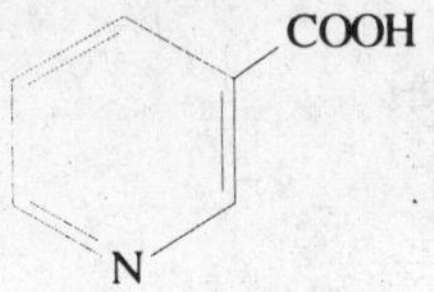

COOH CONH$_2$
N N
Nicotinic Acid 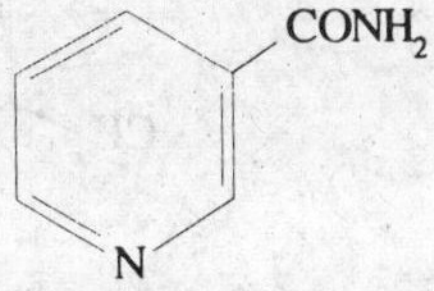 Nicotinamide

Nicotinic acid is sparingly soluble in water. The Nicotinamide is different from the acid in being freely soluble in water. Nicotinamide has odourless white crystalline powder. Nicotinamide has diphosphopyridine nucleotide, triphosphopyridine nucloticle as active coenzymes.

Occurrence or source. The important sources of nicotinic acid are wheat, yeast, liver, barley, maize, rice etc.

Nicotinamide occurs mainly in combined form is coenzyme I and II. The normal daily requirement of this vitamin is between 15 and 20 mg.

Deficiency diseases. Nicotinic acid deficiency causes a disease known as pellagra which is dermatitis and pigmentation of the skin, especially in regions exposed to the sunlight, other symptoms of Pellagra are weakness, digestive disturbance, mental confusion, spinal cord degeneration etc.

Uses : Nicotinic acid is used as a peripheral vasodilator, as it causes peripheral vasodilation by direct relaxation of the vasular smooth muscles. It can be administered both orally as well as parenterally.

Nicotinic acid or amide is used in the treatment of pellagra is daily dose of 50 to 500 mg. Large dose of Nicotinic acid is also used to reduce serum cholesterol levels.

6. Vitamil B$_6$ or Pyridoxine. It is not one compound but is a mixture of pyridoxine, pyridoxal and pyridoxamine which are derivatives of pyridine. These substances are close in their chemical structure

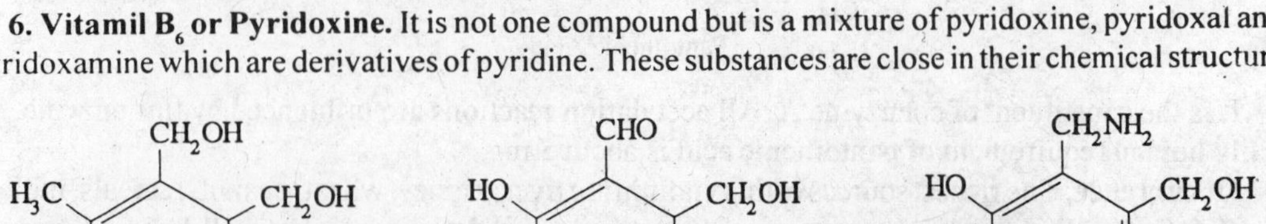

CH$_2$OH CHO CH$_2$NH$_2$
H$_3$C CH$_2$OH HO CH$_2$OH HO CH$_2$OH
H$_3$C N H$_3$C N H$_3$C N
Pyridoxine ⇌ Pyridoxal ⇌ Pyridoxamine

which are as follows and these substances can transform into one another.

Vitamin B_6 groups are thermally stable. Pyridoxal phosphate acts as a coenzyme for amino acid decarboxylases and transaminases. Pyridoxine is produced synthetically. All the three members of B_6 vitamin are active biologically because they are interconvertible, but pyridoxal phosphate is most active as it is obtained by phosphorylation of pyridoxal.

Source : It is found both in animals as well as in plants, the main sources are wheat, maize, fish, liver, milk, egg, cereals, meat, yeast etc.

Deficiency disease. Vitamin B_6 deficiency causes derangement of enzymes controlling carbohydrate metabolism which result in abnormal G.T.T. Deficiency in infants cases abdominal distension, convulsions, failure to gain weight etc. In adults deficiency of this vitamin causes lesion of skin, peripheral neuritis, mental changes etc.

In rats it causes inhibition of growth, reduction in accessory reproductive organs etc.

Uses : It is used in the treatment of convulsions due to pyridoxine deficiency in infants. It is also used in the treatment of pellagra. Pyridoxine in the daily dose of 20 to 100 mg is used in vomiting, of pregnancy and in radiation sickness. It is also used in treating abnormal G.T.T. and mental depressions.

7. β-Biotins vitamin H. It is organic acid which functions as coenzyme for many carboxylation reactions. It has the following structure.

CO

HN NH

HC — CH

H_2C $CH.(CH_2)_4.COOH$

S

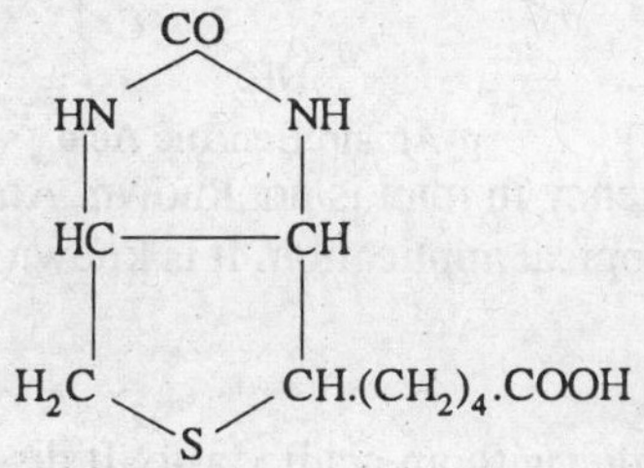

Sources : It occurs in small amounts in all the tissues of higher animals. The main sources are egg-yolk, liver, kidney, milk etc.

Deficiency disease : Biotin deficiency is not encountered naturally in humans. However, it has been found out that a diet containing large amounts of raw egg white or avidin causes both deficiency which results in loss of hair, paralysis, dermatitis etc.

The daily requirement of this vitamin is not known.

Uses : As the clinical deficiency syndrome is not clear this vitamin is not in use therapeutically.

8. Mesoinositol. Chemically it is known as hexahydroxy cyclohexane which has the following structure:

OH

HO OH

HO OH

OH

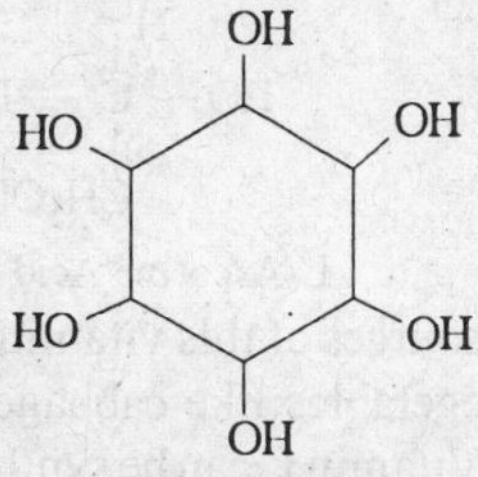

Mesoinositol

There are nine isomers of a hexahydroxy cyclohexane seven of which are optically inactive.

Sources : It is found in all plants and animal tissues. The important source is citrus fruits like lemon, grapes, oranges etc.

Deficiency diseases. Its deficiency in rate causes loss of weight and hair, out of the nine isomeric forms myoinositol is biologically active. In man deficiency of this vitamin is not specific.

Uses : Inositol has been used in the diseases associated with the metabolism of fats.

9. Vitamin B_{12}. It is also known as cyanocobalamin and this vitamin is a cobalt containing complex structure which is related to porphyrins and has a complex structure. It is an odourless, tasteless crystalline powder. It is sparingly soluble in water. It is stable in neutral aqueous solution.

Source : It is found in all animal tissues especially in the liver of fish, pig, sheep etc.

Deficiency diseases : The deficiency of the vitamin causes pernicious anaemia and degradation of the spinal cord. The same symptoms are also seen in the case of folic acid deficiency.

Uses : Vitamin B_{12} is effective against anaemia and B_{12} is therapeutically more beneficial than folic acid. It is available in ampoules. The crystalline drug should be stored in well-closed jars. Ampoules should be stored in a dark place.

10. Amino benzoic acid. This is also included as a member of vitamin B group. It is a white, odourless crystalline powder and gets darkened on the exposure to light. It has the following structure:

COOH

NH_2

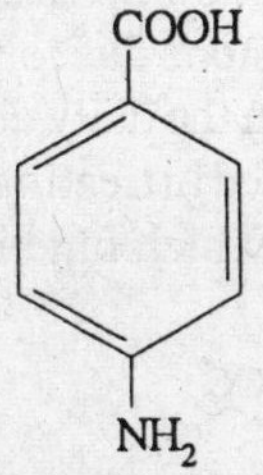

p-Amino Benzoic Acid

Any disease due to its deficiency in man is not known. Aminobenzoic acid is used in some Sun Screen lotions and creams for topical application. It is known to have some adverse side effects like hepatitis, hypoglycenia etc.

Vitamin C or ascorbic acid

It is a white crystalline powder with an acidic taste. It dissolves readily in water and in alcohol though it is insoluble in either benzene or chloroform. It behaves like a monobasic acid. It is decomposed on boiling and by treating with alkali.

It has the following structure:

O
HO C
C
O
C
HO HC
HO—C—H
CH_2OH

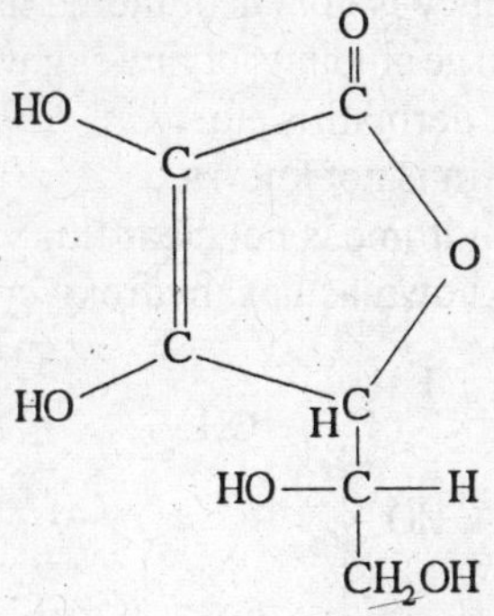

L-Ascorbic Acid

Source : The plants are important sources of this vitamin. It is obtained from the citrus fruits like oranges, lemon, grapes etc., green vegetables like cabbage, tomatoes etc. One ml. of onion juice contains about 0.5 mg of this vitamin. Vitamin C can be synthesised from the juices of citrus fruits and it should be stored in non-metallic containers and should be protected from light.

Deficiency disease. Vitamin C deficiency leads to a disease known as Scurvy, the symptoms of this disease are hemorrhages in subcutaneous tissue, muscles or joints swollen, inflamed, bleeding and spony gums etc. In infants the symptoms are loss of apetite, hemorrhages gum changes etc. Scurvy is usually seen in babies who are given powder milk.

A human milk contains a good amount of vitamin C which is not present in powder milk.

Ascorbic acid is essential for the formation of collagen, intercellular matrix and thus for the development of cartilage, bones and teeth. It also helps in the formation of haemoglobin, erythrocyte maturation and in the conversion of folic acid to tetrahydrofolate. Ascorbic acid also helps in carbohydrate metabolism. It is readily absorbed from gastrointestinal tract.

Uses : It is used in the treatment of scurvy, when it is administered in the dose of 100 to 500 mg per day. Infertile scurvy can be prevented by giving 5 mg of this vitamin per day. Vitamin C helps in absorption of iron from the intestine as it facilitates the conversion of ferric into ferrous ions. Vitamin C is effective against common cold, though its role in the treatment of common cold is still controversial. It is also used in variety of conditions like anaemia, dental infections, haemorrhages etc.

The minimum daily requirement of ascorbic acid is 30 mg for an adult and about 5 mg per kg of body weight for infants.

Vitamin D (calciferols)

These are derivatives of sterols and are formed by irradiation of specific sterols. These sterols are known as pro-vitamins. Until now five vitamins of this group are isolated which are as follows :

Vitamins	*Provitamins*
D_2 (ergocalciferol)	Ergosterol
D_3 (cholecalciferol)	7-dehydrocholesterol
D_4	22, 23 Dihydroergosterol
D_5	7-Dehydrositosterol

Vitamin D_1 is irradiated ergosterol with a mixture of lumisterol.

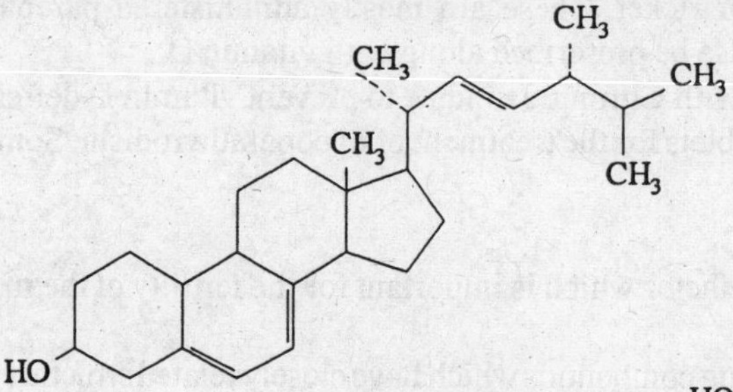

Ergosterol (Vit. D_2)

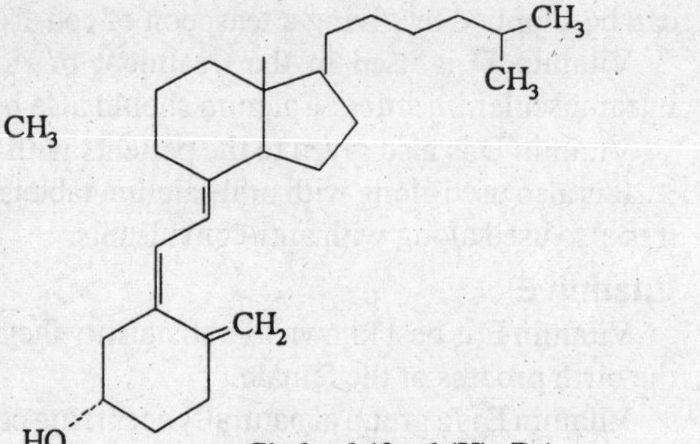

Cholecalciferol (Vit. D_3)

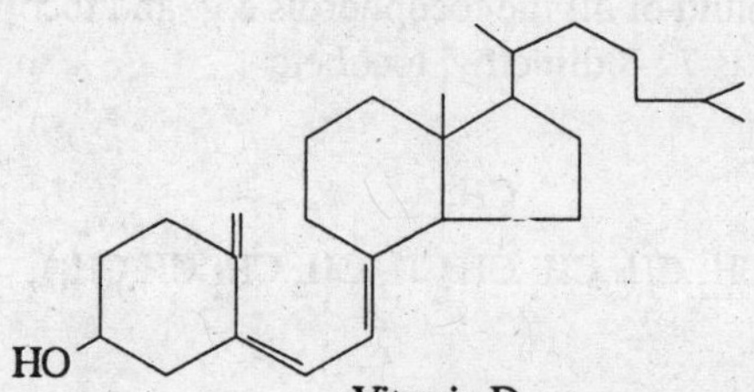

Vitamin D_4

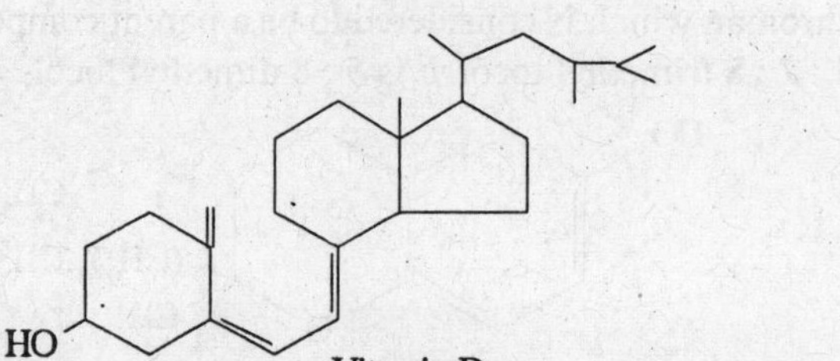

Vitamin D_5

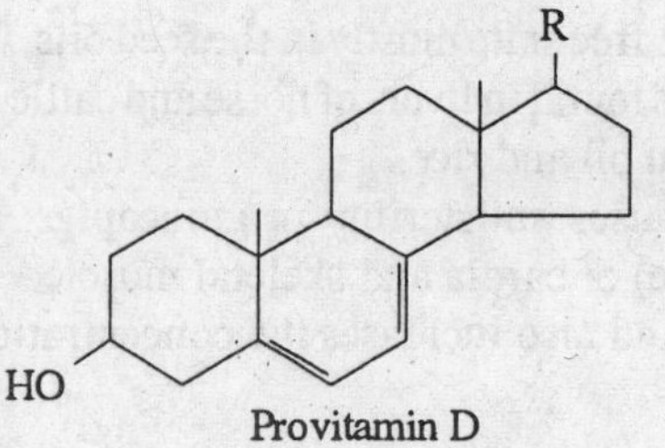

Provitamin D

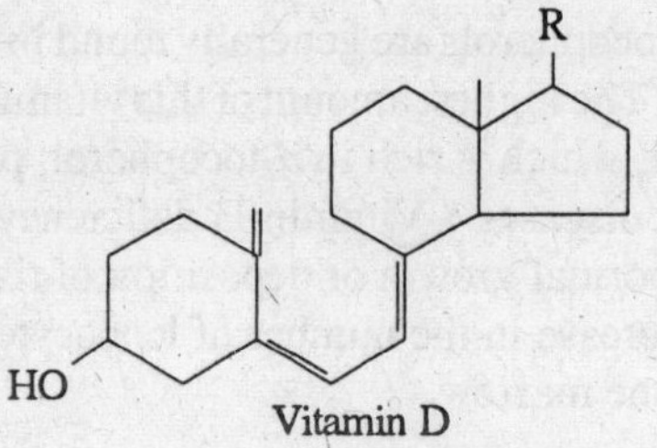

Vitamin D

The vitamin D_4, D_5 and others differ from Vitamin D_2 only in the side chain in the Sterol molecule and are produced in a similar way by irradiation of their provitamin.

Cholecalciferol D_3 occurs as white crystals and is sparingly soluble in water, but it gets dissolved in fixed oil. It is stored in light resistant airtight containers.

Ergocalciferol (D_2) are colourless crystal and are soluble in fixed oils.

The main importance of vitamin D is that it plays a very important role in the absorption of calcium and phosphorus from gastrointestinal tract and for their transport which ensures the mineralization of teeth and bone tissue of a growing organism.

The metabolic actions of vitamin D is due to its active metabolite known as 1,25-dihydroxycholecalciferol which is biosynthesized by the kidney.

Source : The main source of vitamin D is cod-liver oil. It is also obtained from the liver of the seal and other sea animals. It is found in small amounts in the yolk of eggs, butter, milk etc. Vitamin D_2 is an active agent in irradiated yeast.

Deficiency disease : The deficiency of vitamin D results in a disease known as ricket, the symptoms of this disease are that there is softening and irregular growth of the bones, which result in deformity. This disease is mostly found in children from 6 months to 2 years of age.

Uses : Vitamin D is considered both as a vitamin and as a hormone. As a vitamin it is used to prevent rickets by maintaining plasma inorganic phosphate level above a certain threshold, it also maintains calcium above a certain critical level, it also prevents muscle weakness.

The active metabolite of vitamin D acts as a hormone and its major actions are on intestine, kidneys, bones, muscles etc.

The daily requirement of vitamin D in children and adults is about 10 mg. The daily requirement can be supplied by giving a teaspoon of cod-liver oil.

Vitamin D is used in the treatment of ricket. These are mostly administered parentally by intramuscular injection. Calcium should also be prescribed along with vitamin D.

Vitamin D is also given to the patients with Chronic Jaundice to prevent vitamin D deficiency.

It is also used along with oral calcium tablets for the treatment of hypoparathyrodism. Sometimes it is also used along with anti-convulsants.

Vitamin E

Vitamin E is best known as antisterility factor which is important for the fertility of the male and the birth process of the female.

Vitamin E is a group of naturally occurring compounds which have closely related structure. These compounds are collectively known as tocopherols, which are characterised as α, β, γ, σ tocol or chroman which is considered to be a parent compound of all the tocopherols e.g. and tocopherol is 5 : 7 : 8 trimethyl tocol, β is 5 : 8 dimethyl tocol, γ is 7 : 8 dimethyl tocol etc.

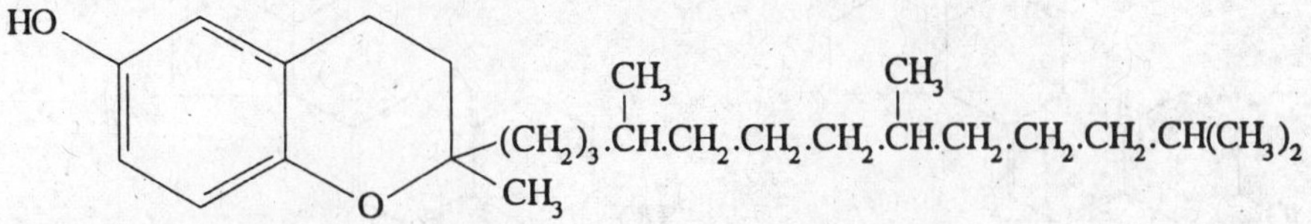

HO

CH_3 CH_3

$(CH_2)_3.CH.CH_2.CH_2.CH_2.CH.CH_2.CH_2.CH_2.CH(CH_3)_2$

O CH_3

Tocol

These are also used in industry as antioxidants.

Sources : Tocopherols are generally found in the free state mostly in the seed oils, wheat germ oil, cotton seed oil. The highest amount of this vitamin is found in livers of horse and cattle. It is also found in soyabean oil, which is rich in *d*-tocopherol, palm oil and rice.

Deficiency diseases : Vitamin E deficiency causes antisterility in guineapigs. It gives rise to dystrophy (abnormal growth or deposition of tissue) of cardia and skeletal muscle.

It causes increase in the number of lencocytes and also increases the concentration of RNA and DNA in the bone marrow.

Uses : Tocopherol acetate is an important drug for treating deficiencies due to vitamin E. This vitamin is used in a variety of disorders such as sterility, myopathy etc. Sometimes it is useful in treating megaloblastic anaemia especially when this anaemia does not respond to vitamin B_{12} and folic acid treatment.

It is also used in the treatment of cystic fibrosis, a disease where there is abnormal amount of secretion in the pancreas and later involves the lungs which results in repeated lung infection and fatty stools.

Vitamin K

There are two important vitamins in this group namely K_1 and K_2. Vitamin K_1 generally occurs as phytomenadione which is 1,4-naphthaquinone derivative. It is a deep yellow, viscous oil. It is sparingly soluble in water but dissolves in fixed oil. It is quite stable in air but on exposure to light it gets decomposed.

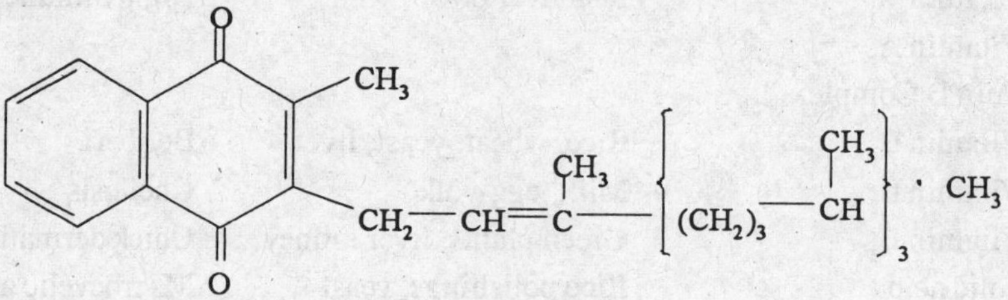

Vitamin K_1

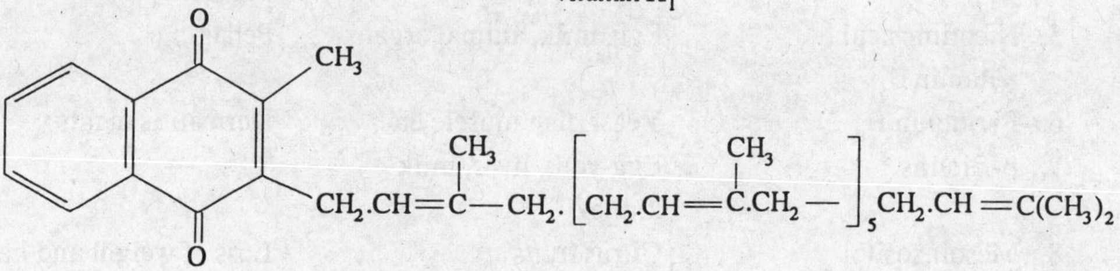

Vitamin K_2

Vitamin K_2 also contains 1,4-napthaquinone, nucleus.

Vitamin K has some important synthetic analogues e.g. Menadione which is 2-methyl-1,4-naphthaquinone, phthiocol which 3 hydroxy-2 methyl 1 : 4 naphthaquinone. These have antihaemorrhagic properties.

Source : Vitamin K is found in all green leafy vegetables, carrot tops, cauliflower, spinach etc. Vitamin K_2 is found in fish, meat, egg, yolk etc.

Deficiency disease : The vitamin K is essential for coagulation of blood, its increases the time of blood clotting and these are also known as antihaemorrhagic vitamins.

Vitamin K helps in blood coagulation by activating the prothrombin for the formation of blood clotting enzyme.

Uses : Vitamin K is a haemotological agent which is used as a blood coagulant. It is also used for the treatment of anticoagulant toxicity. This vitamin is also used in hypoprothrombinemia (deficiency of prothrombosis in the blood).

Vitamin K is administered as tablet of 5 mg potency and as ampoules containing 10 or 50 mg per ml of Vitamin K_1 oxide.

Vitamin P

It is a crude extract containing eriodictyol-*l*-rhamnoside and hesperitin-*l*-rhamnosido-*d*-glucose along with other physiologically active compounds.

Its deficiency causes decrease in capillary resistance and an increase in vascular permeability and thus results in haemorrhagic condition of the skin.

Eriodictyol

S. No.	*Name of the Vitamin*	*Source*	*Deficiency Diseases*
I.	1. Vitamin A_1	Fish liver oil	Night blindness
	2. Vitamin A_2		
II.	Vitamin B Complex		
	1. Vitamin B_1	Rice, wheat, yeast, liver	Beriberi
	2. Vitamin B_2	Milk, egg-yolk	Cheilosis
	3. Vitamin B_3	Green plants, liver kidney	Chick dermatitis
	4. Folic acid	Rice polishings, yeast	Macrocyclic anaemia deficiency of RBC
	5. Nicotinic acid Vitamin B_5	Egg, milk, animal organs	Pellagra
	6. Pyridoxin B_6	Yéast, fish muscle etc.	Dermatitis in rats
	7. β-Biotins	Egg-yolk, liver, milk kidney	—
	8. Mesoinsositol	Citrus fruits	Loss of weight and hair in rats
	9. Vitamin B_{12}	Liver of fish, pig	Pernicious anaemia
III.	Vitamin C	Citrus fruits, green vegetables	Scurvy
IV.	Vitamin D	Fish liver oil	Rickets
V.	Vitamin E	Wheat germ oil, Cottonseed oil	Sterility
VI.	Vitamin K	Cereals, leafy vegetables	Haemorrhage
VII.	Vitamin P	Grapes, oranges etc.	Haemorrhage

Estimation of Thiamine (Vitamin B_1)

1. Thiochrome method

(*a*) **Principle.** When thiamine is treated with a mild oxidising agent like Potassium fericyanide in alkaline medium. It is converted to thiochrome which gives a blue fluorescence in ultraviolet light. Under standard conditions the fluorescence is proportional to the thiochrome present and hence to thiamine in original solution.

(*b*) **Preparation of standard thiamine hydrochloride.** Weigh accurately 25 mg of thiamine hydrochloride. Make up the volume to 1000 ml using dilute alcohol. 10 ml of this solution is diluted to 50 ml with potassium chloride. This is the standard solution. Add 5 ml of standard solution into a tube and add rapidly 3 ml of potassium ferricyanide (1%) within 30 seconds. Add 20 ml of isobutyl alcohol, mix well for 90 seconds. Prepare a blank in another tube instead of the assay solution (vitamin to be estimated) in the same manner (with and without the oxidising agent). To each of the tubes, add 2 ml of dehydrated alcohol. Swirl and measure the fluorescence using a fluorophotometer. The number of mg of thiamine hydrochloride in 5 ml of the assay is given by the formula $(A\text{-}b)/(S\text{-}d)$, where A is the average of the flurophotometer reading with assay preparation without containing oxidising agent, "*b*" is the reading with assay preparation without the oxidising agent. "*S*" is the reading with standard preparation with oxidant and "*d*" is the reading of the standard without the oxidant.

Tablets of vitamin B is estimated by adopting the following procedure.

Powder about 20 tablets, weigh the powder and transfer to a 100 ml conical flask. Add 5 ml of dil. Hydrochloric acid plus 10 ml of water and heat the above to boiling and cool add 1N NaOH to neutralise the acid. Add water to make the volume to 100 ml. Transfer 5 ml of the solution to 50 ml flask and add 10 ml of 50% ethanol. Add 5 ml of diazotate reagent (10 ml of 0.06% *p*-amino acetophenone + 0.2 N HCL at 5° plus 3 ml of 0.2% $NaNO_2$ plus 3 ml 2N NaOH). Place the above solution in a water bath at 20°C, dilute to 50 ml with isopropyl alcohol. Determine, the extinction of the above solution at 520 mm in a photoelectric colorimeter. Perform a blank. Subtract the blank reading. Draw a calibration curve E vs conc of vitamin. Evaluate amount of vitamin by interpolation.

Estimation of Vitamin C

Among the various methods available for the estimation of vitamin C, Titration with a dye called 2.6 dichlorophenol-indophenol is the most widely used one.

Principle : This method is based upon the reduction of the dye by an acid solution of ascorbic acid. The capacity of an extract of the sample to reduce a standard solution of the dye, as determined by titration is directly proportional to the ascorbic acid.

Dye (pink) + Ascorbic Acid → Dye (colourless) + Dehydroascorbic Acid

(*a*) ***Preparation of 2,6 dichloriphenol-indophenol.*** It is obtained in the tablet form. One tablet is crushed in about 40.0 ml of hot water and diluted to 50.0 ml.

(*b*) ***Preparation of standard solution.*** It is prepared by accurately weighing 20 mg of ascorbic acid and making up the solution to 1 litre.

Pipette 5 ml of dilute fruit juice containing about 0.1 mg of vitamin C into a boiling tube, add 1ml of glacial acetic acid, titrate with the dye solution to a faint permanent pink colour. Record the titre value as "*T*". Repeat titration with 5 ml of water for the blank (B1) and add 5 ml of ascorbic acid standard solution (st). Calculate the vitamin C content of the test sample.

Therefore 'x' ml of dye = 0.5 mg of ascorbic acid. In case of 'z' ml of dye is required to titrate 5 ml of unknown lemon extract provided

$$\therefore \text{5 ml of lemon extract} = \frac{0.5 \times 'Z'}{x}$$

$$= \text{1 mg ascorbic acid 15 ml}$$

$$= \text{1/5 mg ascorbic acid/ml of the extract}$$

Notes: (*a*) The stabilising medium metal phosphoric acid—acetic acid mixture is prepared by dissolving 15 mg of metaphosphoric acid in a mixture containing 40 ml of glacial acetic acid and 450 ml of water. The mixture is filtered and stored in refrigerator. It should not be stored for more than ten days.

(*b*) Standard 2, 6 dichlorophenol indophenol solution is prepared by dissolving 42 mg of sodiumbicarbonate and 52 mg of sodium 2, 6 dichlorophenol indophenol in 50 ml of water. The mixute is diluted to 200 ml, filtered and kept in refrigerator. It can be used for three days.

(*c*) *Standard ascorbic acid.* It is prepared by dissolving 100 mg of crystalline ascorbic acid in 100 ml of metaphosphoric-acetic acid. 10 ml of this solution is taken in a 100 ml standard flask and the solution is made up to the mark with metaphosphoric acetic acid (0.1 mg/ml).

Followed by estimation of vitamins as given below.

Estimation of Vitamin A

Many colour reactions have been proposed for estimation of vitamin A. The most important method is that of Carr and Price which is as follows:

Principle. Vitamin A forms coloured compounds with certain agents for example when a solution of antimony trichloride in chloroform is added to the solution of vitamin A it gives a blue colour which shows the maximum intensity after 10 seconds, but then immediately starts fading. The intensity of the colour is measured in a colorimeter. The absorbance of this blue colour at 620 nm is a function of the concentration of vitamin A.

Procedure. Evaporate 5.0 ml of vitamin A solution, dissolve the residue in chloroform. Add a saturated solution of antimony trichloride in chloride known as Carr-Price reagent. Measure the extinction of the blue solution using a photo electric colorimeter (Care should be taken to measure the intensity within a few seconds, as the colour fades immediately).

Repeat this procedure with aliquots of vitamin A in chloroform of known potency and prepare a calibration curve. Determine the potency of the sample from the graph.

Vitamin A is also estimated by another method known as Budowski-Bondi method. Here the alcohol (vitamin A) is converted to anhydrovitamin A in benzene solution using *p*-toluene sulphonic acid as catalyst. The increase in absorption at 399 nm, which results from the dehydration is directly proportional to the amount of vitamin A present. Absorbance is measured in a UV visible spectrophotometer.

Chapter 13

Cancer and Antineoplastic Drugs

Cancer

The cancer is a disease which can affect any individual at any age and it can develop in any organ of the body. The cancer is actually an abnormal development that changes normal cells into cancerous cells. The most obvious feature of many cancers is the development of a new growth, a nodule or a tumour in the tissues of their origin. This tumour develops due to unusual malignant growth or enlargement of a tissue due to uncontrolled repeated division of some cells. Tumours can develop anywhere on the surface or inside the body. All tumours are not malignant in nature. Generally there are two types of tumours.

(*i*) A non-malignant or benign tumour

This type of tumours are more common than malignant tumours. A benign tumour grows slowly and may become quite large but it remains restricted to the place of its origin and does not spread to the other areas of the body. These tumours are non-malignant *i.e.* they do not develop into a cancer.

(*ii*) A malignant or cancerous tumour

It may begin as a small growth and may grow slowly but later on it grows rapidly. The most important feature is that it starts spreading to the neighbouring tissues and gets a grip like a crab. The cancerous cells do not obey the cell restraining mechanism and keep on multiplying un-controllably. Later on these cells of the tumour break away from it and migrate through the blood stream to the different parts of the body and they get accumulated as secondary growth. This stage is fatal.

Some common causes of Cancer

The exact cause of the cancer is not known but there are several factors which cause cancer and these are known as cancer producing or carcinogens, a few of them are as follows:

1. Constant physical irritation of any tissue of the body can cause cancer. For example, heavy smoking causes constant irritation of the lung tissues which may produce cancer. Similarly, chewing pan or tobacco causes continuous irritation in the mouth which may cause mouth cancer.
2. It has been found out that any type of injury continued over a long time often initiate cancer, for instance, the jagged edge of a broken tooth rubbing against the tongue can cause cancer.
3. Ionizing radiations, X-rays, Ultra-violet rays are carcinogenic in nature. Even contact with intense sunlight results in cancer of the skin.
4. Some aromatic amines used in the manufacture of synthetic dyes cause cancer of urinary bladder.
5. There are some types of cancer particularly found in India due to habits and some customs. For instance, in Andhra Pradesh a particular tribal people are in habit of smoking a local cigar with

the burning end inside the mouth. Some of these people develop an ulcer on the palate which often turns into cancer after some years.

Spread of Cancer

Cancer is a dreadful disease and once it is developed it spreads in different ways.

1. The cancer cells multiply at the site of their origin and form a tumour. These cells move to surrounding tissue and get fixed to them and make these cells malignant.
2. The cancer cells which multiply rapidly are loosely attached to one another and some of them are detached and are carried by the lymph stream to the different organs. Wherever these cells are deposited they continue to grow independently and also invade the adjacent tissues. This development is known as secondary growth and their eradication is extremely difficult.
3. Some cancer cells spread through the blood stream and they get scattered much rapidly. The tumours of the connective tissue and bone marrow etc. show this type of tendency.

Treatment of Cancer

The following methods are generally used for the treatment of cancer :

1. **Surgery :** In this method the cancerous cells are removed by surgery. This method is not reliable because if any cancerous cells are left off or gets displaced during the operation then they form secondary growth.
2. **By Radiation :** In this method malignant tumours are exposed to radiation like X-ray, radium, gamma rays etc. to kill the cancerous cells.
3. **Chemotherapy :** Treatment of cancer by drugs is known as chemotherapy. In spite of the tremendous research work during past two decades, the chemotherapy of malignant disease still remains unfruitful as compared to the chemotherapy of other microbial infections.

The drugs which are used in the treatment of cancer are called Antineoplastic agents. These drugs probably destroy the malignant cells (growing tissue) and these drugs should be in such doses as to destroy all the malignant cells. Sometimes this treatment is successful initially but there is always possibility of recurrence.

The antineoplastic drugs could be classified according to their mode of action as follows.

1. Alkylating or Cytotoxic Agents

These drugs act by transferring an alkyl radical to a suitable receptor site and the alkylating agents forms highly reactive quaternary ammonium derivatives which react with the functional groups of the physiologically important molecules in the cells and then these cells are not available for normal metabolic reaction. Alkylating agents are nucleophlic in nature hence they react with nuclei acid bases and inhibit DNA synthesis. They also interfere with cell replication by, cross-linkage of DNA strands. Thus these drugs help in arresting cell division.

Some important examples of alkylating agents are Mechlorethamine, Mustard drugs, Chlorambucil, Triethylenemelamine (TEM), Busulfan etc.

(*i*) **Chloram-Bucil (leukeran).** It is a drug containing nitrogen mustard group [$(ClCH_2CH_2)_2N$] and has the following structure.

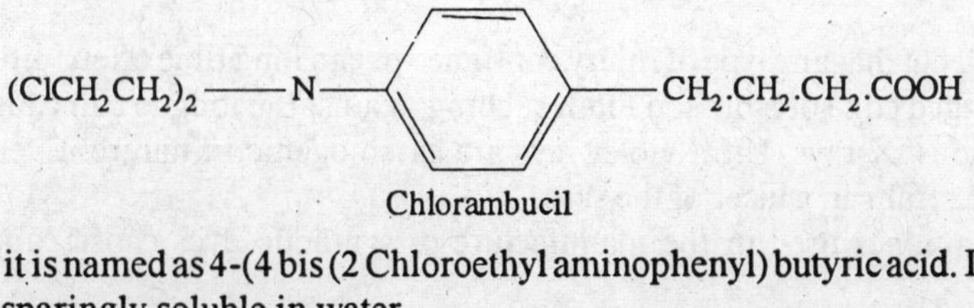

$$(ClCH_2CH_2)_2 — N — C_6H_4 — CH_2.CH_2.CH_2.COOH$$

Chlorambucil

Chemically it is named as 4-(4 bis (2 Chloroethyl aminophenyl) butyric acid. It is a white crystalline powder and is sparingly soluble in water.

It is effective against malignant cells of lymphoid series. It is administered orally and is available as 2 mg tablets. This drug is used in the treatment of chronic lymphocytic leukaemia. It is also useful in carcinoma of breast, ovary and testis. It is used in autoimmune disorders.

Adverse effects : Gastro intestinal irritation, bone marrow depression with leukopenia.

(*ii*) **Mustine :** Chemically, mustine is N, N-bis (2-chloroethyl)-methyl ammonium chloride.

It is a white crystalline hygroscopic powder. It should be stored in well closed containers. It is highly

$$(ClCH_2CH_2)_2\overset{+}{N}H-CH_3\ Cl^-$$

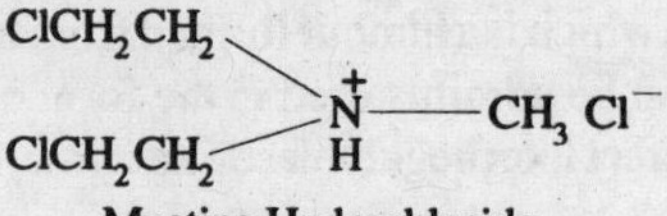

Mustine Hydrochloride

soluble in water. Hodgino—a progressive malignant condition of reticular endoth elial cell.

It is usually administered as intravenous injection. It is used in Hodgkin's disease alongwith vinca alkaloids. It is also useful in lymphomas and in carcinoma of the lung and other solid tumours.

Adverse effect : Nausea and vomiting.

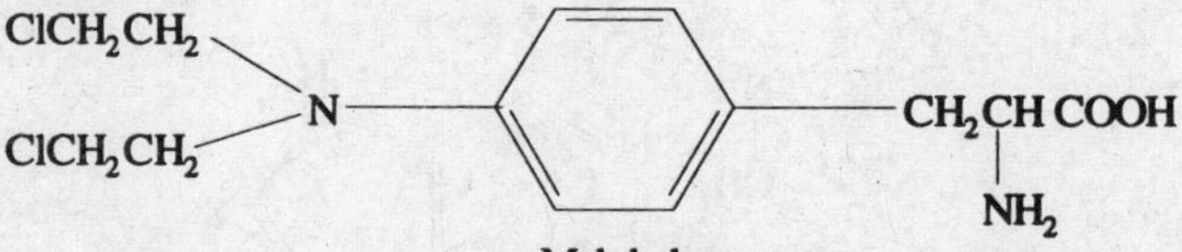

$$(ClCH_2CH_2)_2N-C_6H_4-CH_2CH(NH_2)COOH$$

Melphalan

(*iii*) **Melphalan :** Chemically it is 4-bis (2-chloroethyl) amino-L-phenylalanine.

It is odourless, white powder and is sparingly soluble in water. It should be stored in cool, dark place.

It is used in the treatment of multiple myeloma, carcinoma of the breast and ovary. It is also used in autoimmune disorders.

2. Antimetabolites

Chemical substances which take part in cellular metabolic reactions are called as metabolites and an antimetabolite is a chemical agent which by virtue of its close structural similarity to the metabolite, blocks its action. The antimetabolites check the production of carcinogenic compound produced by the abnormal metabolism of some organic compounds. These antimetabolites act either by preventing the combination of the metabolite with its specific enzyme or it may combine itself with the specific enzyme and thus transform these compounds to an inactive one.

The antimetabolites which are useful carcinogens are structural analogues of substance essential for the synthesis of the nucleic acid bases e.g. folic acid or substances which are incorporated into the nucleic acid chains.

The examples of antimetabolites are :

(*a*) Folic acid antagonists e.g. Methotrexate etc.

(*b*) Purine antogonists e.g. 6-mercapto-purine, Azathioprine

(*c*) Pyrimidine antagonists e.g. Fluoro-uracil, cytosine-arabinose etc.

(*i*) **Methotrexate.** It is a yellowish orange crystalline powder and is sparingly soluble in water. It is soluble in both acids and alkali hydroxides. Its solution is sterlised by filtration and it should be stored in airtight, light resistant containers.

It is an antimetabolite of folic acid and has the adjoining structures.

It is pteridine derivative and its chemical name is N-(4 (2,4 diamino pteridin-6-ylmethyl) methyl amino) Benzoyl) L-glutamic acid.

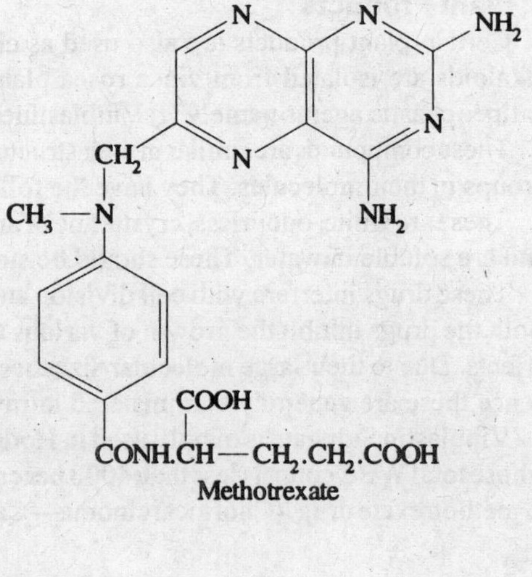

N N NH_2 N N CH_2 CH_3-N NH_2 COOH CONH.CH—CH_2.CH_2.COOH

Methotrexate

Folic acid is essential for the production of the co-enzyme tetrahydrofolic acid (THF). Methotrexate drug competes with folic acid for this co-enzymes and restricts the production of THF. The lack of THF leads to inhibition of DNA synthesis and thus inhibits cell replication and eventually the cells die.

This drug acts on the bone marrow. It is used in acute lymphatic leukaemia especially in children. It is also useful in choriocarcinoma which is a tumour formed from malignant proliferation of epithelium of the charionic villi. This drug can be administered in the form of injections or tablets.

This drug has several adverse effect like megaloblastic anaemia, intestinal ulceration, diarrhoea, liver damage etc.

(*ii*) **Azathioprine.** It is mercaptopurine derivative. Chemically it is named as 6-(1-methyl-4-nitroimidazol-5ylthio)-purine.

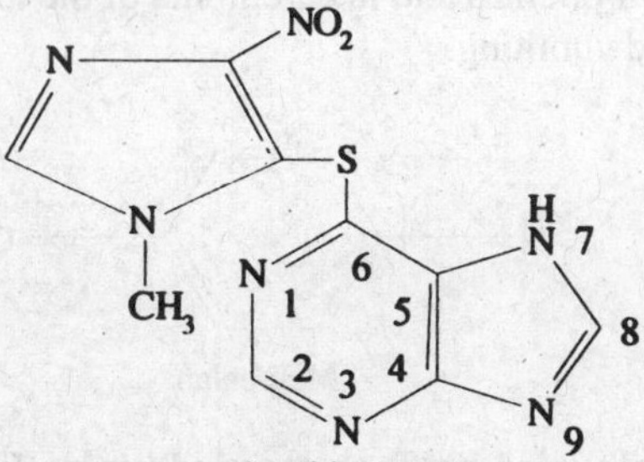

Ayathioprene

It is a pale yellow powder. It is sparingly soluble in water. It is stored in light resistant containers. It is either given orally or by injection as azathioprine sodium. It is used mainly in immunosuppressant to facilitate the survival organ and tissue transplants. It is also useful in rheumatoid arthritis, renal disorders, skin disorders etc.

(*iii*) **Fluorouracil.** Chemically it is named as 5-fluoropyrimidine 2,4-dione. It is an odourless white crystalline powder. It is sparingly soluble in water. It is stored in light resistant, airtight container.

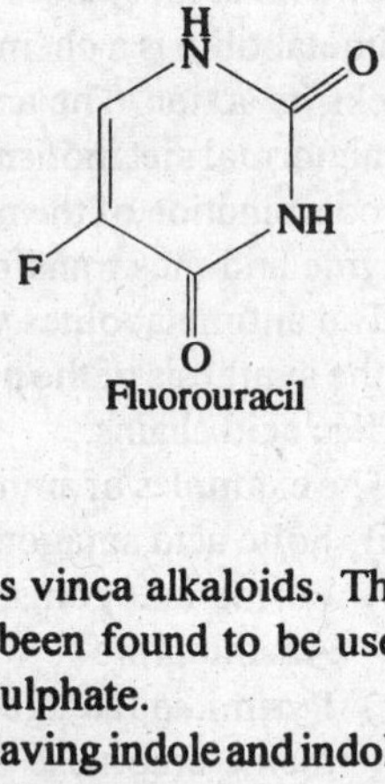

It acts as an antimetabolite of uracil. It is used in inoperable malignant neoplasms, especially those of gastrointestinal tract, breast, liver. It is used topically for treating solar or actinic keratoses and other tumours of the skin.

Fluorouracil

Adverse effect: Nausea, gastrointestinal ulcers, bone marrow, depression.

3. Plant Products

Certain plant products are also used as chemotherapeutic agents such as vinca alkaloids. These alkaloids are isolated from vinca rosea plants. Two vinca alkaloids have been found to be useful antineophastic agents namely (*i*) Vinblastine Sulphate and (*ii*) Vincristine Sulphate.

These compounds are similar in their structure and are dimeric compounds having indole and indoline groups in their molecules. They have the following structure. (See next page.)

These are white, odourless, crystalline or amorphous powders. These are very hygroscopic in nature and are soluble in water. These should be stored in air-tight containers and in cool places.

These drugs interfere with cell division and inhibits miotic spindle formation during cell division. Both the drugs inhibit the growth of various tumours. These drugs also produce certain neurological effects. Due to their large molecular size these drugs are not well absorbed when administered orally hence these are generally administered intravenously.

Vinblastin Sulphate is mainly used in Hodgkin's disease. This drug should not be given to a patient whose total WBC count is less than 4000 per cmm. It is also used in choriocarcinoma which are resistant to methotrexate drug. (Choriocarcinoma—Cancer of outer membrane of the embryo).

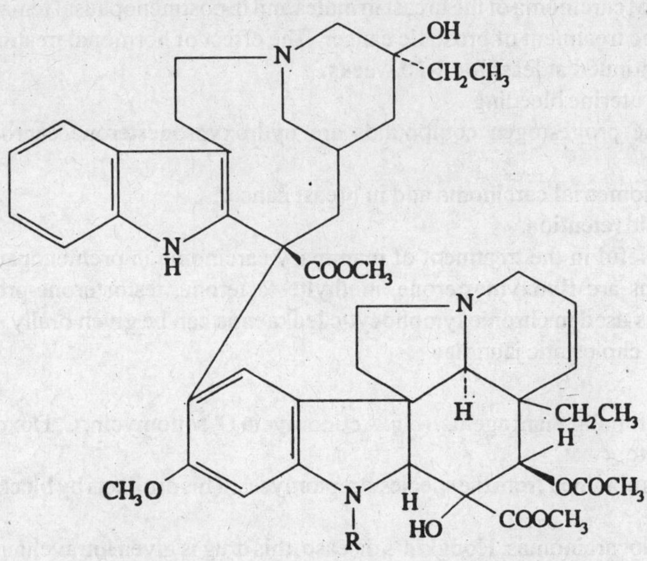

R = CH_3 — Vinblastine
R = CHO — Vincristine

Vincristine is used in the treatment of acute lymphatic leukaemia in children.

Some common adverse reaction of these drugs are vomiting, mental depression, peripheral neuropathy etc.

4. Hormones

Hormone therapy is different from other cytotoxic drugs and are useful neoplastic agents. These drugs do not kill the malignant cells and are slower in their effect. They act by one of the severâl ways.

1. They inhibit the cell growth.
2. They act by stimulating differentiation of cells which responds to the body's controlling and inhibitory influence.
3. They inhibit the circulation of the substance which stimulates the malignant cells.

The hormone therapy is mainly used in the treatment ofbreast and prostate cancer. Some important examples of hormones used in malignancies are Estrogens, Progestins, Adreno-Corticosteroids, Thyroxine etc.

Estrogens : These are mainly used in breast cancer in post menopausal women. The effect of hormonal therapy is usually slow and the treatment should be continued for atleast 8-10 weeks to get some positive result. These drugs are also used in the treatment of prostate cancer.

Some synthetic estrogens are also used as drugs which are as effective as the natural one but are much cheaper in cost—e.g. diethylstillbestrol.

5. Adrenocorticosteroids

(*i*) An example of adrenocorticosteroids is Hydrocortisone, which is a glucocortico.

These drugs are used in acute lymphatic leukaemia in children. It prevents anaemia in disseminated cancer patients. They are used in critically ill cancer patients alongwith cytotoxic drugs to produce relief and a sense of well being. They are also useful in controlling hypercalcenia caused by certain tumours.

(*ii*) Another Aderenocorticosteroid drug i.e. Prednisone. It is given in doses of 60-100 mg daily initially and then reducing to 20-40 mg daily.

Adverse effect : hypertension, fluid retention etc.

(*iii*) **Estrogens :** Some of the estrogenic compounds are diethyl stillbestrol, dimenstrol, oestradiol.

These are useful in the treatment of carcinoma of the breast in males and in postmenopausal females. The diethylstillboestrol is used in the treatment of prostatic cancer. The effect of hormonal treatment is usually slow and it should be continued at least for 8-10 weeks.

Adverse effect : Fluid retention, uterine bleeding.

(*iv*) **Progestins :** Some of the progestogen compounds are hydroxyprogesterone caproate, Medroxyprogesterone.

These are used in metastatic endometrial carcinoma and in breast cancer.

Adverse effect : Occassional fluid retention.

(*v*) **Androgens :** These are useful in the treatment of mammary carcinoma in premenopausal women. Some important androgens are fluoxymesterone, methyltestosterone, testosterone-propionate, generally Fluoxymesterone is used in chronic lymphocytic leukaemia can be given orally.

Adverse effect : Fluid retention, cholestatic jaundice.

5. Antibiotics

Certain antiobiotics are useful antimaliganant agents. (*e.g.* Actinomycin D, Mitomycin. C, Doxorubicin, Mithramycin, Rubidomycin etc.

Actinomycin D. This antibiotic is obtained from the species streptomyces. This drug acts by blocking DNA dependent RNA synthesis.

It is used in Wilm's tumour, choriocarcinomas, Hodgkin's disease, this drug is given intravenously.

6. Radio active isotopes

These are the isotopes of various elements which give radiation. These radiations cause destruction of the cells. The therapeutically useful radioisotopes are Radiogold ^{198}Au, Radiophosphorus ^{32}p, radio iodine ^{131}I.

Radioiodine (^{131}I). This is useful in certain cases of thyroid cancer. Radioiodine acts by giving beta radiation. In some cases surgery followed by radioiodine therapy causes excellent results. Radioactive iodine ^{131}I is available as a solution of Na^{131}I and can be administered orally in patients who undergo this therapy should dispose their urine carefully to avoid spreading of harmful radiation.

7. Miscellaneous Agents

Hydroxy-Urea and urethane are the important examples of miscellaneous agents.

Hydroxy Urea : It is a white odourless crystalline powder and is hygroscopic in nature. It gets decomposed in presence of moisture hence it is stored in a well closed container. It has the following structure

$$HONHCONH_2$$

Hydroxy Urea

It is used in tumours of the ovary and in chronic myeloid leukaemia. It is usually administered orally.

Adverse effect : bone marrow/depression.

(*ii*) **Urethane :** It is ethyl carbamate

$$H_2NCOOCH_2CH_3$$

It finds use in the treatment of chronic myeloid leukaemia. Myeloid leukaemia is a malignant disease in which there is a excessive production of leucocytes in the bone marrers.

Chapter 14

Diabetes and Hypoglycemic Drugs

Diabetes is a disease of metabolism due to deficiency of insulin. Blood sugar level is maintained constant at a value of 70-120 mg of glucose/100 ml. Though several hormones are involved in the maintenance of diabetes, the most important ones are insulin and glucagon. Diabetes is caused as a result of loss of the balanced effect of these hormones, usually due to less insulin production, sugar starts accumulating in the blood and blood sugar level increases and sugar passes in urine along with other minerals. High blood sugar level is known as hyperglycaemia and presence of sugar in the urine is known as glycosuria. Insulin is secreted by the beta cells of islets of Langerhans of pancreas. Insulin is necessary for the burning of sugar in the tissues with the help of oxygen.

Diabetes is of two types: (1) Diabetes insipidus (2) Diabetes mellitus.

(1) **Diabetes insipidus :** This diabetes arises due to lesser secretion of vasopressin (an antidiurtic hormone) by posterior pituitary gland. Symptoms of this type of diabetes is that patient passes larger quantity of urine, mainly water and there is excessive thirst.

This type of diabetes can be cured by injecting pituitrin secreted by posterior pituitary.

(2) **Diabetes mellitus :** This type of diabetes arises by the deficiency of insulin and the blood sugar rises more than normal. This disturbs metabolism of protein and other factors in body. Diabetes mellitus are of two types :

(*a*) Juvenile (*b*) Adult type

Juvenile diabetes mellitus is hereditary and is treated by giving insulin.

Adult type occurs in elderly people and is treated by controlling the diet and oral antidiabetic drugs should be given.

The main symptoms of these type of diabetes are, increased thirst, increased urinary output and ketonaemia and ketouria, i.e. presence of ketone bodies in the blood and urine. Diabetes should be suspected when any of the following symptoms are present. If there is any history of diabetes in one's family then an individual should check for diabetes.

Diagnostic test for diabetes in urine and blood has been discussed in Chapter 4 in detail. Diabetes can be detected by the glucose tolerance test, which divides diabetes according to the sugar level of the patient as normal, mild, moderately severe or severe diabetes (blood sugar values are discussed in Chapter 4 under GTT test).

If diabetes is properly controlled a diabetic patient can live a normal long life.

Control of Diabetes

Diabetes can be controlled by stabilising the body metabolism, It can be controlled by the following ways.

(1) Adjustment of diet to relieve stress on the pancreas.
(2) Insulin should be given to bring the raised blood sugar to near normal level.
(3) Intake of oral antidiabetic drugs to stimulate the pancreas to produce more insulin.
(4) Regular exercise.

Control of diabetes should be continued for years. The diet control is of immense importance in controlling diabetes and should be maintained life long. Diabetic patients should avoid :

(*a*) Concentrated sweets.

(*b*) Substances like potatoes, sweet etc., which contains a large proportion of carbohydrates.

(*c*) Cereals which contain relatively greater percentage of carbohydrates *e.g.* rice, wheat, bajara etc., contain 70 to 80% of carbohydrates. Pulses contain 56 to 60% of carbohydrates.

(*d*) Fruits and vegetables should be eaten according to their sugar content e.g. leafy vegetable contains 3 to 5% carbohydrates, vegetables 6 to 15%, roots and tubers 15 to 25%, sweet fruits like banana contains 35% whereas water melon content of sugar is low i.e. it is about 5%.

(*e*) A lot of vegetable with low carbohydrate content should be taken.

(*f*) Meat, fish etc. do not contain sugar and can be taken in usual quantities. A vegetarian should take two glasses of milk every day.

If the diet of the diabetic is not properly controlled, insulin or oral hypoglycemic drugs will not control the disease properly. A diebetic person should control his or her bodyweight and should avoid overeating.

Regular exercise should be included in daily routine as the muscles live on glucose and exercise helps them to utilise it.

Diabetic patients are more vulnerable to infectious skin diseases like boil, carbuncles etc. These infections must be attended immediately and proper care should be taken.

The drugs which lower the blood sugar and can treat the symptoms of diabetes mellitus are known as hypoglycaemic drugs. These drugs could be categorised as (*i*) insulin and insulin preparation, which are employed only parenterally and (*ii*) oral hypoglycaemic drug which can be administered orally.

1. Insulin : Insulin was first isolated by Banting and Best in 1922 from pancreatic glands of cattle. Later on, it was synthesized by Sanger in the laboratory.

Chemical structure of insulin

Human insulin is a polypeptide, having a molecular weight of about 6000. It consists of two amino-acid chains A and B, which are linked by two disulphide (-s-s-) linkage. The chain A contains 21 amino acids and B contains 30 amino acid, all in a known sequence, hence a polypeptide chain in having 51 amino acids. The disulphide bridges are essential for its biological activity. Beef and pig insulin are structurally similar to human insulin in all respects, except in a few amino acid residues. Normal human pancreas contains about 8-10 mg of insulin.

It is an amphoteric protein and forms salts with weak acids and alkalis. It is soluble in water. It can combine with proteins such as protamine and with zinc without any change in its biological activity.

Insulin is inactivated by digestive enzymes, hence it is not suitable for oral administration. The free amino and aliphatic hydroxyl groups of insulin are not required for biological potency. The solubility of insulin depends on its physical state (*i.e.* amorphous crystalline) on the concentration of zinc and on the nature of buffer in which it is being suspended.

Insulin is prepared from the pancreas of ox and pig. Now-a-days it is commercially prepared by new technology known as recombinant DNA technology.

Preparations and dosage. Plain insulin is commonly injected about half an hour before each major meal. The intermediate and long acting preparations are injected once a day before the breakfast. Different insulin preparations are available which are as follows :

(1) ***Insulin injection*** is a acidic solution of crystalline insulin having potency of 20, 40 or 80 units per ml. It is a clear liquid having a pH of 3.2.

(2) ***Isophane or NPH insuline.*** It is a modified protamine zinc insulin and is a cloudy suspension having 7.2 pH. It is an intermediate acting insulin.

(3) ***Insulin zinc suspension*** (amorphous, semi lente). It is an insulin zinc suspension and is not a clear solution. Its pH is 7.2 and it is a start acting insulin.

(4) ***Insulin Zinc suspension I.P.*** (lente). It is a mixture of 3 volumes of insulin zinc suspension,

amorphous and 7 volumes of crystalline insulin zinc suspension. It is an intermediate acting insulin. Its pH is 7.2.

Insulin mixtures are unstable except lente and must be prepared before use. Addition of zinc prolongs the duration of action of insulin. Now-a-days, some new insulin preparations are in use. These are purer than the older preparation and are neutral in nature. Hence it is possible to mix neutral regular insulin with any of the modified insulins in any proportion. These new preparations are also more stable. Some important preparations are :

1. NUSO—It is bovine insuline in a clear neutral solution.
2. Actrapid—It is also a clear neutral solution of pH 7.0 of recrystalline porcine insulin.
3. Rapitard—It is a cloudy mixture of 1 : 3 actrapid and bovine insulin buffered to pH 7.0. Its onset of action is quick and smooth.

The insulin injection should state the number of units per ml and the source of the insulin.

Uses : 1. It is used as hypoglycomic drug for the treatment of diabetes mellitus.

2. Insulin is also used in Schizophrenia.

3. Glucose—insulin drip is used for the treatment of hyperkalaemia (*i.e.* excess of potassium in the blood).

4. Insulin tolerance test with determination of plasma GH is used as a diagnostic test for hypopituitarism.

5. NPH or lente insulin is used for the treatment of Juvenile diabetes.

6. For routine control of diabetes an intermediate acting preparations is commonly used, *e.g.* NPH.

7. Glucose-insulin-potassium drip is used to treat patient with myocardial infarction.

Oral Hypoglycaemic Agents

Insulins are ineffective through oral administration and hence oral hypoglycaemic drugs were introduced which could control diabetic when administered orally. An ideal antidiabetic drug should be effective by mouth, nontoxic and correct the basic metabolic defects in diabetes. Although the drugs available now do not meet all the requirements, but the most important drugs belongs to two groups namely :

(1) Sulphonylureas

(2) Biguanides

The oral antidiabetic drugs activate the islets of langarhans to secrete insulin and inhibit protein to be converted to glucose.

1. Sulphonyl Ureas. These drugs are chemically related to Sulphonamide and have a basic structure as :

$$R_1—SO_2NHCONH—R_2$$

(basic structure of Sulphonyl urea drugs).

Sulphonyl ureas are useful in treating diabetes which cannot be controlled by diet alone. These drugs can be used to treat mild diabetics during pregnancy. Sulphonyl ureas are quickly absorbed from the intestinal track.

These drugs are ineffective in juvenile diabetes and in presence of Ketosis and during severe stress. Some important drugs of this group are:

(*i*) Tolbutamide, (*ii*) Chlorpropamide, (*iii*) Glybenclamide, (*iv*) Tolazamide etc.

(*i*) **Tolbutamide :** It is a white crystalline powder and is sparingly soluble in water. Chemically it is 1-butyl-3-tosyl urea and has the following structure.

$$CH_3—C_6H_4—SO_2NHC(=O)NH(CH_2)_3CH_3$$

Tolbutamide

It is a substituted urea derivative ($NH_2CO\ NH_2$) in which the hydrogen atom of each nitrogen has been substituted by butyl group (—$CH_2CH_2CH_2CH_3$) and toluene-*p* sulphonyl group or tosyl group—

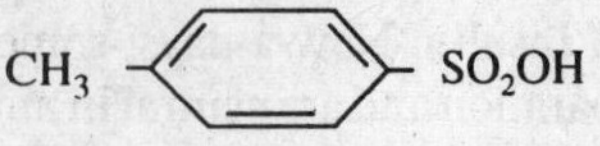

$CH_3-C_6H_4-SO_2OH$

It is the most commonly used drug and is relatively nontoxic. It is also available as aqueous solution (prepared by dissolving 1g in 10 ml of solvent) for travenous injections. Intravenous tolbutamide is used in the diagnosis of diabetes mellitus and it can also predict whether a newly diagnosed diabetic person will respond to sulphonyl urea therapy. Its duration of action is 6-12 hours.

Tolbutamide 0.5 to 1 g tablets are available. Some important commercial products available are Unitolbid, Diabetal, Artosin etc.

Adverse effect : skin rashes.

(*ii*) **Chlorpropamide**. It is a white crystalline powder and is sparingly soluble in water. This is also a substitued urea derivative and it has the following structure.

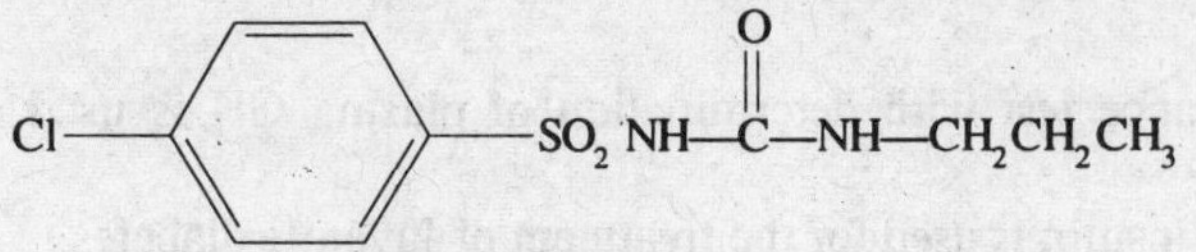

$Cl-C_6H_4-SO_2\ NH-C(=O)-NH-CH_2CH_2CH_3$

Chlorpropamide

Chemically it can be named as (1 (4 Chlorobenzene sulphonyl)-3-propyl urea.

Chlropropamide is effective in the treatment of mild or moderately severe diabetes and is also effective when tolbutamide therapy fails.

These drugs are also used in diabetes insipidus, chlorpropamide controls diabetes in a single daily dose and duration of action is about 60 hours.

Some of the products available are diabigon, diabinese, diamide etc.

Adverse effect. Excess dose may increase the risk of jaundice.

(*iii*) **Glibenclamide.** It is a white crystalline powder and chemically named as 4 (2 (5 chloro-2-methoxybenzamido) ethyl) benzene sulphonyl-3 cyclohexyl urea.

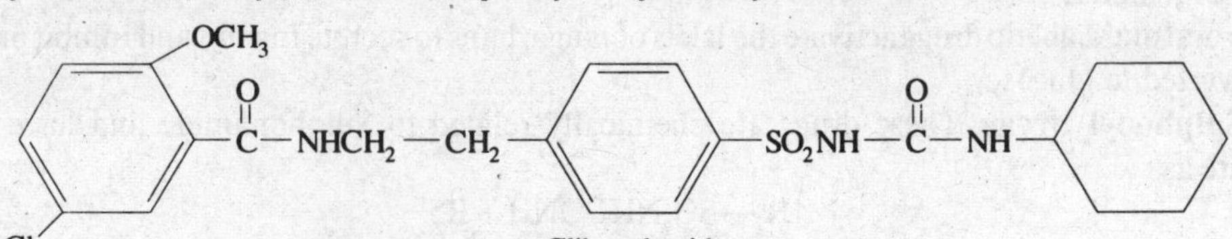

$(Cl)(OCH_3)C_6H_3-C(=O)-NHCH_2-CH_2-C_6H_4-SO_2NH-C(=O)-NH-C_6H_{11}$

Glibenclamide

Glibenclamide is effective in a single daily dose and is effective in patients where the diabetes cannot be controlled by maximum doses of chlorpropamide.

These are available as 5 mg tablets and its duration of action is for 12-15 hours.

2. Biguanides. When two guanide molecules are joined together through a common —NH— link it is known as biguanides and their general chemical structure is as follows:

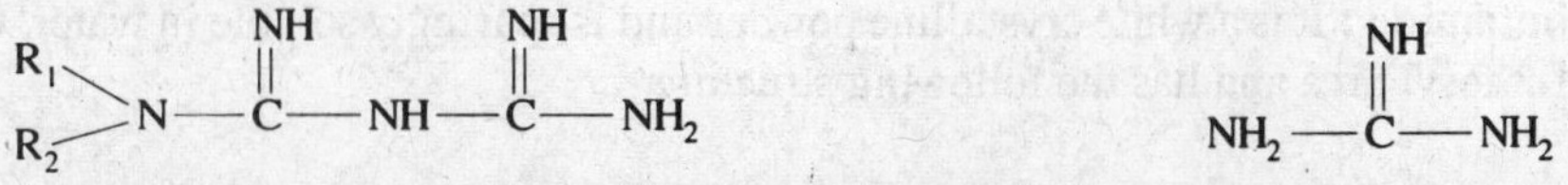

$R_1R_2N-C(=NH)-NH-C(=NH)-NH_2$ $\quad$ $NH_2-C(=NH)-NH_2$

General Structure of Biguanides

Biaguanides control all types of diabetes mellitus. These drugs reduce glucose absorption from the intestine. These drugs can be used to treat mild diabetes during pregnancy. Though these drugs are not used in treating juvenile diabetes, but it is used along with insulin to control juvenile diabetes.

Some important examples of biaguanides are Phenformin and Metformin.

Phenformin : It is available as 25 mg tablets and as 50 mg capsules. It has following structure.

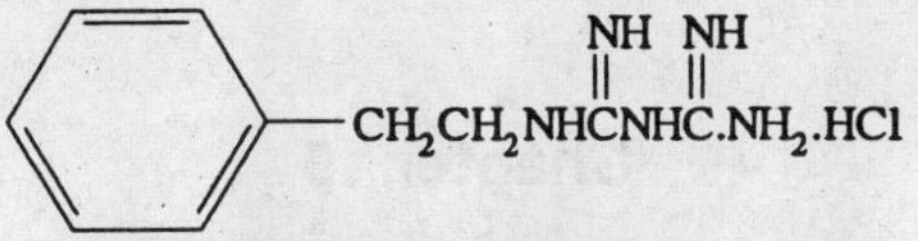

$$C_6H_5-CH_2CH_2NHC(=NH)NHC(=NH).NH_2.HCl$$

Phenformin Hydrochloride

Chemically it is named as 1-phenethyl biguanide. It is a white crystalline powder having a bitter metallic taste. It is more powerful than tolbutamide. It is generally used as its hydrochloride salt. This drug should not be given to patients suffering from kidney, liver diseases or any cardiac disease. Phenformin has shorter duration of action than Metformin.

Metformin. It is chemically, 1,1 dimethyl-biguanide hydrochloride. It is a white, hygroscopic crystalline powder with a bitter taste. It is soluble in water. It should be store in tightly closed containers. Its action and uses are similiar to those of phenformin.

Official product of Metaformin: Metformin hydrochloride, Metformin tablets.

Diabetic patient should take sweetening agents, which are non-carbohydrate substances devoid of food value. These are useful as sugar substitutes in diet and beverages, *e.g.* Saccharin, Sodium Cyclamate.

Chapter 15

Anti-Convulsant Agents

These are the drugs which are used in the treatment of convulsive disorder like epilepsy etc. These drugs can be divided as :

1. **Drugs which cures convulsions.**
2. **Drugs which are administered to prevent the occurence of convulsion e.g. drugs used in epilepsy.**

Most of the anti-convulsant drugs depress central nervous system and are non-specific. The anti-convulsant drugs can be centrally acting e.g. barbiturates or they can act on spinal cord e.g. mephesin or they act by relaxing peripheral skeletal-muscles.

Epilepsy is a disease of central nervous system disorder and there is loss of consciousness with or without convulsions for some time. These convulsions generally occur due to abnormal and excessive electroencephalogram (EEG). Epileptic attack can be of different types such as (*i*) *Major epilepsy* where there is sudden loss of consciousness with major convulsions. (*ii*) *Temporal lobe epilepsy*—In this case, convulsions are absent and the attack consists of a brief period of loss of memory or some abnormal behaviour. (*iii*) *Minor epilepsy*—These attacks consist of sudden loss of consciousness. without convulsions. (*iv*) *Focal epilepsy*—In this case, convulsions may remain confined to a single limb. (*v*) *Petit mal*—This epilepsy may occur in childhood and there is impairment of consciousness (*vi*) *Myclonic type*—In this case there are clonic jerks with characteristic EEG changes. (*vii*) *Hypsarrhythmia*—This epilepsy occurs in infants. In this case EEG is abnormal with myoclonic jerks. The drugs used in epilepsy are classified as

1. Hydantoin derivaties—*e.g.* phenytoin, ethotoin, methoin etc.
2. Barbiturates—*e.g.* methylphenobarbitone, phenobarbitone, metharbitone, primidone etc.
3. Oxazolidinediones—*e.g.* Trimethadione, paramethadione etc.
4. Acetyl Urea derivates—*e.g.* Phenylacetyl urea.
5. Succinimides—*e.g.* Ethosuximide, methsuximide etc.
6. Miscellaneous—*e.g.* Bromides, diazepam, sulthiame etc.

1. Barbiturates. These are derivates of barbituric acid.

H N 1, O, O, 6, 2, 5, 3 NH, 4, O

Barbituric Acid

Barbituric acid is 2,4,6 trioxo-hexahydropyrimidine and is formed by condensation of urea with malonic acid ester. Barbituric acid itself has no central system depressant activity, but substitution by alkyl or aryl group at position-5 makes it physiologically active. Some important barbiturates are discussed here.

(*i*) *Phenobarbitone* : Phenobarbitone is 5 ethyl-5 phenyl barbituric acid. It has the following structure.

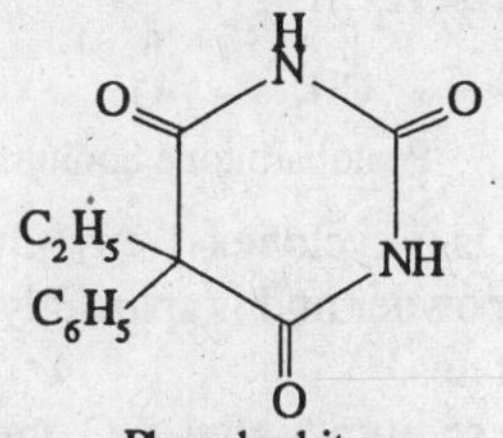

Phenobarbitone

It is a white, odourless crystalline solid with bitter taste. Phenobarbitone is available in 30, 60, 100 mg tablets and is used orally. Its sodium derivative i.e. phenobarbitone sodium is available in injection form. These are long acting barbiturates. Phenobarbitone is hygroscopic in nature hence it should be stored in airtight containers.

This drug can be used in any type of epilepsy except petit mal. The daily dose varies from 60 to 180 mg given in divided doses. This drug is used in convulsion of infants and children.

Use of this drug may cause depression, drowsiness and lethargy. After a prolonged use of phenobarbitone and its sudden discontinuation is dangerous. Hence its doses must be reduced gradually.

(*ii*) *Methyl phenobarbitone* : It is obtained by replacing hydrogen bymethyl nitrogen at position-1 of barbituric acid.

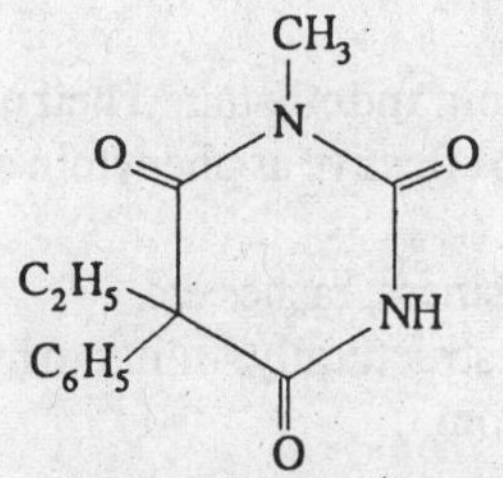

Methylphenobarbitone

It is used as an anti-convulsant drug.

(*iii*) *Metharbitone* (Gemonil) : It is a weaker anti-convulsant than phenobarbitone and mephobarbitone.

(*iv*) *Amylobarbitone* : Chemically it is named as 5 ethyl-5 isopentyl barbituric acid and has the following structure.

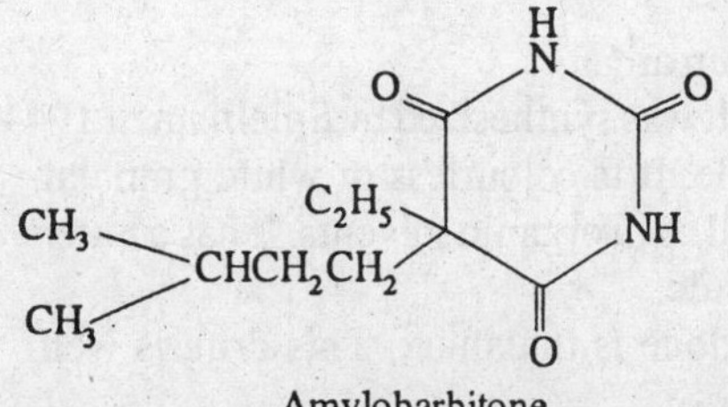

Amylobarbitone

It is used as pre-operative sedative as an anticonvulsant drug.

(*v*) *Pentobarbitone Sodium* : Chemically it is named as 5-ethyl-5-(1-methyl butyl) barbiturate.

It is soluble in water and is available in ampoules. It is an intermediate-acting barbiturate. It is also administered in capsule form and is used as an anticonvulsant drug.

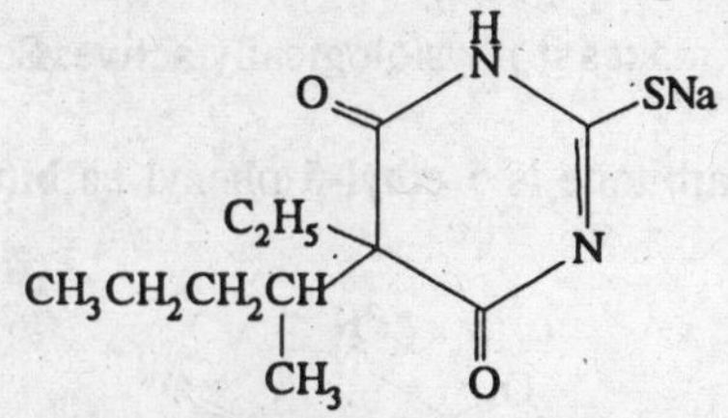

Pentobarbitone Sodium

(*vi*) *Cyclobarbitone* : Chemically it is 5-cyclohex-1-enyl-5-ethyl barbituric acid. It is used as its calcium salt which is yellowish white powder and is sparingly soluble.

It is used in insomnia and anxiety states.

(*vii*) *Quinal barbitone sodium* : It is sodium 5-allyl-5-(1-methyl butyl) barbiturate. It has similar properties like pentobarbitone.

(*viii*) *Secbutobarbitone* : It has ethyl and a secondary butyl group at position 5. It is a white crystalline powder having bitter taste and is soluble in water

2. Hydantoin. Hydantoins are imidazoline 2,4 dione heterocyclic system. The most important drugs are as follows :

(*i*) *Diphenyl hydantoin* (Phenytoin). It is diphenyl substituted hydantoin and has the adjoining structure.

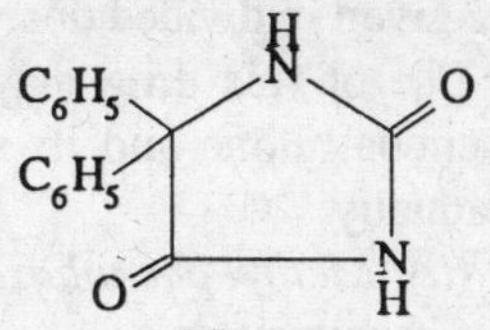

It is white odourless powder and is insoluble in water. Phenytoin sodium is available as 50 to 100 mg tablets. Phenytoin is slowly absorbed from the intestine.

It is used in the treatment of major epilepsy. This drug is also used in focal epilepsy and in temporaral lobe epilepsy. These drugs are also used as cardiac arrhythmias.

Phenytoin

Other important hydantoins are methoin and ethotoin. Their pharmacological properties are similar to those of phenytoin, but they are not so effective as phenytoin and are used in the treatment of major epilepsy and temporal lobe epilepsy.

Adverse effect : nervousness, drowsiness, rashes etc.

3. Oxazolidinediones. This group is structurally similar to hydantoins, except at position-1 where instead of NH group there is oxygen atom.

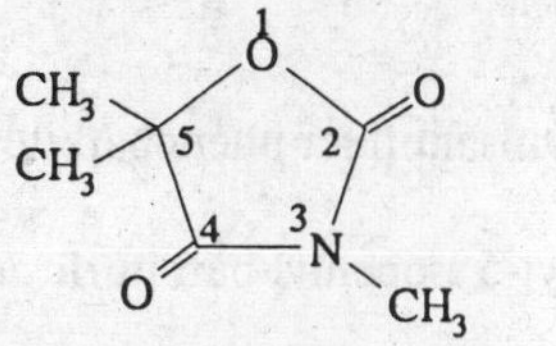

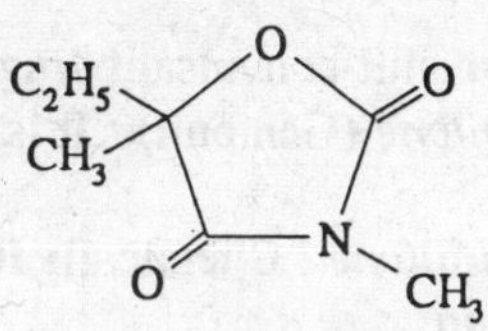

Some important drugs of this group are :

(*i*) *Troxidone* (Trimethadione) : It was synthesized by Spielman in 1944. Chemically it is named as 3,5,5-trimethyloxazoline 2,4-dione. It is colourless or white granular substance. It is soluble in water as well as in organic solvents. It has a bitter taste and is available in 250 mg capsule.

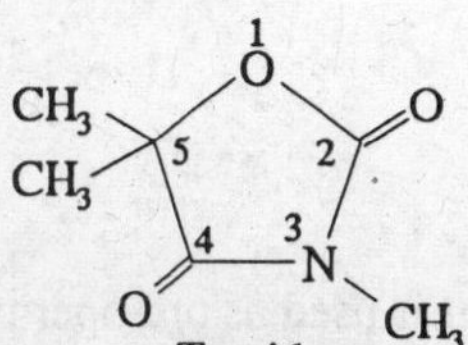

The main site of action of troxidone is thalamux. This drug is well absorbed from the intestinal tract.

This drug is used in Petit mal epilepsy and psychomotar epilepsy. In patients suffering from both petit mal and major epilepsy troxidone is given along with phenobarbitone. Its structure is as given alongside.

Troxidone

It has some adverse effects like drowsiness, blurring of vision in bright light, skin rashes, exfaliative dermatitis etc.

(*ii*) *Para methadione*: Chemically it is 3,5 dimethyl, 5-ethyl oxazoline, 2,4-dione. It is a clear, colourless liquid at room temperature. It has similar therapeutic uses as that of troxidone.

4. Acetyl-Urea derivatives e.g. (*i*) *Phenacemide (Phenyl Acetyl urea).*

It is a stright chain chemical analogue of 5-phenyl-hydantoin.

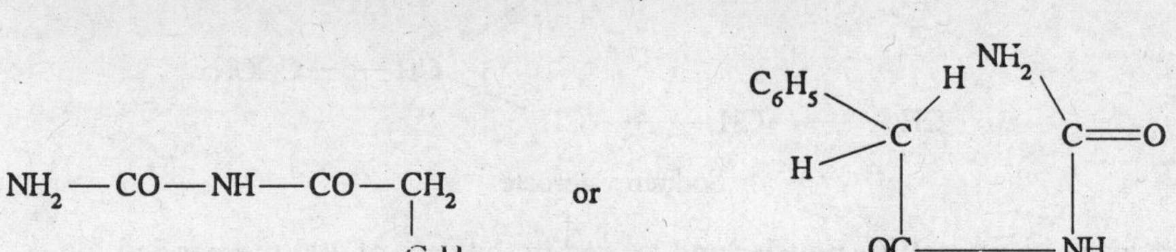

NH2—CO—NH—CO—CH2
C6H5
or
NH2
C6H5
H
C
C=O
H
OC
NH

Phenacemide

It is absorbed from the gastrointestinal tract and is completely metabolized by the liver.

It is a wide spectrum anti-epileptic drug and the action of a single dose persists for 4 to 5 hour. This drug is mainly used in major psychomotor and petit mal epilepsy. This drug can be used in combination with other drugs like diphenyl hydantoin and phenobarbitone.

This drug has toxic side effects like mental changes, skin rashes, gastro-intestinal signals, hepatitis, headache etc.

(*ii*) *Pheneturide* : It is better drug than Primidone as it is less toxic and now-adays it is generally used in place of primidone.

5. Succinimides. Many derivates of succinimides are known for its anti-convulasant activity. The most important drugs are Ethosuximide, methsuximide, phensuximide.

(*i*) *Ethosuximide* : This is the most frequently used succinimide.

It is a white odour-less powder and has a bitter taste. It is available as 250mg capsule or as a syrup containing 250 mg per 5 ml. It is soluble in water and has the following structure.

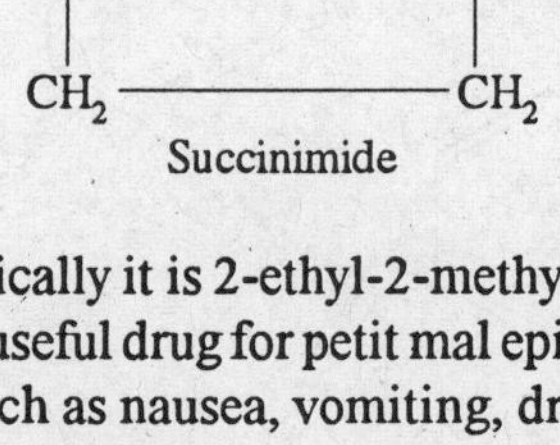

NH
OC
CO
CH2
CH2

Succinimide

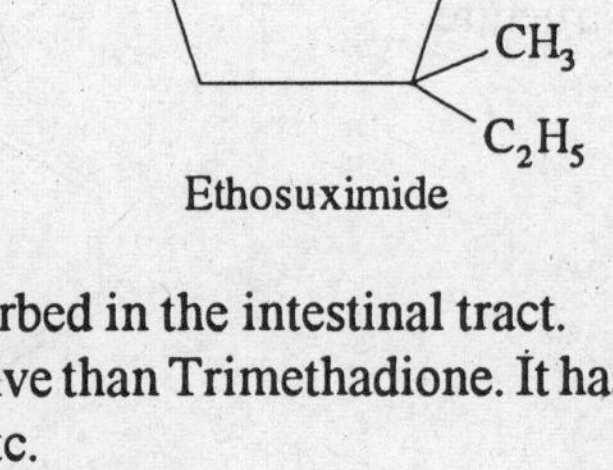

H
N
O
O
C
C
CH3
C2H5

Ethosuximide

Chemically it is 2-ethyl-2-methyl succinimide. It is absorbed in the intestinal tract.

It is a useful drug for petit mal epilepsy and is more effective than Trimethadione. It has many side effects such as nausea, vomiting, drowsiness, skin rashes etc.

(*ii*) *Methsuximide* : Its pharmacological actions are similar to that of ethosuximide. This drug can also be used for temporal lobe epilepsy. It is used in the dose of 500-600 mg daily.

6. Miscellaneous drugs. There are many other compounds having various chemical structure and are used as anticonvulsant drugs. Some important miscellaneous drugs are : Bromides, Sulthiame, Diazepam, Amphetamine, Carbamazepine, Acetazolamide, Premidone, Sodium Valproate etc.

(*i*) *Diazepam (valium):* It is a yellowish white odourless crystalline powder. It is slightly bitter in taste and is sparingly soluble in water and is soluble in alcohol. It is used in the treatment of petitmal

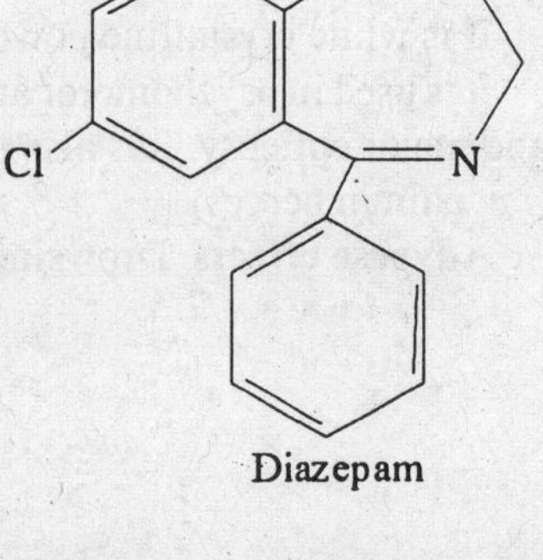

CH3
O
N
Cl
N

Diazepam

and status epilepsy. Sometimes it is used along with phenobartitone and diphenyl hydantoin for the treatment of major epilepsy.

(*ii*) *Sodium valproate:* It is sodium dipropyl acetate and has the following structure.

$CH_3—CH_2—CH_2$ \
$CH—COONa$ /
$CH_3—CH_2—CH_2$

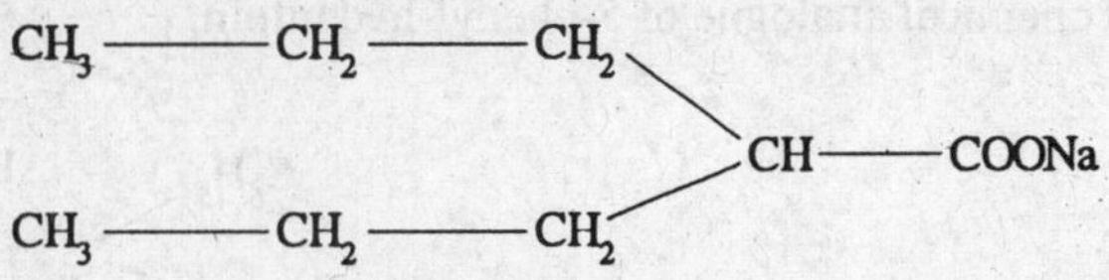

Sodium valproate

It occurs as white crystalline powder and is used in the dose of 200 mg per day.

It is a useful anticonvulsant drug and is used in petit mal, myodonic major epilepsy, and temporal lobe epilepsy. This drug has less adverse side effects and has minimum sedative effect.

(*iii*) *Sulthiamine:* It is a benzene sulphonamide derivate, where the hydrogen of para-amino group, is substituted by saturated ring structure and has the following structure.

O O S N —(benzene ring)— SO_2NH_2

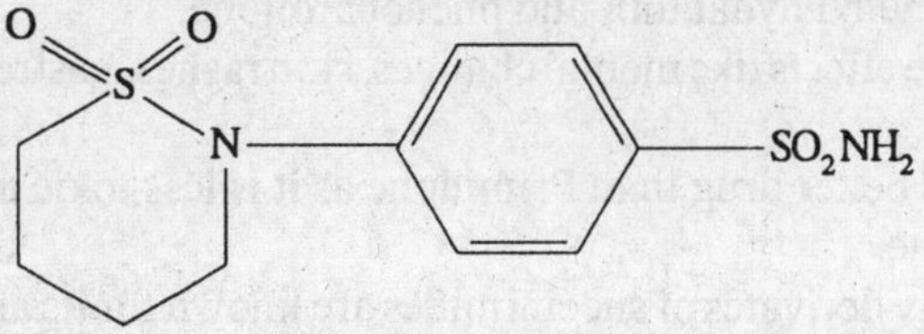

Sulthiamine

It is a white powder which is sparingly soluble in water and is administered in the dose of 100-600 mg per day. It is absorbed from intestinal tract.

It is mainly used in temporal lobe epilepsy. It is also useful drug for treating major epilepsy and myodonic seizures.

It has adverse side effects, such as blurring of vision, weight loss, kidney damage, psychotic reactions etc.

(*iv*) *Carbamazepine:* It is a dibenzazepine derivate which has structural resemblance to imipramine, an antidepressant drug, carbamazepine has a carbamoyl group attached to nitrogen atom and has the following structure.

N
$O{=}C—NH_2$

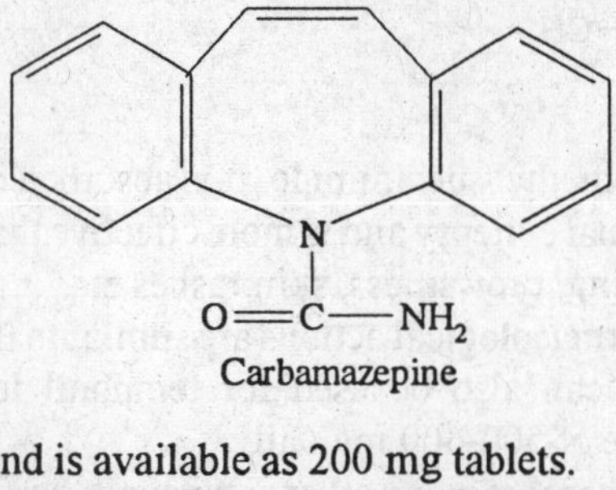

Carbamazepine

It is white crystalline powder and is available as 200 mg tablets.

It is used in psychomotor and grand mal epilepsy. It is also useful in the treatment of temporal lobe and major epilepsy. It is also used as an analgesic to control severe pain due to trigeminal neuralgia (*i.e.* pain in nerves).

Adverse effects. Drowsiness, blurred vision, depression etc.

Chapter 16

Analgesics, Antipyretic and Anti-inflammatory Agents

Analgesics

Analgesics are the drugs which relieve pain by acting on the central nervous system and they reduce pain without loss of consciousness. Analgesics can be divided into two main groups namely :

(A) **Narcotic analgesics.** These drugs produce depression of the central nervous system and are mainly of two types.

(*i*) Natural analgesics *e.g.* morphine, codeine etc.

(*ii*) Synthetic analgesics *e.g.* pethidine, methadone etc.

(B) **Non-narcotic analgesics.** These drugs do not produce significant depression of the central nervous system. Unlike narcotic analgesics, the non-narcotic analgesics possess anti-inflammatory and antipyretic effect *e.g.* Salicylates and related compounds.

A. Narcotic analgesics

(Naturally Occurring) Morphine and its derivatives

Morphine. Morphine is an opium alkaloid obtained from unripe seed capsule of Papaver somniferum (poppy plant). Morphine is generally used as sulphate or hydrochloride. Morphine sulphate is a white odourless crystalline compound while its hydrochloride is obtained as a white powder), both are soluble in water. Morphine is a phenanthrene-isoquinoline alkaloid and has the following structure.

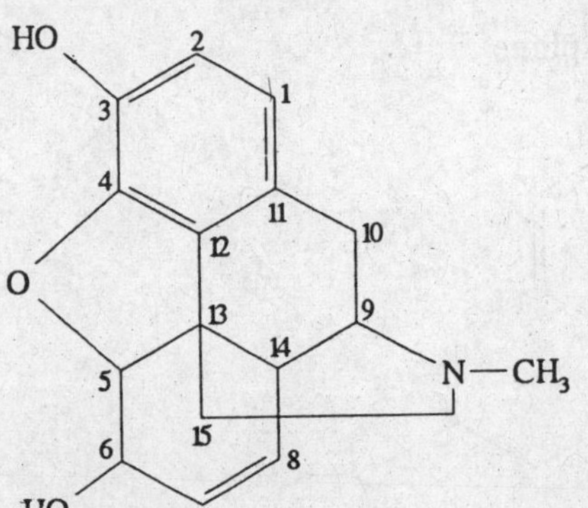

Morphine

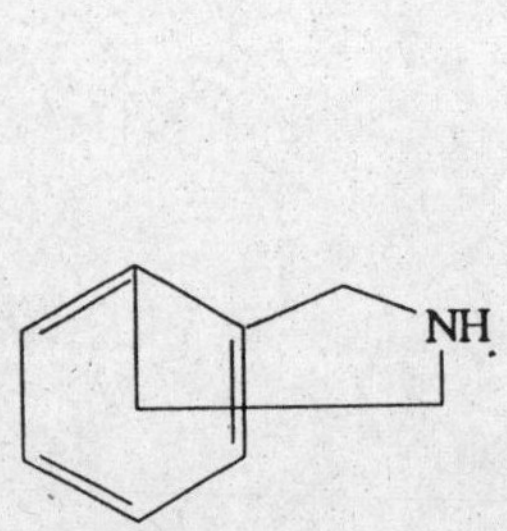

Isoquinoline

The solutions of morphine are sterilised at 98°C to 100°C for 30 minutes with a bactericide or by filtration.

Analgesic action of Morphine. Morphine is considered a powerful analgesic. It produces relief of pain in a dose which usually does not alter the other functions of central nervous system. The use of morphine does not lead to any abnormal behaviour or slurred speech. In large doses morphine can relieve all types of pain. Morphine reduces pain by increasing the threshold of pain which helps in producing an euphoria, *i.e.* a feeling of well being and it also modifies the emotional reaction to pain. Thus morphine does not remove pain completely, but in therapeutic doses, it helps the patient to tolerate the pain and the pain no longer remains as a source of concern. Morphine induces analgesia by acting on receptors situated both in the higher centers and the spinal cord.

Morphine produces a depression of respiration and it also produces miosis which results into excessive contraction of pupils and intake of morphine is characterised by pin-point pupils.

Uses. The main therapeutic use of morphine is as an analgesic for relieving the pain. Morphine is used to reduce pain in conditions such as acute myocardial infarction, fractures of bones, burns, pleurisy etc. Morphine can be administered intravenously for prompt relief of pain. It is also used to reduce post operative pain Morphine also shows sedative effect and is given to patient with internal bleeding etc. It also finds use in acute left ventricular failure. It is invaluable in the treatment of acute pulmonary-edema. Morphine is also used as premedication before surgery.

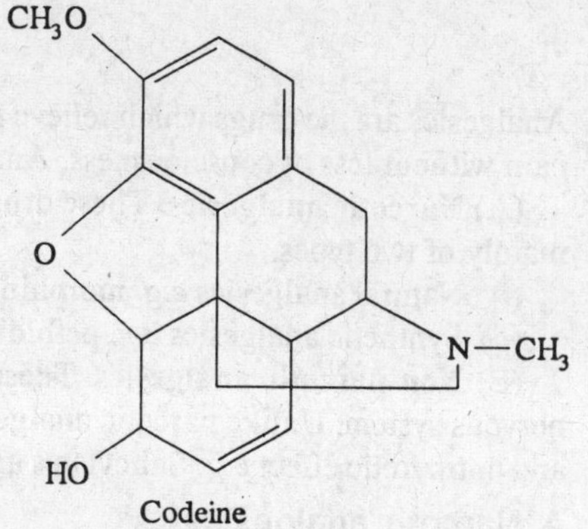

The major setback of morphine is that it is a drug of addiction due to its euphoriant effects. An overdose of morphine produces acute morphine poisoning. Other adverse side effects are dryness of mouth, mental clouding, dysphoria, vomiting, headache, fatigue, constipation etc. Other phenanthrene alkaloids of opium are codeine and buprenorphine.

CH_3O

O

$N—CH_3$

HO

Codeine

Codeine occurs as white crystalline powder and is sparingly soluble in water. It has some bitter taste.

It is much less potent analgesic than morphine and unlike morphine it is much better absorbed when administered orally.

Derivates of Natural Morphine

The important derivates of morphine are Heroin and Apomorphine, Dihydroxy morphine, Methyl-dihydromorphinone, Oxymorphone etc.

1. Heroin : It is a diacetyl derivative of morphine

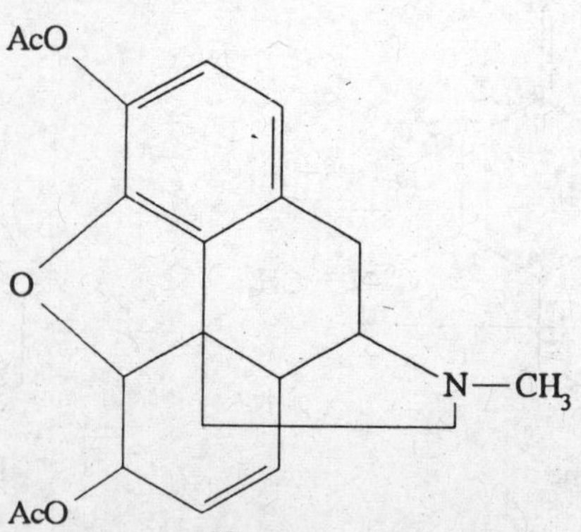

AcO

O

$N—CH_3$

AcO

Heroin

It is more powerful analgesic than morphine and produces more euphoria. The tendency of addiction to this drug is much more than morphine and hence this drug is rarely used as analgesic.

2. Apomorphine : This drug is obtained by the acid-catalyzed re-arrangement of morphine.

It is greyish white crystalline powder. It is sparingly soluble in water and alcohol. This drug is used as an emetic and is a stimulant of chemoreceptor triger zone. Apomorphine induces vomiting within a few minutes. It is administered parentally.

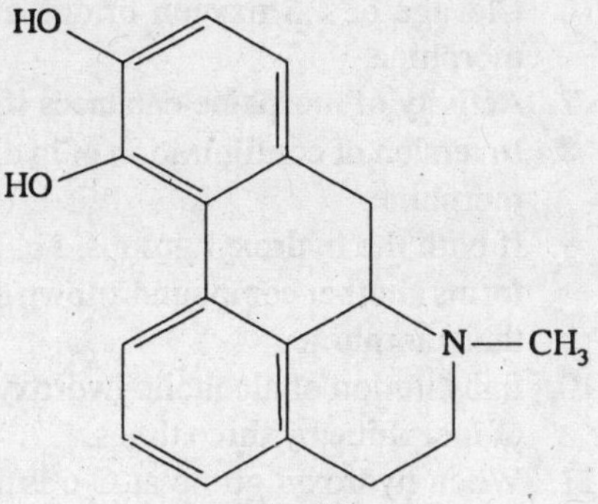

Apomorphine

Other important derivates of morphine are :

(1) Dihydroxy morphine,
(2) Oxymorphone,
(3) Methyl dihydromorphinones and
(4) Ethyl morphine hydrochloride.

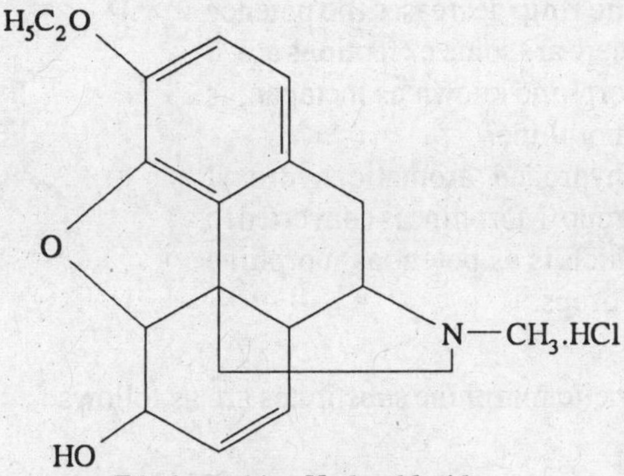

Ethylmorphine Hydrochloride

These drugs are effective analygesics and their duration of action lasts for 4-5 hours. Their toxicity is similar to that of morphine. Ethylmorphine hydrochloride is used in opthalmology in the form of drops and an ointment and as an anti-inflammatory agent.

Structure activity relationship of Morphine

1. The physiological activity of morphine is attributed to the presence of phenanthrene skeleton.
2. The activity of morphine depends to a great extent on its stereochemical nature *e.g.* the natural morphine which is laevo rotatary shows analgesic activity whereas the synthetic morphin which is dextro-rotatory do not exhibit any activity.
3. If the alcoholic hydroxyl group is oxidised to a keto function group the activity of the morphine increases, but toxicity also increases.
4. When the phenolic hydroxyl (*i.e.* OH gp attached to the aromatic nucleus) of morphine is masked, the activity of morphine decreases.
5. If the alcoholic hydroxyl group at C-6 is shifted to C-8 position as in pseudocodeine the activity of morphine decreases.

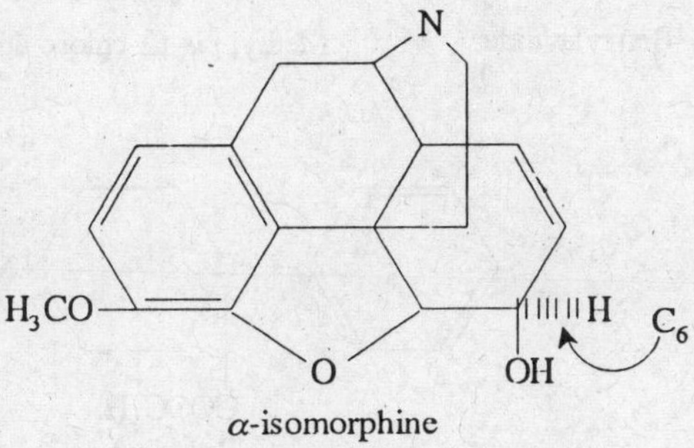

α-isomorphine

6. Clevage of 4,5 oxygen bridge and substitution in the aromatic ring lowers the activity of morphine.
7. Activity of morphine enhances if there is saturation of 7.8 double bond.
8. Inversion of configuration of hydroxyl group at C-6 as in isomorphine increase the potency of morphine.
9. If both the hydroxyl groups, *i.e.* phenolic and alcoholic group of morphine are acetylated. It forms another compound known as Heroin, which is a powerful analgesic and is more addicting than morphine.
10. Substitution of alcoholic hydroxyl group by chlorine (chloromorphine) increases the potency of morphine by three times.
11. When hydroxyl group at C-6 is substituted by a hydrogen (dihydrodexoxy morphine D) the potency of morphine increases about ten times.
12. Substituting phenyethyl group in place of methyl group on nitrogen (N—CH3) yields a compound which is more active than morphine.
13. Generally introducing new substituents in the aromatic or alicyclic ring, decreases the potency of morphine, but there are some exceptions *e.g.* 5 methyl dihydro morphine known as metapan, is more potent then morphine.
14. Substituting the hydrogen aromatic hydroxyl group of an ethyl group. Morphine is converted to ethyl morphine which is as potent as morphine and is used as eye drops.

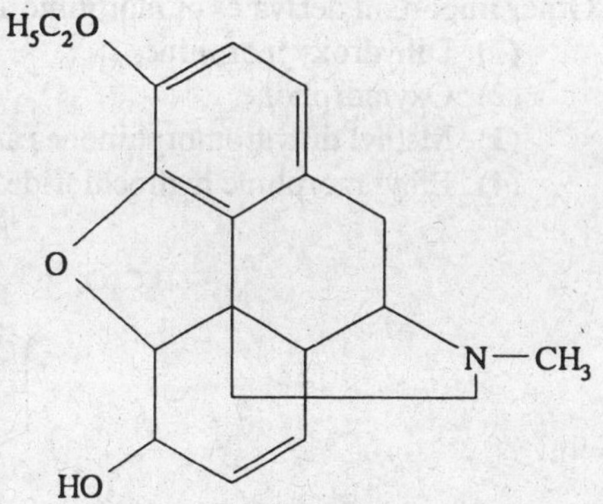

H_5C_2O O N—CH_3 HO

Ethyl morphine

Synthetic analgesics

Some important synthetic morphine substitutes are as follows :

(*i*) Pethidine,
(*ii*) Methadone,
(*iii*) Morphinan and
(*iv*) Benzomorphan.

(*i*) **Pethidine (Meperidine, Demeral).** Pethidine is a white, crystalline substance with a bitter taste. It forms salts with acid. It is synthesised as follows:

Its action is similar to that of morphine. Pethidine is a useful analgesic especially when short duration of action is required.

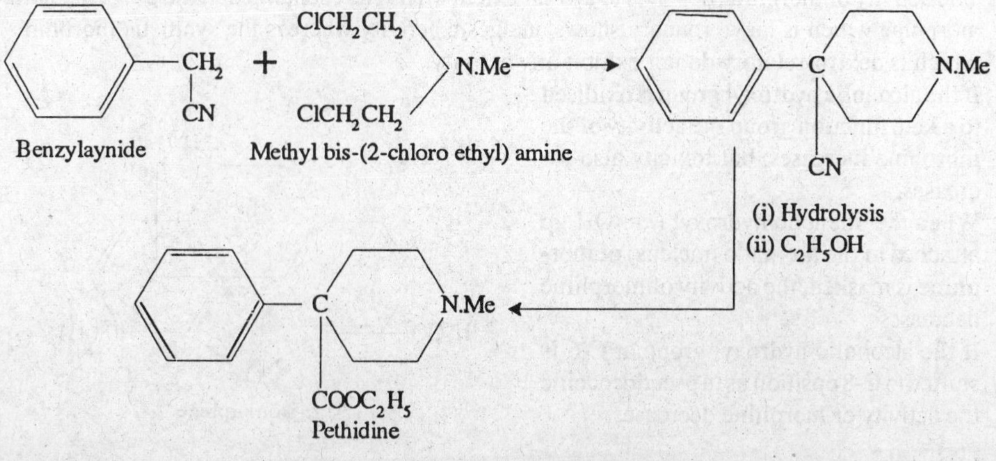

CH_2 CN + $ClCH_2CH_2$ $ClCH_2CH_2$ N.Me ⟶ C N.Me CN

Benzylaynide Methyl bis-(2-chloro ethyl) amine

(i) Hydrolysis
(ii) C_2H_5OH

C N.Me $COOC_2H_5$

Pethidine

Pethidine can be used in place of morphine as an analgesic in myocardial infarction, burns etc. It is a respiratory depressant like morphine. It is also used for the treatment of shock.

(*ii*) **Methadone.** (Amidone, Dolephine). It is a synthetic compound whose potency is slightly greater than morphine. It is synthesised as follows :

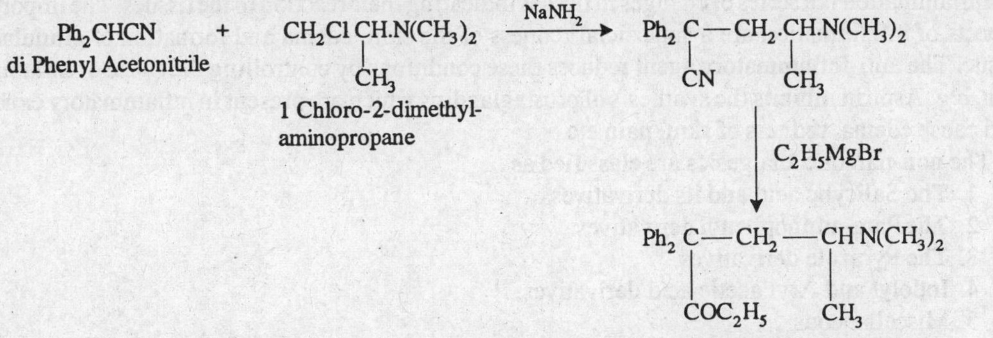

$$\underset{\text{di Phenyl Acetonitrile}}{Ph_2CHCN} + \underset{\text{1 Chloro-2-dimethyl-aminopropane}}{CH_2Cl\,CH(CH_3).N(CH_3)_2} \xrightarrow{NaNH_2} Ph_2C(CN)-CH_2.CH(CH_3).N(CH_3)_2$$

$$\xrightarrow{C_2H_5MgBr} Ph_2C(COC_2H_5)-CH_2-CH(CH_3)N(CH_3)_2$$

Amidone or Methadone

Methadone is available as Methadone hydrochloride tablet doses 5 to 10 mg. It is also available in ampoules in injection form. It is a racemic mixture and the *l*-isomer is more active than *d*-isomer.

It is used as a substitute for morphine and pethidine for relief of severe viceral pain. Methadone is also respiratory depressant.

(*iii*) **Morphinan.** This is also synthetic analgesic. It has the following structure.

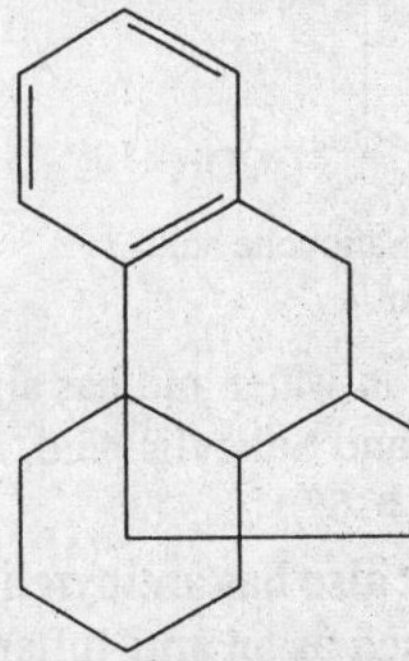

N—H

Morphinan

HO

N—CH₃

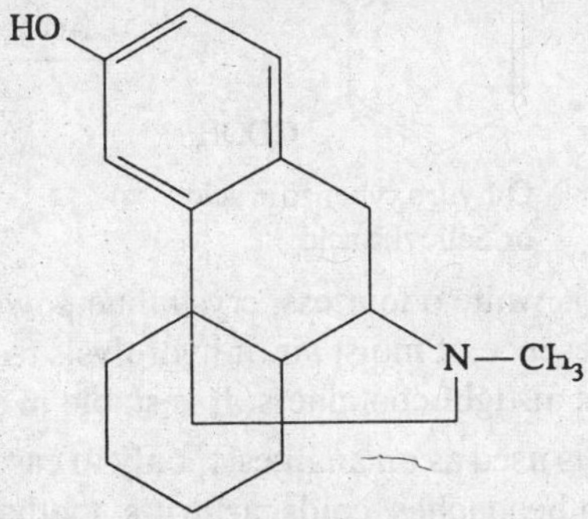

Levorphanol

An example of morphinan compounds is levorphanol. It is more potent analgesic than morphine and is well absorbed when administered orally.

(*iv*) **Benzomorphan compounds** *e.g.* Phenazocine, Pentazocine etc. It is a new synthetic agent and is much more potent than morphine. It is generally used intramuscularly in the dose of 2 to 3 mg.

Narcotic anatogonist

The drugs which are used as an antidote for the narcotic analgesics are called narcotic anatogonist. It antagonizes the effects of morphine, *e.g.* N-allyl morphine, Naloxone, Nalorphine, Levallorphan etc.

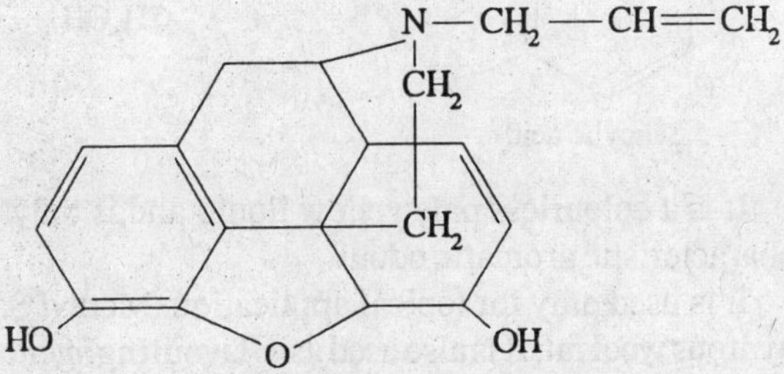

N—CH_2—CH=CH_2

CH_2

CH_2

HO O OH

Nalorphine

B. Non-narcotic analgesics

These analgesics also possess antipyretic and anti-inflammatory property.

Antipyretic agents reduces the elevated body

temperature. The hypothalamus (situated at the base of the brain) plays an important role in regulating the body temperature. It regulates the body temperature by cutaneous vasodilation and by excessive sweating. In fever, the thermostatic mechanism is set at higher level and these antipyretic agents induce changes in the central nervous system in the region of the anterior of the hypothalamus and help to bring the thermostatic mechanism at the normal level.

Inflammation is a series of changes in tissues indicating their reaction to the tissues. The important aspects of inflammation are a superficial redness of the skin, edema and formation of granulation tissue. The anti-inflammatory agent reduces these conditions by controlling the cause of inflammation, *e.g.* Aspirin inhibits the synthesis of prostaglandins which are present in inflammatory exudate and cause edema, redness of skin, pain etc.

The non-narcotic analgesics are classified as

1. The Salicylic acid and its derivatives.
2. The Para-aminophenol derivatives.
3. The Pyrazole derivatives.
4. Indolyl and Aryl acetic acid derivatives.
5. Miscellaneous.

Salicylic Acid Derivatives

The derivative of Salicylic acid are well-known for their analgesic and antipyretic action. The most important drugs are aspirin (acetyl salicylic acid), sodium salicylate etc.

(*i*) **Aspirin (Acetyl Salicylic Acid).** It is obtained by the acetylation of Salicylic acid in the following manner :

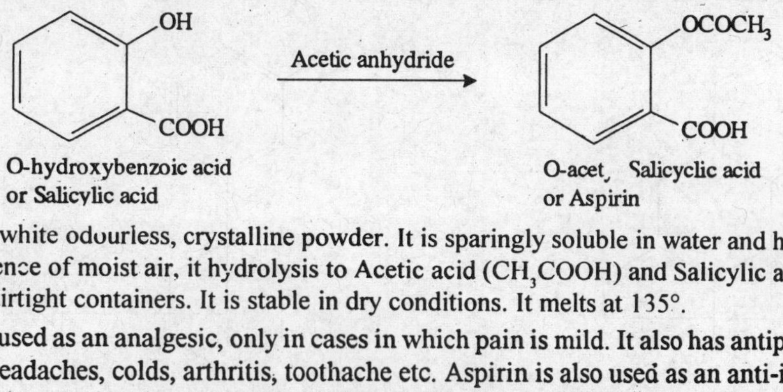

OH, COOH —Acetic anhydride→ $OCOCH_3$, COOH

O-hydroxybenzoic acid or Salicylic acid — O-acet, Salicyclic acid or Aspirin

Aspirin is white odourless, crystalline powder. It is sparingly soluble in water and has slight acid taste. In presence of moist air, it hydrolysis to Acetic acid (CH_3COOH) and Salicylic acid. It should be stored in airtight containers. It is stable in dry conditions. It melts at 135°.

Aspirin is used as an analgesic, only in cases in which pain is mild. It also has antipyretic action. It is used in headaches, colds, arthritis, toothache etc. Aspirin is also used as an anti-inflammatory agent. It helps in reducing edema, tissue swelling etc. It is also useful as antirheumatic drug and gives relief from signs and symptoms of inflammation. Aspirin has also shown beneficial effect in radiation diarrhoea.

(*ii*) **Methyl salicylate** (oil of winter green). Chemically it is O-hydroxy benzoate.

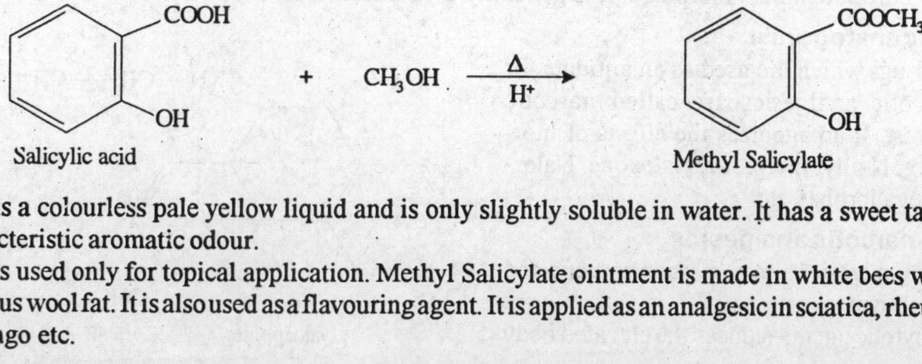

COOH, OH + CH_3OH —Δ, H^+→ $COOCH_3$, OH

Salicylic acid — Methyl Salicylate

It is a colourless pale yellow liquid and is only slightly soluble in water. It has a sweet taste and characteristic aromatic odour.

It is used only for topical application. Methyl Salicylate ointment is made in white bees wax and hydrous wool fat. It is also used as a flavouring agent. It is applied as an analgesic in sciatica, rheumatic, lumbago etc.

(*iii*) **Sodium Salicylate.** It occurs as colourless small crystals or as crystalline powder. It has characteristic unpleasant taste and is soluble in water.

It is usually taken in a mixture form with alkali and is mainly used for integumental pain and acute rheumatic fever.

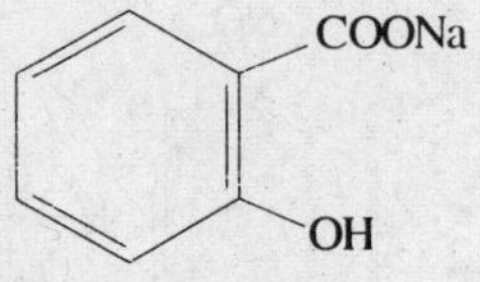

COONa
OH
Sodium Salicylate

(*iv*) **Salicin.** It is obtained from bark of willows. It is also synthesised by the action of acetobromo-glucose and Salicylaldehyde.

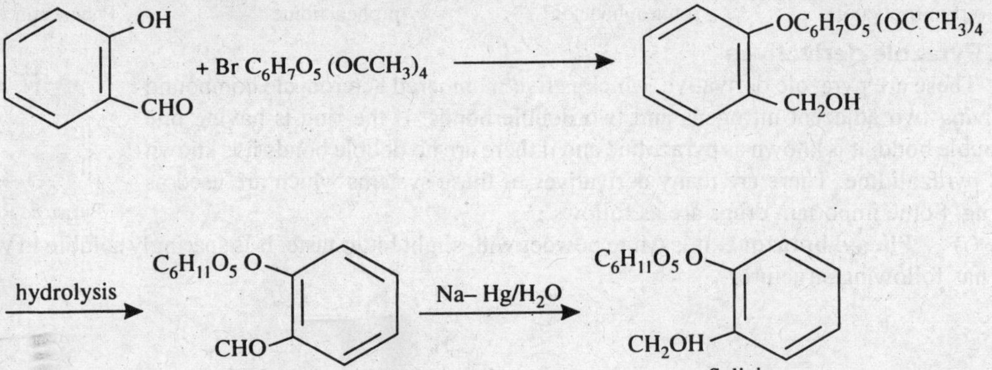

OH, CHO + Br $C_6H_7O_5\ (OCCH_3)_4$ ⟶ $OC_6H_7O_5\ (OC\ CH_3)_4$, CH_2OH

$\xrightarrow{\text{hydrolysis}}$ $C_6H_{11}O_5\ O$, CHO $\xrightarrow{Na-Hg/H_2O}$ $C_6H_{11}O_5\ O$, CH_2OH

Salicin

It is used in rheumatic pain and fever.

(*v*) **Diethylamine Salicylate.** It is a diethylamine salt of salicylic acid. It is a white crystalline substance and is soluble in water. It is used only topically and is available as creams and is used in rheumatic and muscular pain.

2. The para-aminophenol derivatives

These compounds have very pronounced analgesic and antipyretic effects. They are not useful anti-inflammatory drugs. The commonly used drugs are phenacetin (acetophenatidin) and para-acetamol (acetamainophen).

(*a*) **Para acetamol.** It is a white crystalline solid. It is 4-hydroxy Actanilide and is prepared as follows :

It is odourless and highly soluble in water. The total daily dose should not exceed 2.5% in adults.

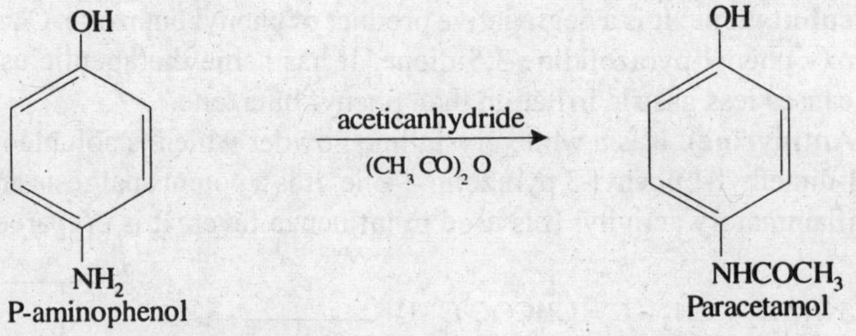

OH, NH_2 $\xrightarrow[(CH_3\ CO)_2\ O]{\text{aceticanhydride}}$ OH, $NHCOCH_3$

P-aminophenol   Paracetamol

Paracetamol is more potent antipyretic than phenacetin. It is used for relief of pain and fever.

Adverse effect : Sweating, nausea, vomiting etc.

(*b*) **Phenacetin.** It is a white odourless, crystalline powder and is sparingly soluble in water. It has a slight bitter taste. It is prepared from *p*-nitrophenol as follows :

It is often used with aspirin for relief of integumental pain. It is more toxic than aspirin. Phenacetin is hydrolysed to paracetamol in vivo hence it is believed that phenacetin's, antipyretic-analgesic action is due to its hydrolysis to paracetamol.

Adverse effect : More toxic than aspirin.

p-nitrophenol $\xrightarrow[-\,HBr]{C_2H_5Br}$ p-nitrophenetole $\xrightarrow[\text{Reduction}]{\text{Fe-HCl}}$ p-phenetidine $\xrightarrow{(CH_3CO)_2O}$ Phenacetin

(Structures: p-nitrophenol: NO_2 / OH; p-nitrophenetole: NO_2 / OC_2H_5; p-phenetidine: NH_2 / OC_2H_5; Phenacetin: $NHCOCH_3$ / OC_2H_5)

3. Pyrazole derivatives

These are pyrazole derivative, which is five-membered heterocycle compound having two adjacent nitrogens and two double bonds. If the ring is having one double bond, it is known as pyrazoline and if there are no double bonds it is known as pyrazolidine. There are many derivatives of these systems which are used as drug. Some important drugs are as follows :

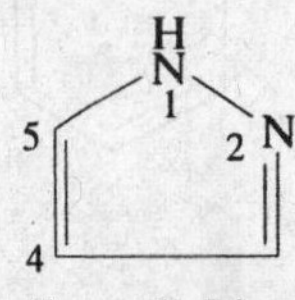

Pyrazole Ring

(*i*) **Phenyl butazone.** It is white powder with slight bitter taste. It is sparingly soluble in water. It has following structure.

(Structure: pyrazolidine ring with C_6H_5 on N1, C_6H_5 on N2, O at C5 and C3, $CH_3CH_2CH_2CH_2$ at C4)

Phenyl Butazone

Chemically it is named as 4-butyl-1,2-di-phenyl pyrazolidine-3,5-dione. It is available as tablets whereas its sodium derivative is available in ampoules containing 200 mg of phenyl butazone sodium per ml.

It is used for relieving pain in acute gout. It is also useful in the treatment of spondylitis and in rheumatoid arthritis. It produces variety of side effects, hence it should not be employed routinely as an analgesic or antipyretic.

(*ii*) **Oxyphenbutazone.** It is a degradative product of phenyl butazone. Chemically it is 4-butyl-2-phenyl-1-hydroxy phenyl-pyrazolidine-3,5-dione. It has some therapeutic uses as that of phenyl butazone, but it causes less gastric irritation than phenyl butazone.

Phenazone (Antipyrine). It is a white crystalline powder which is soluble in water. Chemically it is named as 2,3-dimethyl-1 phenyl-3 pyrazolin-5-one. It is a potent analgesic and antipyretic. It also possesses anti-inflammatory activity. It is used in influenza fever. It is prepared as follows :

$$C_6H_5NHNH_2 + CH_3 - \underset{OH}{\underset{|}{C}} = CHCOO\,C_2H_5 \longrightarrow$$

Phenyl hydrazine; Acetoacetic ester

(Product: pyrazolone ring with O, C_6H_5–N, N–H, CH_3)

$\xrightarrow{CH_3Cl}$ (ring with O, C_6H_5–N, N–CH_3, CH_3)

Phenazone

Analgin. It is a phenazone derivative, having $N(CH_3)CH_2SO_3Na$ substituent at position-4. It is a yellowish white powder and is soluble in water. It has slight bitter taste. It has potent analgesic and antipyretic action. It has toxic side effects hence it should not be used routinely to relieve pain and fever. It is prepared as follows :

Amino Phenazone $\xrightarrow[\text{Formal dehyde, Sodium bisulphite}]{HOCH_2SO_3Na}$ Melubrin $\xrightarrow{(CH_3)_2SO_4}$ Analgin

4. Indolyl and aryl acetic acid derivates

The important drugs of this group are as follows :

(*i*) **Indomethacin.** It is an indole acetic acid derivative. It has following structure.

Indole   Indomethacin

Chemically it is named as 1-(4-chlorobenzoyl)-5-methoxy-2-methyl-3 indolylacetic acid. It is brownish yellow crystalline powder. It is insoluble in water, but it is soluble in organic solvents. It is available n 25 mg capsules.

This drug has inflammatory, analgesic and antipyretic ons. It is effective in rheumatic disorders. It is par- larly useful in the treatment of gout.

has some adverse side effects like nausea, vomiting, shes, diarrhoea etc.

Sulindac. It is a fluorinated derivative of hacin. Chemically it is [5-fluoro-2 methyl-1 (4- lphinyl-benzylidene) inden-3 yl] acetic acid. s as yellow crystalline powder and is insoluble It has longer duration of action than cin. It used in the treatment of rheumatic and eletal disorders.

Sulindac

5. Miscellaneous drugs

(*i*) **Ibuprofen (Brufen).** It is a synthetic compound and it is isobutylphenyl-propionic acid. Chemically it is named as 2-(4-isobutylphenyl) propionic acid.

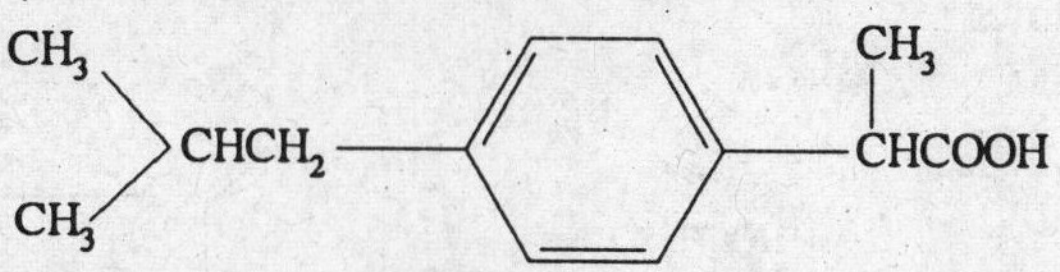

Ibuprofen

It is a white crystalline powder and is insoluble in water.

It has analgesic and weak anti-inflammatory action. It is used in rheumatoid disease. Its analgesic action is less than that of aspirin.

(*ii*) **Ketoprofen.** Chemically it is 2-(3-benzoyl phenyl) propionic acid.

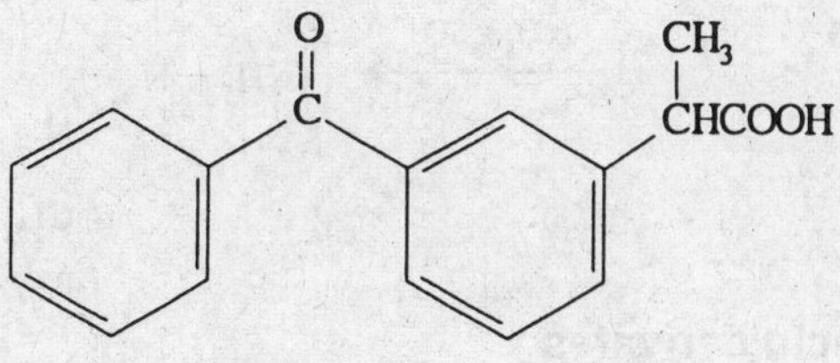

Ketoprofen

It is white crystalline powder and is insoluble in water. It is more potent than Ibuprofen and is used in the dose of 150 mg. It has anti-inflammatory, analgesic, and antipyretic actions.

(*iii*) **Naproxen.** It is a propionic acid derivative and has methoxy-2-naphthyl group as substituent. It is a white crystalline powder and is insoluble in water. It is the dextrorotatory isomer.

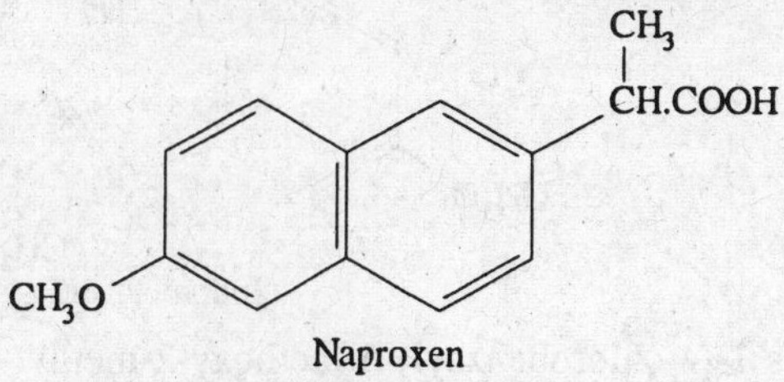

Naproxen

It is used in the treatment of rheumatoid arthritis.

(*iv*) **Tromaril.** This is a anthranilic acid derivative. Chemically it is N-phenyl ethyl anthranilic acid.

It has analgesic, antipyretic and anti-inflammatory properties. It is used orally in a dose of 1200-2400 mg per day.

Chapter 17

Anaesthetics

The term anaesthesia means loss of sensation. It is derived from a Greek word "anaesthesia", which means insensibility. Anaesthetics produces insensibility to the vital function of all types of cells and especially those of nervous system. The most important aspect of anaesthetics are that the loss of consciousness is reversible i.e. as soon as the concentration of the anaesthetic is decreased, the effected organs return to the normal state. Priestley discovered the first inhalation anaesthetic, nitrous oxide in 1776. Later on in 1846, Marton showed the use of anaesthesia in surgical operations. It produces temporary insensibility to pain or feeling in the body or a particular organ which has to undergo surgical operation. The discovery of anaesthetics proved to be a great boon to mankind as the patient has no agony during the operation.

An ideal anaesthetic agent should have the following characteristics :

1. It should be able to give enough relaxation time for the surgery to be performed.
2. It should be inert, poten, non-flammable, and should be non-irritating to the mucous membrane.
3. It should not have any toxic effect on heart, brain, liver, kidney etc.
4. It should be stable to light and heat, soda lime.
5. It should not reduce the blood pressure drastically.
6. It should not induce vomiting or nausea.
7. It should not give any adverse effect with the administration of other drugs.
8. It should be economical.
9. It should have analgesic action and muscle relaxation.

It is difficult to find an anaesthetic agent having all these qualities, but these can be combined with suitable ancillary drugs to make it an ideal anaesthetic.

Some drugs are used before the administration of an anaesthetic agents. These drugs are known as pre-anaesthetic medication and are used to make anaethesia safer and more agreeable to the patient. Some common pre-anaesthetic drugs are morphine, pethidine, barbiturates etc.

An anaesthetic may produce unconsciousness all over the body or in a particular organ. The anaesthetics could be classified as follows according to their mode of action :

I. General anaesthetics and

II. Local anaesthetics.

I. General Anaesthetics. The anaesthetics belonging to this group produce unconsciousness all over the body, by depressing the central nervous system to such an extent that the sensitivity to pain or feeling is lost.

II. Local Anaesthetics. These agents abolish the sensation of pain in localized area, i.e. they do not affect the whole body, but only a part of the body becomes insensitive to any pain or feeling. These drugs affect the peripheral nervous tissues and block nerve conduction and thus abolish all sensation in the part supplied by the nerve.

General Anaesthetics

There are many theories regarding the mechanism of action of general anaesthetics. According to lipid theory, it is said that efficiency of a general anaesthetic is directly proportional to its affinity for the lipids. The central nervous system is rich in lipids and chemicals with high lipid solubility can easily penetrate into the central nervous system and produce anaesthesia.

According to surface tension theory, the general anaesthetic can reduce the surface tension of the neuronal cell membrane and thus produces anaesthesia.

General anaesthetics are capable of depressing all the functional elements of central nervous system, but it is postulated that it inhibits the ascending reticular activating system, which maintains the state of consciousness.

The general anaesthetics are further classified as :

1. Volatile general anaesthetics.
2. Non-volatile general anaesthetics.

1. Volatile General Anaesthetics

This group includes the drugs which are volatile liquids or gases, e.g. ether, halogenated compound, cyclopropane and nitrous oxide etc.

The behaviour of volatile or gaseous general anaesthetics in the body are governed by the basic laws of gases such as Boyle's law, Dalton's law of partial pressure, Graham's law of diffusion etc.

The anaesthetic activity of volatile general anaesthetics can be divided into four stages which are different from each other. These stages are :

(*a*) *Stage of analgesia* : This stage stretches from the beginning of inhalation to loss of consciousness.

(*b*) *Stage of delirium* : This stage is from the loss of the consciousness to the beginning of surgical anaesthesia.

(*c*) *Stage of surgical anaesthesia* : At this stage patient starts deep breathing and passes into the third stage when slowly reflex activity is lost.

(*d*) *Stage of respiratory paralysis* : This stage is characterised by severe depression of medullary centres. General anaesthetics can be applied by open method through a simple mask like Schimmel busch mask covered with six to ten layers of gauge or close method in which a chemical agent such as soda lime is used to reabsorb carbondioxide present in the expired air and needs a complicated apparatus.

Other methods are semi-open or semi-closed method.

Some important volatile liquids are as follows :

(*i*) **Ether.** Anaesthetic ether is 96-98 per cent Diethyl ether. It is a colourless volatile liquid with a pungent smell with the boiling point of 35°C. It is soluble in organic solvents like alcohol, chloroform, benzene etc.

Ether is prepared by Williamson's synthesis in the following way.

$$\underset{\text{Ethyl bromide}}{C_2H_5Br} + \underset{\text{Sodium ethoxide}}{NaOC_2H_5} \longrightarrow \underset{\text{Diethyl ether}}{C_2H_5OC_2H_5} + NaBr$$

It can also be prepared from an alcohol in the presence of a dehydrating agent.

$$\underset{\text{Ethyl alcohol}}{2C_2H_5OH} \longrightarrow \underset{\text{Ether}}{C_2H_5OC_2H_5} + H_2$$

Ether should be stored in well-closed, light resistant bottles as when it is exposed to air or light it forms peroxides or acetic aldehydes, which are irritant, hence ether is marketed in sealed amber coloured bottles. Ether is inflammable. The stabilisers like sodium pyrogallate, propyl gallate etc. are added in anaesthetic ether. To avoid these complications ether is stored in a container which is copper plated internally.

Ether is eliminated through lungs only, a small fraction about 8 to 9 per cent is oxidized in the body.

Advantages of ether

1. It induces surgical anaesthesia without any pre-anaesthetic medication.
2. It is a safe anaesthetics.
3. It produces muscular relaxation and has analgesic actions also.
4. It does not modify blood pressure.
5. It can be administered without a complicated apparatus.
6. It is economical and is quite stable, if stored properly.

Disadvantages

1. The induction of ether is slow.
2. It boils at low temperature and it is inflammable and its vapours are irritating to mucous membrane.
3. Nausea and vomiting appear especially during the recovery from the anaesthesia.
4. The recovery is slow.
5. Ether convulsions may occur in children.

Derivatives of ether have been found to be less irritating than ethyl ether and are found to be more powerful. Examples:

(*a*) *Methyl-n-propyl ether:* It is found to be more powerful than ethyl ether and shows less adverse side effects.

(*b*) *Vinyl ether:* It has essential features of ethyl ether and vinyl ether anaesthetics. It was synthesized by Major and Ruigh from ethylene chlorohydrin in the following way.

$$2ClCH_2CH_2OH \xrightarrow{H_2SO_4} \begin{matrix} ClCH_2CH_2 \\ \quad\quad\quad\quad > O \\ ClCH_2CH_2 \end{matrix}$$

2, 2' dichloro diethyl ether

$$\downarrow C_2H_5OH + KO$$

$$\begin{matrix} CH_2{=}CH \\ \quad\quad\quad\quad > O \\ CH_2{=}CH \end{matrix}$$

Vinyl ether

It is a colourless liquid having a characteristic odour. It is miscible with alcohol, ether and chloroform. It should be stored in a well closed container and should be stored in a cool, dark place.

Advantages

1. It is more potent than ether with a speedy recovery.
2. It is useful for dental procedures.
3. It can be used for minor operations of short duration.

Disadvantages

1. It damages lever.
2. It is inflammable.

(*ii*) **Chloroform**. It is a powerful volatile liquid anaesthetic. It is prepared from bleaching powder and ethyl alcohol as follows :

$$CaOCl_2 + H_2O \longrightarrow Ca(OH)_2 + Cl_2$$

$$C_2H_5OH + Cl_2 \longrightarrow \underset{\text{Actaldehyde}}{CH_3CHO} + 2HCl$$

$$\underset{\text{Acataldehyde}}{CH_3CHO} + Cl_2 \longrightarrow \underset{\text{Chlorol}}{CCl_3CHO} + 3HCl$$

$$2CCl_3CHO + Ca(OH)_2 \longrightarrow \underset{\text{Chloroform}}{CHCl_3} + (HCOO)_2Ca$$

It is colourless volatile liquid having a characteristic odour. In chloroform 1 to 2 percentage of V/V ethyl alcohol is added to prevent the formation of poisonous carbonyl chloride. Ether also might react with any carbonyl chloride, thus formed.

It is miscible with ether and ethanol. Chloroform should be protected from light as it forms toxic phosgene.

Advantages

Chloroform's induction is prompt and its vapours are not flammable. It is used for obstetrical anaesthesia.

Disadvantages

Chloroform has several toxic effects, particularly liver and cardiac toxicity. Due to its toxicity chloroform is no longer in use as anaesthetics.

(*iii*) **Halothane.** It is a fluorinated volatile liquid and its structure is similar to chloroform. It has the following structure.

$$CF_3\text{—}CH(Br)(Cl)$$

Halothane (2 bromo- 2 chloro- 1, 1, 1 trifluorothane)

It is a heavy colourless liquid which is non-flammable. It has a characteristic sweet fruity odour and its boiling point is 50°C. It is miscible with chloroform, ether, trichloroethylene etc. It should be stored in air tight containers at low temperature and it should be protected from light. It is stable in presence of alkalis.

Advantages

1. It is not flammable and does not irritate the respiratory passage.
2. It has a nice pleasant odour.
3. It is a potent anaesthetic.
4. Post operative vomiting is low.
5. It relaxes bronchial muscle, hence it is safe to use for patients having bronchial asthma.
6. Halothane is used as an anaesthetic during plastic surgery.

Disadvantages

1. It causes cardiovascular depression hence blood pressure decreases.
2. It cannot be used for intro-abdominal operation.
3. Halothane depresses respiration.
4. It is a poor analgesic,
5. It is expensive and it has to be given through a special apparatus.
6. It causes hepatic damage.

(*iv*) **Trichloroethylene**. Its chemical structure is as follows :

$$\underset{Cl}{\overset{H}{>}}C=C\underset{Cl}{\overset{Cl}{<}}$$

Tri-chloroethy lene

It is prepared by the following method.

$$\underset{\text{Tetrachloroethane}}{CHCl_2.CHCl_2} + \underset{\text{Alkali}}{NaOH} \xrightarrow{\text{decomposition}} \underset{\text{Trichloroethylene}}{CCl_2 = CHCl} + NaCl + H_2O$$

It is a clear, colourless or blue coloured volatile liquid, its boiling point is 86°C. It has a characteristic odour. Thymol is added in it as a preservative. It is not miscible with water but is miscible with alcohol, chloroform and ether. It should be stored in airtight and light resistant containers and should be kept in cool places.

Advantages

1. It is a potent analgesic with rapid onset of action.
2. It is recommended for short surgical procedures such as orthopaedic manipulations, dental extractions etc.
3. It is used as an analgesic for migraine pain, angina pectoris and child birth.

Disadvantages

1. It breaks down in presence of soda lime to highly toxic products and hence it cannot be used in closed circuit.
2. It produces poor muscle relaxation.
3. It may be hepatotoxic and produce tachypnoea. Tachypnoea = shallow respiration.

(*v*) **Ethyl chloride**. It is a non-irritating high volatile and inflammable liquid with a boiling point of 12°C. At ordinary temperature and pressure, it is a gas. It has a pleasant characteristic odour. It is miscible

$$\underset{\text{Ethylchloride}}{CH_3CH_2Cl}$$

with organic solvents like alcohol and ether, but it is slightly miscible in water. It should be kept at temperature below 12°C in airtight containers and should be protected from light. When sprayed on skin, it evaporates rapidly, cooling the skin and it produces transient paralysis of cutaneous sensory nerve endings. Its advantage is that : (1) it can be used as a local anaesthetic for minor surgeries such as incision etc. (2) As general anaesthetic it induces anaesthesia within 1 to 2 minutes and recovery also occurs within 2 to 3 minutes. Due to its quick induction property it is mainly used for induction of general anaesthesia.

It has many disadvantages, hence this drug is not in much use now.

1. It may damage liver and induce cardiac arrhythmias or cardiac arrest.
2. It does not produce adequate muscle relaxation.
3. It is difficult to maintain steady depth of anaesthesia.
4. Its margin of safety is low.

Some important gaseous anaesthetics are as follows :

(*vi*) **Nitrous oxide**. It is obtained by heating ammonium nitrite upto 200°C as follows:

$$NH_4NO_2 \xrightarrow{\Delta} \underset{\text{Nitrous oxide}}{N_2O} + 2H_2O$$

It is colourless inorganic, non-irritating gas. It has a sweet taste. It is non-flammable. It is available in steel cylinders at a pressure of 650 to 800lbs per square inch. The cylinders should be kept at a temperature below 37°C. It is also known as laughing gas. Nitrous oxide is not altered within the body and is eliminated through lungs with 2 to 5 minutes after its withdrawal.

Advantages

1. It is one of the safest anaesthetic agent and does not induce post-operative nausea and vomiting.
2. It is non-flammable and is non-irritating.
3. It provides rapid induction and also rapid recovery.
4. It has a pronounced analgesic action and is employed for tooth extraction, obstetrical (child birth), painful procedures like changing and cleaning burn wounds etc.

Nitrous oxide is used with oxygen to induce anaesthesia which is then maintained by ether. This technique is called gas-oxygen-ether technique.

Disadvantages

1. It is not a potent anaesthetic and pre-anaesthetic medication has to be given.
2. It may produce violent excitation.
3. Hypoxia (a state of diminished oxygen) may be produced during prolonged nitrous oxide anaesthesia.
4. It may increase cardiac irregularities.

(*vii*) **Cyclopropane**. It is obtained from 1,3-dichloropropane as follows:

$$ClCH_2.CH_2.CH_2Cl + \underset{\text{Metal}}{Zn} \longrightarrow \underset{\text{Propane}}{\text{cyclo-}(CH_2)_3} + ZnCl_2$$

It is a colourless sweet smelling gas. It is administered in a closed circuit. It is available in metal cylinders in compressed form. Cylinders should bear the name or chemical symbol of the gas. It should be kept in a cool place.

Advantages

1. It is a potent anaesthetic and has a fast induction. Recovery is also rapid and smooth.
2. It maintains blood pressure and cardiac contractility even on prolonged administration, hence it is an ideal anaesthetic agent in haemorrhagic shock.
3. It does not produce irritation of respiratory passage.
4. It is a good muscle relaxant.
5. It is a valuable anaesthesia for major thoracic and abdominal surgeries in patients with pulmonary infections or heart disease, as it produces quiet breathing, hence there is enough supply of oxygen.
6. It does not produce any visceral damage.

Disadvantages

1. If induction is not done carefully, it produces tachypnoea breath-holding, coughing etc.
2. It may produce cardiac irregularities.
3. It depresses respiratory centre.
4. Cyclopropane may increase capillary oozing.
5. Anaesthetic mixture of cyclopropane and oxygen is explosive and this is a great disadvantage.

Intravenous Anaesthetics or Non-volatile Anaesthetics

These anaesthetics produce unconsciousness when they are administered parenterally or intravenously.

They induce anaesthesia rapidly. There are some ultra short acting barbiturates like thiopental sodium and some non-barbiturates such as eugenol derivative (*e.g.* Prapanidiol), phencyclidine, derivative (*e.g.* ketamine), steroid (*e.g.* althesin), etomidate. These are some of the important drugs which are used as non-volatile general anaesthetics.

(*i*) **Thiopental sodium**. It is a yellowish white hygroscopic powder and has a characteristic odour and a bitter taste. It is soluble in water and its solution is freshly prepared before using it in injection form. It is marketed in sterile form in a sealed container, in the specified amount of water for injection.

Thiopentone sodium has gained a reputation as a "truth serum". It is believed that under its effect people speak freely and truthfully. It is also used in psychiatric treatment.

It is a derivative of barbituric acid and is obtained from malonic ester in the following:

$$CH_2(COOC_2H_5)_2 \text{ (malonic ester)} \xrightarrow[C_2H_5I]{C_2H_2ONa} (C_2H_5)(H)C(COOC_2H_5)_2$$

$$\xrightarrow[\text{(ii) } CH_3CH_2CH_2CHI\,CH_3]{\text{(i) } NaOC_2H_5} (C_2H_5)(CH_3(CH_2)_2CH(CH_3))C(COOC_2H_5)_2 \xrightarrow{\text{Urea}} (C_2H_5)(CH_3(CH_2)_2CH(CH_3))C{<}\begin{matrix}CO-N=\\ CO-HN-\end{matrix}{>}C-ONa$$

$$\xrightarrow[C_2H_5ONa]{(H_2N)_2C=S} (C_2H_5)(CH_3(CH_2)_2CH(CH_3))C{<}\begin{matrix}CO-N=\\ CO-HN-\end{matrix}{>}C-SNa$$

Thiopental Sodium

Chemically it is named as 5-ethyl-5 (1-methyl butyl)-2 thio-barbiturate.

When administered intravenously, it produces complete anaesthesia of short duration. The patient passes through the stages of hypnosis and deep sleep to anaesthesia. First the consciousness is lost and then reflex activity is lost and lastly the medullary centres are lost.

It is used as anaesthetic agent for operations of short duration such as fracture, dental surgeries etc. It is also used as an anti-convulsant.

Advantages

1. It is non-explosive and is easy to administer.
2. Its induction is rapid and without any unpleasant feeling or irritation.
3. The tendency of nausea, vomiting and excitement is much less.
4. It does not effect respiration and recovery is fast.
5. Generally there are no post-anaesthetic complications.

Disadvantages

1. It needs constant supervision to avoid an overdosage.
2. It depresses vasomotor centre and myocardium.
3. It is a poor analgesic.
4. It does not produce adequate muscular relaxation.
5. Sometimes during induction, it may produce even fatal reaction like cessation of respiration etc.

(*ii*) **Methohexitone**. It is a methylated derivative of oxybarbiturates and chemically it is named as 5-allyl, 5(1-methyl butyl) barbiturate.

It is a white crystalline powder and is only sparingly soluble in water. It is used as its sodium salt and methohexitone is more potent than thiopentone.

It is used as 1 per cent solution in injections and is less irritant to veins than thiopental.

Its uses, advantages and disadvantages are similar to other barbiturates like thiopental sodium.

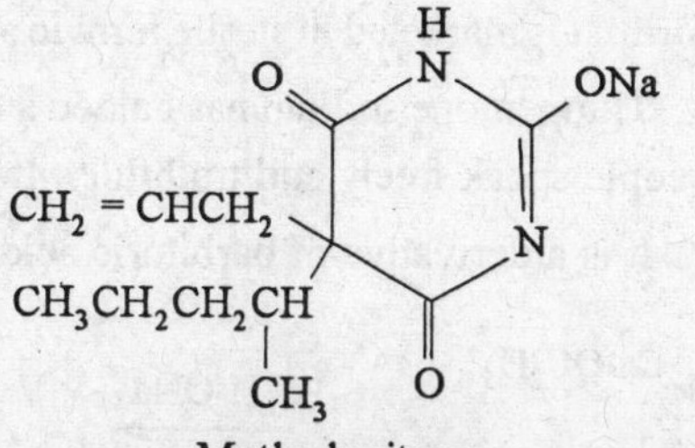

H N O ONa $CH_2 = CHCH_2$ N $CH_3CH_2CH_2CH$ CH_3 O

Metho hexitone

(*iii*) **Propanidid**. It is eugenol derivate and chemically it is named Propyl 3-methoxy 4 [N, N diethyl acetamidoxy] phenyl acetate.

It is colourless or greenish yellow oily liquid. It has a faint odour and is miscible with alcohol, chloroform and ether. It is less potent than thiopental.

It is a short acting anaesthetic and analgesic.

It has many disadvantages like a fall in blood pressure, respiratory depression and may produce involuntary movements such as tremors etc. It also produces nausea, vomiting, headache, excessive salivation etc.

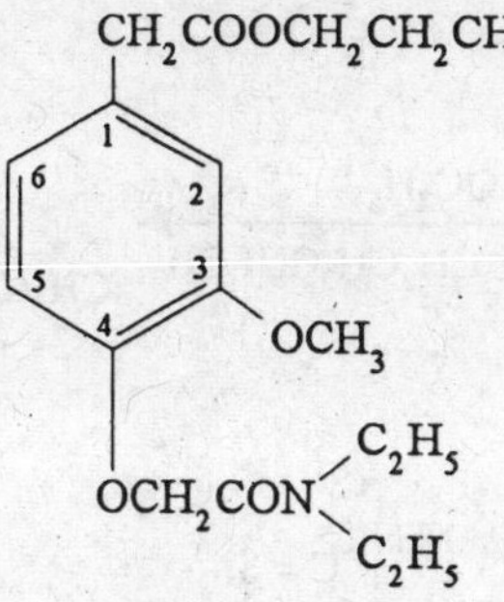

$CH_2COOCH_2CH_2CH_3$ 1 2 3 4 5 6 OCH_3 OCH_2CON C_2H_5 C_2H_5

Local Anaesthetics

These substances are applied topically or by injection and abolishes sensation in localized area. They are directly applied to peripheral nervous tissue and block nerve conduction and are generally applied to somatic nerves, cell body, etc.

Local anesthetics should have the following requisites :

(*i*) It should be non-irritant.
(*ii*) It should not be painful.
(*iii*) It should have low toxicity.
(*iv*) It should be stable.
(*v*) It should produce anaesthesia for sufficient time, to carry out the operation.
(*vi*) It should not damage nerve cells.
(*vii*) Sterilisation of these drugs should be easy.

Local anaesthetics could be divided into the following groups.

1. Natural *e.g.* cocaine.
2. Synthetic nitrogenous compounds.
   (*a*) Derivatives of para aminobenzoic acid *e.g.* procaine, benzocaine, orthoacine.
   (*b*) Derivative of acetanilides *e.g.* lignocaine.
   (*c*) Derivatives of Quinoline *e.g.* cinchocaine.
3. Synthetic non-nitrogenous compounds, *e.g.* benzyl alcohol, propanediol.
4. Miscellaneous drugs *e.g.* clove oil, phenol, chlorpromazine etc.

Local anaesthesia is produced in many ways. Such as

(*a*) *Surface anaesthesia*—These are generally used as ointments to relieve pain due to burns, ulcers etc.

(*b*) *Infiltration anaesthesia* is produced by direct exposure to drug and the drug is administered subeutaneously.

(*c*) It can also be administered through *spinal anaesthesia* when the drug is injected within the dual membrane, surrounding the spinal cord.

The local anaesthetics can also be classified into two broad categories. (*i*) ester (C_6H_5—COO—R) and (*ii*) amides (C_6H_5—CONH—R).

The Esters

1. Cocaine. It is an alkaloid obtained from the leaves of Erythroxylon Coca (Coca-tree). It is methyl benzoyl ester of Ecgonine, which is a tropane derivative. This compound is basic in nature and is levorotatory.

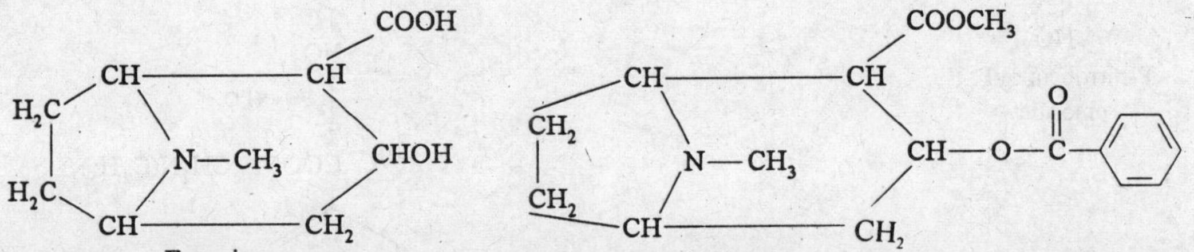

Ecgonine

Cocaine

Cocaine is a white crystalline powder having a bitter taste. It is sparingly soluble in water, but its hydrochloride is quite soluble in water and is hygroscopic in nature.

Advantages

It is a surface anaesthetics and is used in the surgery of the nose, throat and eye. Besides its local anaesthetic action it also stimulates central nervous system.

It has many disadvantages. It is a drug of addiction and dilates the pupils and raises the blood pressure. It produces euphoria and fatigue. It is poisonous in nature and allergic reactions to this drug are frequent.

The structural activity of Cocaine suggested that its anaesthetic activity is due to the presence of

$$C_6H_5\text{—COO}(CH_2)_n\text{—N}\begin{matrix} R \\ R' \end{matrix}$$

Keeping this structure in mind, some *p*-amino benzoic acid derivates were prepared and tested for their activity. Among the derivatives of *p*-amino benzoic acid some important drugs used as local anaesthetics are as follows.

2. Benzocaine. It is synthesized from *p*-nitrotoluene in the following:

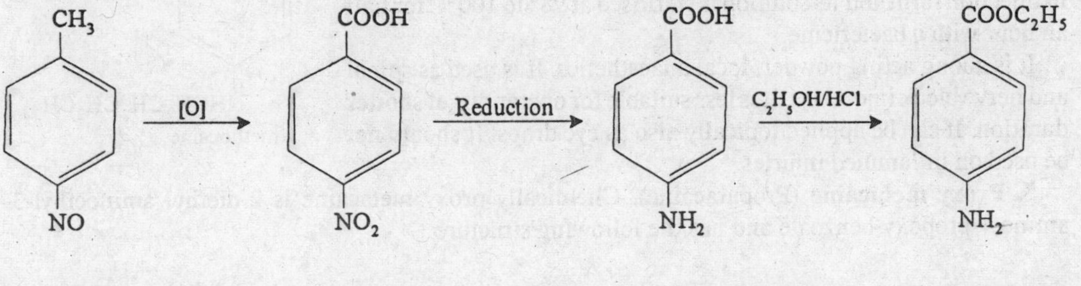

$$\text{(}CH_3\text{, NO)} \xrightarrow{[O]} \text{(COOH, }NO_2\text{)} \xrightarrow{\text{Reduction}} \text{(COOH, }NH_2\text{)} \xrightarrow{C_2H_5OH/HCl} \text{(}COOC_2H_5\text{, }NH_2\text{)}$$

Benzocaine is an ester derivative and chemically it is ethyl-4 aminobenzoate.

It is a white crystalline powder. It is stored in air tight and light resistant containers. It is sparingly soluble in water, but its salt with *p*-phenol sulphonic acid is soluble in water. It is a weak base. It possesses low toxicity. It is also available under the name anaesthesin. It is useful as a dusting powder. It is also used in throat lozenges. Orally it is used for relieving pain of gastric ulcer. It is also available in ointment form and is a surface anaesthetic.

3. Procaine (novocaine, seurocaine). It is diethyl amino, aminoethyl ester of para aminobenzoic acid and is synthesized in the following way.

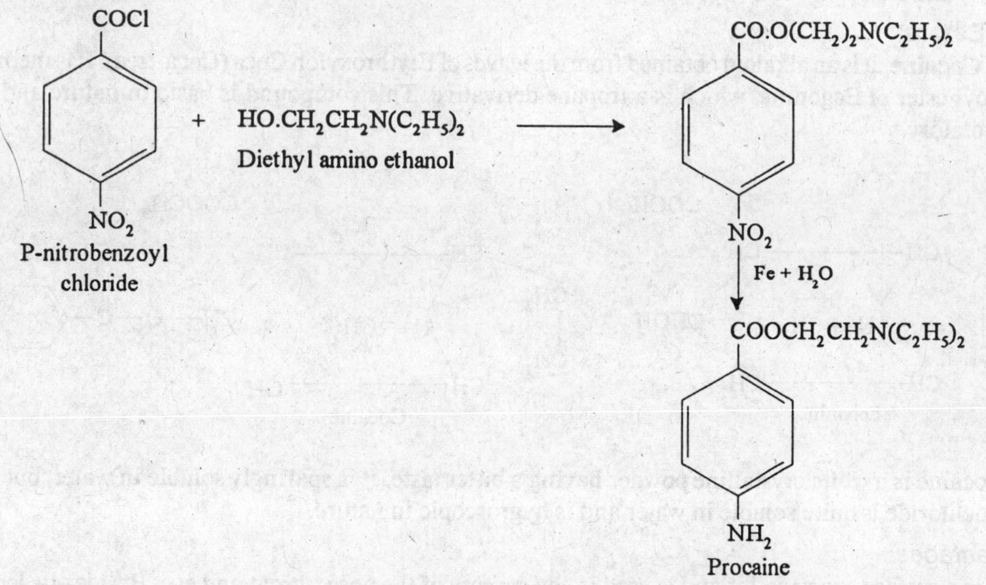

COCl (benzene ring) NO_2
P-nitrobenzoyl chloride
+ $HO.CH_2CH_2N(C_2H_5)_2$
Diethyl amino ethanol
→ $CO.O(CH_2)_2N(C_2H_5)_2$ (benzene ring) NO_2
Fe + H_2O
↓
$COOCH_2CH_2N(C_2H_5)_2$ (benzene ring) NH_2
Procaine

Chemically it is named as 2-diethyl-amino ethyl-*p*-aminobenzoate. The free ester is an oil but it is used as the hydrochloride salt, which is stable and is soluble in water and alcohol. It is an odourless, white crystalline powder. Procaine is rapidly hydrolysed by certain enzymes in the plasma and liver.

It is ineffective in topical use and is generally available in injection form. It is a nerve block and spinal anaesthetic. It has vasodilator property. Its activity is much less than Cocaine, but it is an important local anaesthetics because it is non-irritant and non-habit forming drug.

Its main disadvantage is that it is poorly absorbed from mucous membrane and hence has no topical use.

4. Amethocaine (Anethaine pontocaine). It is a derivative of dimethyl amino ethyl benzoate. The *p*-amino group is substituted by an butyl group.

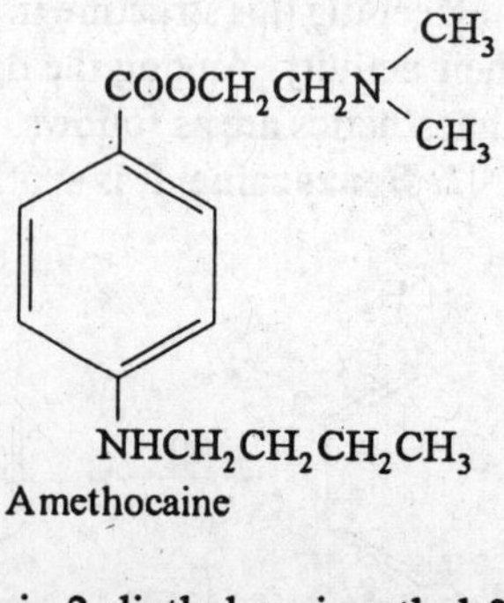

Chemically it is 2-dimethyl amino ehtyl-p-butyl amino benzoate. Its hydrochloride salt is used as the local anaesthetics. It is available in injection form and its solution is sterilised at 98° to 100°C for half an hour with a bactericide.

$COOCH_2CH_2N(CH_3)_2$ (benzene ring) $NHCH_2CH_2CH_2CH_3$
Amethocaine

It is a long acting powder, local anaesthetics. It is used as spinal and nerve block injections. It is less suitable for operations of shorter duration. It can be applied topically also as eye drops. It should not be used on inflammed injuries.

5. Proxy metacaine (Proparacaine). Chemically proxy metacaine is 2-diethyl aminoethyl-3-amino 4-propoxy-benzoate and has the following structure.

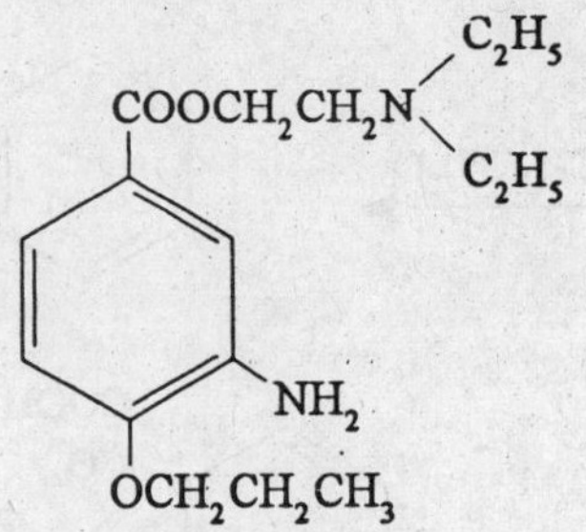

$$\text{COOCH}_2\text{CH}_2\text{N}(\text{C}_2\text{H}_5)_2$$

NH_2

$OCH_2CH_2CH_3$

Proxymetacaine

It is a derivative of procaine, generally its hydrochloride salt is used and is white or buff colour powder, which is soluble in water and gives acidic solution.

It is mainly used as topical local anaesthetic in ophthalmic preparations. It is more potent than amethocaine.

The amides

The acetanilide and methyl acetanilide groups are used as antipyretic and analgesic drugs and by keeping this structure in view, the $COCH_3$ group of these amides were elaborated to $COCH_2NR_2$ and the resulting compounds were tested and were found to have local anaesthetic properties. Some of the important drugs are as follows :

1. Lignocaine (xylocaine, lidocaine). Lignocaine is the most important and commonly used drug of this group. It is synthesized in the following way.

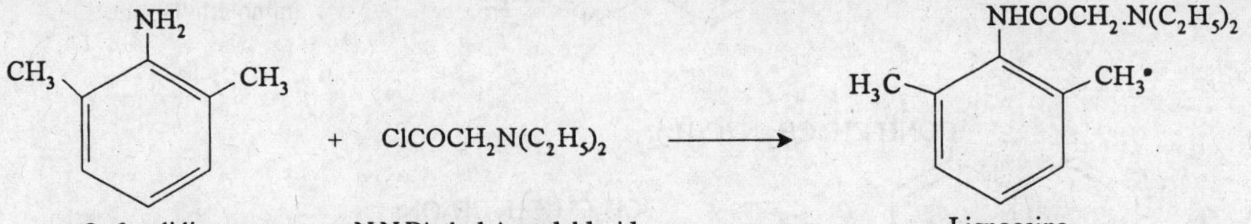

NH_2, CH_3, CH_3 (2, 6 xylidine) + $ClCOCH_2N(C_2H_5)_2$ (N,N Diethyl Acetylchloride) ⟶ $NHCOCH_2.N(C_2H_5)_2$, H_3C, CH_3 (Lignocaine)

2, 6 xylidine — N,N Diethyl Acetylchloride — Lignocaine

Chemically it is named N-diethyl amino acetyl-2,6-xylidine. It is a crystalline solid and is sparingly soluble in water. It is used as its hydrochloride salt. Its main advantage is that it is extremely resistance to hydrolysis. It is very stable and can be stored for a long time at room temperature.

It is used topically as surface anaesthetics and as ointment. It is also available in injection form as spinal nerve block injections. It has a quick onset of action. Addition of adrenaline prolongs its action for about two hours. It is also a good analgesic agent and is used for dental analgesia. It is also used in cardiac arrythmias. It is also used for the patients who are allergic to procaine.

Its toxicity is similar to that of other local anaesthetics and it may cause drowsiness.

2. Cinchocaine hydrochloride (nupercaine). It is an amide of quinoline derivative and is a potent anaesthetics. It is synthesized from Isatin in the manner shown on the following page.

Cinchocaine. Chemically it is named as 2-butaoxy-4 (N (2-diethyl amino ethyl) carboxamide) quinoline. It is a crystalline substance having slight bitter taste. It is soluble in water. On exposure to light, it gets darkened, hence it should be stored in light resistance containers. It is used locally in the ointment form and it is used as solution form as injection for spinal anaesthesia.

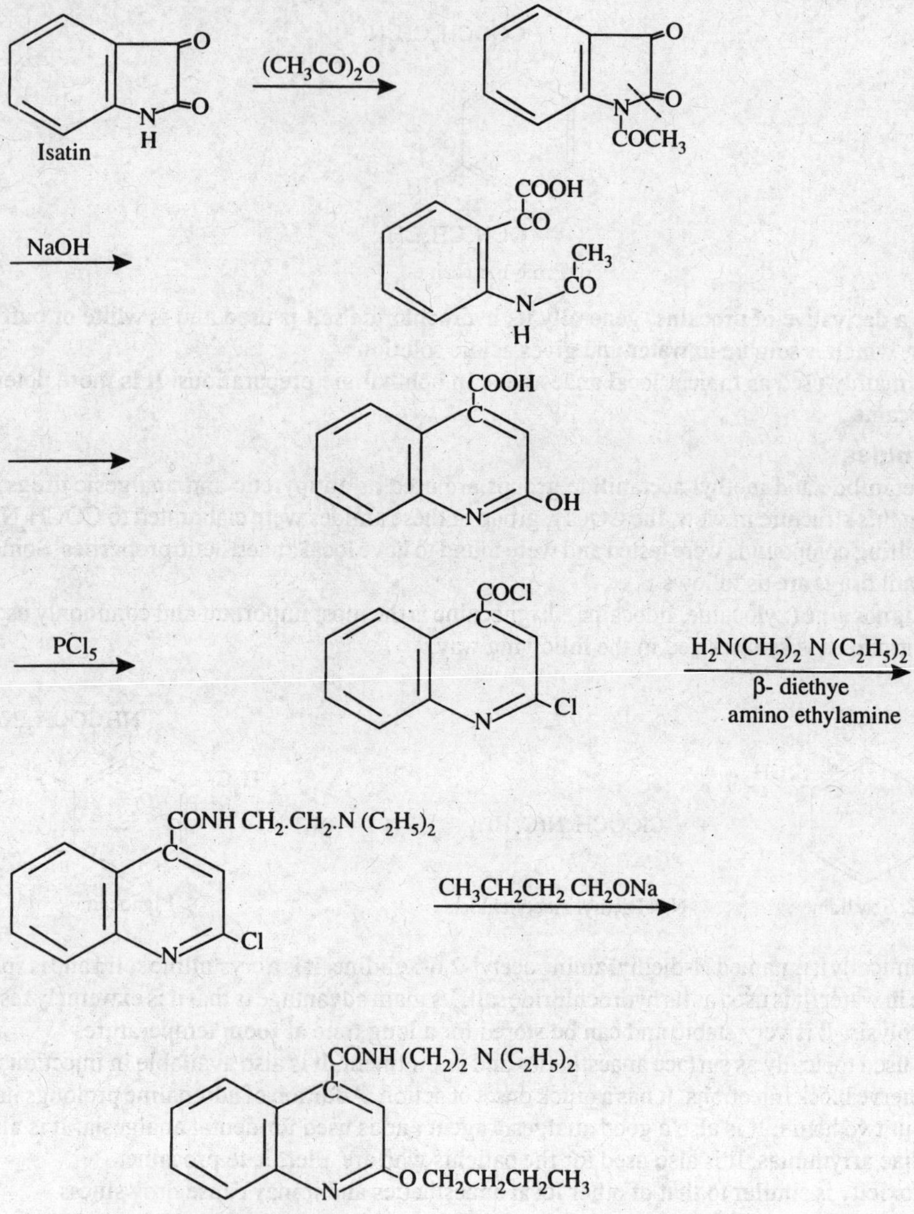

Isatin $\xrightarrow{(CH_3CO)_2O}$ N-acetyl isatin (N–$COCH_3$)

$\xrightarrow{NaOH}$ $COOH$–CO–C_6H_4–NH–CO–CH_3

$\longrightarrow$ 2-hydroxyquinoline-4-carboxylic acid ($COOH$, OH)

$\xrightarrow{PCl_5}$ 2-chloroquinoline-4-carbonyl chloride ($COCl$, Cl)

$\xrightarrow[\beta\text{- diethye amino ethylamine}]{H_2N(CH_2)_2\ N\ (C_2H_5)_2}$

$CONH\ CH_2.CH_2.N\ (C_2H_5)_2$ (2-chloroquinoline, Cl)

$\xrightarrow{CH_3CH_2CH_2\ CH_2ONa}$

$CONH\ (CH_2)_2\ N\ (C_2H_5)_2$

$O\ CH_2CH_2CH_2CH_3$

Cinchocaine

Chapter 18

Anti-Septics and Disinfectants

The term 'Septic' is derived from a Greek word *septikos* which means putrify or rot, and antiseptic and disinfectants are the chemical substances which are used to kill or inhibit the growth of pathogenic micro-organisms. Antiseptics and disinfectants are the drugs which are widely used by the lay public. These agents are used in their original form or are often added to easily available everyday essential articles like soaps, toothpaste, after shave lotions etc. An ideal antiseptic and disinfectant agent should have a wide antimicrobial spectrum and should destroy bacteria, viruses, spores, protozoa, fungi etc. If these are used for disinfection of surgical instruments, it should not corrode the metal. It should have quick action, high penetrating power and chemical stability.

Distinction between Disinfectants and Antiseptics

Antiseptics are those chemical compounds which prevent the sepsis of wounds by inhibiting the action of micro-organism either by stoping their reproduction or by killing them. These are anti-infective agents which are applied to living tissues and are mainly bacteriostatic.

On the other hand disinfectant are the chemical compounds which are applied to non-living or inanimate objects and they kill the organisms immediately.

An antiseptic prevents the growth of micro-organism as long as it remains in contact, while a disinfectant kills the organism. Thus an antiseptic action is milder but more prolonged whereas the action of disinfectant is immediate but is of short duration.

Disinfectants are mainly used in public health sanitation and for cleaning water etc., whereas antiseptics are used for wounds, for sterilising surgical apparatus and are used by surgeons, nurses and dentists for washing their hands.

Sterilisation is a process which is defined as complete destruction of all living micro-organisms and freeing an article, a medium or surface of all micro-organisms. Sterilisation can be achieved by various methods such as physical agents e.g. heat, radiation, or by chemical means, e.g. acids, alcohols, phenols, dyes, heavy metals etc.

Standardization of Disinfectants and Antiseptics

These drugs are standardised in terms of phenol coefficient or Rideal-Walker coefficient.

Phenol coefficient is defined as the dilution at which it will kill bacteria in a given time (which is generally 10 minutes) compared to the dilution required for phenol.

$$\text{Phenol coefficient} = \frac{\text{Germicidal dilution of the compound}}{\text{Germicidal dilution of Phenol}}$$

Standardization of these drugs are done by Rideal-Walker method. In this method, first the minimum concentration of phenol which is required to kill a 24 hours culture of B. typhosus in a known duration of time is determined and then the concentration of the unknown disinfectant is determined which produces the same effect under the same condition. The phenol co-efficient is determined by substituting these values in the above formula.

Phenol coefficient values of some important disinfectants are mentioned in the Table-1 given below.

Table 1

	Name of the Disinfectant	*Phenol Co-efficients (activity against E. Typhosus)*
1.	Phenol	1.0
2.	Chlorophenol	3.6
3.	2, Chlorophenol	3.9
4.	2, 4, 6 Trichlorophenol	23.0
5.	4-n Butyl Cresol	100.0
6.	4-n Amyl Cresol	280.0
7.	Thymol	30.0

Various Antiseptic and Disinfectant could be studied under the following headings :

1. Phenols and related compounds.
2. Halogen compounds.
3. Dyes.
4. Organic mercurials.
5. Formaldehyde and its derivates.
6. Nitrofuran derivates.
7. Cationic surface active agents.

1. Phenols

It is also known as carbolic acid and is one of the oldest antiseptic. It was introduced in 1867 by Lister as an antiseptic in surgery.

Chemically phenol is hydroxy benzene.

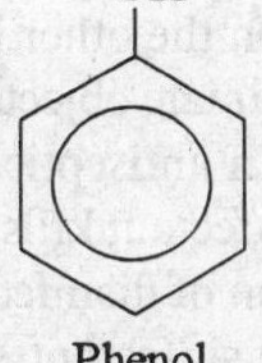

Phenol

It is a colourless or pinkish needle-shaped deliquescent crystals having a characteristic smell. It is soluble in alcohol, water, glycerine, fixed oil etc. On exposure to air it turns pink.

It is a potent bactericidal and fungicidal agent and acts by denaturing the bacterial proteins. Hence it is called a general proto-plasmic poison injuring the micro-organisms and the cells of the body tissue. Due to this reason, it is used as a standard against which the efficiency of the other antiseptics are measured. If phenol is administered orally, it causes severe gastrointestinal corrosion, pain, vomiting, shock and it may cause convulsion which sometimes become fatal.

Some important preparations of phenols and their uses are as follows :

(*i*) Liquified phenol has 80% W/W of phenol in water. It is effective against Gram positive and Gram-negative bacteria.

(*ii*) Its dilute solutions (0.2 to 1.0%) are bacteriostatic where as stronger concentration are bactericidal in action.

(*iii*) Its 20% solution is used as dressings for small wounds, liquified phenol is used in dentistry due to its anaesthetic effect.

(*iv*) Phenol glycerin contains 16 per cent W/W of phenol in glycerin. These are used as ear drops and mouthwash.

(*v*) Its solution in glycerol is used as an analgesic and antiseptic in mouth ulcers and tonsilitis. This agent has been occasionally used as a 5% solution in almond oil for sclerosing haemorrhoids.

Phenol is no longer used as an antiseptic as its derivatives have been found to be less toxic and better antiseptic drugs. Some of the important derivatives of phenols are as follows:

(*a*) **Alkyl substituted phenols**. Introduction of an alkyl group in the phenol increases its germicidal properties. Alkyl substituted phenols are as follows:

(*i*) *Cresol* : It is a mixture of ortho, para, meta derivaties of phenol.

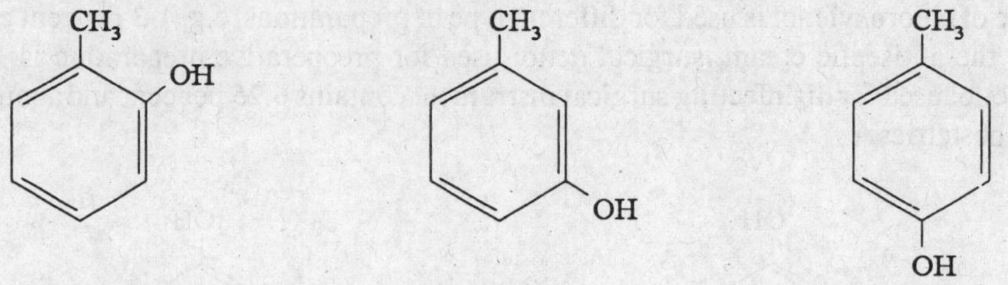

O-Cresol M-Cresol p-Cresol

It is found as pale yellow liquid and has the characteristic smell of phenol. It is soluble in water and alcohol and is used as 2 per cent solution.

Cresol is much more active than phenol but its toxicity is nearly equal to that of phenol. It is used as a general disinfectant.

(*ii*) *Lysol* : It is a saponified preparation of cresol with a vegetable oil or mixed fatty acids. It contains 50% V/V of cresol. Lysol is a disinfectant which is very useful for domestic and hospital use. It is also used for disinfecting inanimate things and animal waste products.

(*iii*) *Thymol* : It is a naturally occurring antiseptic and is obtained from the oils of thyme as follows :

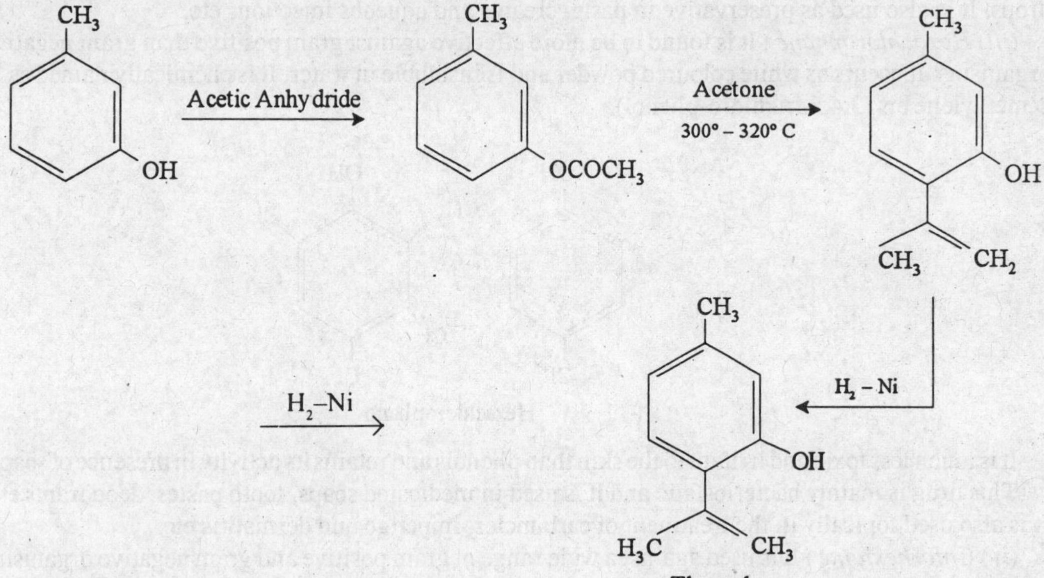

Acetic Anhydride; Acetone 300° – 320° C; H_2 – Ni

Thymol

It is 2 isopropyl-5 methyl phenol and it is a white crystalline solid with a characteristic pungent smell. It is sparingly soluble in water, alcohol and vegetable oils. It is stored in light resistant and tight closed containers.

Thymol has both antiseptic and antifungal properties. These are used in antiseptic lotions, mouth washes, pain balm, tooth ache drops etc. It is also used as an antioxidant.

(*b*) **Chlorinated Phenols**. Introduction of chlorine in phenols increases its antiseptic property. Important chlorinated phenols are as follows :

(*i*) *Chloro Xylenol* : It is popularly known as Dettol, chemically it is named as 4 chloro-3, 5 Xylenol.

It is a white crystalline substance and has a characteristic odour. It is soluble in alcohol and only sparingly soluble in water. It is effective against gram positive and gram negative organisms. Different percentage of chloroxylenol is used for different type of preparations, e.g. 1-3 per cent concentration is used in the antiseptic cream, surgical dettol used for preoperative preparation is 1.44 per cent whereas dettol used for disinfecting surgical instrument contains 6.25 per cent and about 2.5 per cent is used in obstetrics.

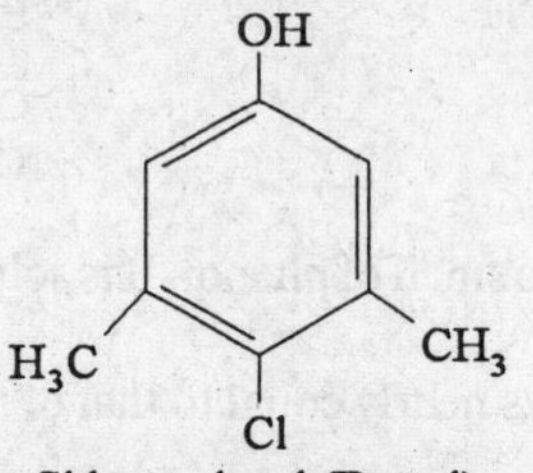

OH, H_3C, CH_3, Cl

Chloroxylenol (Dettol)

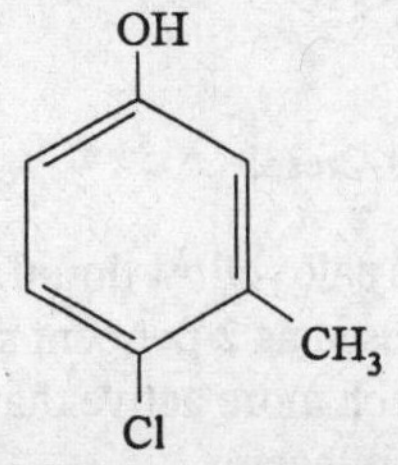

OH, CH_3, Cl

Chlorocresol

(*ii*) *Chlorocresol* : It is 4-chloro-3-methyl phenol and is a white crystalline solid with a phenolic odour.

It is soluble in alcohol but only sparingly soluble in water.

It is a potent bactericide with low toxicity. Its 0.05 per cent solution is used for preservation of eye drops. It is also used as preservative in paste, creams and aqueous injections etc.

(*iii*) *Hexachlorophane* : It is found to be more effective against gram positive than gram negative organisms. It occurs as white coloured powder and is insoluble in water. It is chemically named as 2, 2'methylene bis (3,4,6 trichloro-phenol).

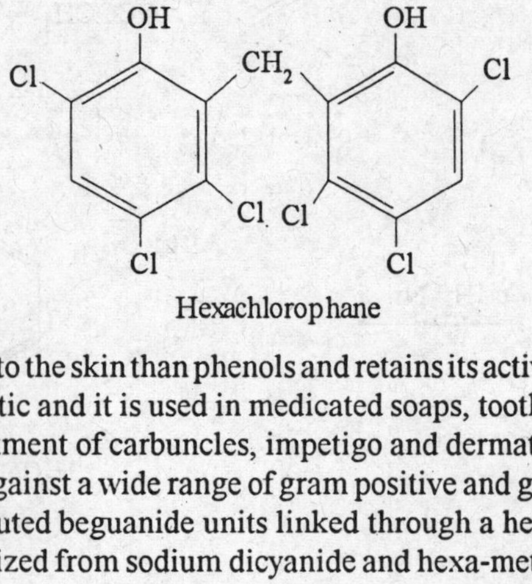

OH, OH, Cl, CH_2, Cl, Cl, Cl, Cl, Cl

Hexachlorophane

It is much less toxic and irritant to the skin than phenols and retains its activity in presence of soaps.

This drug is mainly bacteriostatic and it is used in medicated soaps, tooth pastes, deodorants etc. It is also used topically in the treatment of carbuncles, impetigo and dermatitis etc.

(*iv*) *Chlorhexidine* : It is used against a wide range of gram positive and gram negative organisms (it has two *p*-chlorophenyl substituted beguanide units linked through a hexamethyline unit) gram negative organisms and is synthesized from sodium dicyanide and hexa-methylene diamine and has the following structure.

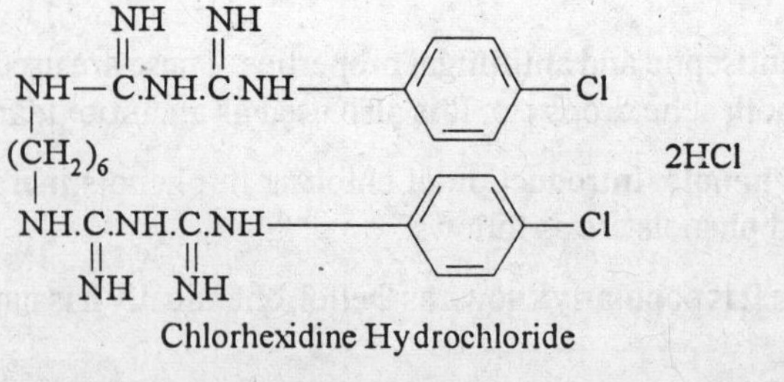

NH NH
NH—C.NH.C.NH—(C₆H₄)—Cl
$(CH_2)_6$ 2HCl
NH.C.NH.C.NH—(C₆H₄)—Cl
NH NH

Chlorhexidine Hydrochloride

It is a non-toxic and non-irritant antiseptic. It is used along with cetrimide for the preparation of Savlon. Its gluconate solution is used for the treatment of burns and wounds. It is also used as an ingredient in dusting powders, creams, ointments, lozenges etc.

2. Halogen Compounds

Halogens and halogen compounds are extensively used as disinfectants and as antiseptics.

Chlorine is one of the most commonly used halogen which is used for purification of water. Chlorine exerts a bactericidal action against a variety of gram positive and gram negative organisms in very low concentrations. The antibacterial activity of chlorine is mainly due to the formation of hypochlorous acid.

Other important halogen compounds are :

(*i*) **Chloramines**. These are N-Chloro compounds and liberate hypochlorous acid in contact with water. Its anti-septic property is greatest at pH 7 and its action decreases as the solution becomes more alkaline or acidic. This is an organic chloride and 2% solution is used as disinfectant for wounds and its 0.2 to 0.5 per cent solution is used for mouth wash etc.

(*a*) *Chlorarmine-T or Chlorozone:* Chemically it is named as Sodium-p-toluene sulphone chloramide and is synthesized form *p*-toluene sulphonyl chloride as follows:

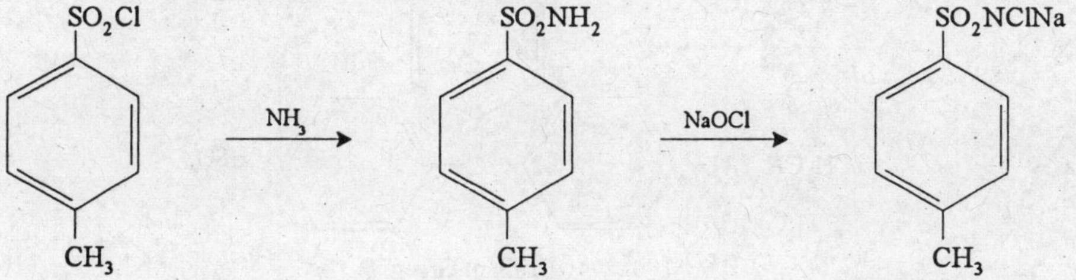

SO_2Cl (p-toluene sulphonyl chloride, CH_3) $\xrightarrow{NH_3}$ SO_2NH_2 (CH_3) $\xrightarrow{NaOCl}$ SO_2NClNa (CH_3)

(Chloramine-T)

It occurs as trihydrate crystalline yellowish white powder and has a faint smell of chlorine. It is soluble in water and alcohol.

It is less irritant and is used as a disinfectant lotion in some infectious diseases like measels, scarlet fever etc. It is also used as mouth wash.

(*b*) *Dichloramie T:* It has the structure as given alongside.

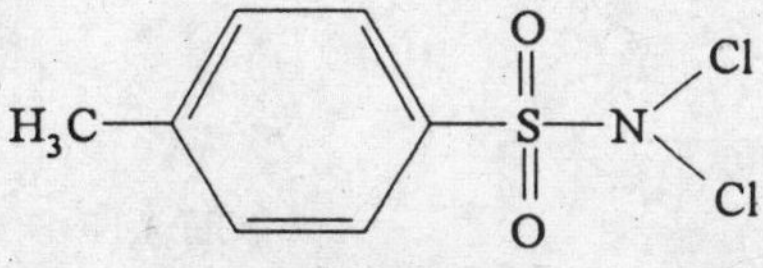

$H_3C-C_6H_4-SO_2-NCl_2$

Dichloramine-T

It is soluble in alcohol and it is used as disinfectant in 5% solution for dressing wounds and its 1% solution is used as ointment for mucous surfaces.

(*ii*) **Halozone.** Chemically it is p-dichloro sulphamoyl benzoic acid.

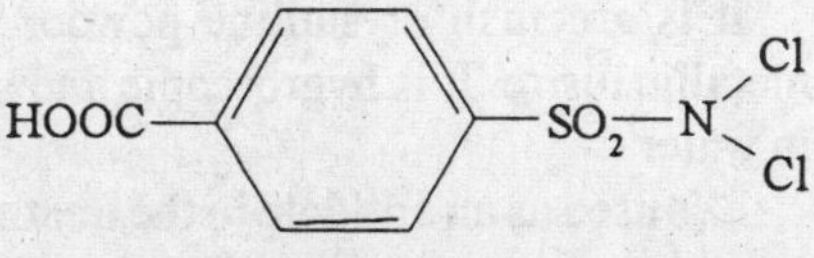

It contains 25% of active chlorine and is mainly used for purification of drinking water.

$HOOC-C_6H_4-SO_2-NCl_2$

(*iii*) **Chlorinated lime or Bleaching powder.** This is also extensively used for purification of drinking water.

3. Dyes

Natural dyes have been used since ancient times in foods, medicines, cosmetics etc. Now-a-days these natural dyes have been replaced by more stable synthetic dyes and these are used as antiseptics, chemo-therapeutics, colouring agents etc.

Some important dyes are as follows :

(*i*) **Crystal Violet (gentian violet).** It occurs as a dark green powder and has metallic lustre. It is sparingly soluble in water and has the following structure.

It is a triphenyl methane dye and is named as 4-4, 4' (dimethyl amino-benzhydrylidene) cyclohexa-2,5 dien-1 yliedene-dimethyl ammonium chloride. It gives deep blue colour with water.

It is a potent antiseptic and is effective against gram positive organisms. Its aquous solution is applied topically in the treatment of burns, boils, skin infections, chronic ulcers etc. It is also used as anthelmentic agent.

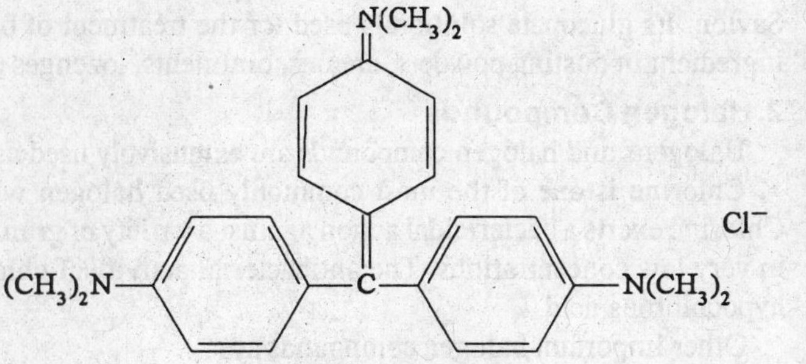

$N(CH_3)_2$ Cl^- $(CH_3)_2N$ C $N(CH_3)_2$

Crystal Violet

(*ii*) **Brilliant green.** This is also triphenylmethane dye and chemically named as 4 (4-diethylaminobenzhydrylidene) cyclohexa-2, 5-dien-1-ylidenediethyl ammonium hydrogen sulphate.

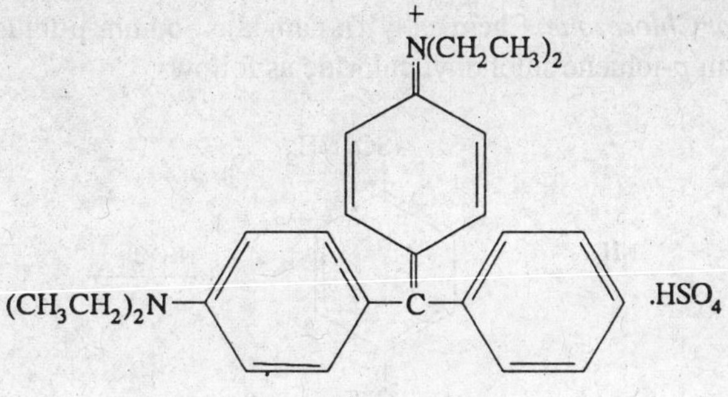

$\overset{+}{N}(CH_2CH_3)_2$ $(CH_3CH_2)_2N$ C $.HSO_4$

Brilliant Green

It occurs as golden crystals and is soluble in water and alcohol.

Its 1% aqueous solution is used as an antiseptic for wounds, burns, impetigo, minor injuries. Its 0.02% solution in water is used by aerosol inhalation.

(*iii*) **Methylene Blue.** It is a thiazine dye and has a following structure.

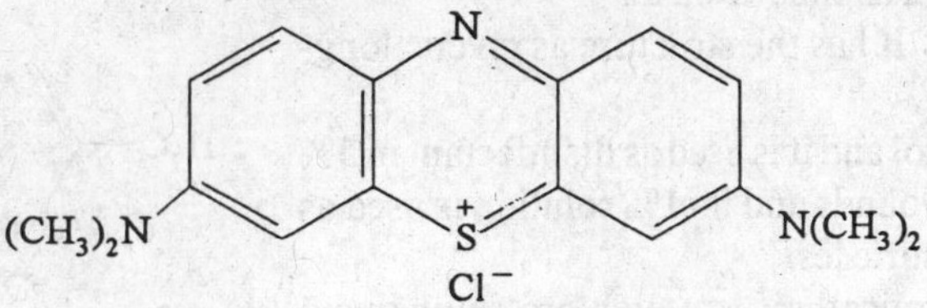

N $(CH_3)_2N$ $\overset{+}{S}$ $N(CH_3)_2$ Cl^-

Methyelene Blue

It is greenish crystalline powder with a metallic lustre. It is hygroscopic and soluble in water.

It is used as an antidote in the treatment of cyanide poisoning. It is weak antiseptic and is used as an intestinal and urinary antiseptic.

(*iv*) **Mercurochrome.** It is important Xanthene dye and is a mercurial fluorescein dye. It has the adjoining structure.

It is used as an antiseptic in 1 to 5% solution for skin and mucous membrane. It is a non-irritating, but a weak skin disinfectant.

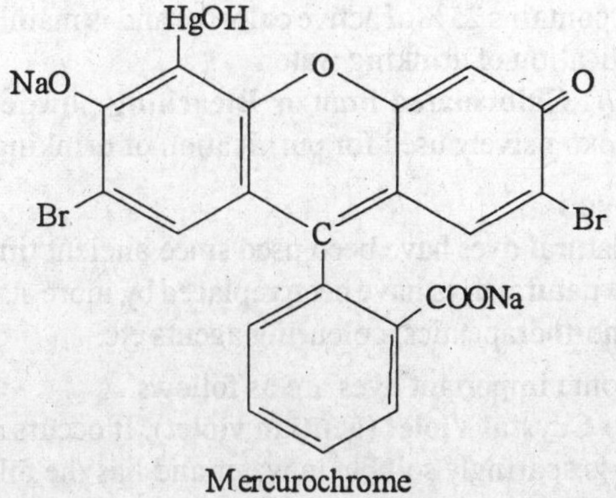

HgOH NaO O O Br C Br COONa

Mercurochrome

4. Organic Mercurials

Many inorganic mercury compounds are well known for their potent germicide activity. These are mainly bacteriostatic and act by inhibiting the sulphydryl enzymes of bacteria. The commonly used inorganic mercurials are yellow mercuric oxide, mercuric oxide eye ointment etc. These are mainly used as eye ointments.

Some of the organic compounds having mercury have been found to be potent antiseptic agents. In these compounds mercury is found to be covalently bonded to the organic molecule. These compounds are found to be bacteriostatic in action and are less toxic than inorganic mercurials. Some of the important mercurials are as follows :

(*i*) **Thiomersol.** It occurs as cream coloured powder and is soluble in water. It is affected by light and should be marketed in light resistant containers. Chemically it is named as Sodium (O-carboxy-phenylthio) ethyl mercury.

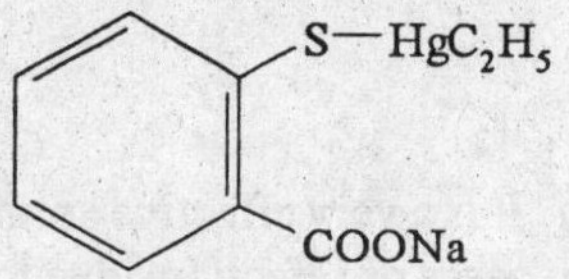

S—HgC_2H_5

COONa

Thiomersol

Its dilute aqueous solutions are used for disinfecting skin before surgical operations and is also used for sterilizing surgical instruments. It is also used for preservation of biological products. Dilute solution of Thiomersal is also used as antiseptic for irrigation purpose.

(*ii*) **Nitromersol.** This is a mercurial compound.

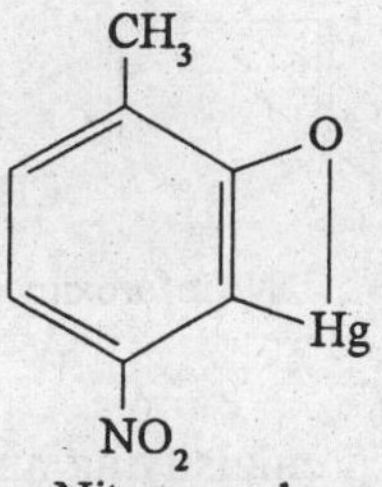

CH_3

O

Hg

NO_2

Nitromersol

It is a powerful antiseptic and its application is similar to Thiomersol. It is used in ophthalmology. It is used as an eye ointment employed in conjuctivitis, corneal ulcers etc.

(*iii*) **Phenyl mercuric nitrate, acetate and borate.** These compounds are used in many commercially available spermicidal creams. These have antibacterial and antifungal properties and are used as preservatives for injections.

5. Formaldehyde and its derivatives

(*i*) **Formaldehyde** is a gas (HCHO) and its 34-38 per cent aqueous solution is used as fumigant. Its solution is a colourless liquid and has a characteristic odour and burning taste.

Its solution in methanol is known as formalin and is highly effective against several bacteria, viruses and fungi. Formalin is used as preservative for pathological specimen. 2-5% solution of formaldehyde is used for disinfecting surgical gloves and instruments. It is also used in dentistry. It is extremely irritant to the mucous membranes and precipitates tissue proteins when it is applied to the skin it hardens the epidermis. Its 3% solution is used for removing warts on the palms and soles.

(*ii*) **Gluteraldehyde.** It is a staurated aldehyde.

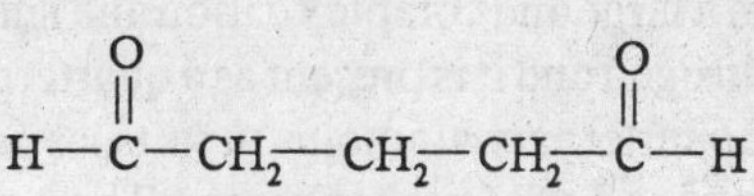

$$H-\overset{O}{\overset{\|}{C}}-CH_2-CH_2-CH_2-\overset{O}{\overset{\|}{C}}-H$$

Glutaraldehyde

It is soluble in water and forms acidic solution.

It is a potent bacteriocide and fungicide. It is less irritant than formaldehyde. It is mainly used for sterilizing surgical instruments.

6. Nitrofuran derivatives

Many nitrofuran derivates are used as antiseptics. Some important derivates are as follows :

(*i*) **Nitrofurazone (Furacin).** Nitrofurazone is chemically 5-nitro-2 furaldehyde semicarbozone. It occurs as yellow crystalline powder and it is soluble in water.

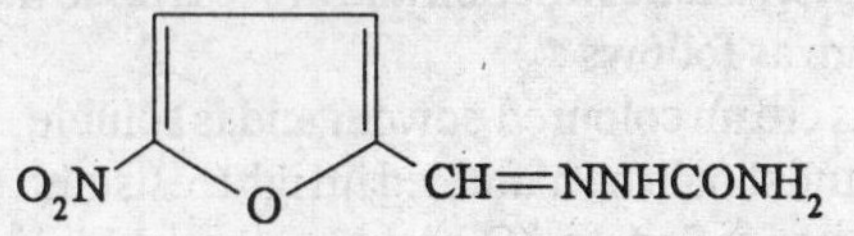

O_2N–(furan ring, O)–$CH{=}NNHCONH_2$

Nitrofurazone

It is a very effective antiseptic against both gram-positive and gram negative organisms. It is used in cases of infections due to burns, ulcers, wound and other skin diseases.

Nitrofurazon blocks an energy transfer by the organism necessary for cell division and its antibacterial action is due to its inhibition of bacterial respiratory enzymes.

(*ii*) **Nitrofuroxime.** It is a bacteriocidal compound and is used in vaginal infections.

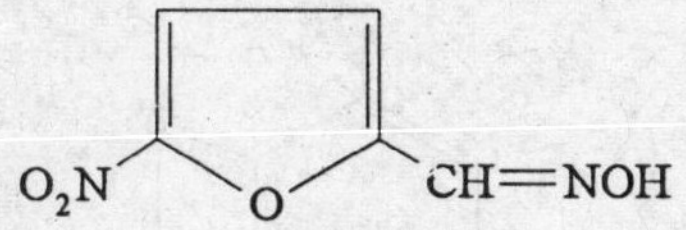

O_2N–(furan ring, O)–$CH{=}NOH$

Nitrofuroxime

7. Cationic Surface Active Agents

Cationic surface agents have marked emulsifying and detergent action and have bactericidal activity against both gram-positive and gram-negative organisms. These compounds modify the characteristics of the interface be'ween the liquids or a liquid and gas or a liquid and a solid or any general surface. These are most effective in neutral solution and their bactericidal activity decreases in the acidic medium. Surfactants are characterised by the presence of hydrophilic (having affinity for water) and hydrophobic [lacking affinity for water or lipophilic (fat soluble) group]. When the lipophilic group is a cation, these surfactants are classified as cationic surfactants, e.g. lauryl triethyl ammonium chloride.

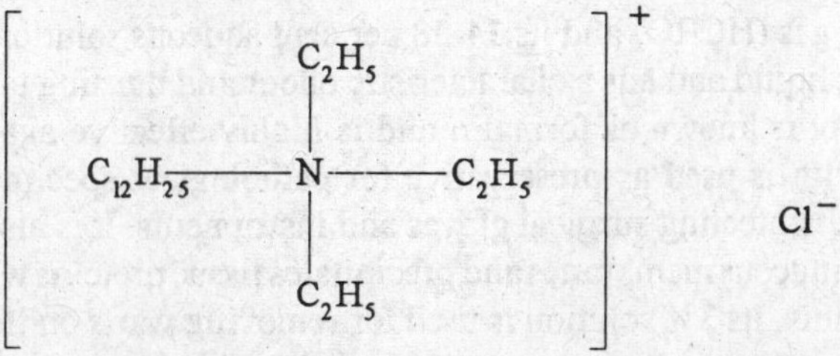

$$\left[C_{12}H_{25} - \overset{\displaystyle C_2H_5}{\underset{\displaystyle C_2H_5}{\overset{|}{\underset{|}{N}}}} - C_2H_5 \right]^+ Cl^-$$

The lipophilic groups i.e. C_2H_5 and $C_{12}H_{25}$ are cations, hence it is cationic surfactant.

These surfactants dissociate into a large and complex cation and smaller inactive anion. The cation generally contains a pentavalent nitrogen and it is present as a quarternary ammonium, pyridinium or piperidinium group or as aliphatic quarternary ammonium salts, as quarternary picolinium salts etc. The hydrocarbon chain usually contains 8 to 16 carbon atoms. These solutions of cationic surfactants are colourless, odourless, non-irritating and non-toxic in nature. These are soluble in water and are non-corrosive to metallic and rubber surgical appliances. These compounds readily combine with proteins and their activity decreases in presence of serum, pus and other organic matter. Quarternary compounds are incompatible with anionic surfactants like soap.

Surfactants are antimicrobial agents and it is believed that they act by lowering surface tension and thus damage the cell membrane resulting in leakage of the cellular metabolites, which is fatal to the bacteria.

Surfactants are widely used as skin disinfectant, and as an disinfectant for clothes, they are also used as creams, ointments, powders which are used in hospitals and also for domestic purposes. At present surfactants are most widely used antiseptics. Some important cationic surfactants are benzalkonium chloride, cetrimide, cetylpyridinium chloride, domiphen bromide, octaphonium chloride, dequalinium chloride etc.

(*i*) **Benzalkonium chloride.** Chemically, it is alkyl dimethyl benzyl ammonium chloride in which the alkyl groups are having eight to eighteen carbon atoms.

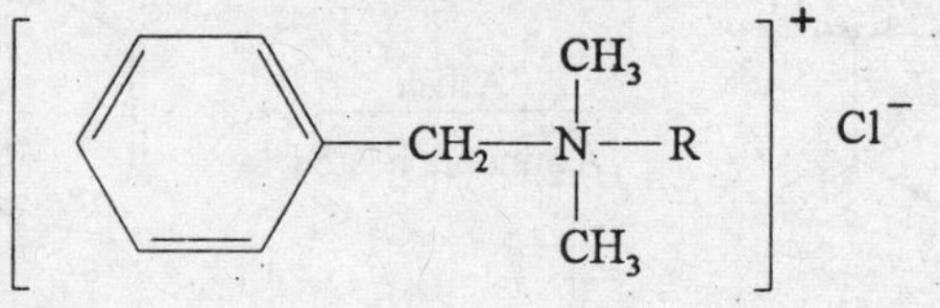

$$\left[C_6H_5-CH_2-\overset{CH_3}{\underset{CH_3}{N}}-R \right]^+ Cl^-$$

where $R = C_8H_{17}$ to $C_{18}H_{37}$

It occurs as white or yellowish white amorphous powder or as gelatinous mass and has a typical aromatic odour.

It is highly soluble in water and its aqueous solution is pale yellow in colour. It is used as detergent, keratolytic, and as an emulsifying agent. These are used for the pre-operative disinfection of unbroken skin. Its dilute solution is as an antiseptic in wounds, burns and in obstetrics. It also finds its use for disinfecting surgical instruments and baby diapers to prevent napkin rashes. Benzalkonium chlorides lozenges are used for the treatment of pharyngeal and it is used as a supplementary for vaginal infections. It is also used as preservative for eye drops.

(*ii*) **Cetrimide.** It contains tetradecyl bromide and some small quantities of inorganic salts such as sodium bromide water etc. Tetradecyl bromide is obtained by the action of trimethyl amine or alkvl bromides in the following way :

$$CH_3.[CH_2]_{13}.Br + \underset{\text{Trimethyl amine}}{N(CH_3)_3} \longrightarrow \underset{\text{Tetradecyl bromide}}{\left[CH_3.[CH_2]_{13}N^+(CH_3)_3 \right] Br^1}$$

It is a creamish, hygroscopic powder, which is readily soluble in water and alcohol. It has a faint odour and soapy taste. It is marketed under the trade name of Cetavlon.

Its applications are similar to that of benzalkonium chloride. Its dilute solution is used as a preservative for eye drops. It is a less toxic antiseptic and detergent and finds use for the treatment of wounds, burn, sterilizing the surgical instruments etc.

(*iii*) **Cetylpyridinium chloride.** Chemically it is 1, hexadecylpyridinium chloride monohydrate.

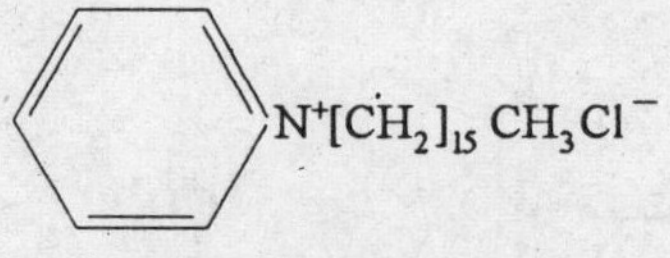

$$C_5H_5N^+[CH_2]_{15}\,CH_3Cl^-$$

Cetylpyridinium chloride

It is a white powder having slight odour and is soluble in water. It is used both as an antiseptic and germicide. It is also used as lozenges and mouth wash for treatment of superficial infections of the mouth and throat. Its solution is used for minor wounds and burns and also for cleansing the skin before operation.

(*iv*) **Domiphen bromide.** It is consisting of dodecyldimethyl-2-phenoxyethyl ammoniaum bromide. It occurs as light yellow crystalline substance and is soluble in water.

It is used as an antiseptic detergent and its 0.02% aqueous solution is used for wounds and for irrigation of the bladder and urethra. It is also used for surgical dressing. It is used as lozenges for disinfecting mouth and throat.

(*v*) **Dequalinium chloride.** Chemically it is 4,4' diamino-2,2' dimethyl N,N decamethylene di (quinolinium chloride). It is prepared from isotin in the following way :

It is white colour powder and is only slightly soluble in water.

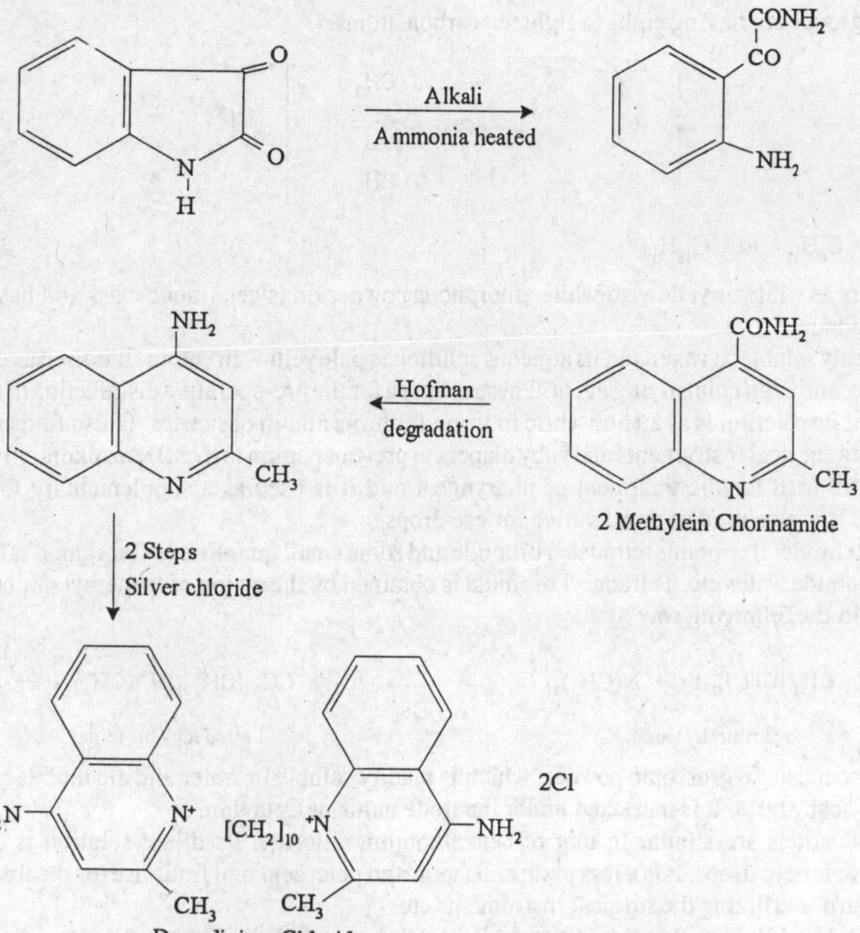

Alkali, Ammonia heated → ($CONH_2$, CO, NH_2)

Hofman degradation

2 Methylein Chorinamide

2 Steps, Silver chloride

$$H_2N-\text{(quinolinium, }CH_3)-N^{+}[CH_2]_{10}{}^{+}N-\text{(quinolinium, }CH_3)-NH_2 \quad 2Cl$$

Dequalinium Chloride

It possesses antibacterial and anti-fungal properties and is used as lozenges and paints for mouth and throat infections.

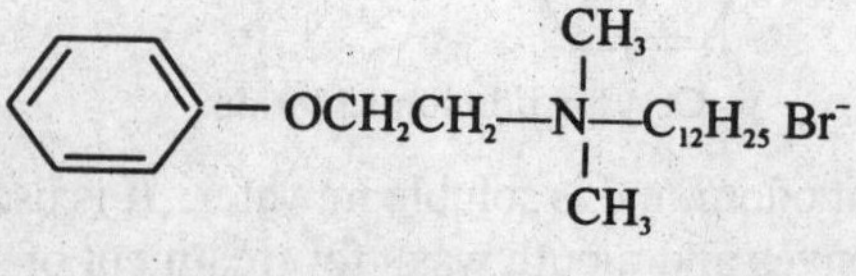

$$C_6H_5-OCH_2CH_2-N(CH_3)_2-C_{12}H_{25}\ Br^-$$

Chapter 19

Organic Pharmaceutical Aids

Many organic substances do not have therapeutical value but they are very useful in the manufacture or compounding of various pharmaceutical preparations. These substances are called pharmaceutical aids. The various examples of pharmaceuticals aids can be classified as follows :

I. (A) Preservatives, (B) Antioxidants and (C) Sesquestrants
II. Emulsifying agents
III. (A) Colouring, (B) Flavouring and (C) Sweetening agents
IV. Stabilizing and suspending agents
V. Ointment bases and related agents
VI. Solvents
VII. Miscellaneous substances.

I (A) Preservatives

Preservatives are antimicrobial compounds which are capable of inhibiting microbial growth in preparations. Preservatives should have the following properties :

1. They should not react with the ingredient of the formulation or with the containers.
2. They should not interfere with the official assay method.
3. They must be non-toxic and should be effective under pH conditions of the formulation.
4. They must be soluble in water when used in water and oil emulsion.

Some important preservatives used in pharmaceutical preparations are benzoic acid, esters of P-hydroxy benzoic acids, cinnamic acid, sorbic acid, phenylmercuric nitrate, phenyl mercuric borate, chlorbutanol, methyl and propyl paraben, benzalkonium etc.

(*i*) **Benzoic Acid** (C_6H_5COOH). It is a colourless needle-like crystalline solids. It is readily soluble in boiling water. It is soluble in fats and oils. These are incompatible with ferric salts and salts of heavy metals.

A preparation of 0.1% of benzoic acid is an effective preservative for pharmaceutical preparations. It is effective at a range of pH 5.

These are used as food preservatives but are not used as preservative for ophthalmic preparations, benzoic acid ointment are used for fungus infections of the skins.

(*ii*) **Hydroxy Benzoates** (HOC_6H_4COOR). These are effective against fungal growth but are less active against bacterial growth.

These are also used as food preservatives, numerous cosmetics and pharmaceutical preparation.

(*iii*) **Sodium Benzoate** (C_6H_5COONa). It is soluble in water and is generally used as preservative in place of benzoic acid. This is also effective when pH range is below 5.

These are mainly used as food preservatives.

(*iv*) **Benzyl Esters of Benzoic Acid.** The methyl, ethyl, propyl, butyl, benzyl esters of P-hydroxy benzoic acid are used as preservatives for pharmaceutical preparation. These are odourless, tasteless, nontoxic crystalline solids.

Their water solubility decreases with increase in molecular weight. They are soluble in alcohol, glycerol, fats and oils. The hydroxy benzoate are active at pH 7 to 9.

Methy hydroxy benzoate, propyl hydroxy benzoate and butyl hydroxy benzoate—These are used as water soluble sodium salts.

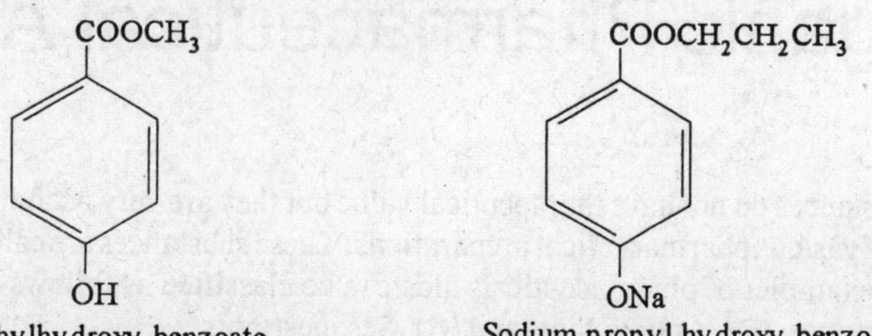

Methylhydroxy benzoate   Sodium propyl hydroxy benzoate

(*v*) **Chlorinated Tertiary Alcohol** e.g.: Chlor-butol [$Cl_3CC(CH_3)_2OH$]
(1, 1, 1, -trichloro-2-methyl propan-2-ol)

It is white crystalline compound and has a characteristic odour and taste. It is slightly soluble in water but is readily soluble in alcohol and glycerol. It is decomposed by alkalis.

It is used as preservative in injections and eye drops.

(*vi*) **Phenyl mercuric compounds.** Phenyl mercuric nitrate ($C_6H_5HgNO_3$), Phenyl mercuric hydroxide ($C_6H_5HgOH_3$), Phenyl mercuryorthoborate, etc. are also used as preservatives for pharmaceutical preparations. These are white as crystalline powder and soluble in glycerol and fixed oils.

Phenyl mercuric nitrate in concentration of 0.001% as preservative for injection solution and in concentration of 0.002% of for eye drops. It is spermicidal and is used in some contraceptive preparations.

(*vii*) **Benzalkonium Chloride.** In a concentration of 0.01% solution it is used as preservative for eye drops.

(*viii*) **Methyl-propyl Paraben.** These are generally used as preservative for pharmaceutical preparation against fungus contamination.

(*ix*) **Cetrimide.** It is used in a concentration of 0.005% as preservative in eye drops.

Besides these compounds chloroform and cresol, ethanol are used as preservatives for various injections etc.

I (B) Antioxidant

The pharmaceutical preparations which contain vitamins, fats, unsaturated oils, phenolic compounds etc. undergo deterioration on storage because of atmospheric oxidation. Antioxidant agents are added to such preparation to prevent oxidation and thus prevents subsequent spoilage of such preparations.

The important examples of antioxidants are esters of gallic acid, butylated derivatives of P-hydroxyanisole, or P-hydroxy toluene etc.

(*i*) 3, 4, 5, tri-hydroxy benzoic acid is known as gallic acid. Various derivatives like ethyl, propyl, octyl, dodecyl esters of gallic acid are used as antioxidants. They are also used to stabilise compounds like ethers and paraldehyde by preventing peroxide formation. They are also added to oils and fats to prevent rancidity.

Alkyl gallates are also important antioxidants. These occurs as creamy white odourless crystalline powders. Ethyl, octyl, dodecyl gallates are tasteless but propyl gallate is bitter in taste. These alkyl

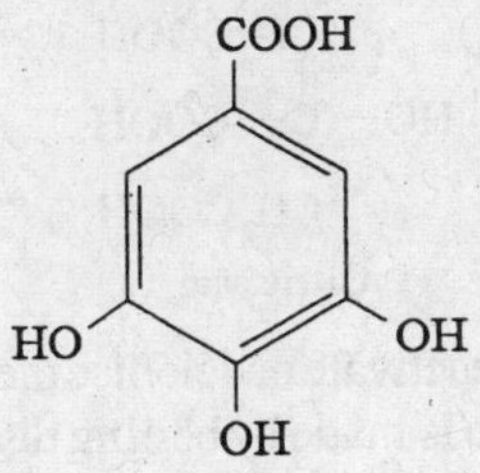

Gallic acid

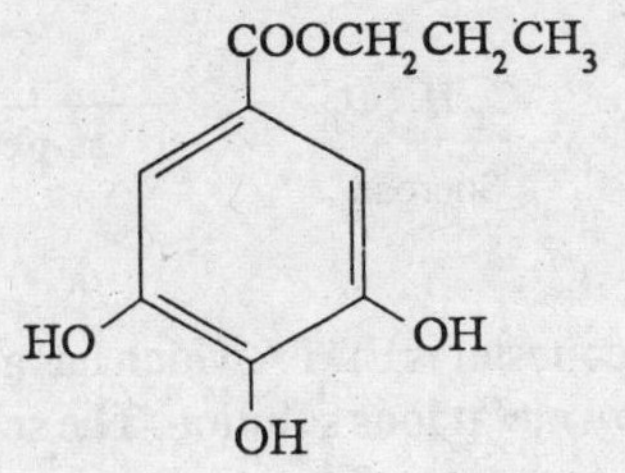

Propyl gallate

gallates differ in their solubility in water for e.g. ethyl gallate is slightly soluble in water, propyl gallate is very less soluble in water but octyl and dodecyl esters are insoluble in water. These are readily soluble in alcohol and acetone.

(*ii*) Butylated hydroxy anisole which is chemically 2-t-butyl-4-methoxy phenol and butylated hydroxy toluene which is chemically 2, 6-di-t-butyl-p-cresol. These are crystalline solids and are insoluble in water but are soluble in organic solvents like alcohol, glycol, ether, chloroform etc., and in many oils. These should be protected from light.

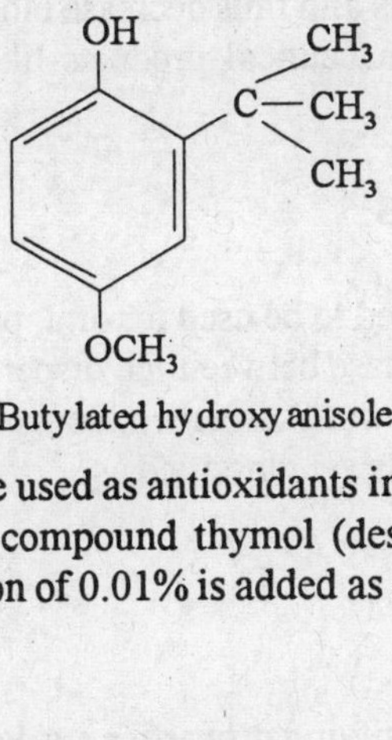

Butylated hydroxyanisole

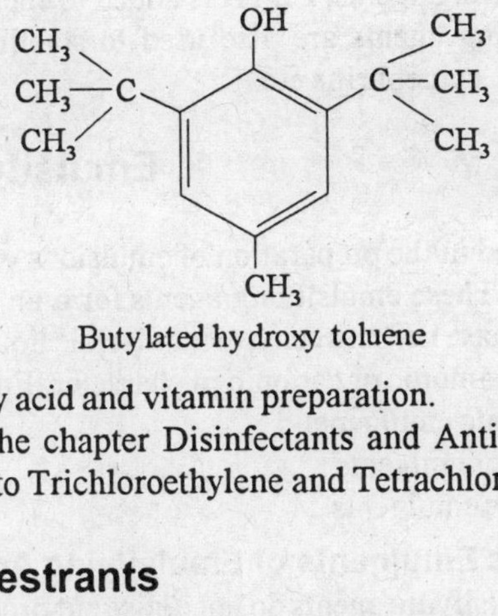

Butylated hydroxy toluene

These are used as antioxidants in fixed oils, fatty acid and vitamin preparation.

Another compound thymol (described under the chapter Disinfectants and Antiseptics) in a concentration of 0.01% is added as an antioxidant to Trichloroethylene and Tetrachloroethylene.

I (C) Sequestrants

These are another type of preservative. It has been found out that oxidation of food is catalysed by traces of metals which are naturally present in the food, for e.g.: potatoes become black when pealed and exposed to air or an apple when it is cut and exposed to air it turns brown. The trace metals may get into food from the processing machinery. The spoiling of food by this type of oxidation can be prevented by adding a sequestering agent. Some important sequestrants are ethylene diamine tetraacetic acid, citric acid, hydroxy acids, etc.

Sequestration is the state of combining a cation in a large complex soluble ion and thus prevent the original cation from exhibiting its usual properties. Some common cations are Ca^{2+}, Mg^{2+}, Mn^{2+}, Sb^{3+}, Bi^{3+}, Fe^{2+}, Fe^{3+} etc. The sequestrants used are tartaric and citric acid, glycerin, sucrose etc. Citric acid is a natural sequestering agent and employed in cookery when lemon juice is rubbed on to certain vegetables to prevent their discoloration.

When citric acid, is used along with another antioxidant known as butylated hydroxy anisole (BHA), the citric acid increases the antioxidant power of BHA. These metals get into food from the soil and from machinery during harvesting. Cu^{2+}, Fe^{2+} and Ni^{2+} catalyse the oxidation of fats. Citric acid wraps around the metal ion and thus make them ineffective as a catalyst.

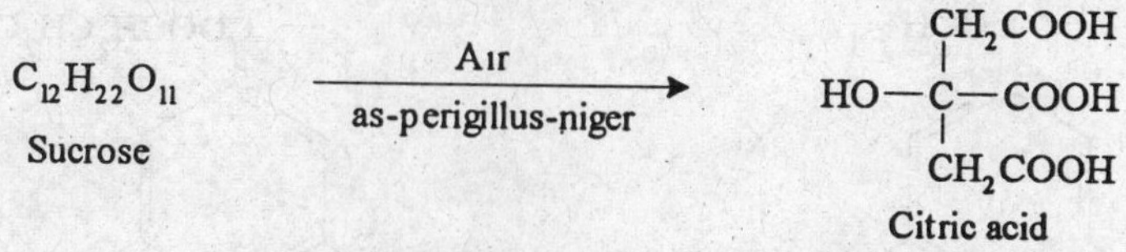

$$C_{12}H_{22}O_{11} \xrightarrow[\text{as-perigillus-niger}]{\text{Air}} HO-C(CH_2COOH)_2-COOH$$

Sucrose — Citric acid

Another sequester is EDTA which has a unique property to deactivate or deionise the alkaline earth ions and heavy metal ions solution. The sodium calcium edetate is made by boiling disodium edetate with calcium carbonate. The calcium complex of EDTA has the following structure.

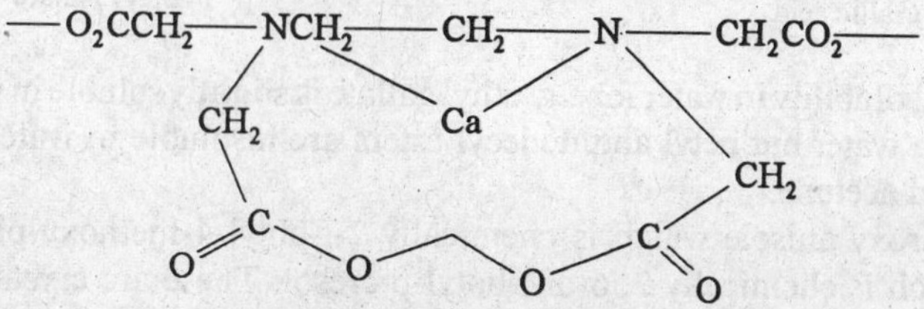

$$-O_2CCH_2-NCH_2-CH_2-N-CH_2CO_2-$$

(Ca chelated by the two N atoms and two $CH_2-C(=O)-O$ groups)

Calcium complex of EDTA

The emulsions of alkaline earth ions are stabilized by the addition of relatively small amounts of these complexing agents, EDTA is added to chelate calcium ions and thus decrease blood calcium.

Sequestering agents are also used to stabilize many pharmaceutical products like pencillin, ascorbic acid, epinephrine etc.

II. Emulsifying Agents

These are used in the preparation of emulsions which are intended to be used for oral, parenteral or external use. These emulsifying agents form an intermediate phase between the dispersed and the continuous phase i.e. between two immiscible liquids. These agents are surface active agents and may be non-ionic, anionic or cationic in character. Emulsifying agents are classified as

(*i*) Non-ionic emulgents
(*ii*) Anionic emulgents
(*iii*) Cationic emulgents

(*i*) Non-ionic Emulgents or Emulsifying Agents

These emulsifying agents do not ionise but produce their emulsifying properties *e.g.* Polyethylene glycol, esters of fatty acids, cetomacrogol, polysorbates, sorbitan esters, glyceryl esters etc.

(*a*) Examples of sorbitan esters include sorbitan monolaurate, sorbitan mono-oleate, sorbitan monostearate.

(Tetrahydrofuran ring bearing HO, OH and $-CH(OH)CH_2O.CO.C_{11}H_{23}$)

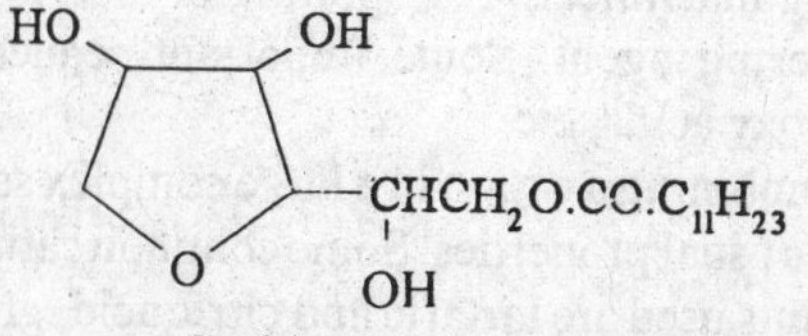

Sorbitan Monolaurate

These are the mixtures of partial esters of sorbitol and its mono and dihydrides with lauric, oleic and stearic acid respectively. The sorbitan mono-laurate and mono oleate occur as amber coloured oily viscous liquids while sorbitan monostearate is a yellow waxy solid.

(*b*) Glyceryl monosterate is a glycerol ester. It is a poor water-in-oil emulsifying agent yet it is a good stabiliser of water-in-oil and oil-in-water emulsions which are used internally as well as externally. It is a white hard waxy mass and is insoluble in water.

(*c*) *Polysorbates :* These are the complex mixture of partial fatty acid esters of sorbital and its mono and dianhydrides copolymerised with almost 20 moles of ethylene oxide for each mole of sorbital and its anhydrides. These complex structures have both hydrophobic and hydrophilic properties, which makes it an active emulsifying agent. These emulsifying agents are unaffected by high concentration of electrolytes or by changes in pH. These are yellow coloured oily liquid and are miscible in all proportions with water and alcohol. It is insoluble in liquid paraffin and fixed oils.

(*d*) *Cetomacrogal 1000 :* It is used in making ectomacrogal emulsifying wax. This wax is used in the making of cetomacrogol emulsifying wax which is in turn used for making water-in-oil emulsions. These are useful in the preparation of cetomacrogal cream and emulsifying ointments.

(*ii*) Anionic Emulgents

The emulgents in which the anion contains the characteristic group which confers the emulsifying properties of the compound e.g. soaps such as sodium-oleate, sulphonated compounds such as sodium lauryl sulphate etc.

Sodium Lauryl Sulphate. It is a mixture of sodium alkyl sulphates having mainly sodium dodecyl sulphate ($C_{12}H_{25}OSO_3Na$). It is a useful emulsifying agent for detergent and as a wetting agent in shampoos. It also forms an ingredient of emulsifying wax, paraffin ointments and simple ointments. Emulsifying ointments generally contain 3 parts by weight of emulsifying wax, 5 parts of white soft paraffin and 2 parts of liquid paraffin.

(*iii*) Cationic Emulgents

These emulgents in which the cation contain the characteristic group that produces the emulsifying properties are called cationic emulgents or cation-active emulgents.*e.g.* Quarternary ammonium compounds, (Cetyl trimethyl ammonium bromide having cetyl as cation), cetrimide has been discussed under disinfections and antiseptics (Chapter 18).

An emulsion can be prepared by wet gum method or dry gum method.

In wet gum method acacia gum is mixed with equal quantity of water and then required quantity of oil is mixed and stirred till a homogenous mixture is formed.

In dry gum method the required quantity of oil is mixed with dry powdered acacia and it is mixed well. Then twice the quantity of water as that of acacia is added at once to the contents until a thick white cream is formed.

III. Colouring, Flavouring and Sweetening Agents

(A) Colouring Agents

Colouring agents may be defined as compounds employed for the purpose of imparting colours. These agents can be classified in various ways such as inorganic and organic colouring agents or natural and synthetic colouring agents. The members of these groups are used as colours for cosmetics, pharmaceutical preparations, foods, bacteriological and diagnostic agents.

Natural colouring agents are obtained from plants and animal sources. These are mainly used for foods, drugs, cosmetics, antiseptic purposes etc.

Colours obtained from plants are used to colour lotions, cosmetics and other preparations which are generally used for external application. The plant pigments can be extracted and used as colouring agents e.g. Chlorophyll, β-Carotene, Alizarin, Indigo etc.

Natural colouring agents can be obtained from animal sources also. e.g. *Cochineal*—which is the dried female insect known as Dactylopus coccus costa, containing egg and larvae. This is used as tinfurents or solution as a red colouring agent.

Caramel or burnt sugar is used as a colouring agent to give dark brown colour to the syrups and other preparations.

Synthetic colouring agents are also very useful as colouring agents for various pharmaceutical preparations. e.g. Amaranth which is synthetic azo dye and it can be named chemically as trisodium-3-hydroxy-4 (4-sulphonate-1-naphthyl azo)-naphthalene-2,7 disulphonate. It is a reddish brown powder and dissolved in water, gives magenta-red colour. It is mainly used for colouring medicines. It is also used for dyeing wool and silk and is also used in colour photography.

HO SO_3Na N=N SO_3Na

Amaranth

(B) Flavouring Agents

These agents are used to make a material peasant smelling by imparting acceptable flavour. Technically "flavour" refers to a mixed sensation of taste, touch, smell etc. to a preparation. These are mostly natural in origin. However, synthetic flavours have also been prepared but they are best imitation of natural flavours.

These flavours are generally obtained from the essential oils obtained from plants. e.g. Aniseoil, caraway oil, clove oil, cinnamon oil, coriander oil, lemon oil, orange oil, nutmeg oil, peppermint oil, spearment oil etc.

Some flavouring agents are obtained from crude drugs such as liquo-rice, ginger are also used.

Synthetic flavouring agents are also very useful such as, vanillin, benzaldehyde, methyl salicylate etc.

(*i*) **Vanillin :** It is 4-hydroxy-3 methoxy-benzaldehyde. It is a white or cream colour crystalline needle-like substance. It has characteristic taste and odour of vanilla.

It is obtained from vanilla fruits but for commercial use it is mainly prepared synthetically. It is soluble in alcohol, volatile oils etc. and should be stored in closed light resistant containers.

CHO OCH_3 OH

Vanillin

(*ii*) **Benzal dehyde**—(C_6H_5CHO) : It is a light yellow colour liquid with the characteristic smell of bitter almond oil.

(*iii*) **Methyl Salicylate** is a local analgesic and is also used as flavouring agent.

(*iv*) **Chloroform** in the form of chloroform spirit or chloroform water is used as flavouring and preservative agent.

Some officially permitted flavours are as follows :

Amyl acetate (banana)	Allylcaproate (Pine apple)	Carvon (spearmint)
Aceta anisole (hay like)	Citral (lemon)	Ethyl formate (Rum)
Ethyl Propionate (fruity)	Methyl salicylate (winter green)	Eugenol (spice)
Ginger oil (ginger)	Menthol (peppermint)	Geraniol (Rose)

Selection of flavours for various pharmaceutical preparations are done by expert consultants While selecting a flavour for an oral dosage, the flavour is selected according to its taste for example if the preparation is sour in taste raspberry flavour is preferred or for bitter taste cocoa flavour is generally used or salty taste cinnamon flavour is preferred.

(C) Sweetening Agents

Taste is a chemical sense and is a result of the contact between the taste possessing material and the taste buds of the tongue. There are four primary tastes namely sweet, bitter, sour and saline. From olden time substances with offensive taste were administered by mixing with pleasant tasting materials like honey etc. These agents which make the preparation sweetish are called sweetening agents. In modern times the following sweetening agents are used. Such as sugars (sucrose, lactose etc.), sorbitals, saccharin, cyclamates, glycerin etc.

(*i*) **Sucrose.** Sugars have been used as sweetening agents from the beginning like sugar coated pillas etc. Sucrose is used as a sweetening agent. It is extracted from sugarcane and sugar beet. It is a disaccharide and is chemically named as B-D frutofuranosyl-α-D gluco pyranoside.

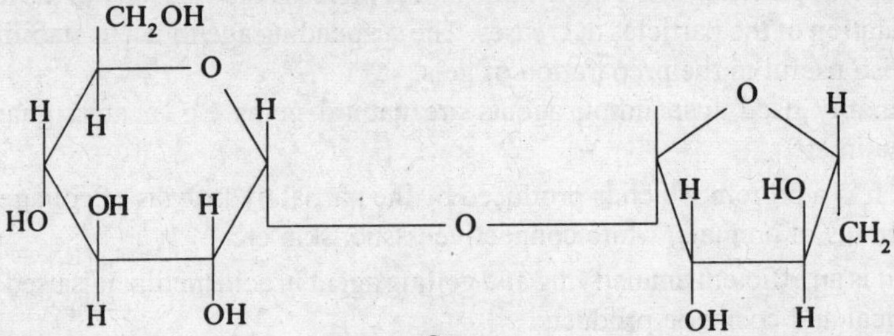

Sucrose

It is colourless crystalline solid having a sweet taste. It is readily soluble in water. It is used in Lozenge and for making syrups.

(*ii*) **Sorbital.** It is a reduction product of the monosaccharide glucose. It occurs as white crystalline powder and has sweet taste. It is hygroscopic and is soluble in water.

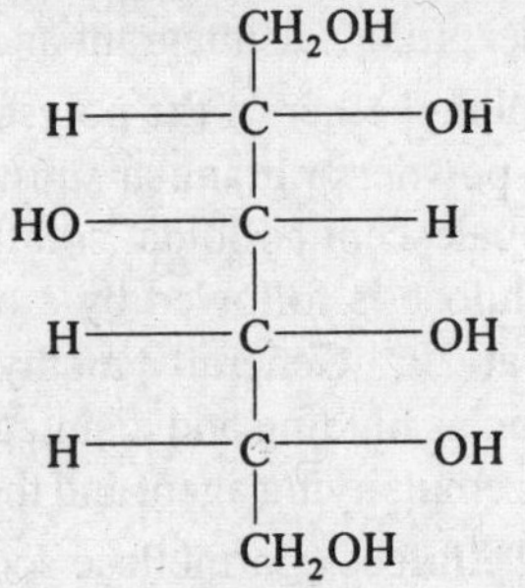

Sorbital

It is used as a substitute for glycerol in pharmaceutical preparation. Sorbital solutions are used as humectant in pharmaceutical and cosmetic creams. It is also used as a plasticizer in making gelatin capsules. It is also used in tooth-pastes.

(*iii*) **Saccharin.** It is prepared synthetically, chemically it is 1, 2 benzisothiazolin-3-one 1, 1 dioxide. It occurs as odourless, white crystalline powder.

It has a faint aromatic odour with very intense sweet taste. Saccharin is also used as saccharin sodium which is more palatable and free from the unpleasant after-taste of saccharin. Saccharin is prepared from toluene when it is heated with chlorosulphonic acid to form ortho toluene sulphonyl

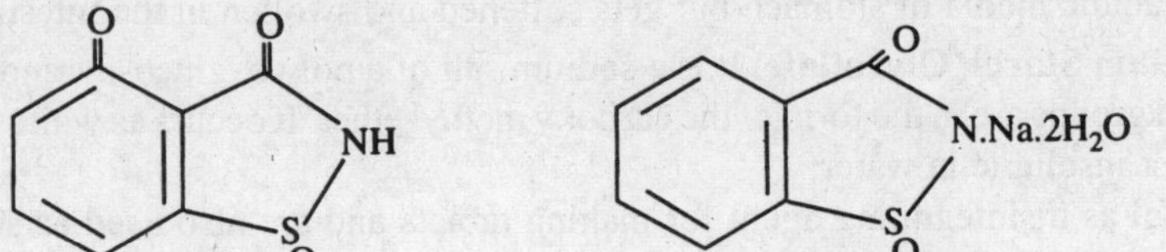

Saccharin

Saccharin Sodium

chloride. This is then reacted with ammonia to form sulphonamide, the methyl group is then oxidized with dichromate yielding ortho sulphamyl benzoic acid which when heated forms saccharin.

It has 550 times more sweetening power than sucrose. These are used as substitute for sucrose. It is used by diabetic patients as a substitute for sugar.

IV. Stabilizing and Suspending Agents

A suspension is defined as a preparation in which the insoluble material is dispersed in the liquid medium in finely divided form. The suspending agent is added to such a preparation to prevent the crowding of disperse particles into a solid mass and to increase the viscosity of the medium so that the rate of sedimentation of the particles decreases. The suspending agents act as stabilizers in emulsions.

These are also useful in the preparation of gels.

Some commonly used suspending agents are natural gums e.g. acacia, guar, gum, sterculia, tragacanth, gelatin etc.

(*i*) **Gelatin.** It is a protein which is produced by the partial hydrolysis of collagen which in turn is derived from bones of animals, white connective tissue, skin etc.

(*ii*) **Pectin.** It is an efficient emulsifying and gelling agent in acid media. It is used in the preparation of pharmaceutical and cosmetic products.

In pharmaceutical suspensions generally cellulose derivatives such as methyl cellulose, carboxy methyl cellulose, hydroxy propyl methyl cellulose, gelatine synthetic polymers etc. are used.

(*iii*) **Cellulose.** These are obtained as microcrystalline, purified, partially depolymerised cellulose from fibrous plant products, with mineral acids.

It is white or cream coloured crystalline powder. It swells in water which when dispersed, produces a white opaque dispersion or gel. It is used in various pharmaceutical preparations. The non-dispersible grade is used as a binder, filler, disintegrant and lubricant in preparation of tablets.

(*iv*) **Methyl Cellulose.** It is a methyl ether of the polysaccharide cellulose which contains about 29% of methoxyl group. It is white powder or granular substance. These are odourless and are slowly soluble in water and the viscous opalescent colloidal solution is stable in acid or alkaline medium. Generally the name of methyl cellulose is followed by a number which indicates the approximate viscosity of a 2% W/V solution at 20°. Generally methyl cellulose 2500 and 4500 are used as thickening agents in suspensions and as binding and disintegrating agents in tablet. The low viscosity grade methyl cellulose 20 is used as emulsifying agent and the medium grade are used as bulk laxative.

(*v*) **Sodium Carboxy Methyl Cellulose** (Carmellose sodium). It is a sodium salt of poly carboxy methyl ether of cellulose. It occurs as cream or white coloured powder or granules. It is hygroscopic in nature and is soluble in water.

These are mainly used as suspending agent for insoluble powders in aqueous preparation for oral, parenteral and external application. It is also in use as emulsifying agent for oil-in-water emulsions. These are also used in food industry.

(*vi*) **Cellulose Acetate Phthalate.** These are cellulose derivatives in which some of the hydroxyl groups of cellulose get esterified by acetyl groups and other hydroxyl groups are esterified by hydrogen phthalate. It occurs as colourless or white free flowing powder and is insoluble in water. It may be odourless or has a faint odour of acetic acid.

It is useful as an enteric-coating material for tablets and capsules and cellulose acetate phthalate are affected by acidic media of stomach but gets softened and swollen in the intestinal fluid.

(*vii*) **Sodium Starch Glycollate.** It is a sodium salt of a poly α-gluco-pyranose in which some of the hydroxyl groups are in the form of the carboxy methyl ether. It occurs as white free flowing powder and is almost insoluble in water.

It is useful as disintegrating agent for making tablets and are also used as suspending agent for extemporaneous preparations.

Some purely synthetic suspending agents are :

(*viii*) **Carbomer.** It is a cross linked, high molecular weight polymer of acrylic acid. It occurs as a white fluffy powder with a characteristic odour. It is slightly soluble in water, alcohol and glycerol.

These are used as neutralised gel, suspending agent in pharmaceutical preparations for internal and external use. It is also used as binding agent in tablets and in the formulation of prolonged-action tablets.

(*ix*) **Polyvinylpyrrolidone.** It is a synthetic polymer which is having linear 1-vinyl-2 pyrrolidone group. It can be chemically named as poly-(2-oxopyrrolidin-1-yl ethylene)

$$\left[-CH-CH_2- \right]_n$$

Poly viny lpy rrolidone

It is a white colour solid which is hygroscopic in nature. It is soluble in water.

It is used as dispersing agent for dyes in making coloured tablet. It is also used as a stabilizing agent in vitamins, enzymes etc. salicylate preparations.

V. Ointment Bases and Related Agents

Ointments are semisolid preparations for application to the skin and generally contain a medicament. They are used as emollient protective or for other surface effects or they may contain medicaments which are intended to be absorbed systematically. The drug is incorporated in soft hydrocarbons or other synthetic substances called the ointment bases.

The physical nature of the ointments should be such that when it is applied to the skin it should soften but not melt necessarily. The ointment includes substances which are used as bases for the preparation of cream, emulsions and suppositaries.

The bases used for ointments are generally animal, vegetable or mineral fats and oils or even synthetic materials. Thus, an ointment can be defined as "solution, suspension or emulsion of medicament in the base".

Ointments are used either as inert carriers for drugs or for their various emulsifying and emollient properties and are also useful for the improvement of the texture, stability or water repellent properties of the final preparation. There are various ointment bases in the use and there can be no ideal ointment base since different types are required for different purposes. A base suitable for certain conditions may not be suitable for some other conditions. Various bases include greasy base, anhydrous emulsifying bases, emulsion bases, aqueous non-greasy bases etc.

Types of ointment bases:

1. Hydrocarbon bases.
2. Absorption bases.
3. Water miscible bases.
4. Water soluble bases.

1. Hydrocarbon Bases

These are obtained from petroleum product known as paraffin. Paraffin may be soft paraffin or hard paraffin and liquid paraffin.

Soft Paraffin. Soft paraffin is purified semisolid hydrocarbon from petroleum. This may be white or yellow coloured. White base is used for colourless medicament. Both (yellow and white) paraffins are transluscent, soft greasy, odourless, mp 36°. This paraffin is not used for eye ointment. This is soluble in volatile oils, light petroleum etc. but is insoluble in water.

Hard Paraffin. It is a mixture of solid hydrocarbon obtained from petroleum. It is odourless, colourless or white translucent wax like substance. It is slightly greasy and it is solidified at 50-57°C. It is almost insoluble in water.

Liquid Paraffin. It is a mixture of liquid hydrocarbons from petroleum. It occurs as transparent, colourless, odourless oily liquid. It is available only as liquids which on storage tends to oxidise with the formation of peroxide. Tocopherol (10 ppm) or butylated hydroxy toluene are generally used as antioxidant.

Paraffin ointment contains a mixture of hard paraffin, white soft paraffin, white bees wax and cetostearyl alcohol.

Characteristics of hydrocarbon bases

(*a*) They are not absorbed by the skin.
(*b*) They restrict the loss of moisture from the skin, hence keep the skin soft, but on prolonged application it may lead to water logging with continuous maceration of skin.
(*c*) They are generally immiscible with water hence their removal is difficult.
(*d*) They are sticky.
(*e*) They are inert as they are saturated hydrocarbons.
(*f*) They are suitable for sterilized preparation.
(*g*) They do not encourage microbes hence preservatives are not needed.
(*h*) They are cheap.
(*i*) They retain body heat.
(*j*) Water absorption is low (i.e. only upto 5–10%) in aqueous or hydrophylic medicine which can be incorporated suitable for dry skin.

2. Absorption Bases

These are hydrophylic bases and can absorb considerable amount of water (unlike hydrocarbon bases). These are of two types namely (*i*) non-emulsified bases and (*ii*) emulsions.

(*i*) **Non-emulsified Bases.** They consist of a mixture of sterol type emulgent with one or more paraffin, hence they can absorb water or aqueous solution to produce emulsions. They are good emollients and are easier to spread, they arrest oil soluble medicament to penetrate the skin. Examples are wool fat, wool alcohol, bees wax, Cholesterol etc.

(*a*) *Wool Fat* : It can absorb 50% weight of water and is known as anhydrous lanolin. It is an important constituent of official ointment bases. These are used in simple ointments and in eye ointment.

(*b*) *Hard Fat* : It is a mixture of mono, di and triglycerides of the saturated fatty acids $C_9H_{19}COOH$ to $C_{17}H_{35}COOH$. It is used for the preparation of suppositories.

(*c*) *Bees Wax* : It is obtained from the honey comb of the common honey bee. It is a mixture of waxy esters mainly myricyl palmitate. It melts above 60°. These are used as an ingredient of paraffin ointment.

(*d*) *Hard Wax—Theobroma Oil* : Theobroma oil (Cocoa butter) is obtained from the seeds of theobroma cocoa or chocolate plant. At ordinary temperature it is yellowish white hard wax like substance. It is a mixture of glyceryl esters of stearic, palmitic, oleic and other acids. It is a useful ingredient of emollient bases. Its main use, however, is a suppository base.

(*e*) *Wool Alcohol* : These are prepared by saponification of the wool grease of sheep and have about 30% of cholesterol. It is a golden brown solid. It is used in making wool alcohol ointments and also used as emulsifying agent for water-in-oil emuslions.

(*ii*) **Emulsions :** There properties are almost same but they absorb more water. Examples hydrous wool fat, oily cream base etc.

(*a*) *Hydrous Wool Fat (Lanolin)* : These are prepared by 70% wool fat and 30% of water. They are used either alone as emollient or in several ointment bases.

(*b*) *Oily Cream Base* : It is mixture of equal parts of wool ointment and water. Although these bases are hydrophilic in nature it is difficult to remove them. It adheres well to skin.

3. Water Miscible Bases

These ointments can be washed easily with water after use and have many other advantages such as it has no interference with the skin functions, it contains surfactant contents which have good contact with skin. It is preferred by patients due to high cosmetic acceptability. If applied to hairs, it can be easily removed. Some examples are emulsifying agents, cetrimide emulsifying ointment, cetomaerogol emulsifying ointment, etc.

Cetomacrogol Emulsifying Wax. It is a mixture of solid aliphatic alcohols having mainly stearyl and cetyl alcohols. It is used to improve emollient properties of parraffin ointment.

Polyethylene Glycols. These are liquids having 200–700 molecular weight. Those with molecular weight over 1000 exist as wax-like solids. These are water soluble and inert and are used as water miscible bases for ointments and stabilisers of emulsions.

The water miscible bases are generally used in the preparations of aqueous creams where easy removal from the skin is desirable.

4. Water Soluble Bases

These bases are soluble in water non-volatile, inert agents and hence can be removed by washing. They are easily absorbed by skin hence facilitates penetration of medicament. They are anhydrous, hydrophilic, water soluble, greaseless and help in absorption of medicament. They are good solvents and hygroscopic in nature. They are found to be compatible with many dermatalogical medicaments e.g. Macrogels.

Macrogels are polyethylene glycols having high molecular weight. These are condensation polymeric compounds and are obtained by the reaction of ethylene oxide in presence of sodium hydroxide at high temperature and pressure. Their general formula is $CH_2OH(CH_2OCH_2)_nCH_2OH$. They contain both ether and alcohol functions. Their molecular weight is represented by a number following the name. Example Macrogel 200, 300, 400 (Viscous liquid).

Macrogel 1500 (greasy semisoild) and macrogel 1500, 4000, 6000 (waxy soilds). These clear liquids and have faint characteristic odour and are miscible in water. The solids are white, creamy and hard lumpy flakes. These are used as water soluble bases for ointments and stabilisers of emulsion. These are also used in suppositories, pessaries, cosmetic creams etc.

A Few Important Ointments

1. Salicyclic Acid Ointment. It is prepared as follows :

Salicyclic acid (10 grs). Zinc oxide (2 grs), Starch powder (2 dr), Vaseline upto 1 oz.

It is used for fungus infections of the skin especially ring worm.

2. Sulphur ointment. Flowers of sulphur (50 grs), Vaseline upto 1 oz.

It is used in scabies infection. It is applied all over the body (below neck) after bath.

3. Sulphanilamide ointment. Sulphanilamide (50 grs), White Vaseline 1 oz.

It is used in boils and pyogenic infections.

4. Zinc oxide ointment. Zinc oxide (75 grs), Yellow soft paraffin 1 oz.

It is used in dermatitis.

5. Benzyl benzoate emulsion. Benzyl benzoate (1.5 dr), Stearic acid (10 grs) Triethnanol+ amine (3 minim), Water upto 1 oz.

It is also used in scabies. For children it should be diluted with equal amount of water.

6. Mercury ointment. Calomel (10 grs), Lead acetate (10grs), Zinc oxide (10 grs), ointment of mercuric nitrate upto 1 oz.

It is an antiseptic ointment.

7. Boric eye ointment. Boric acid (30 grs), Soft white paraffin (4 dr), Anhydrous lanolin (4 dr).

These are used for eye infections.

8. Calamina Lotion. Prepared calamina (25 grs), Zinc oxide (25 grs), Glycerine (30 minim), Lime water (1 dr), distilled water upto 1 oz.

Used in eczema and other forms of dermatitis.

VI. Solvents

Some organic liquids are used in pharmaceutical preparations as solvents or vehicles. The important solvents which are used in pharmaceutical preparations are alcohol, isopropyl alcohol, glycerol, propylene, glycol, acetone solvent, ether, chloroform, ethyl oleate, diethyl phthalate etc.

(*i*) **Ethyl Alcohol or Ethanol** (CH_3CH_2OH). It is a transparent, colourless, mobile, volatile liquid. It has a burning taste and characteristic odour (spirit like). It is hygroscopic and flammable. It is miscible with water, solvent ether, chloroform etc. Ehtyl alcohol should be stored in tightly closed container and should be kept at dark and cool places.

Alcohol which is officially used are dilute ethanol solutions. It is used for the preparation of tinctures, spirits, and also as solvent and preservative for pharmaceutical preparations.

For industrial uses, methylated spirit is used which is a mixture of 19 volumes of alcohol and 1 volume of approved wood naphtha. It occurs as colourless, mobile, volatile liquid having same characteristic odour that of alcohol and odour of wood of naphtha. These are used as solvent for surface coating, plastics, toilet preparations etc. It should not be used for internal use.

(*ii*) **Isopropyl Alcohol** $(CH_3)_2CHOH$. It occurs as clear colourless inflammable liquid with a burning and slightly bitter taste. It is a solvent for cosmetics, lotions and non-aqueous agents. It is also useful as a non-aqueous agent for tablet granulation. It is used externally for preoperative skin cleaning. It is also used in lotions.

(*iii*) **Glycerol (Glycerin).** Chemically it is propane 1, 2, 3 triol.

It occurs as clear colourless, odourless liquid. It has a sweet taste and a sensation of warmth. It is soluble in water and alcohol. It is a useful solvent for ear drops. It is also used as an emollient and water retaining agent in many creams and jellies.

$$\begin{array}{c} CH_2OH \\ | \\ CH_2OH \\ | \\ CH_2OH \end{array}$$

The properties of propylent glycol $CH_3CH(OH)CH_2(OH)$ resembles with those of glycerol. It is used as solvent for a number of drugs which are insoluble in water and are unstable in aqueous solution.

(*iv*) **Acetone** (CH_3COCH_3). It is mainly used as a solvent for fats, resins, and many other organic substances. It is also a useful solvent for preparing film coating of tablets. It is a flammable liquid which is miscible in water.

(*v*) **Diethyl ether** ($C_2H_5OC_2H_5$). It is also known as solvent ether. It is used as solvent for oils, resins and other substances. It is also used for cleaning skin before carrying out surgical operations.

(*vi*) **Chloroform** ($CHCl_3$). It is used as a solvent for resins, alkaloids, fats, fixed and volatile oils and rubber. It is used in liniments as a counter irritant.

(*vii*) **Diethyl phthalate.** It occurs as clear liquid which is insoluble in water. It is used as a denaturant of alcohol, as solvent and plasticier.

(*viii*) **Dimethyl Sulphoxide** (DMSO) (CH_3SOCH_3). It occurs as a colourless viscous liquid. It is hygroscopic in nature. It is miscible in water. It is used as highly polar solvent as it can dissolve in both inorganic and organic compounds. It is used as a solvent for topically applied drugs.

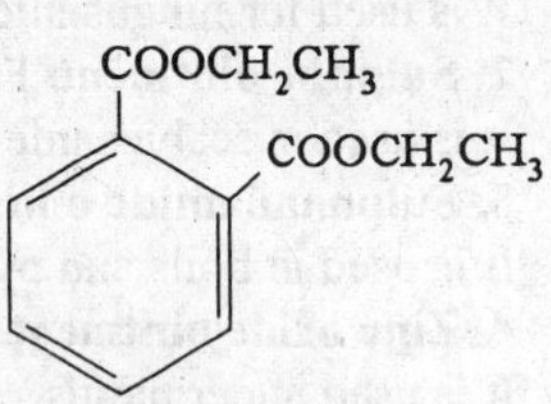

$COOCH_2CH_3$

$COOCH_2CH_3$

Diethyl phthalate

VII. Miscellaneous Substance

There are certain other compounds which are used in making pharmaceutical preparations. The most important among them are lactose, starch, stearic acid, magnesium stearate, organic bases such as ethylenediamine, meglumine, phenol-sulphophthalein, evans blue, phenolphthalein, sodium fluorescein, etc.

(*i*) **Meglumine.** It is an amino-sugar. Chemically it can be named as 1-deoxy-1-methylamino-D-glucital. It occurs as a white micro crystalline powder having a bitter taste. It is soluble in water.

$$\begin{array}{c} CH_2NHCH_3 \\ | \\ (CHOH)_4 \\ | \\ CH_2OH \end{array}$$

Meglumine

It is used for the preparation of salts with iodinated organic acids, which is used as media diagnostic agents.

(ii) **Lactose.** It is also known as milk sugar. It is a disaccharide and is chemically named as 4-0-β-galactopyranosyl -α-D-glucopyranose monohydrate. It occurs as white crystalline powder having a slight sweet taste.

It is soluble in water. The main use of lactose is as a diluent to provide bulk to powders and as a diluent in compressed tablets, sterilised lactose is used as a diluent for antibiotic powders.

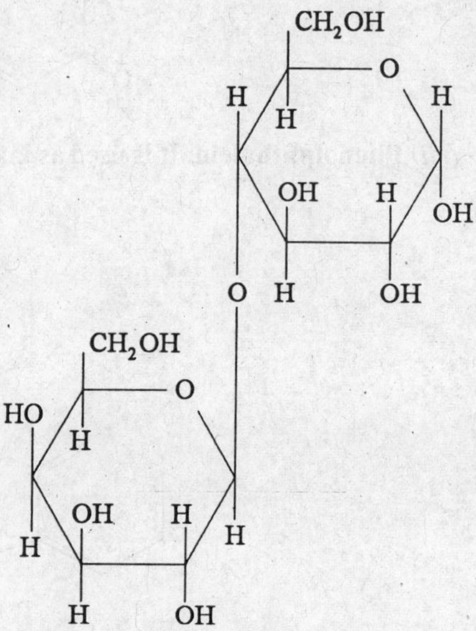

(iii) **Starch.** It is polysaccharide and is obtained from maize or rice or wheat or from tubers of potatoes. It occurs as a fine white powder.

It is sparingly soluble in cold water but gets dissolved in hot water. It should be kept in air tight containers and should be stored in dry, cool and dark places.

Starch is an absorbent and is used in dusting powders either alone or mixed with zinc oxide etc. It is also useful as a binder and as a disintegrating agent in many tablets. It is also useful as a filtering medium.

Lactose

(iv) **Steric Acid.** Officially used stearic acid is a mixture of fatty acids mainly of stearic $C_{17}H_{35}COOH$ and palmitic acid $C_{15}H_{31}COOH$. It occurs generally as a white crystalline powder and is sparingly soluble in water.

It is used as a lubricant in formation of compressed tablets and as an enteric coating for pills and tablets. Its soap, such as magnesium stearate, is used as lubricant in tablet making. Magenesium stearate is also used as a dusting powder in cosmetics.

(v) **Phenol Sulphophthalein.** It is red in colour at pH 8.4. It is available as sodium salt which is used as in intravenous or intramuscular injections.

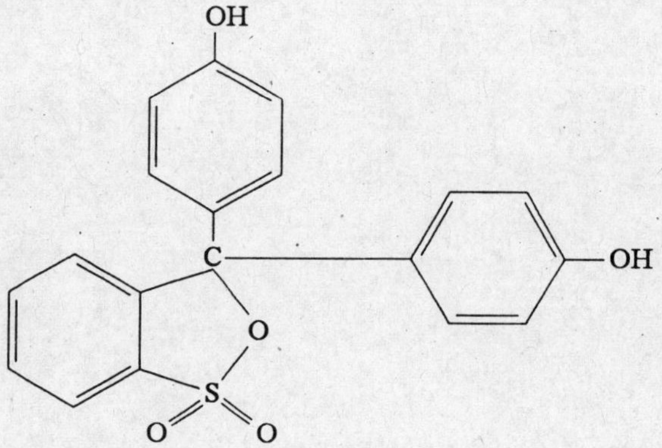

(vi) **Sodium Fluorescein.** It is used as opthalmologic diagnostic agent.

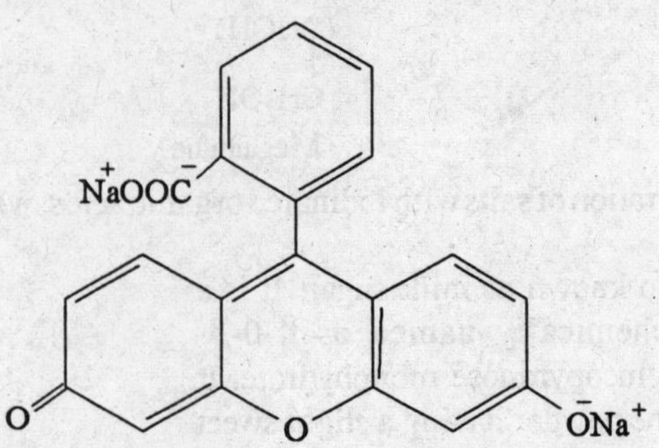

(vii) **Phenolphthalein.** It is used as laxative.

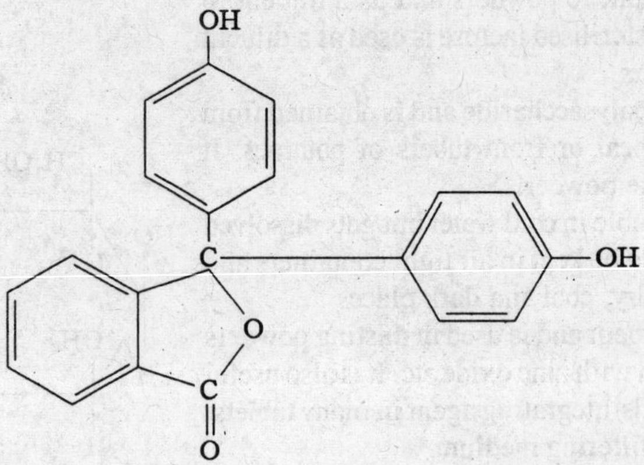

Chapter 20

Organic Diagnostic Agents

Diagnostic methods have become increasingly complex and frequently involve the use of drugs. Diagnostic agents are used for clinical diagnosis of a disease. A variety of organic substances have been employed for visualization of the liver, gall bladder, urinary tract, bronchi, lungs, heart, blood vessels and spinal canal. Drugs of varying chemical structure have been used for the study of blood flow, blood volume and for the diagnosis of diseases such as pernicious anemia. A number of radio active substances have been introduced to measure organic function. According to their useage these drugs are divided into the following groups.

I. Drugs used as X-ray contrast media.

II. Drugs used to test organ function

III. Drugs used to determine blood volume and hemopoietic function

IV. Drugs used for miscellaneous diagnostic tests.

(I) **X-ray contrast Media Radioopaque.** Certain salts of heavy metals such as barium, bismuth and compounds containing iodine or bromine are opaque to X-ray. By administering these preparations in various ways it is possible to visualize with the fluoroscope or to photograph by X-ray. Various structures such as the gastrointestinal tract, gall bladder, ducts, ureters, kidney, pelvis, uterus and fallopian tubes. Some of these important drugs are discussed here.

(i) **Barium sulphate.** It contains 98.0 to 100.5% of barium sulphate. It is fine, white, odourlers, tasteless, bulky powder. Its suspension in water is neutral to litmus paper.

It is used for the purpose of making the intestinal tract opaque to the X-ray, so that it may be photographed. Soluble salts of barium are highly poisonous hence it is highly important to be sure that the sulphate used for diagnostic purpose is of high quality.

(ii) **Sodium Diatrizoate injection.** It is a sodium salt of diatrizoic acid and it is named chemically as 3, 5 diacetamida 2, 4, 6 tri-iodobenzoic acid. It is a clear colourless to pale yellow solution. The viscosity of which increases with the concentration. It is soluble in water and is sterilised by autoclaving. It is used intravenously usually in a dose of 30 ml of a 50% solution.

It is used for radiological examination of different organs and especially of kidney in advanced renal diseases.

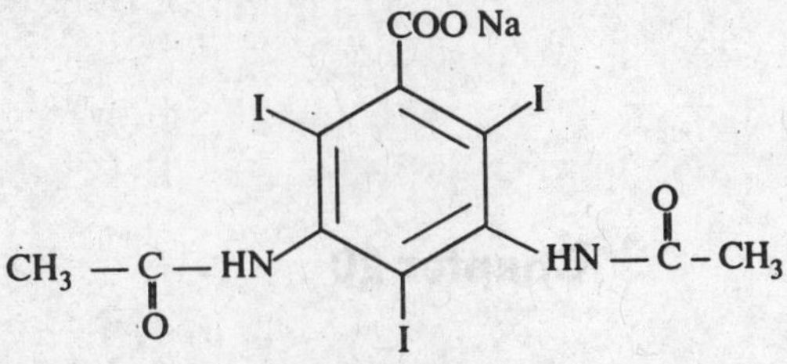

COO Na
I I
CH_3 — C — HN HN — C — CH_3
O O
I

Sodium Diatrizoate

(*iii*) Iothalamic Acid

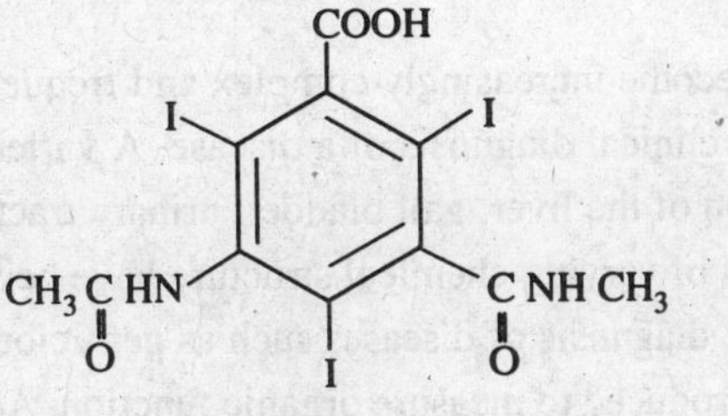

COOH
I I
CH_3 C HN C NH CH_3
O I O

Iothalamic acid

It is the structural isomer of diatrizoic acid having one of the acetamido group replaced by methyl-carbamoyl group. Chemically it is named as 5-Acetamido, 2, 4, 6-triiodo-N methyl isophthalamic acid. It is used as meglumine and sodium salts. It may contain small amounts of suitable buffers and of calcium disodium edetate or disodium edetate as a stabilizer. It is a clear, colourless solution, the viscosity of which increases with the concentration.

It is used for intravascular angiocardiography. It is rapidly distributed throughout the vascular system and promptly exereted by the kidney. It should not be given to the patients known to be sensitive to Iodine preparations and in patients with severe cardiovascular disease. Meglumine Iothalamate is used for intravenous urography (Radiography of any part of urinary tract).

(*iv*) Iopanoic Acid

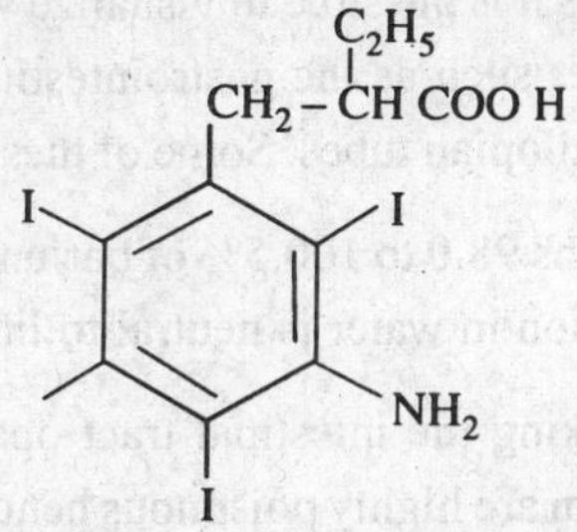

C_2H_5
CH_2 – CH COO H
I I
NH_2
I

Iopanoic acid

It is cream coloured powder, tasteless and has faint characteristic odour. It gets affected by light and metls with decomposition between 152°C amd 158°C, Chemically it is named as 2 (3 amino 2, 4, 6 triiodobenzyl) butyric acid, it is insolube in water and is soluble in alcohol, chloroform, ether and in alkali hydroxides, carbamates etc.

It is administered orally as a radiopaque medium in cholecystography (X-ray photography of the gall bladder after it has been made opaque by administration of radio-opaque dyes) It is promptly absorbed from the gastrointestinal tract, concentrated in the gall bladder and subsequently excreted in the bile.

(*v*) **Iodipamide**

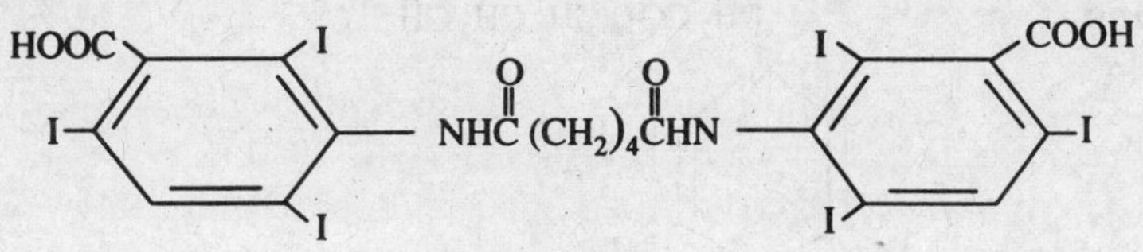

HOOC, I, I (ring) — $NH\overset{O}{\overset{\|}{C}}(CH_2)_4\overset{O}{\overset{\|}{C}}HN$ — (ring) I, COOH, I, I

Iodipamide

It has two 3 amino 2, 4, 6 triiodobenzoic acid units which are linked through their nitrogen atom through the adipoyl (– CO CH_2 CH_2 CH_2 CH_2 CO] group. Chemically it is named as 3, 3 adipoyl diamino bis- (2, 4, 6 triiodobenzoic acid.) It is sparingly soluble in water and is generally used as meglumine and sodium salt. It is a clear, colourless viscous liquid and its viscosity increases with concentration.

Sodium Iodipamide is used for cholecystography and cholangiography (photography of gall bladder, the cystic and common bile duct.) Meglumine Iodipamide is used in non surgical visualization of the biliary ducts.

(*vi*) **Iophendylate**

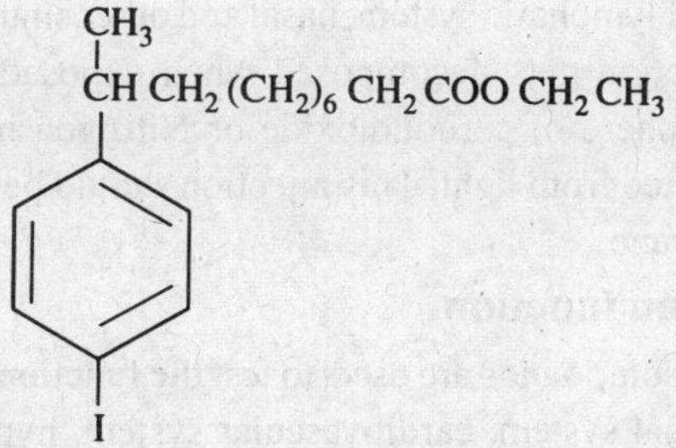

CH_3
|
$CH\ CH_2\ (CH_2)_6\ CH_2\ COO\ CH_2\ CH_3$ (on benzene ring, para-I)

Iophendylate injection is a sterile mixture of isomers of ethyliodophenyl undecanoate. It is a odourless, colourless to pale yellow viscous liquid. It is sparingly soluble in water and freely soluble in alcohol, benzene, chloroform and ether.

This injection is used as radiopaque medium for myelography (Radiographic examination of the spinal cord). It is commonly used for visualization of tumors of the intervertibral disc. It is also used for visualising the factors in amniotic fluid before intrauterene blood transfursion

(*vii*) **Diodone**

It is an iodinated pyridone derivative which is used as radiopaque compound.

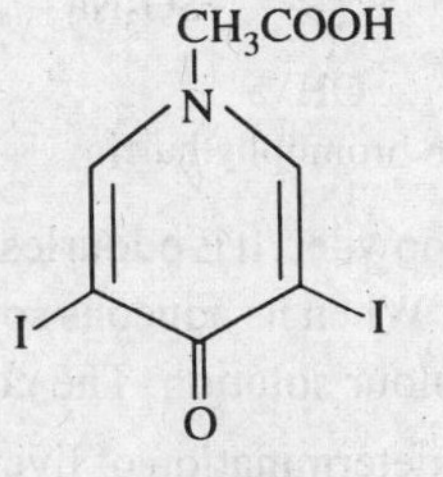

CH_3COOH — N (pyridone ring), I, I, O

Diodone

It is a clear or pale yellow colour liquid and is mainly used for examination of renal pelvis and ureter.

(*viii*) Propyliodone

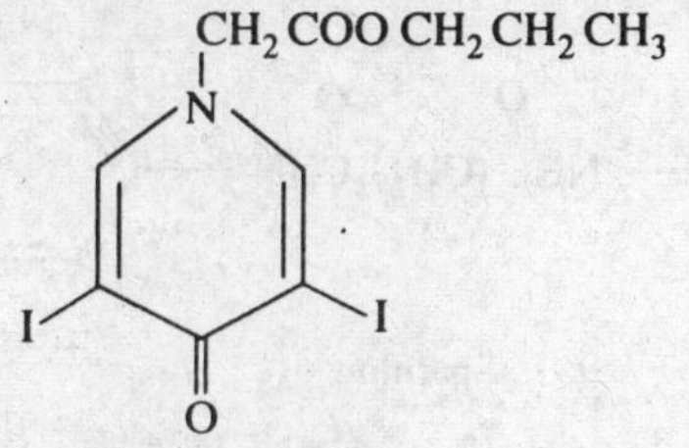

$CH_2COOCH_2CH_2CH_3$

Propyliodone

It is the propyl ester of diodone and is chemically named as propyel 3, 5 di iodo - 4 oxo- 1 (4H) pyridin acetate. It is a odourless, white crystalline powder. Melting range is 187° to 190°C. It is sparingly soluble in water and is soluble in organic solvents like acetone, alcohol and ether.

It is used for bronchography (X-ray photography of the bronchial tree after introduction of an iodised oil opaque to the X-ray).

(*ix*) Iodised Oil Fluid

Iodised oil injection is a sterile iodine addition product of the ethyl esters of fatty acid from poppy seed oil. It is a straw coloured viscous oily liquid.

It is used for examination of lymphatic system, nasal and other sinuses. It is also used for examining the bronchial tract. The injection gets decomposed when exposed to air and sunlight hence this injection is kept in an atmosphere of carbondioxide or Nitrogen in a previously sterilised scaled containers. It should be protected from light. This injection should be administered in plastic syringes which are made with polystyrene.

II. Drugs used to test organ function

Different kinds of chmical compounds are used to test the functioning of various body organs such as kidney, liver, gastrointestinal system, cardiovascular system, hypothalamus, pituitary etc. Many dyes are used as diagnostic agents. Some important dyes used as drugs are discussed here.

(*i*) Sulphobromophthalein sodium

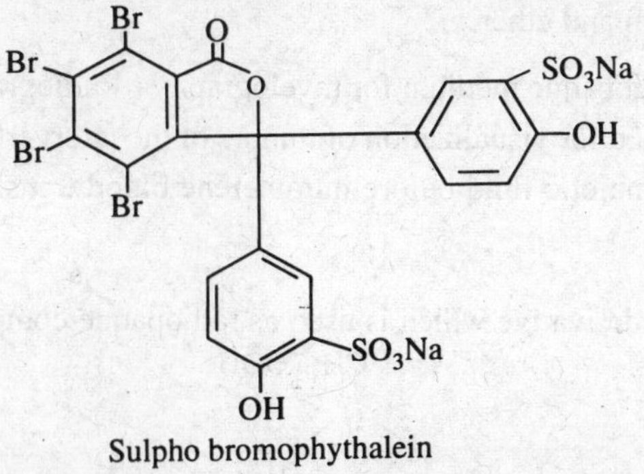

Sulpho bromophythalein

It is a white hygroscopic crystalline powder. It is odourless and has a bitter taste. It is soluble in water but is insoluble in alcohol and acetone. When its aqueous solution is treated with N-sodium hydroxide solution, it yields an intense purple colour solution. The colour disappears on adding acid.

It is used as a diagnostic aid for the determination of liver function. It is injected intravenously and after 80 minutes a sample of blood is drawn and the dye content of the alkalinized serum is compared with a series of standards. The normal liver excretes most of the dye within a period of 30 minutes. Usual dose given intravenously is 5 mg per kg of body weight.

(*ii*) **Fluorescein Sodium.**

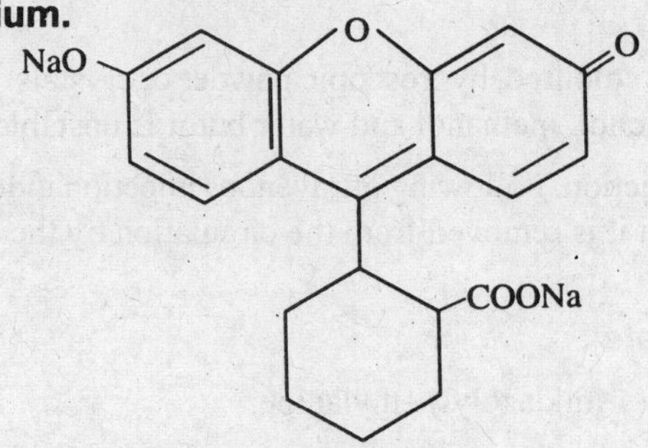

Fluorescein Sodium

It is orange red, odourless, hygroscopic powder. It is highly soluble in water and should be stored in well closed container. Its aqueous solution is strongly fluorescent even in extreme dilution. The fluorescence disappears when the solution is made acidic and reappears when the solution is again made alkaline.

Its 2% aqueous solution is used as an ophthalmic diagnostic aid. It is applied topically for the diagnosis of corneal lesions and detection of minute foreign bodies embedded in the cornea. It has the ability to strain the foreign bodies selectively and can be observed as a green ring surrounding the foreign bodies. Fluorescein also reveals defects or disease of the endothelium of the cornea. It is also used in fitting the hard contact lenses. This dye can be administered intravenously for measuring circulation time. It is also used intravenously to visualise the bile duct and the gall bladder before surgery.

(*iii*) **Inulin**

It is a polysaccharide which is closely allied to starch and is used to test kidney function. It is found in some of the plants belonging to the family composite *e.g.,* Inula, Taraxacum, Lappa, Dahlia variabilis etc. It differs from starch in the following properties *i.e.,* it is cloured yellow by Iodine, does not gelatinize with water and is not found in plants in the form of granules having concentric layers like starch. When hydrolysed with water it produces fructose.

It is used as a diagnostic agent for evaluation of glomerular filtration.

(*iv*) **Sodium benzoate**

It is administered orally or by intravenous injection and is used in a test for liver function. It is exereted as hippuric acid.

(*v*) **Indocyanine Green**

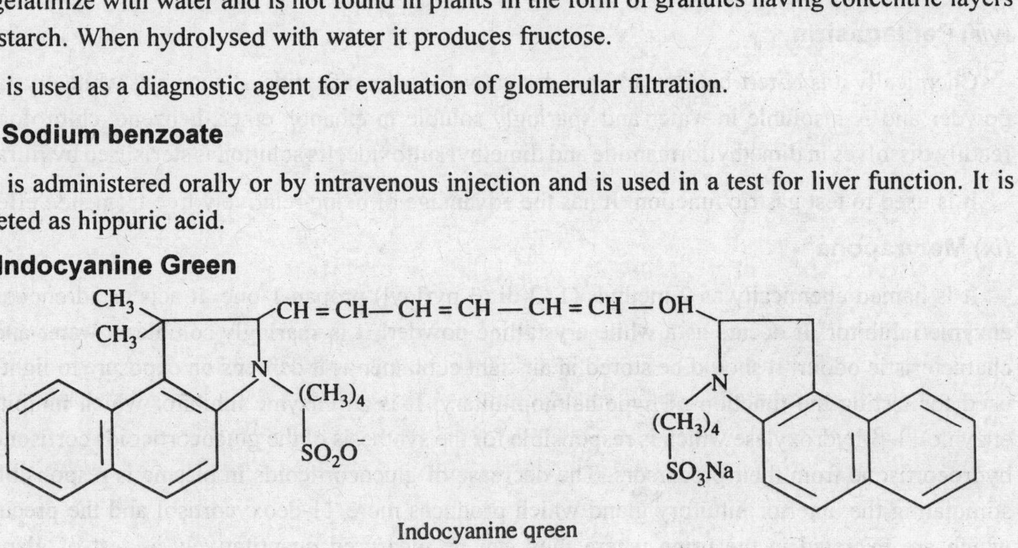

Indocyanine qreen

It is odourless, tasteless, green coloured, hygroscopic powder or crystals. It decomposes gradually above 200° C. It is soluble in alcohol, methanol and water but it is unstable in water.

It is used to determine liver function. Following intravenous injection indocyanine green is rapidly bound to plasma protein and then it is removed from the circulation by the liver and excreted in the bile in unconjugated form.

(*vi*) Mistamine Acid Phosphate

Chemically it is named as 2– (4 imidazolyl) ethylanine.

H
N
$\cdot 2\ H_2PO_4 \cdot H_2O$
N
$H_2NCH_2CH_2$

It is used as diagnostic agent for testing gastrointestinal system. It can be given parenterally for testing the secretory function of the stomach.

(*vii*) Mannitol

It is a hexahydric alcohol which is related to mannose. It occurs as white crystalline powder and is soluble in water and has a swete taste.

$$\begin{array}{c} CH_2\,OH \\ HO - C - H \\ HO - C - H \\ H - C - OH \\ H - C - OH \\ CH_2\,OH \end{array}$$

It is used intravenously for renal function determination.

(*viii*) Pentagastrin

Chemically it is N-tert-butoxycarbonyl derivative of a pentapeptide. It occurs as white crystalline powder and is insoluble in water and sparingly soluble in ethanol, ether, benzene, chloroform. It readily dissolves in dimethylformamide and dimethyl sulfoxide. Its solution is sterislised by filtration.

It is used to test gastric function. It has the advantage of being relatively free from side effects.

(*ix*) Metyrapone

It is named chemically as 2 methyl (1, 2 di (3 pyridyl) propan-1-one. It acts as adrenocortical enzyme inhibitor. It occurs as a white crystalline powder. It is sparingly soluble in water and has characteristic odour. It should be stored in air tight container as it darkens on exposure to light. It is used for testing the function of hypothalmopituitary. It is an enzyme inhibitor which inhibits the enzyme 11-β-hydroxylase which is responsible for the synthesis of the glucocorticoids cortisone and hydrocortisone from their precursors. The decrease of glucocorticoids in plasma is responsible for stimulating the anterior pituitory gland which produces more 11-deoxycortisol and the precursors which are excreted in the urine where they can be measured quantitatively as test of glandular functions.

(*x*) **Xylose.** It is also known as wood sugar.

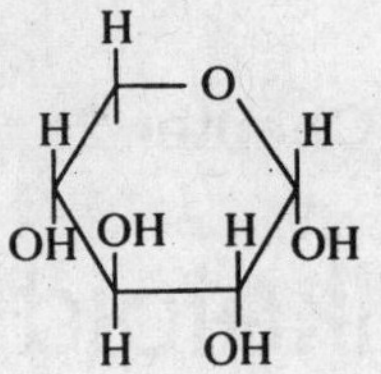

It is administered by mouth for diagnosis of gastrointestinal tract.

III. Drugs used to Determine the Blood volume and Hemopoietic Funtions

The estimation of blood volume is important in detecting impending shock and as a guide to the amount of plasma or other fluids to be used in order to avoid inadequate or excessive dosage. The determination of the circulating red blood cells and the measurement of the absorption of vitamin B_{12} are valuable in the diagnosis of megaloblastic anemia. Diagnostic agent used for this purpose is discussed here.

(*i*). **Evans blue**

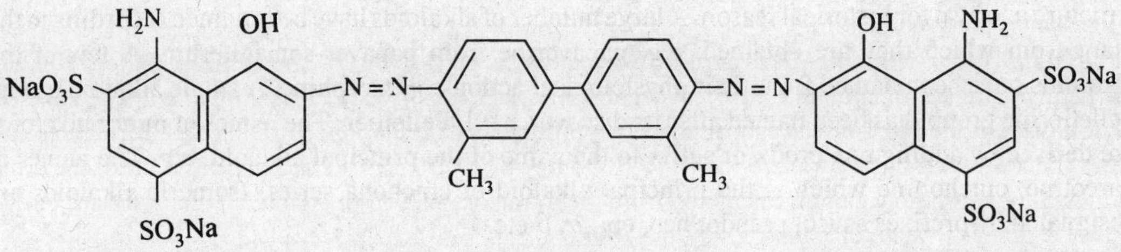

Evans Blue.

It is odourless, green to bluish green or brown powder. The dried product is hygroscopic and is soluble in water, sparingly soluble in alcohol but insoluble in benzene, ether, chloroform etc.

Evans blue is a diagnostic agent employed in blood volume estimation. A known quantity of the dye is quantitatively injected into the blood stream. After some time withdrawal of a sample of blood and determination of the concentration of the dye in the sample indicates the volume of blood in tne patient.

Chapter 21

Alkaloids

The term alkaloids or alkali-like was defined by a pharmacist Meissnei in 1819, as the basic, nitrogenous organic compounds of plant origin. The alkaloids are now generally defined as physiologically active basic compounds of plant origin, having one or more basic nitro atoms in heterocyclic ring system, which induce pronounced physiological activity in animals and man. This definition of alkaloid is not fully satisfactory and does not cover all the compounds classed as alkaloids.

Nomenclature

No attempt has been made for systematic nomenclature of these alkaloids due to their complex structure and also for historical reason. A large number of alkaloids have been named according to the plant from which they are obtained. *e.g.* papaverine from papaver-someniferum. A few of the alkaloids have been named from their physiological action *e.g.* morphine. A single alkaloid group, Pelletierine group, has been named after its discoverer P.J. Pelleitier. The names of minor alkaloids are derived by adding one prefix or suffix to the name of the principal alkaloid. *e.g.* The names of narcotine, cinchonine which is the principal alkaloid of cinchona series, Isomeric alkaloids are designated by prefixes as iso, pseudo, neo, epi, α, β etc.

Classification

A large number of alkaloids are known and consist of such a variety of molecular structure that their rational classification is very difficult. Alkaloids are often classified according to plant from which they are isolated. *e.g.* Cinchona, Opium etc. Nowadays they are classified according to the heterocyclic ring system, which is common to a group of alkaloid. They are classified as follows:

(*i*) Phenylethylamine alkaloids.
(*ii*) Pyrrolidine alkaloids *e.g.* Hygrine etc.
(*iii*) Piperidine *e.g.* coniine
(*iv*) Pyridine—Pyrrolidine *e.g.* Nicotine.
(*v*) Quinoline *e.g.* Quinine
(*vi*) Tropane *e.g.* Atropine.
(*vii*) Isoquinoline *e.g.* Papaverine
(*viii*) Phenanthrene isoquinoline *e.g.* Morphine
(*ix*) Indale *e.g.* Strychnine.
(*x*) Tropolone.

Occurrance

Alkaloids are chiefly found in plants of dicotyledons families and occur in leaves, roots, seeds and bark etc. The alkaloid content of the plant varies with season, age and its locality, closely related alkaloids generally occur together in the same plant *e.g.* nearly twenty alkaloids have been isolated from opium. It is generally observed that a given genus or related genera possesses the same or structurally related alkaloids e.g. seven different genera of the family solanaceae contain alkaloid hyoscyamine. They occur mainly as salt of common plant with acids such as acetic, oxalic, citric, malic, lactic, tartaric, tannic, aconitic acid since alkaloids are basic in nature.

Some alkaloids are found to occur as free amides e.g. piperine, as esters e.g. tropine, and as cocaine or organic acids. Generally the structurally related alkaloids are present as salts of the same acid e.g. cinchona alkaloid with quinine acid, aconitealkaloids with aconitic acid and opium alkaloids with meconic acid.

Functions of Alkaloids

The function of alkaloids in plants are uncertain and they are regarded as by products of plant metabolism. They may act as reservoir of protein synthesis. Alkaloids are also believed to act as protective substance against the animal or insect attack. They may function as plant stimulant or regulators like hormones and regulates activities like growth, metabolism and reproduction. Alkaloids may function as detoxicating agents by methylating, condensing and cyclising the compounds.

Isolation

Alkaloids are isolated by a general procedure, recommended by Manske, is described here.

The dried and powdered plant material is first extracted with petroleum ether to remove soluble fats. The plant residue is then extracted with methyl alcohol to remove cellulosic and other insoluble material and the filtrate so obtained is evaporated. The evaporated mass is dissolved in water and it is acidified to pH 2 and finally steam distilled to remove methyl alcohol. The dark residue solution is either allowed to stand for several days in a refrigerator or heated with molten paraffin to remove suspended impurities. The filtrate is extracted with ether or chloroform to remove water soluble non-basic organic material and then steam distilled, when the steam volatile alkaloids are separated. The solution of the rest of the alkaloid salt is made alkaline and again extracted with ether or chloroform and the ethereal layer obtained after this extraction is evaporated to give crude alkaloid. The general scheme for the extraction of alkaloids is illustrated as follows:

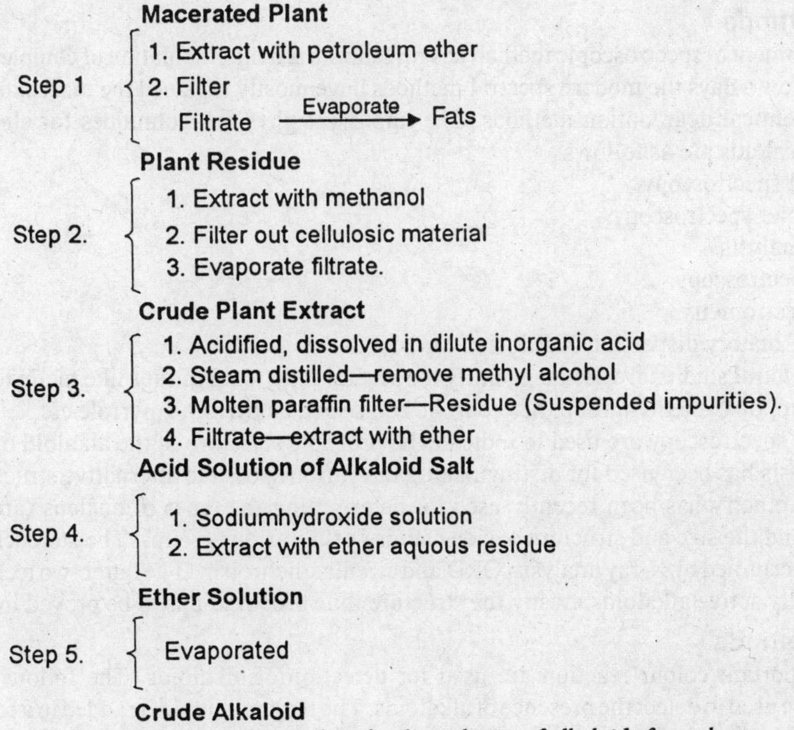

Table showing general procedure for the isolation of alkaloids from plants.

Now a days a new technique has developed for the isolation of alkaloids which is based on the ion exchange principle. The alkaloids react with H^+ ion to form a covalent bond by exchange with a positive ion and then these are absorbed on a hydrogen ion exchanger. Then they are removed from the exchanger with alkali together with a suitable solvent as ethanol or acetone.

Properties of Alkaloid

The important general properties of alkaloids are as follows :

1. Alkaloids are generally colourless, crystalline, nonvolatile solids though a few exceptions are there which occurs as liquids *e.g.* coniine etc.
2. Alkaloids are insoluble in water but are soluble in most of the organic solvents like ethanol, ether, chloroform etc. Liquid alkaloids are soluble in water.
3. Alkaloids are optically active compounds and a majority of them are laevo-rotatory. Although a few dextro-rotatory alkaloids are known *e.g.* coniine, while a few are optically inactive *e.g.* papaverine.
4. Alkaloids are bitter in taste.
5. Most of them possess marked physiological activity and are of great value in medicine. *e.g.* morphine, atropine, quinine etc.
6. They are poisonous in nature.
7. Alkaloids are distinctively basic in character and contains one or more basic nitrogen atoms. Most of the alkaloids are tertiary mono acid bases while a few are secondary bases. Thus they form crystalline salts with mineral acids.
8. Alkaloids are precipitated by common alkaloidal reagent *e.g.* tannic acid, picric acid, perchloric acid etc.

Detection of Alkaloids

Alkaloids are detected by physical methods as well as by chemical methods.

Physical Methods

The development of spectroscopic method has revolutionized the elucidation of complete structure of alkaloids. Now a days the modern spectral methods have mostly replaced the most extensive time consuming chemical degradation methods. The important physical techniques for elucidation of structure of alkaloids are as follows:

(*i*) Infra red spectroscopy.
(*ii*) Ultra violet spectroscopy.
(*iii*) X-ray Analysis.
(*iv*) NMR spectroscopy.
(*v*) Mass spectrometry.
(*vi*) Optical Rotatory dispersion (ORD) and circular dichroism.

Infra red spectral studies are used to identify the several functional groups like methylene groups, N-methyl group, or C-methyl group, heterocyclic rings such as pyridine, pyrrole etc.

Ultra violet spectroscopy are used to indicate the probable structure of the alkaloid molecule

X-ray analysis has been used for distinguishing the various possible alternative structures.

Mass spectrometry has been recently used for determining the types of nucleus (aromatic and heterocyclic) and the size and structure of side chains in the various groups. The stereochemistry of alkaloids is determined by X-ray analysis, ORD and circular dichroism. The latter two techniques are only on optically active alkaloids. Lastly the structure thus proposed has to be proved by synthesis.

Chemical Methods

Several important colour reaction are used for detection of alkaloids. The following general reagents may be used to detect the presence of alkaloids. The reagent should be added to a cold solution of the substance in the minimum quantity of dilute mineral acid. These reagents precipitate the alkaloid from aqueous solution. In many cases the presence of the trace of alkaloids can be detected by adding a suitable reagent to its solution.

1. Mayer's Reagent—(Potassium Mercuri Iodide Solution). It is prepared by adding potassium iodide solution to a solution of mercuric chloride until the precipitate of mercuric iodide redissolves. When alkaloid solution is added to the above reagent a pale yellow or white precipitate is formed. Purine group and a few other alkaloids do not give precipitate with this reagent. This reagent is used in many official alkaloid assays as a test for complete extraction of alkaloids.

2. Dragendorff's Reagent. This is prepared by adding excess of potassium iodide to a solution of bismuth nitrate. When alkaloid solution is added to the above reagent orange or reddish orange precipitate is formed.

3. Wagner's Reagent. It is a solution of iodine in potassium iodide and gives a brown or reddish brown precipitate with most common alkaloids. This reagent is used in official alkaloid assay as a test for complete extraction of colchicine or emetine.

4. Hager's Reagent. It is a saturated solution of picric acid in cold water.

It gives characteristic precipitates which can be identified by microscopic appearance and have sharp melting points.

Quinine

Quinine is a cinchona alkaloid. It occurs along with other cinchona alkaloids like quinidine, cinchonine, cinchonidine etc., in the cinchona bark. Among these alkaloids quinine is the most important. All the alkaloids are present in the bark in combination with various complex organic acids such as quinic acids, cinchotannic acid etc.

Quinine is widely used as an antimalarial drug. Malaria is a protozoan infection caused by plasmodium faiciparun, plasmodium vivax, plasmodium Malaria and plasmodium ovale. The chief agents employed in Malaria chemotherapy are chloroquine, pamaquine, quinine, quinacrine etc.

Sources. It is obtained from Cinchona Ledgeriana and its hybrids.

Extraction. The bark is powdered and mixed with sodium hydroxide solution and slaked lime and the liberated alkaloids are extracted with hot petroleum. The separated extract is shaken with hot dilute sulphuric acid. The alkaloid passes into the aqueous layer as sulphate. The solution is then cooled and the sparingly soluble quinine sulphate crystallizes.

Assay

Quinine Hydrochlorides. The alkaloid is liberated by the addition of sodium hydroxide and extracted with successive quantities of chloroform. Each chloroform solution is washed with water and it is evaporated and residue is dried and weighed.

Quinine-Sulphate. Quinine Sulphate tablets are extracted with water and dilute hydrochloric acid and from the filtered solution the alkaloid is liberated by the ammonia and extracted with chloroform. The chloroform solution is washed with water and the solution is evaporated and the residue is dried and weighed.

Structure of Quinine ($C_{20}H_{24}N_2O_2$)

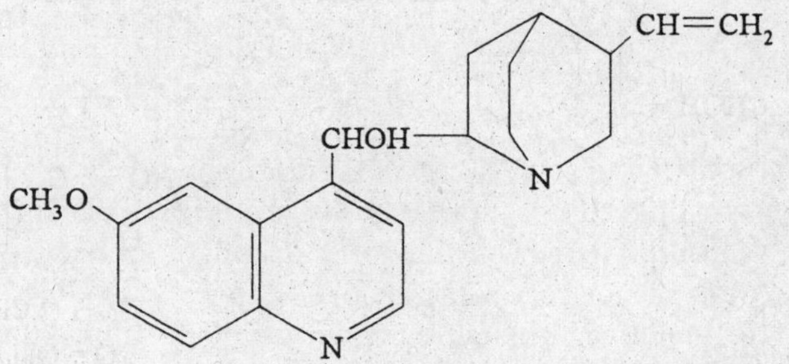

CH=CH$_2$

CHOH

N

CH$_3$O

N

1. It is a derivative of quinoline and contains 6 methoxy quinoline nucleus with a substituent attached to the 4' position.

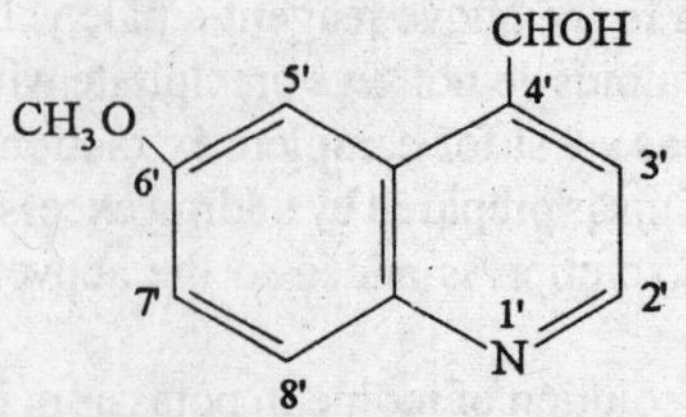

CHOH, CH_3O, 5', 4', 6', 3', 7', 2', 1', N, 8'

2. It has a secondary alcoholic group at position 4 of the quinoline nucleus and 3-vinyl quinuelidine unit is linked to the quinoline ring through this secondary alcohol.

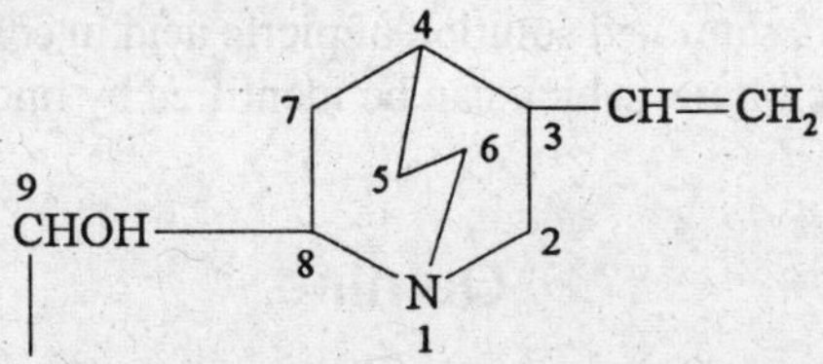

4, 7, $CH{=}CH_2$, 6, 3, 5, 9, CHOH, 2, 8, N, 1

This fromula contains four chiral centres 3, 4, 8 and 9. Nitrogen atom is tertiary and all three valencies form parts of the ring systems and it cannot oscillate. Since the bridge forms a cis fusion between 1 and 4 these centres behave as one chiral unit and hence in all there are four chiral centres. The absolute configuration of quinine are as follows :

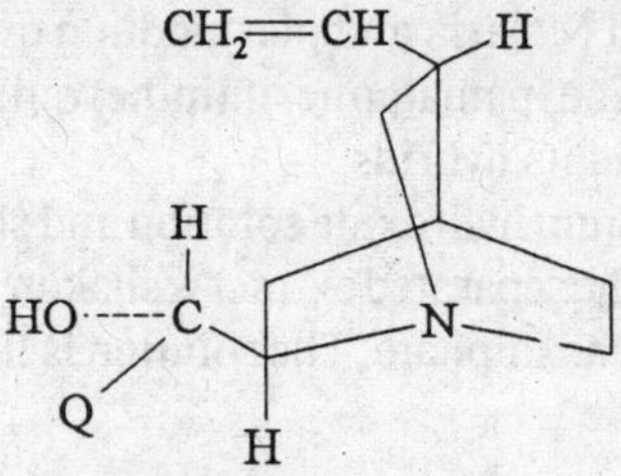

$CH_2{=}CH$, H, H, HO----C, N, Q, H

(+) Quinine

Q = Quinoline nucleus

$CH_2{=}CH$, H, H, N, C, HO, Q, H

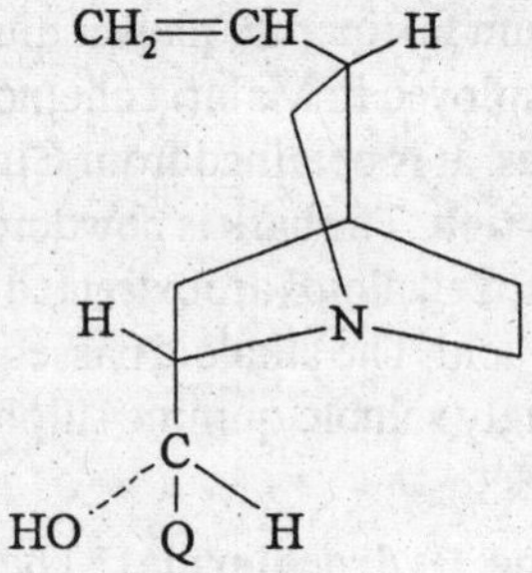

(–) Quinine

Other three cinchona alkaloids have the following structure.

(*i*) **Quinidine ($C_{20}H_{24}N_2O_2$) :** It is a dextro-rotatory stercoisomer of quinine. The two alkaloids could be distinguished by means of polarimeter.

(*ii*) **Cinchonine ($C_{19}H_{22}N_2O$) :** It is identical in structure with quinine except that it does not contain the CH_3O-group like quinine.

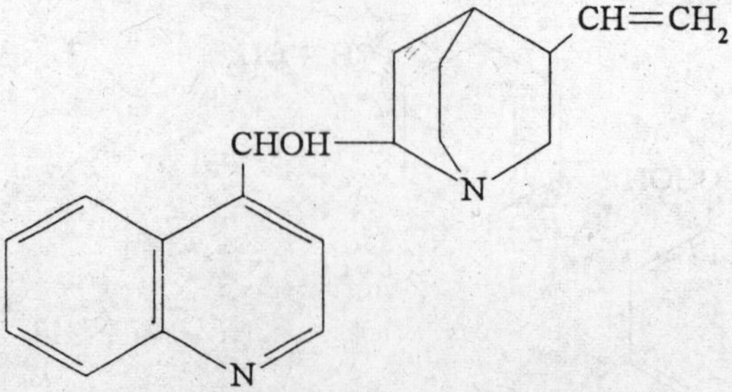

$CH{=}CH_2$, CHOH, N, N

Cinchonine

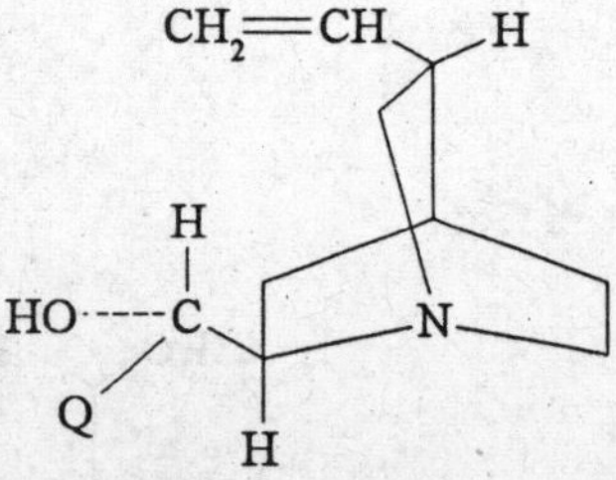

$CH_2{=}CH$, H, H, HO----C, N, Q, H

(+) Cinchonine

Q = Quinoline nucleus

(*iii*) **Cinchonidine ($C_{19}H_{22}N_2O$) :** It is a stercoisomer of cinchonine.

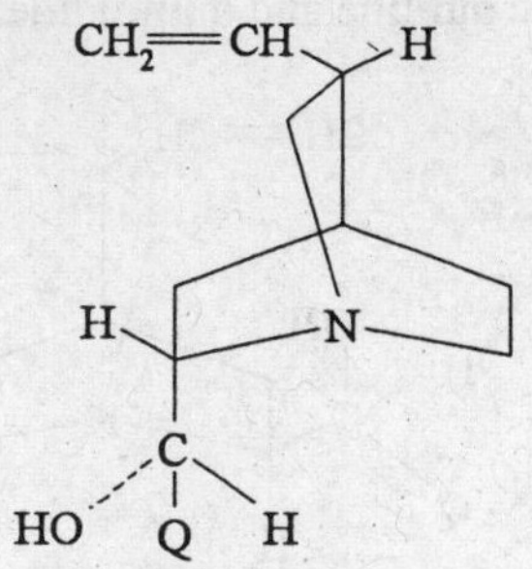

CH_2=CH H

H N

C

HO Q H

(–) Cinchonidine

Properties of Quinine

Some important properties of quinine are as follows :

1. Qunine crystallizes as colourless needles contaning $3H_2O$. It is sparingly soluble in water but readily soluble in most organic solvents. It is bitter in taste and is laevo-rotatory.
2. It is derivative of quinoline.
3. It is a di-acid tertiary base—and forms salts with two equivalents of an acid. The dihydrochloride is more soluble than monohydrochloride. The monosulphate is sparingly soluble in water.
4. On oxidation quinine gives quininic acid.

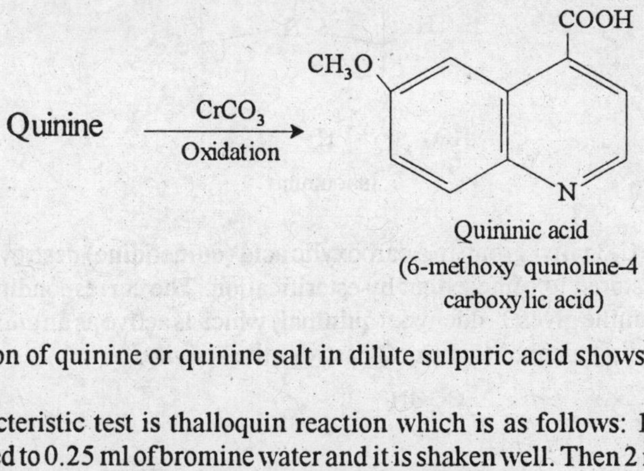

$$\text{Quinine} \xrightarrow[\text{Oxidation}]{CrCO_3}$$

COOH

CH_3O

N

Quininic acid
(6-methoxy quinoline-4 carboxylic acid)

5. A dilute solution of quinine or quinine salt in dilute sulpuric acid shows a characteristic blue fluorescence.
6. Another characteristic test is thalloquin reaction which is as follows: 10 ml of quinine salt solution is added to 0.25 ml of bromine water and it is shaken well. Then 2 ml of dilute ammonia solution is added to it when it gives bright green colouration.

Structure and Activity of Quinine and Related Antimalarials

1. The antimalarial activity of the cinchona alkaloids has been studied. Depending on the infection the biological activity of quinine, quinidine, cinchonidine, cinchonine differ both qualitatively and quantitatively.
2. Quinine is very active against asexual erythrocytic forms of all species of plasmodia that cause human malaria.
3. Quinidine is more active than quinine against plasmodium falciperum plasmodium vivex. Cinchonine is less active.
4. Quinine has many adverse side effects such as blurred vision, deafness etc. It requires dosing and quinine sulphate is poorly absorbed from the gastro intestinal tract. Quinine is still used extensively in the treatment of malarial infections.
5. Reduction of the C-3 vinyl group of quinine has no effect on antimalarial activity of quinine and related alkaloids.

6. **The dihydro compounds are somewhat more active than quinine against P. gallinaceum and P. relictum, epiquinine and epiquinidine differing in stereochemical configuration at C-9 are inactive compared to epimeric quinine and quinindine.**

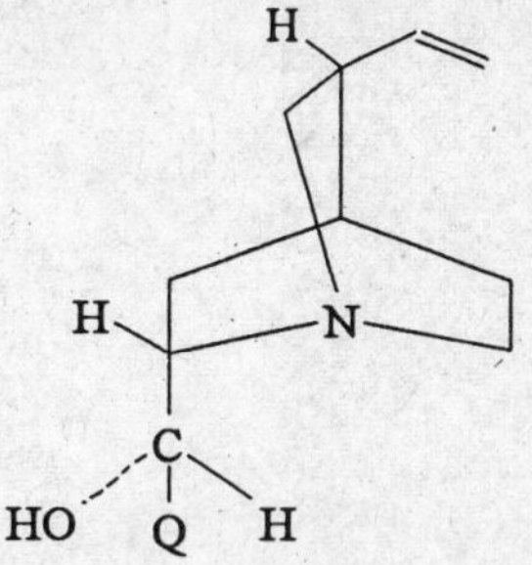

Epihydroquinine

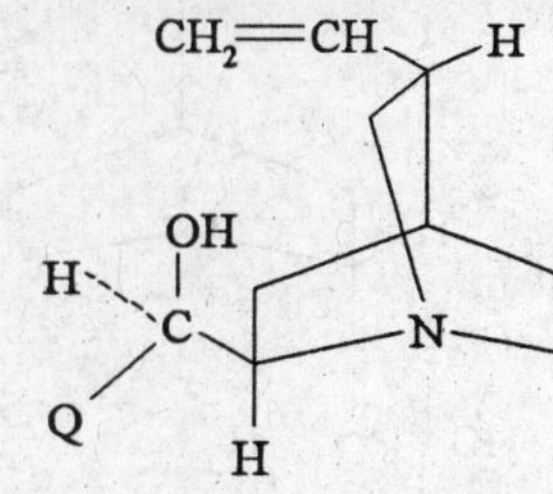

Epiquinidine

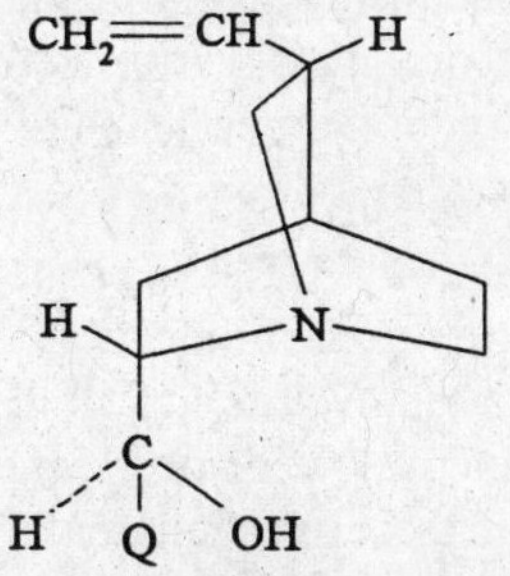

Epiquinine

7. **α, β and isoquinine in which asymmetry at C-3 is destroyed with inversion are as active as quinine against plasmodium relictum.**

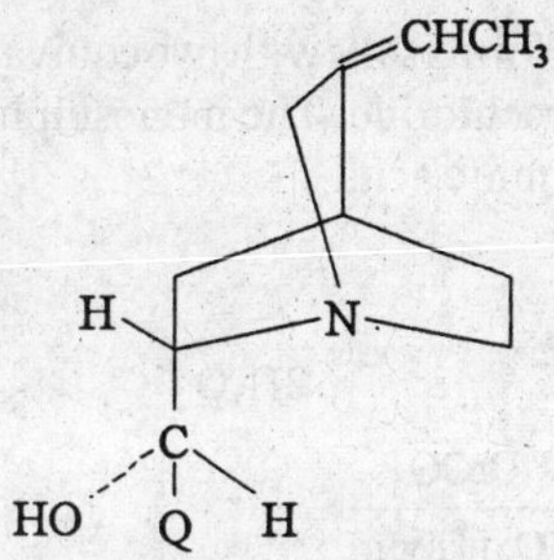

Isoquinine

8. **Oxidation of the C-3 vinyl group to a carboxylic acid (quinetidine) destroys all activity, though the activity is restored to some extent by esterification. The corresponding amide is inactive. Ozonolysis of quinine gives 3-aldehyde (quininal) which is active as an antimalarial. Reduction to the corresponding alcohol (quininol) destroys the activity.**

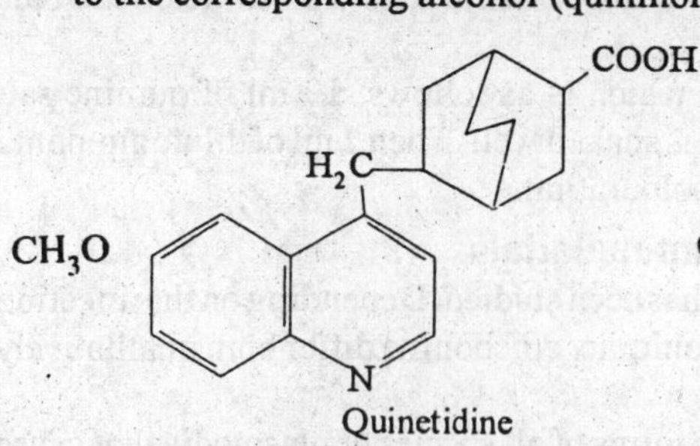

Quinetidine

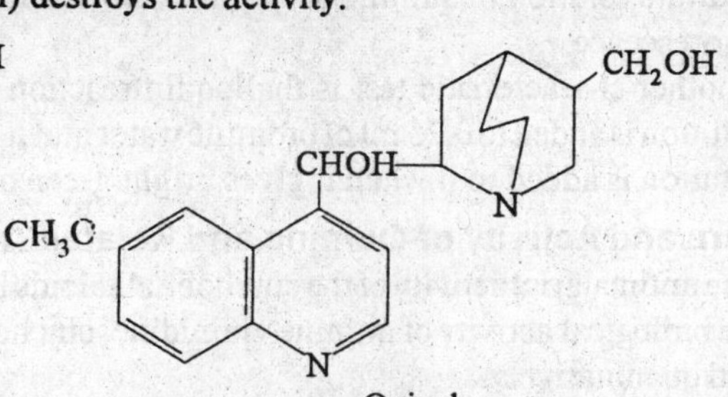

Quinol

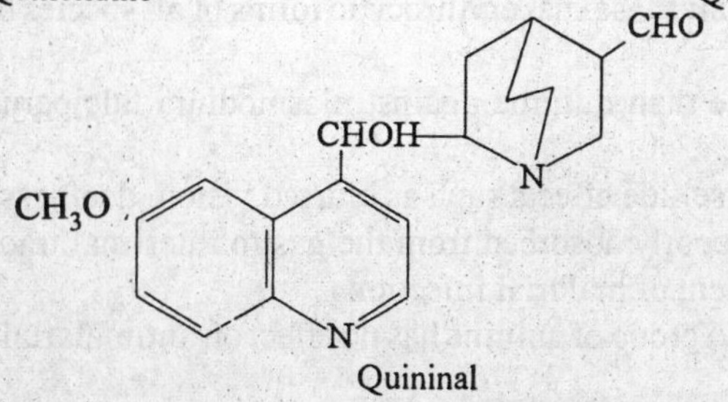

Quininal

9. Ozonolysis of isoquinine does not give the expected 6'- methoxy ruban-3-one 9-ol but, gives ketone-3 acetyl 6' methoxy ruban-9-ol which is inactive.
10. Replacement of C-9 hydroxyl group with halogen or hydrogen results in a loss of antimalarial activity.
11. Replacement of 6' methoxy (—CH_3O) group by 6' alkyl ethers increases its activity.
12. The presence of amine-methyl-4 quinoline methanol moiety is necessary for antimalarial activity.

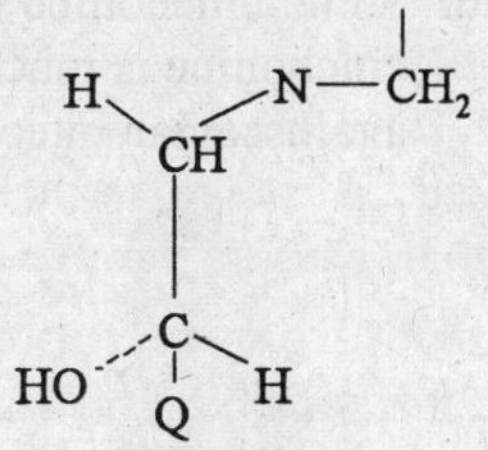

4-Quinolinyl

13. Another similar compound 6' methoxy quinoyl-2-piperidyl carbinol is found to be active against P. relictum.

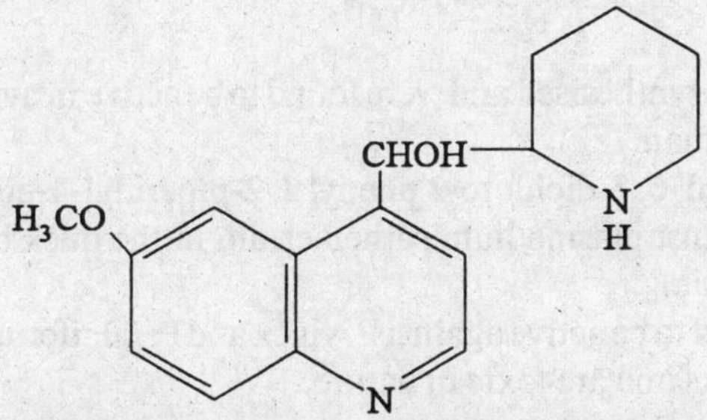

14. The analogous compounds related to cinchonine without the 6' methoxyl group were found to be inactive.
15. Quinine is largely localised in the liver and is oxidised to 2' hydroxy carbo-styril derivative by an enzyme.

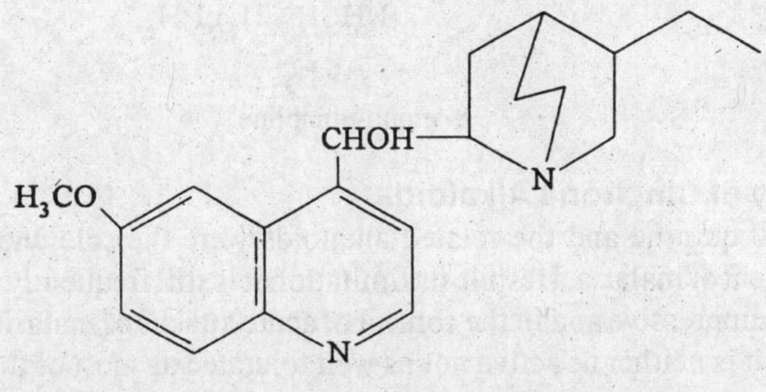

In human beings 2' hydroxy derivative is converted to 2,2' dihydroxy derivative. A compound isolated from urine of human has the following structure.

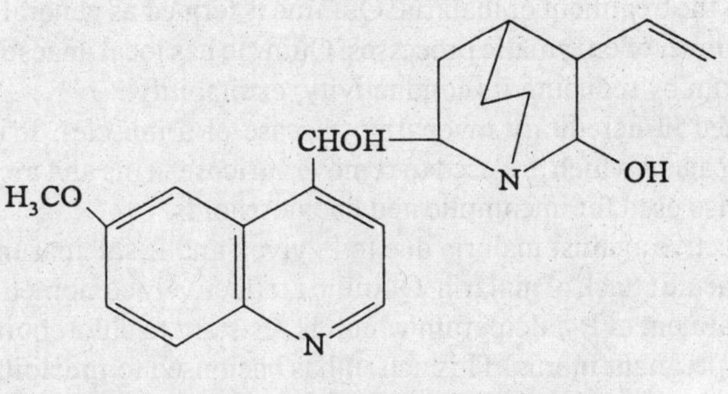

If oxidation at 2' position is blocked antimalarial activity increases.

16. Some cinchonine derivatives were studied for their activity and it was found that though cinchonine is much less active but 2'-hydroxy cinchonine is more active than chinochonine against avian infection.

C_2H_5

CHOH

N

N

C_6H_5

17. Quinolyl carbinolamine was synthesised and were found to be active in avian malarias but these are not useful clinically for man.
18. Another quinoline compound, 6, 8-dichloro-2 phenyl 2, 2-piperidyl-4-quinoline methanol has been found to be active against plasmodium cathemerium in the duck but it is not useful for treatment of malaria in human.
19. 8-amino quinolines are found to be active against P. vivex and P. falciferum but are not effective against acute malarial attacks and are toxic in nature.

H_3CO

N

$NHCH(CH_2)_3NH_2$

CH_3

8-amino-quinoline

Therapeutic Activity of Cinchona Alkaloids

Until the early 1930 quinine and the related alkaloids were the sole chemotherapeutic agents available for the treatment of malaria. Despite its limitation it is still frequently used because of its low cost. It is effective as a suppressive and in the control of acute attacks of malaria due to P. falciperum, P. vivax etc. However, it is neither as active nor as well tolerated as most of the newer synthetic anti malarial drugs.

Therapeutic Uses of Quinine

It is mainly used in the treatment of malaria. Quinine is termed as general protoplasmic poison. It can depress a large number of enzymatic processes. Quinine has local anaesthetic activity. It can also depress the myocardium by reducing its conductivity, excitability.

Quinine is also used in hereditary myopathy (disease of a muscle). It is also used along with urethane as sclerosing agent which are used to remove varicose veins and are also used for closure of hernial ring and are also used for uncomplicated haemorrhoids.

Chloroquine is effective against malaria due to P. vivax and P. falciparum. These are also useful in the treatment of an acute attack of malaria. Quinine is effectively combined with sulphadiazine and tetracycline in the treatment of P. falciperum which is resistant to chloroquine. In therapeutic doses quinine contracts the pregnant uterus. This action has been used to precipitate labour. Quinine has analgesic and antipyretic properties. Intravenous injection of quinine has a relaxant effect on vascular smooth muscle causing a fall in blood pressure.

Morphine

Morphine is one of the oldest narcotic analgesic. It is obtained from opium. There are three important opium alkaloids namely morphine, codeine and thebaine which contain phenanthrene nucleus.

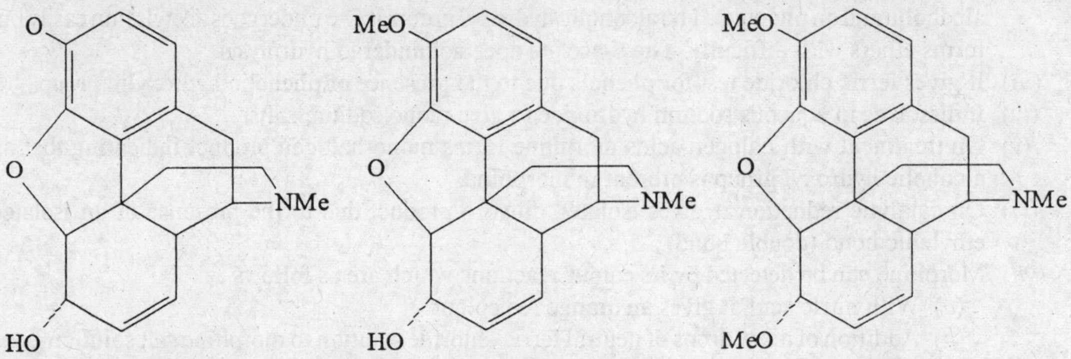

Morphine Codeine Thebaine

Morphine is the chief alkaloid obtained from opium which is the dried sap of the unripe fruit capsules of papaver somniferum.

Structure of Morphine

Morphine is $C_{17}H_{19}NO_3$ and contains a phenanthrene nucleus and has a tertiary nitrogen. It has a secondary hydroxyl group at position 6 and a phenolic group at position 3. There is a ethylene double bond between C-7 and C-8. There is a methyl group at tertiary nitrogen atom. In morphine there are five rings and it has five chiral centres i.e. 5, 6, 9, 13 and 14. It has been found out that the hydrogen atoms at C-5, C-6, C-14 are all cis and the bridge at C-9 and C-13 is also cis. The stereochemistry of morphine is confirmed by X-ray analysis and it shows that rings I, II and oxide bridge lie in a plane and rings III and IV are almost perpendicular to this plane. The cyclohexane ring has chair conformation.

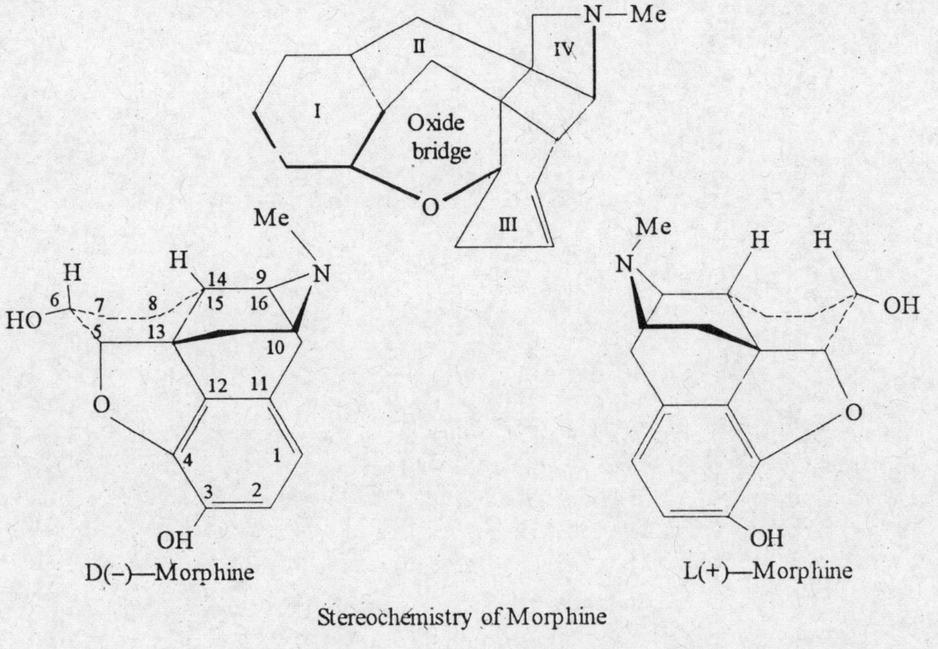

D(–)—Morphine L(+)—Morphine

Stereochemistry of Morphine

Properties of Morphine

(i) It is a solid having its m.p at 254°C and natural morphine is laevo rotatory, has a specific rotation of –131°C. It crystallises from aqueous ethanol and is soluble in solvents like ethanol, pyridine, ethyl acetate, diethyl amine etc. It is insoluble in water and benzene.

(ii) Morphine forms a diacetate salt (heroin) showing the presence of two hydroxylic group i.e. on alcoholic and on phenolic. The alcoholic hydroxyl group at C-6 undergoes acetylation easily but forms ethers with difficulty. The 6-acetate does not undergo hydrolysis.

(iii) It gives ferric chloride test for phenols due to the presence of phenolic hydroxylic group.

(iv) It dissolves in aqueous sodium hydroxide to give mono sodium salts.

(v) On treatment with halogen acids morphine forms mono-halogen product indicating that an alcoholic hydroxyl group is present in morphine.

(vi) On catalytic reduction it gives isolated dihydro product due to the presence of an isolated ethylenic bond (double bond).

(vii) Morphine can be detected by its colour reactions which are as follows :

  (a) With nitric acid, it gives an orange red colour.

  (b) Addition of a few drops of neutral ferric chloride solution to morphine salt solution gives blue colour which disappears on the addition of alcohol or carboxylic acid.

  (c) With potassium ferricyanide solution and neutral ferric chloride solution, morphine gives a blue colour.

  (d) With potassium iodate and dilute sulphuric acid it gives a brown colouration which darkens on addition of ammonia.

  (e) With sulphuric acid it gives very faint pink colour that changes to green and on heating the colour changes to brown.

(viii) Morphine has stimulant and depressant effects on central nervous system and produces a sense of euphoria, and is a valuable analgesic.

The therapeutic uses, action and structure activity relationship of morphine is discussed in chapter 16 (Analgesics) of this book.

Chapter 22

Psychopharmacology

The drug therapy and other organic treatment for mental disorders can be defined as ***psychopharmacology***. The drugs used to modify or to correct pathological behaviour, psychotic disorders, moods or thoughts are known as ***Psychotropic drugs***, which can be classified as follows:

I. Antipsychotic drugs (Major Tranquilizers)
II. Anti anxiety drugs
III. Psychotogenic or Psychedelic drugs
IV. Antidepressant drugs

I. Antipsychotic Drugs

The antipsychotic drugs mainly consist of dopamrines, receptors, antagonists. Antipsychotics are considered as Major Tranquilizers except for a few newly developed drugs. The tranquilizers have been defined as drugs which gives "peace of mind" without causing hypnosis or anaesthesia. Originally, the term tranquilizer was used to describe the psychic effect of reserpine and chlorpromazine, which have the ability to calm a patient suffering from psychotic disorders. The major use cf antipsychotics is to treat schizophrenia and other psychiatric and organic disorders. These drugs are further classified as follows:

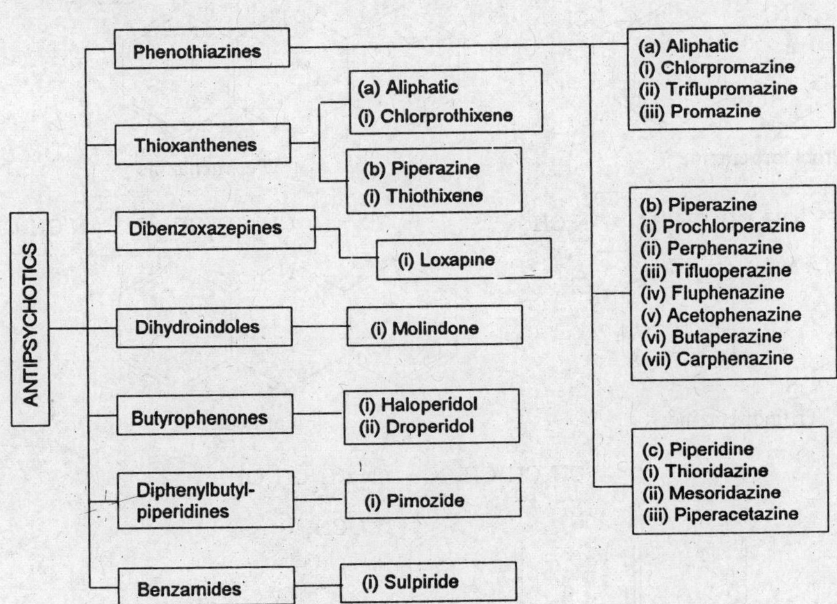

1. Phenothiazines

All the phenothiazines have the same three ring phenothiazine nucleus, but differ in the side chains joined to the nitrogen atom of the middle ring. These drugs are typed according to the aliphatic, piperazine, piperidine nature of the side chain.

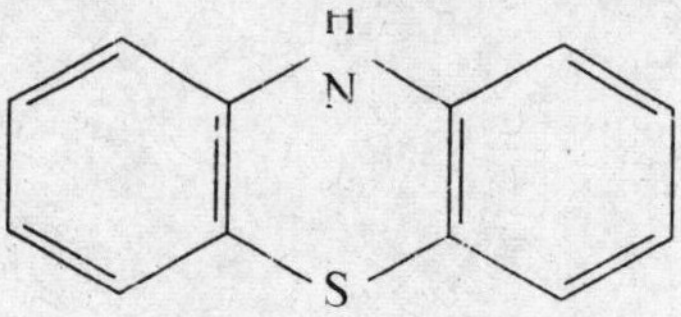

Phenothiazine nucleus

(a) **Aliphatic**

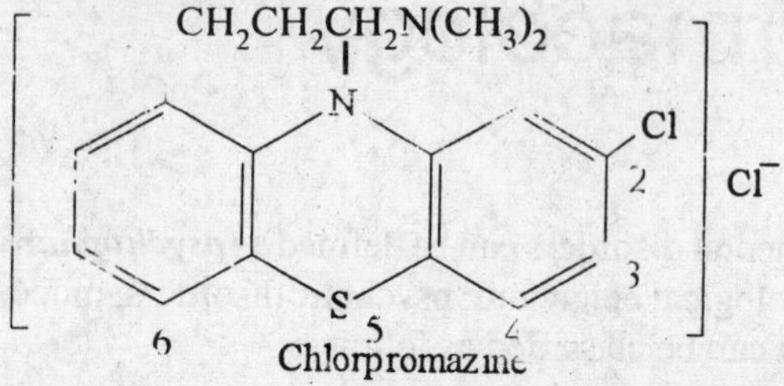

[Chlorpromazine: $CH_2CH_2CH_2N(CH_3)_2$ on N; Cl at 2] Cl^-

Chlorpromazine

$CH_2CH_2CH_2N(CH_3)_2$

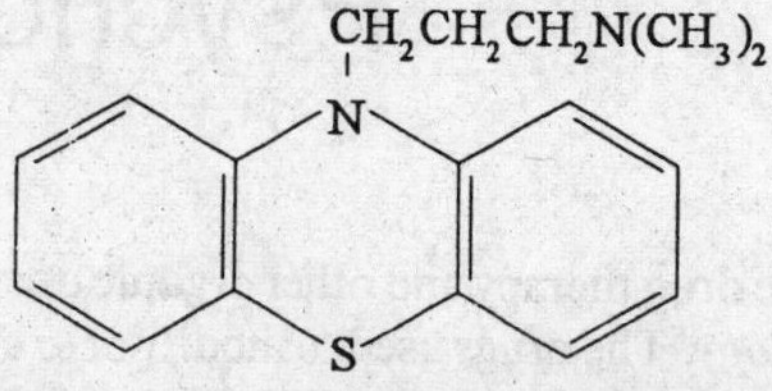

Promazine

$CH_2CH_2CH_2N(CH_3)_2$, CF_3

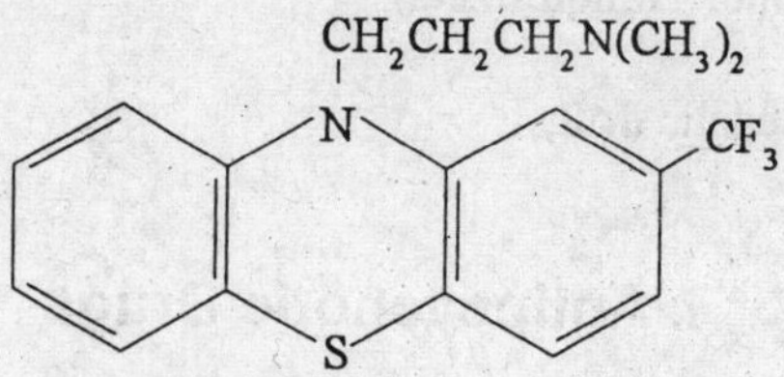

Triflupromazine

(b) **Piperazine**

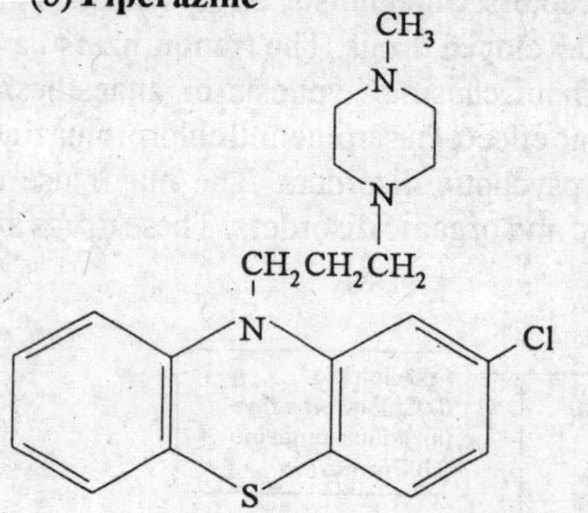

CH_3, $CH_2CH_2CH_2$, Cl

Prochlorperazine

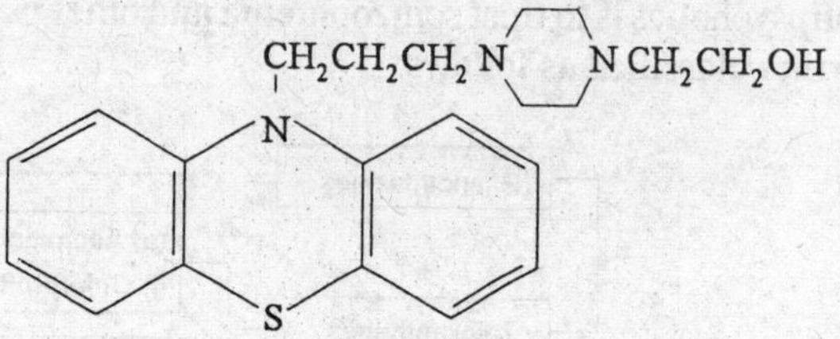

$CH_2CH_2CH_2$ N N CH_2CH_2OH

Perphenazine

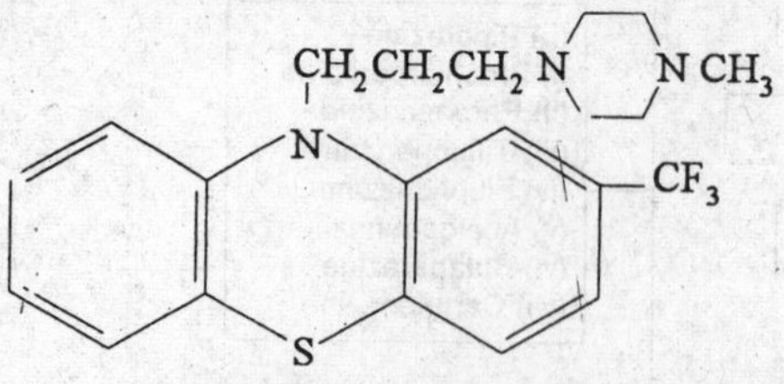

$CH_2CH_2CH_2$ N N CH_3, CF_3

Trifluoperazine

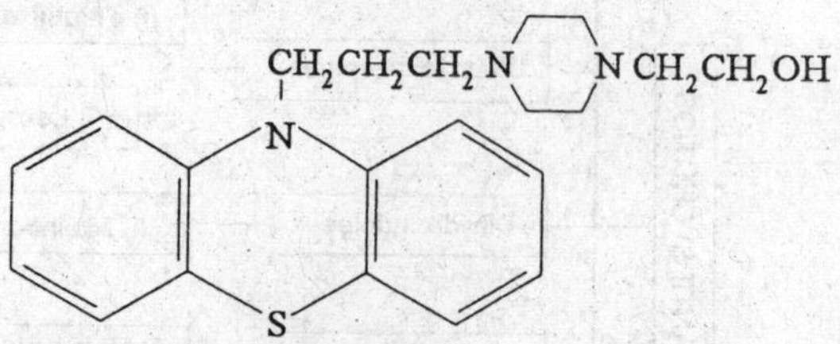

$CH_2CH_2CH_2$ N N CH_2CH_2OH

Fluophenazine

$CH_2CH_2CH_2$ N N CH_2CH_2OH, $COCH_3$

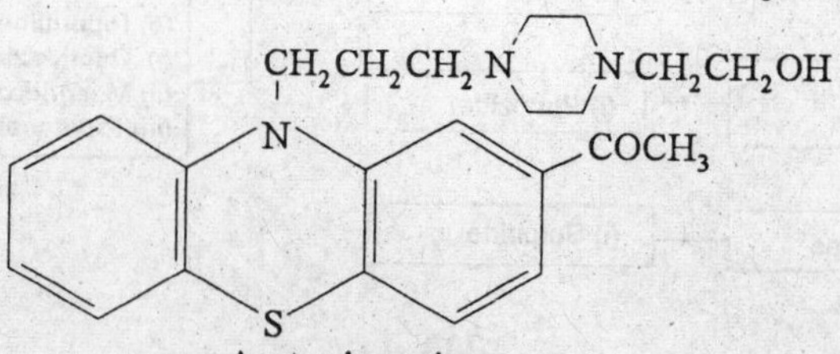

Acetophenazine

(c) Piperidine

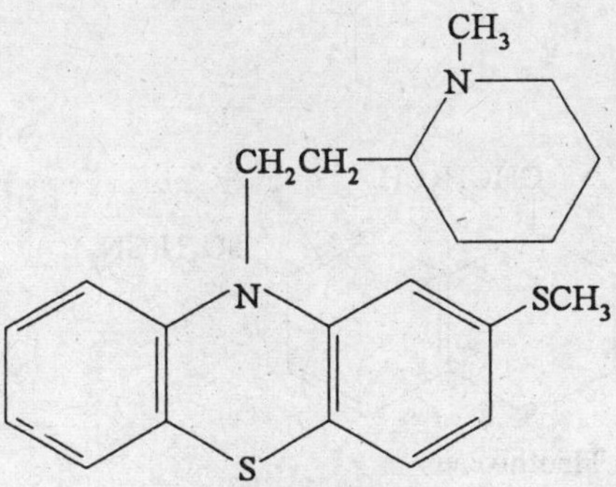

Thioridazine

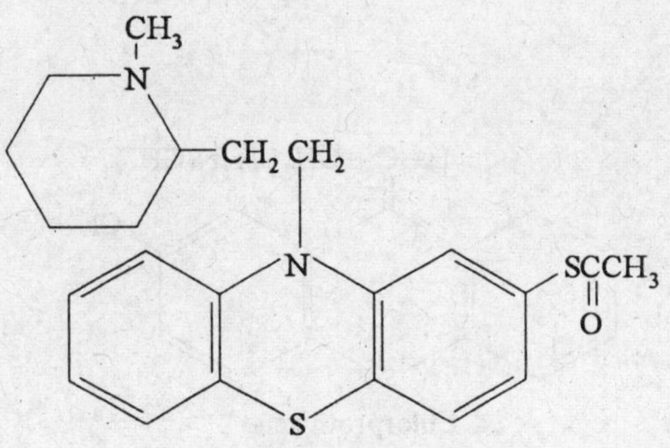

Mesoridazine

The following few drugs are discussed here:

(*i*) *Chlorpromazine,* chemically it is named as 2-chloro-10-(3-dimethyl aminopropyl) phenothiazine hydrochloride. It occurs as white or creamy white powder or as an oily liquid with a bitter taste. It is soluble in water and is stored in tight, closed, light resistant containers. It is available as tablets containing 10, 25, 50, 100 mg of the drug and as intramuscular injection and also as a syrup containing 10 mg per teaspoonful of the drug. It is one of the most extensively used antipsychotic drug. It acts by effecting three integrating systems of the brain, i.e. they reduce the incoming sensory stimuli and then modify the limbic or marginal system and lastly they decrease central sympathetic activity by blocking noradrenaline, dopamine 5 hydroxy tryptamine Transmission in the brain. Chlorpromazine depresses the chemoreceptor trigger zone and thus it is a powerful antiemtic drug and is used as antiemtic drug in radiation sickness, vomiting of pregnancy etc. It also possesses alpha adrenergic blocking effect.

It is an effective drug for psychotic patients suffering from schizophrenia. It also is used to control very aggressive and destructive behaviour in children. It is also used as a preanaesthetic medication.

(*ii*) *Prochlorperazine,* chemically it is named as 2-chloro-[10(3-{-4methyl peperazin-1-yl}propyl)] phenothiazine.

It occurs as pale yellow liquid and is sparingly soluble in water.

It is used as anti psychotic drug.

(*iii*) *Perphenazine,* chemically it is named as 2-chloro-[-10-{3-(4-(2-hydroxyethyl)piperazin-1-yl)}] phenothiazine.

It occurs as white powder and is used as antipsychotic drug in the treatment of severe anxiety and tension.

(*iv*) *Trifluoperazine,* chemically it is named as 10-(-3-{-4 methylpiperazin-1-yl} propyl)-2 trifluoromethyl phenothiazine. It is known as Siquil tablets and is mainly used as antipsychotic, antiemetic drug.

(*v*) *Thioridiazine,* chemically it is named as 10-(2-{1-methylpiperid-2-yl}ethyl)-2-methyl thiophenothiazine.

It is a white crystalline powder and is soluble in water. It is used in the treatment of schizophrenia.

2. Thioxanthenes

The thioxanthene three ring nucleus differs from the phenothiazine nucleus by the substitution of a carbon atom for the nitrogen atom in the middle ring. The two available thioxanthenes have either aliphatic (e.g. Chlorprothixene) or a piperazine (e.g. thiothixene) side chain.

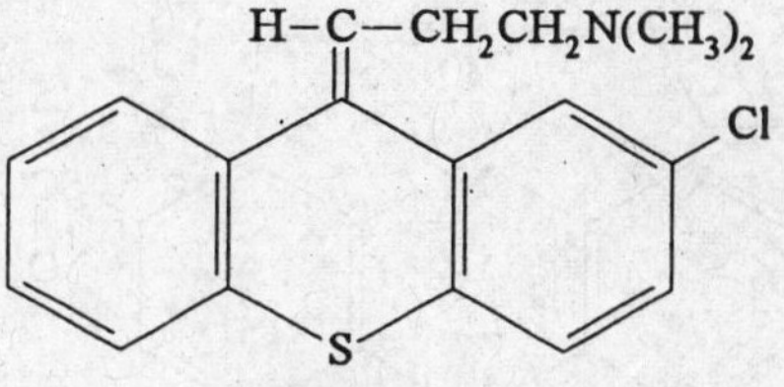

H−C−$CH_2CH_2N(CH_3)_2$ ... Cl ... S

Chlorprothixene

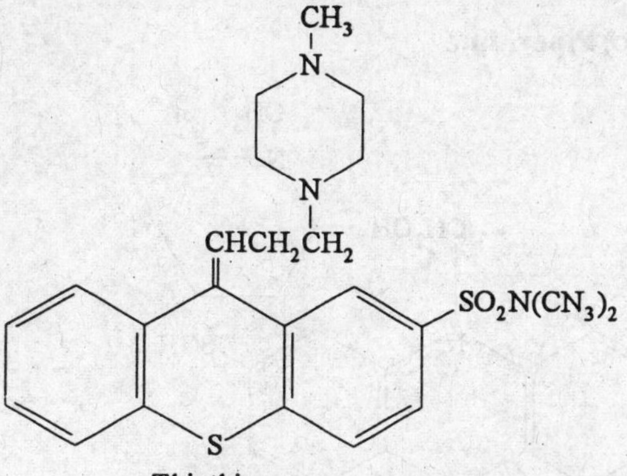

CH_3 N N $CHCH_2CH_2$ ... $SO_2N(CN_3)_2$ ... S

Thiothixene

These are useful drugs for the treatment of schizophrenia.

?. Dibenzoxazepines

The chemical structure of these drugs are based on another modification of the three ring phenothiazine nucleus. Loxapine is a drug belonging to this group, which has a piperazine side chain.

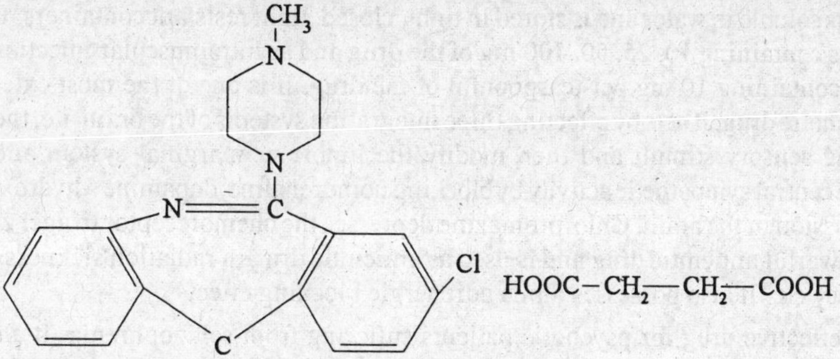

CH_3 N N N=C Cl C HOOC−CH_2−CH_2−COOH

Loxapine

The usual dose of Loxapine is 60-100 mg per day for usual adults. This drug is used in schizophrenias, organic brain psychoses, psychotic depressions.

4. Dihydroindoles

The example of the drug belonging to this group, is molindone. This drug has some unusual properties such as not inducing weight gain and is less epileptogenic than the phenothiazines. It has the following structure:

H N CH_3 O NCH_2 CH_2CH_3 O

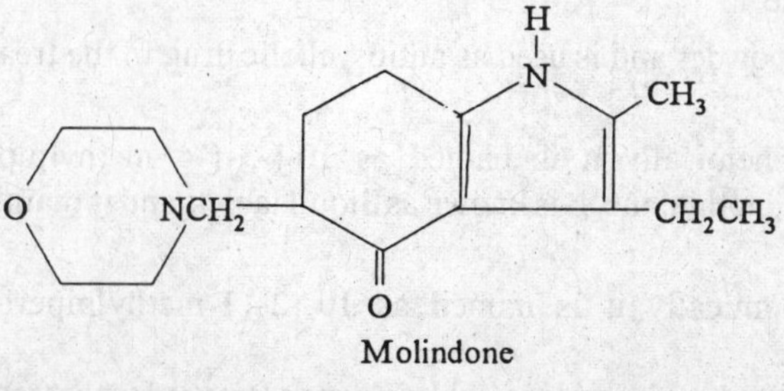

Molindone

5. Butyrophenones

Examples of the drugs are Haloperidol and Droperidol. These have the following structures :

Haloperidol is most widely used antipsychotic drug. Droperidol is used as an adjuvant in anesthesia and sometimes it is used as intravenous antipsychotic drug in emergencies.

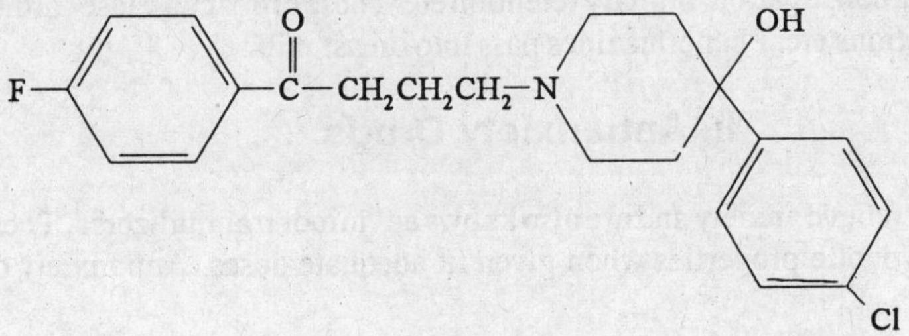

Haloperidol

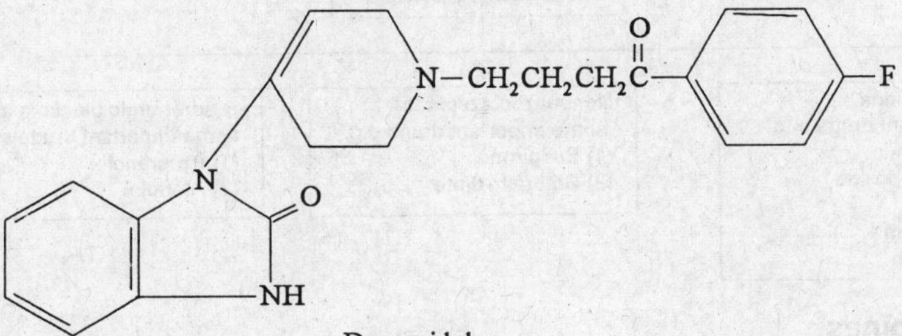

Droperidol

6. Diphenylbutylpiperdines

These drugs are structurally similar to butyrophenones. The most important drug of this group is pimozide. It has the following structure:

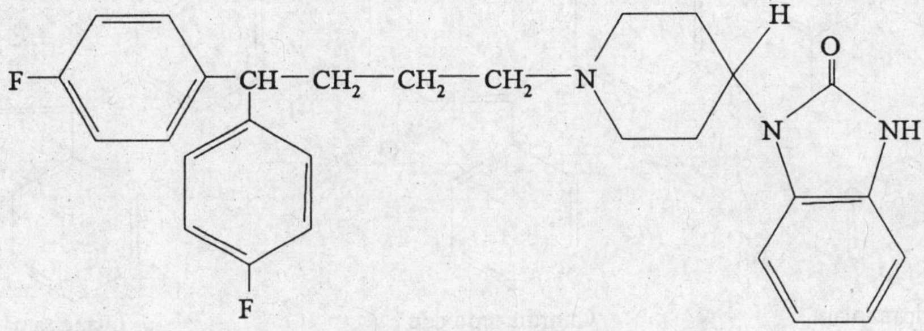

Pimozide

7. Benzamides

Example of the drug is Sulpiride, having the following structure:

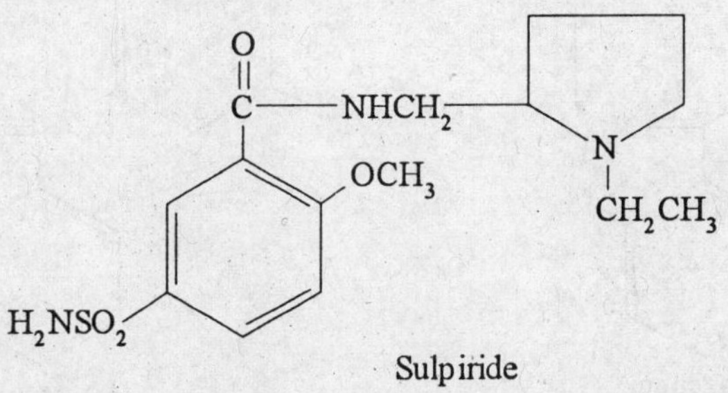

Sulpiride

It is an effective antipsychotic drug with fewer neurological side effects.

Adverse Effects of Antipsychotic Drugs

These drugs cause postural hypertension, cause peripheral anticholinergic effects like dry mouth and nose, blurred vision, constipation, uninary retention etc. These drugs can cause cardiotoxic effect, weight gain, skin erruptions etc. Phenothiazines pass into breast milk.

II. Antianxiety Drugs

These drugs are used to relieve anxiety and are also knows as "minor tranquilizers." These drugs also exhibit sedative and hypnotic properties when given in adequate doses. Antianxiety drugs can be calssified as follows:

ANTIANXIETY DRUGS

- Benzodiazepiines some important drugs e .g.
  (1) Alprazolam
  (2) Chordiazepoxide
  (3) Diazepam
  (4) Flurazepam
  (5) Lorazepam
- Nonbenzodiazepines some important drugs e.g.
  (1) Buspirone
  (2) Chloralhydrate
- B-adrenergic blocking agents some important drugs e.g.
  (1) Propranolol
  (2) Atenolol

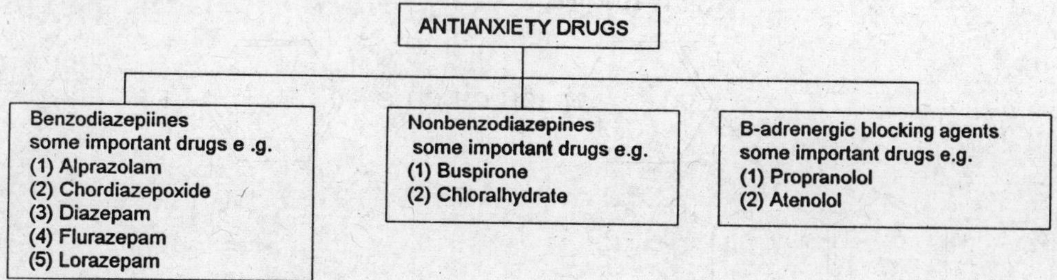

1. Benzodiazepines

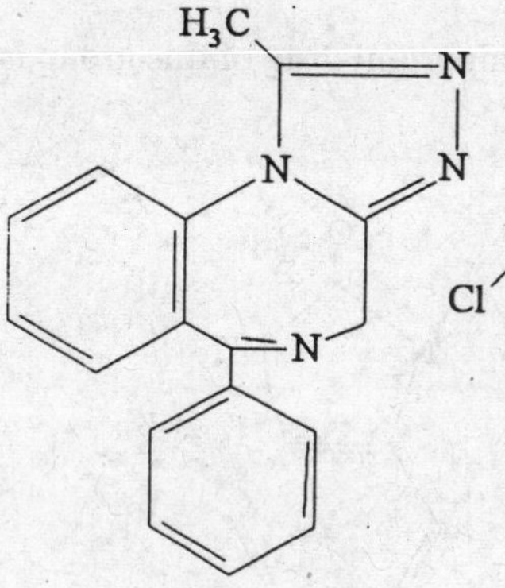

Alprazolam

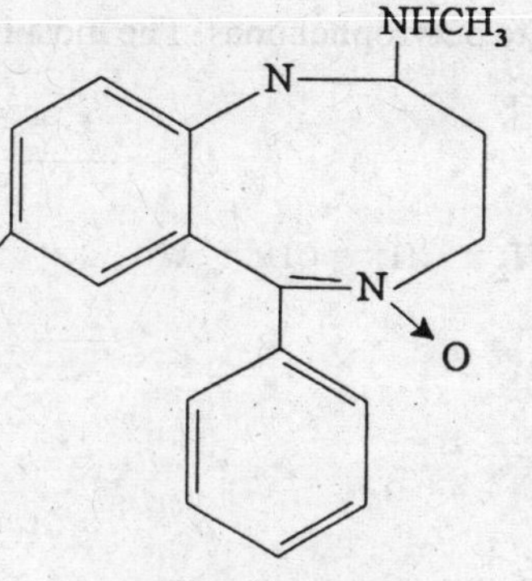

Chordiazepoxide

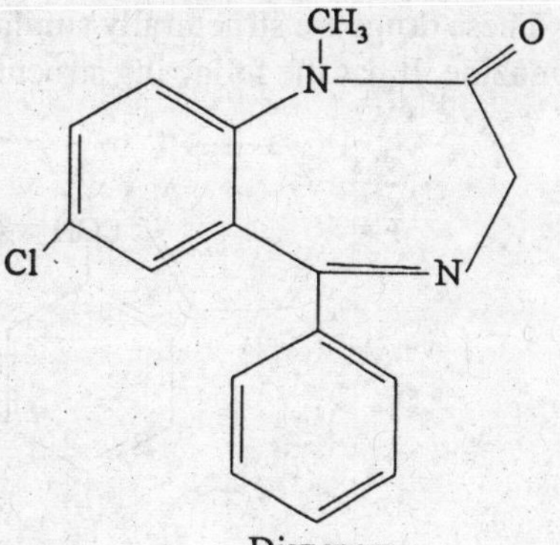

Diazepam

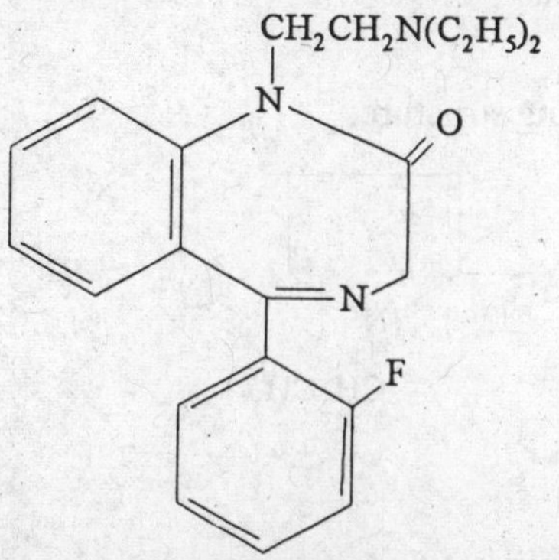

Flurazepam

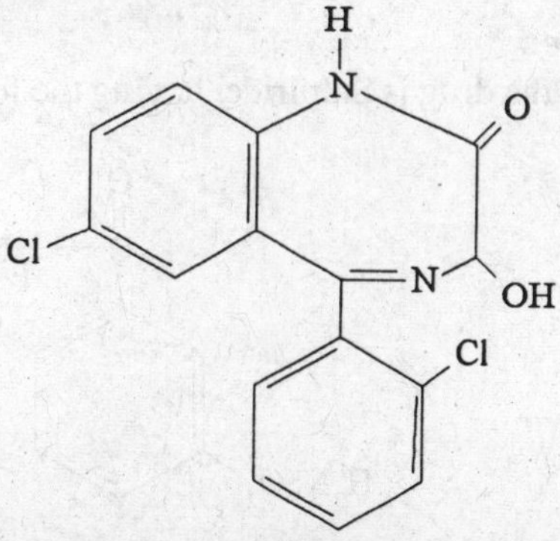

Lorezepam

These drugs are usually effective in moderate doses, have a wide margin of safety and are long acting. All of the benzodiazepines are similar in chemical structure and clinical properties. These drugs are marketed as minor tranquilizers and are used for the treatment of anxiety.

Adverse effects. Ataxia, hypotension, excessive dose can cause coma, disinhibition, dysarthria, etc.

2. Non-benzodiazepines Some important drugs are :

Chloralhydrate. Its chemical structure is $CCl_3CH(OH)_2$. It is one of the oldest sedative hypnotic drug which is used in the treatment of anxiety.

Adverse effects are: nausea, vomiting, diarrhoea, hepatic renal damage. It should not be used during pregnancy.

Buspirone : It is available in 5 and 10 mg tablets. It is useful in the treatment of anxiety, especially in patients who have not received benzodiazepines in the past. Adverse effects are: headache, nausea, dizziness. It should be used with caution for patients with hepatic or renal disease and for pregnant women.

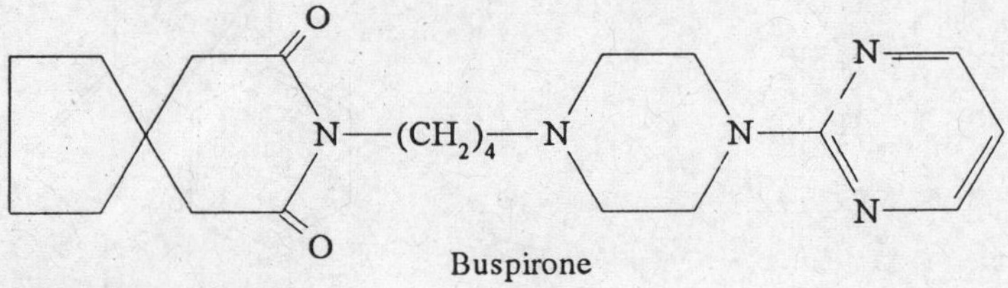

O, N—(CH$_2$)$_4$—N, N—, N, N, O

Buspirone

Aderenergic Blocking Agents

Aderenergic drugs are very useful in the treatment of psychiatric disorders, lithium induced tremor, and in somatic symptoms of anxiety.

Atenolol

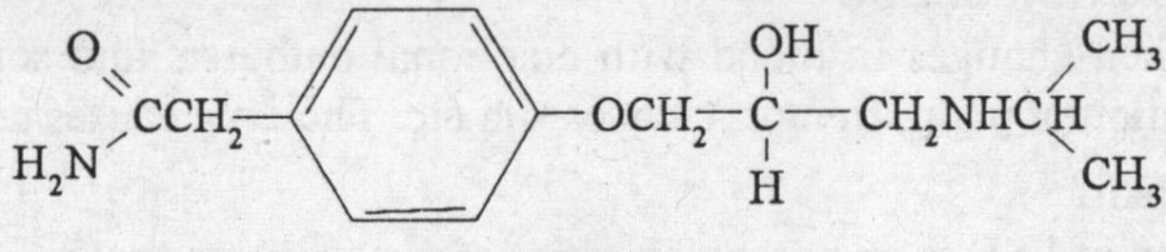

O, H_2N, CCH_2, $OCH_2—C—CH_2NHCH$, OH, H, CH_3, CH_3

Atenolol

OH, $OCH_2—C—CH_2NHCH$, H, CH_3, CH_3

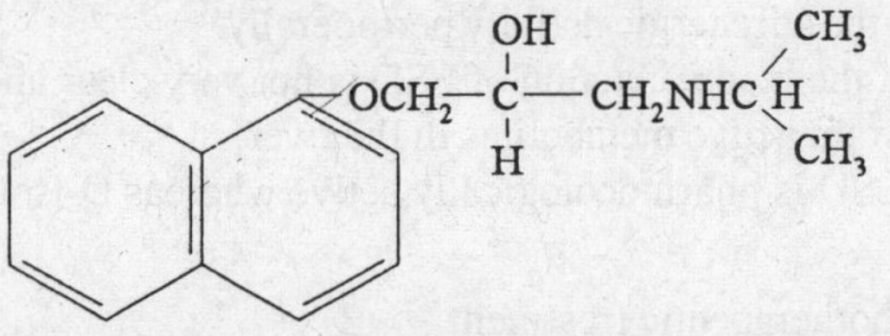

Propranalol

These drugs are effective in treating the aggressiveness and violent behaviour of organic mental diseases, schizophrenia, etc. Propranolol is useful in reducing the peripheral manifestations of anxiety, which are associated with social phobia.

Adverse effects. These drugs can elevate blood glucose by inhibiting insuline release. Some other adverse effects are : hypotension, nausea, vomiting, dizziness, fatigue, insomnia.

III. Psychotogenic Drgus or Psychotomimetic Drugs or Psychedelic Drugs

These drugs produce depersonalization, changes in mood and learned behaviour and also produces hallucination, hence, these are also known as *hallucinogens.* These drugs could be classified as follows :

(*i*) Those with indol ring and show peripheral adnergic activity e.g. lysergic acid diethylamide (LSD), bufotenine etc.

(*ii*) Those without indole ring e.g. cannabis (marijuana) showing miscellaneous action.

(*iii*) Drugs showing anti cholinergic action e.g. atropine etc.

Lysergic Acid Diethylamide (LSD)

LSD is an amine alkaloid and is synthesized from ergot. It has the following structure.

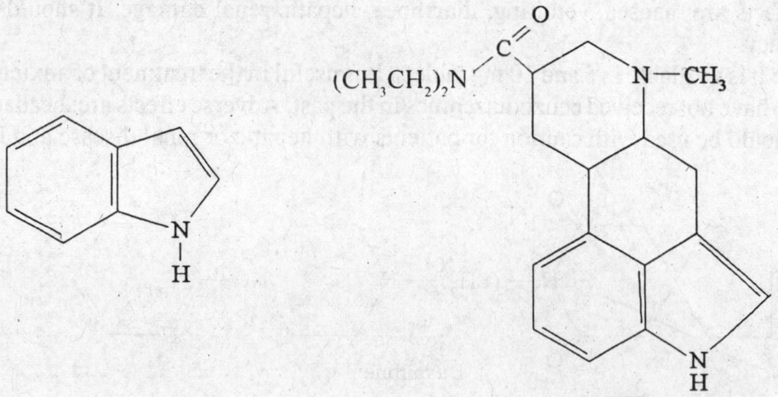

$(CH_3CH_2)_2N-C(=O)-$; $N-CH_3$; N–H

LSD

It is an indole alkaloid and the indole ring is fused to reduced quinoline ring with N-methyl group. Its psychotic properties of LSD was discovered by Hopmann in 1943.

Pharmacological Action of LSD

LSD exhibit marked changes in mood with emotional outbursts and sometimes shows other symptoms like dilatation of pupil, tremor, tachycardia etc. The drug causes psychic dependence.

Mechanism of Action

1. The main action of LSD is on norepinephrine present in the central nervous system.
2. LSD can block the peripheral action of 5-hydroxy tryptamine.
3. LSD can increase the adrenergic activity peripherally.
4. The exact site and the mode of action of LSD is not very clear and most probably the drug is metabolised to certain active metabolites in the liver.
5. The L-isomer of LSD is pharmocologically active whereas D-isomer is not active.

Therapeutic Uses

1. It is used for psychotherapeutic treatment.
2. It helps to activate repressed memories which is useful in psychoanalysis.
3. This drug is used as a research tool to produce experimental psychosis.
4. LSD is used as an analgesic as it helps the patient to bear the chronic pain.

Adverse Effects of LSD

The treatment with LSD should be done very carefully under strict medical supervision as this drug has serious side effects. It produces suicidal tendencies and an overdose of LSD may also be fatal due to respiratory failure. The psychotic changes produced by LSD are generally reversible but some times it is capable of producing permanent psychosis.

The drug causes psychic dependence. It sometimes causes nausea, increases blood pressure. LSD can effect the foetus of a pregnant woman and an expectant mother may deliver a child with clubbed feet etc. Progesterone, phenothiezine, diazepam are antagonistic to LSD.

Source and Preparation

d-lysergic acid is obtained from ergot alkaloids. These ergot alkaloids are product of a filamentous fungus called Claviceps purpurea. This fungus grows parasitically on rye and grasses and cereals etc. These ergot alkaloids yeild lysergic acid on hydrolysis and then LSD is prepared synthetically.

LSD can be detected by a colour test *i.e.* when the LSD is treated with dilute sulphuric acid and traces of ferric chloride it gives blue-violet colour.

Marijuana (Hashish)

It is one of the oldest herbal medicine and is obtained from hemp plants called cannabis sativa and Cannabis indica. The active ingredient is present in the resinous mass of the female plants. It is also known as Hashish or Charas or Bhang, Kif etc. Actually the term Marijuana is used to describe any plant extract which contains the active principle of hashish.

The main constituent and active principle of cannabis are the isomeric tetra-hydro cannabinols having the following structure.

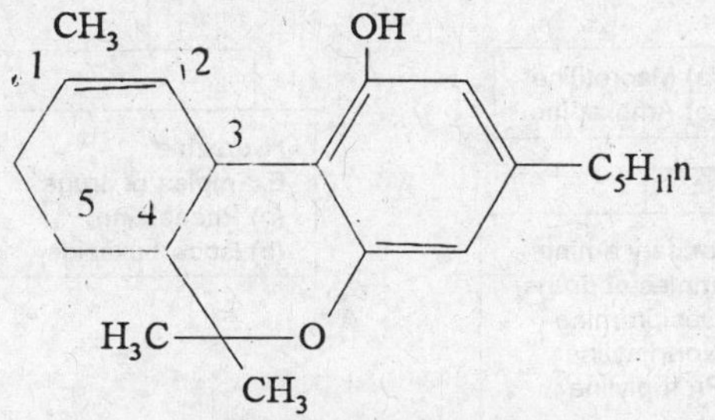

CH_3 OH 1 2 3 $C_5H_{11}n$ 5 4 H_3C O CH_3

Tetra hydro cannabinols

Structure Activity Relationship

1. Tetra hydrocannabinols have other isomers which differ in the location of double bond and also in the stereochemical arrangment of C-1, C-3, C-4.
2. When nC_5H_{11} alkyl side chain of phenyl group is replaced by nC_6H_{13} or nC_7H_{13} groups the activity of cannabinols increases.
3. The racemic synthetic cannabinol are much less active than the mixture of (-) tetrahydro cannabinol.
4. If n-amyl group is replaced by 1, 2 dimethylheptyl group, the activity of cannabinols increases remarkably.

Pharmacological Action of Hashish

These drugs produce some behavioural changes. The drug mainly acts on central nervous system. Unlike LSD it has a sedative effect. It is well absorbed when administered orally. The effect of cannabis can be seen after half an hour or one hour after its ingestion and the effect may persist for 3-5 hours. Some effects of Marijuana are :

1. The drug produces a dreamy state, feeling of well being, joy etc. The subject enters a state of euphoria and may suffer a physiologically depressed state.
2. Drug may produce nausea, vomiting etc.
3. It is found to increase appetite and hunger in humans.
4. It has a sedative effect and there is generally an increase in heart rate and reddening of eyes occur.

Therapeutic Uses of Cannabis

It is used in certin psychotic disorders, in the treatment of glaucoma and is also used as an antiemetic for cancer patients. Sometimes these are also used in the treatment of acute asthma. Its therapeutical uses are quite restricted as it is illegally used in many parts of the world as charas, bhang or ganja for its euphoriant and behavioural effect.

IV. Antidepressant Drugs

The discovery of these drugs were found to be very important in the field of psychopharmacology. These are also known as mood elevators. These drugs are classified as follows :

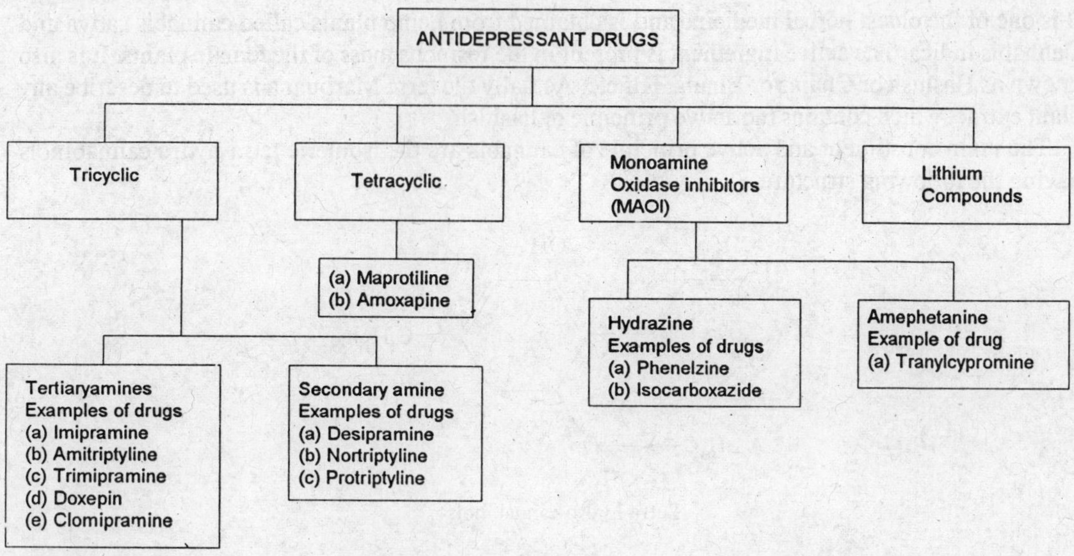

1. Tricyclic

All the Tricyclics have a three ring nucleus in their molecular structure. Imipramine, Amitriptyline, Trimipramine, Doxepin, Clomipramine are called tertiary amines as they have two methyl group on the nitrogen atom of the side chain.

The secondary amines have only one methyl group on the nitrogen atom.

The tertiary amines are metabolized into their corresponding secondary amines in the body.

The chemical structure of Tricyclics are as follows:

Tertiary Amines

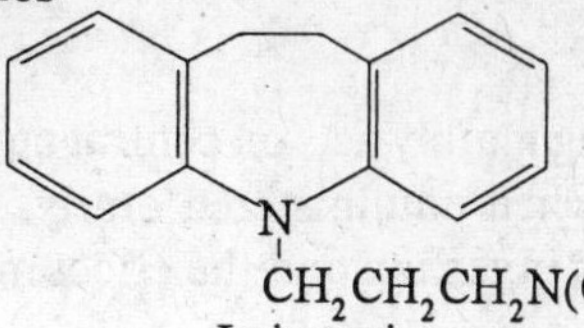

N
$CH_2CH_2CH_2N(CH_3)_2$
Imipramine

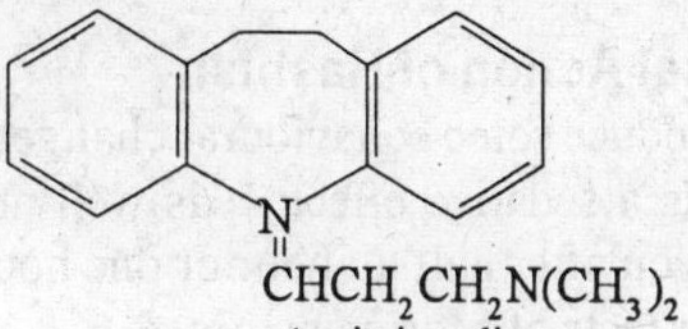

N
$CHCH_2CH_2N(CH_3)_2$
Amitriptyline

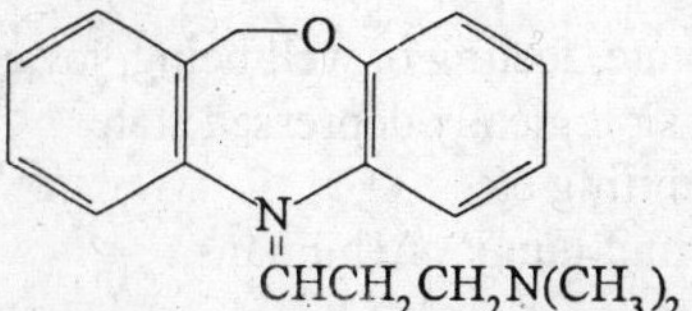

O
N
$CHCH_2CH_2N(CH_3)_2$
Doxepin

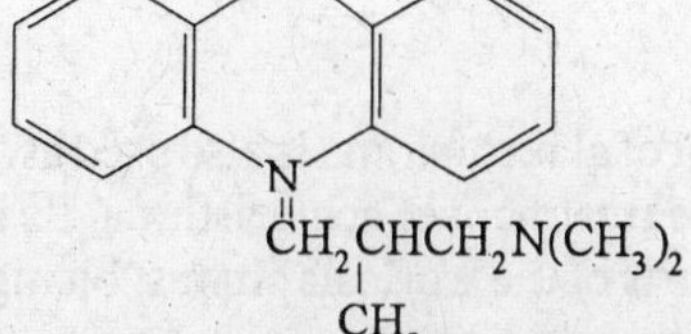

N
$CH_2CHCH_2N(CH_3)_2$
CH_3
Trimipramine

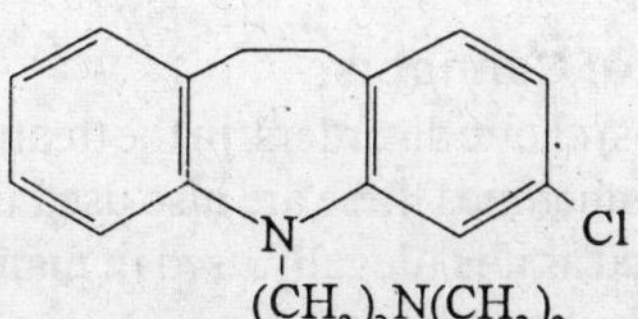

N Cl
$(CH_2)_3N(CH_3)_2$
Clomipramine

Secondary Amines

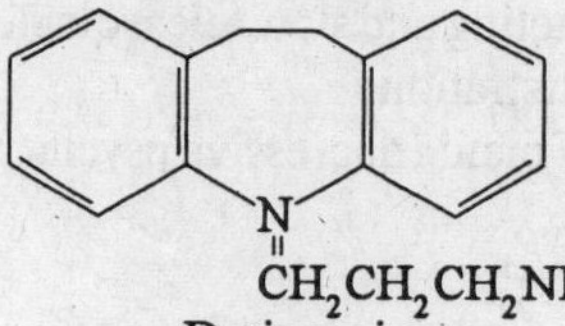

$CH_2CH_2CH_2NHCH_3$
Desipramine

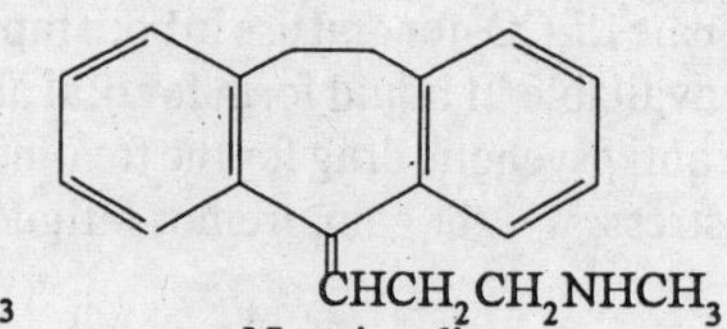

$CHCH_2CH_2NHCH_3$
Nortriptyline

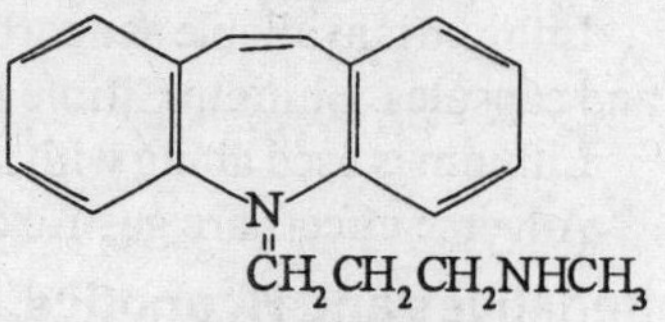

$CH_2CH_2CH_2NHCH_3$
Protriptyline

2. Tetracyclic

These drugs along with tricyclic antidepressants are widley known as TCA's.

The classification of these drugs are based on a gross count of the number of rings in their moleculer structures.

Amoxapine is a dibenzoxazepine which is a derivative of anti psychotic drug, Loxapine. It has cyclic side chain of three ring nucleus, for a total of four rings.

Maprotiline is a tetracyclic with the same side chain as desipramine and its fourth ring bridge the centre of the standard tricyclic nucleus.

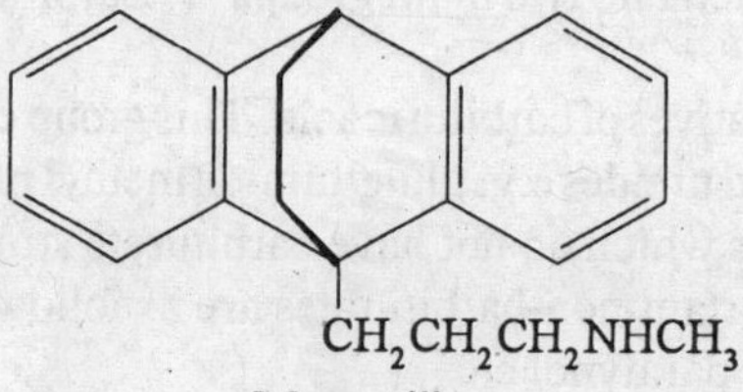

$CH_2CH_2CH_2NHCH_3$
Maprotiline

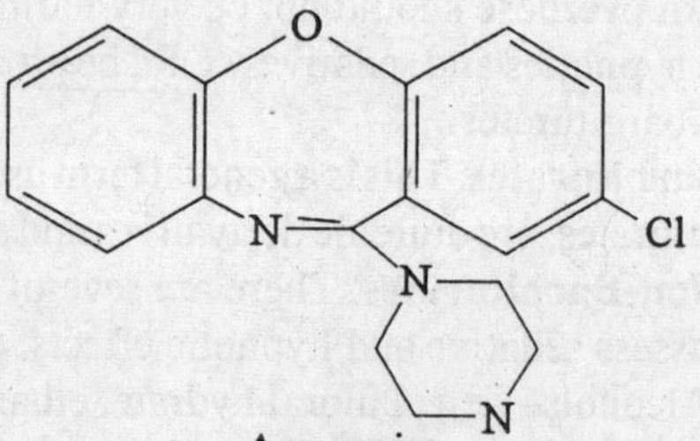

Amoxapine

Tetracyclics also reduce the reuptake of norepinephrine and serotonin. The other therapeutic uses and adverse effects of tetracyclines are similar to that of tricyclics.

3. Monoamine Oxidase Inhibitors

These are as effective as tricyclic antidepressants in treating depression. The commonly used MAOI's are: hydrazine derivatives (*e.g.* Phenelzine and Isocarboxazide) and amphetamine derivative (e.g. Tranylcypromine).

The chemical structure of these drugs are as follows:

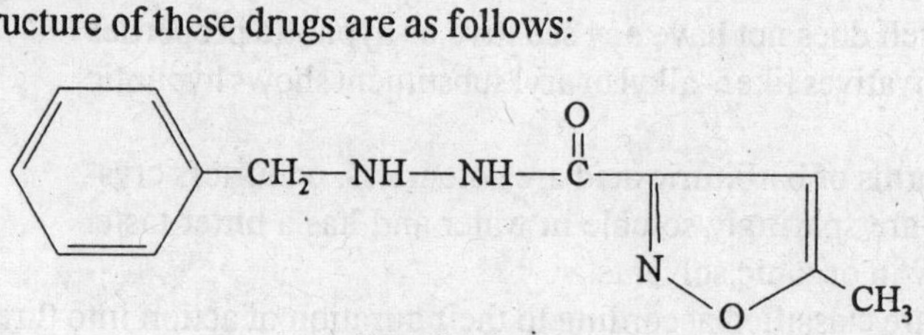

$-CH_2-NH-NH-C(=O)-$ ... CH_3
Isocarboxazide

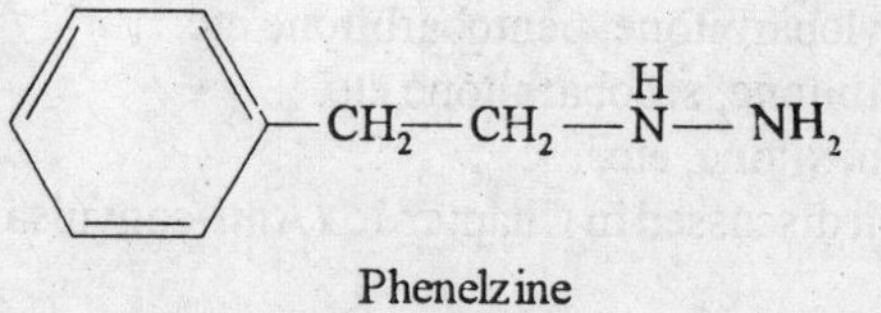

$-CH_2-CH_2-NH-NH_2$
Phenelzine

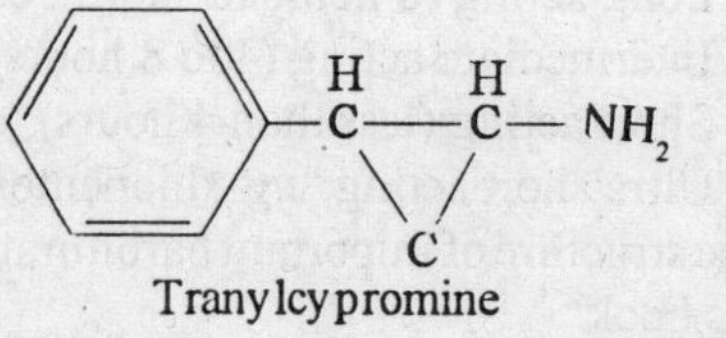

$-CH-CH-NH_2$
Tranylcypromine

These drugs are particularly effective in the treatment of depression, hypersomnia, anxiety, etc. These drugs are readily absorbed when administered orally. The hydrazine MAOI's are metabolized by acetylation.

Adverse effects are hypotension, weight gain, oedema, insomnia, etc.

4. Lithium Compounds

Lithium is available as a carbonate Li_2CO_3 for oral use in both rapidly acting and slow release tablets and capsules. Lithium Citrate is available in liquid form for oral administration.

Lithium is used along with an anti psychotic drug for the treatment of manic depressive psychosis.

Adverse effects are gastric distress, weight gain, tremor fatigue, etc.

Sedatives and Hypnotics

Hypnotic drugs produces sleep resembling natural sleep. Whereas sedatives are the drugs which reduce excitement. Both the drugs are central nervous system depressants and this difference is merely quantitative. Sedative and hypnotics when given in large doses may produce anaestheisa poisoning which may be fatal in some cases.

A ideal hypnotic agent should have the following properties :

1. They should be effective when administered orally.
2. These drugs should not produce addiction, tolerance, or habitation.
3. It should be non-toxic, non-irritant and should not produce hang-overs.
4. It should be cost effective,
5. An overdose should not be very harmful.

The hypnotics and sedatives could be broadly classified into two main groups 1. Barbiturates and 2. Non-barbiturates.

1. Barbiturates. This is a general term used for derivatives ofbarbituric acid. This group comprises ofbarbiturates, are diureide derivatives and some related ureides e.g. glutethimide methyl pryone etc.

2. Non-Barbiturates. There are several compounds which do not have barbiturate structure but they possess sedative and hypnotic effects. Some important non-barbiturates are as follows :

(*a*) Alcohols—*e.g.* Chloral hydrate, ethanol, methyl parafynol etc.
(*b*) Aldehydes—*e.g.* Paraldehyde.
(*c*) Acetylated Carbinols *e.g.* ethylchlorvynol ethinamate etc.
(*d*) Benzodiazepines.
(*e*) Inorganic ions *e.g.* Bromide.
(*f*) Miscellaneous *e.g.* Methaqualone, scopolamine etc.

1. Barbiturates

These are derivatives of Barbituric acid which is obtained as a condensation product of urea and malonic ester.

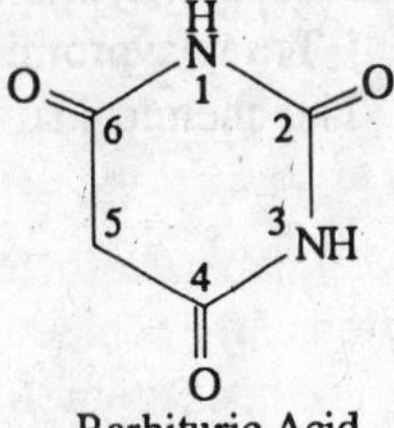

Barbituric Acid

Barbituric acid itself does not have any sedative or hypnotic properties but its substituted derivatives like 5-alkyl or aryl substituent shows hypnotic and sedative activity.

The substituted forms of barbituric acid are colourless, odourless crystalline solids. These are sparingly soluble in water and has a bitter taste. They dissolve readily in organic solvents.

The barbiturates are classified according to their duration of action into three groups.

(*a*) Long acting (8 hours or more): *e.g.* barbitone, methobarbitone, mephobarbitone etc.
(*b*) Intermediate acting (4 to 8 hours): *e.g.* amylobarbitone, pentobarbitone etc.
(*c*) Short acting (less than 4 hours): *e.g.* herbarbitone, secobarbitone etc.
(*d*) Ultra short acting: *e.g.* thiopentone, methohexitone, etc.

The structure of important barbiturates has been discussed in Chapter 15 (Anti-convulsant agents) of this book.

Mechanism of Action of Barbiturates

These drugs cause depression of the activity of the central nervous system. Barbiturates used as sedatives reduce anxiety and the hypnotic doses of barbiturates cause motor in co-ordination and the sleep induced by hypontic doses cause depression of the respiratory system. Barbiturates depresses polysynaptic and monosynaptic reflexes of the spinal cord. The mechanism of barbiturates action on central nervous system is attributed to various biochemical changes like oxidative phosphorylation, increase of concentration of acetylcholine in the brain, inactivation of sodium carrier mechanism etc.

It depresses the myocardium and vasomotor centres and thus produce hypotension.

The distribution of barbiturates in the body is dependent on its lipid solubility, degree of protein binding and the extent of ionization. The short acting barbiturates are highly lipid soluble. As the biological membranes are lipoidal in nature these compounds have a rapid onset of action. An increase in pH of blood and urine increases the ionization of the barbiturates causing efflux of barbiturates from the tissue in the plasma and their excretion.

Other related ureides like glutethimide, methylprylon has the following structure.

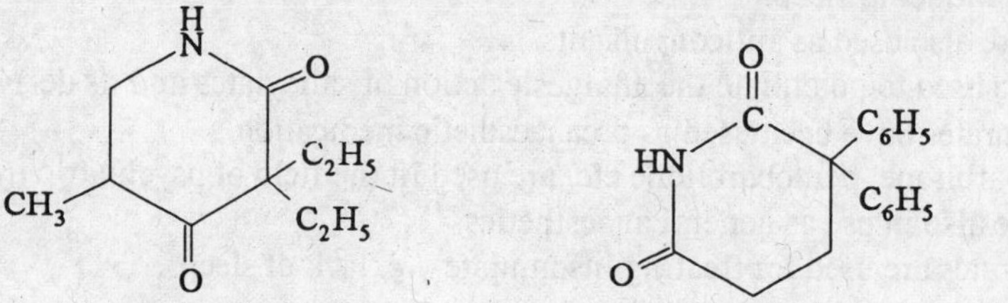

CH_3, C_2H_5, C_2H_5, HN, O, O

Methylprylon

HN, O, C, C_6H_5, C_6H_5, O

Glutethimide

Structure Activity Relationship

The hypnotic properties and the duration of action of barbiturates are modified by the changes in their chemical structures.

1. The substituted barbiturates are obtained by the action of dialkyl diethylmalonate with urea in presence of sodium ethoxide.

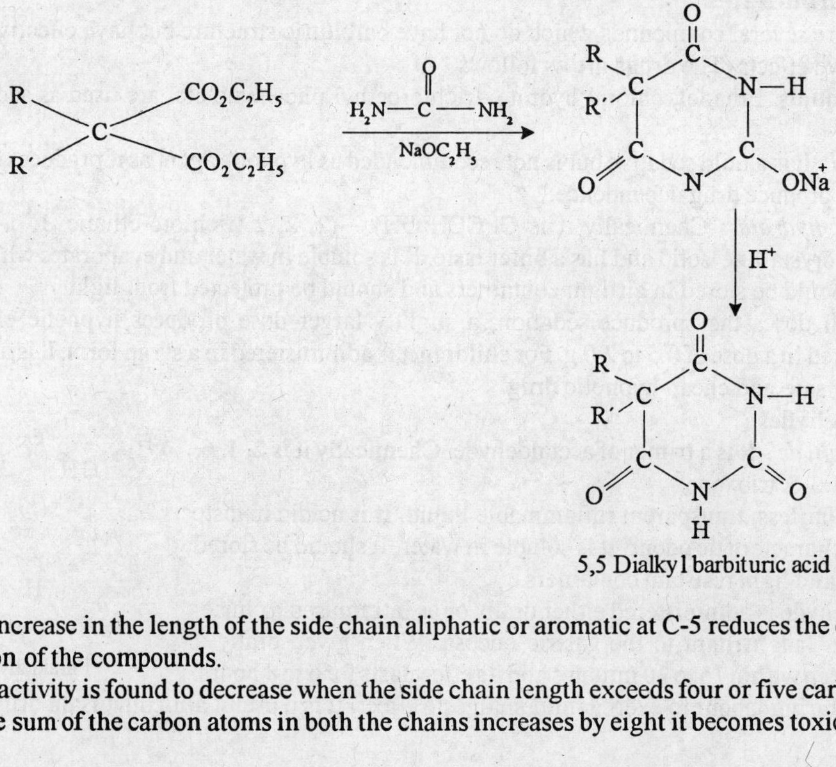

$$\underset{R'}{\overset{R}{>}}C<\begin{matrix}CO_2C_2H_5\\CO_2C_2H_5\end{matrix} \xrightarrow[NaOC_2H_5]{H_2N-\overset{O}{\overset{\|}{C}}-NH_2}$$

R, R′, C, C, N—H, C, C, N, O, O, $O\overset{+}{Na}$

H^+

R, R′, C, C, N—H, C, C, N, H, O, O, O

5,5 Dialkyl barbituric acid

2. An increase in the length of the side chain aliphatic or aromatic at C-5 reduces the duration of action of the compounds.
3. The activity is found to decrease when the side chain length exceeds four or five carbon atoms. If the sum of the carbon atoms in both the chains increases by eight it becomes toxic in nature.

4. Introduction of a phenyl group in the side chain at C-5 reduces the hypnotic properties and the compound shows anticonvulsant effect *e.g.* phenobarbitone.
5. Replacement of oxygens at C-2 with sulphur enhances the lipid solubility of these compounds e.g. thiobarbiturates.

The structural modifications play an important role in deciding the action of the drug as some slight structural change may turn a sedative into hypnotic and an hypnotic into an anaesthetic or into an anti convulsant.

Therapeutic Uses of Barbiturates

1. They are used as sedatives *e.g.* phenobarbitone, amylobarbitone, butobarbitone etc. They are used to relieve tension, anxiety and are also used as in hypertension, peptic ulcer etc.
2. They are used as hypnotics *e.g.* phenobarbitone, secobarbitone, pentobarbitone etc. These are used for inducing sleep.
3. These are also used as anticonvulsants.
4. They are used to potentiate the analgesic action of salicylates and its derivatives.
5. Barbiturates have been used as preanaesthetic medication.
6. Amylobarbitone, pentobarbitone etc. are used in the field of psychiatric treatment.
7. They are also in use as general anaesthetics.
8. Barbiturates are used for treating insomniate. *i.e.* lack of sleep.

Glutethimide

It is absorbed from the gastro intestinal tract and is metabolized in the body. It is used as an hypnotic and its hypnotic effect starts in about 30 minutes, and it is long lasting. The drug is capable of inducing drug dependence.

Methyprylone

It is chemically related to glutethimide and is used as hypnotic drugs and its effective dose is about 200-400 mg at bed time.

2. Non Barbiturates

There are several compounds which do not have barbituric structure but have effective hypnotic and sedative effects. The drugs are as follows.

(*a*) **Alcohols**. Ethanol, chloral hydrate, trichloroethyl phosphate etc. are used as sedatives and hypnotics.

Ethanol : It is a mild sedative but is not recommended as hypnotic agent as it produces excitement and might produce drug dependence.

Chloral hydrate : Chemically it is $Cl_3CCH(OH)_2$—(2, 2, 2 trichloro-ethane, 1, 1-diol). It is colourless crystalline solid and has a bitter taste. It is soluble in water and evaporates when exposed to air. It should be stored in airtight containers and should be protected from light.

In small doses they produce sedation, a slightly larger dose produces hypnotic effects. It is administered in a dose of 0.5 to 2.0 g. For children it is administered in a syrup form. It is an effective, reasonably safe and cheap hypnotic drug.

(*b*) **Aldehydes**

Paraldehyde : It is a trimer of acetaldehyde. Chemically it is 2, 4, 6, trimethyl 1.3.5. trioxane.

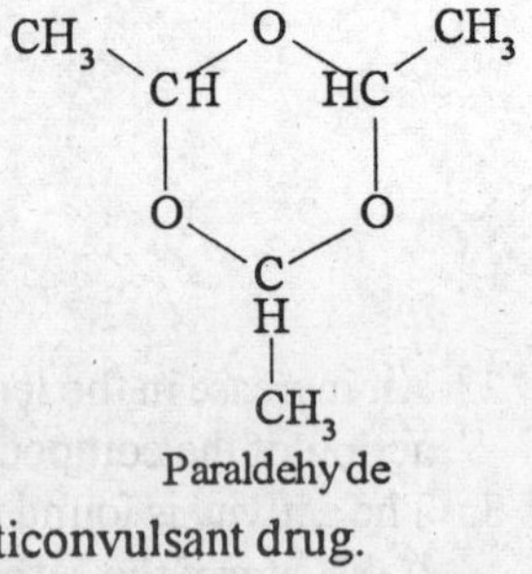

Paraldehy de

It is colourless, transparent inflammable liquid. It is acidic in taste and has a characteristic odour. It is soluble in water. It should be stored in airtight and light resistant containers.

Paraldehyde is administered either orally or as intramuscular injection forms. It is irritant to the gastric nucosa. When given orally it induces sleep within 15 to 30 minutes and its effect lasts for 6 to 8 hours. It is used during labour to exert a mild analgesic effect. It is a useful anticonvulsant drug.

(c) **Acetylated Carbinols**

Ethylchlorvynol : It is an oily liquid and is used for its hypnotic and sedative effects. It is generally administered orally in capsule forms.

(d) **Benzodiazepines**. These are a composed system of a benzene ring fused with a seven-membered heterocyclic ring having two nitrogen atoms and double bonds. These are named as 1,4 benzodiazepine according to the position of nitrogen at position 1 and 4.

Various modifications in the structure of the ring system give rise to many physiologically active compounds. Some important benzodiazepine drugs are as follows.

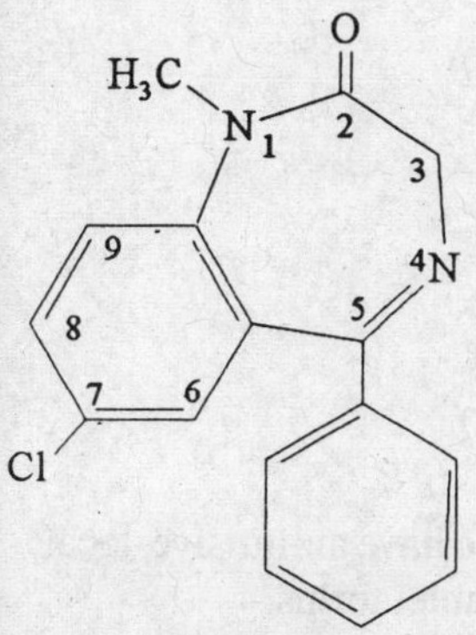

Diazepam

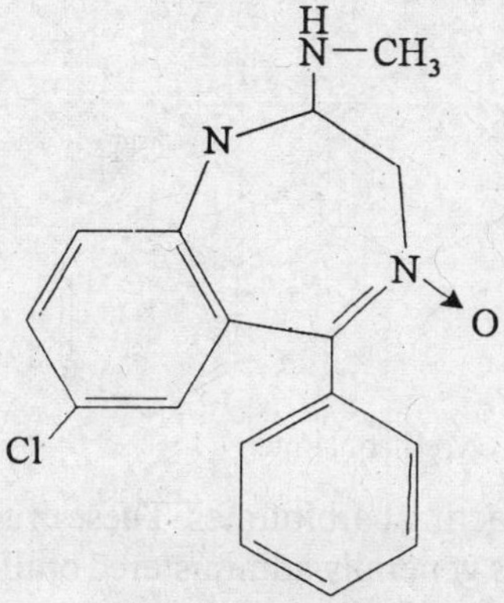

Chlordiazepoxide

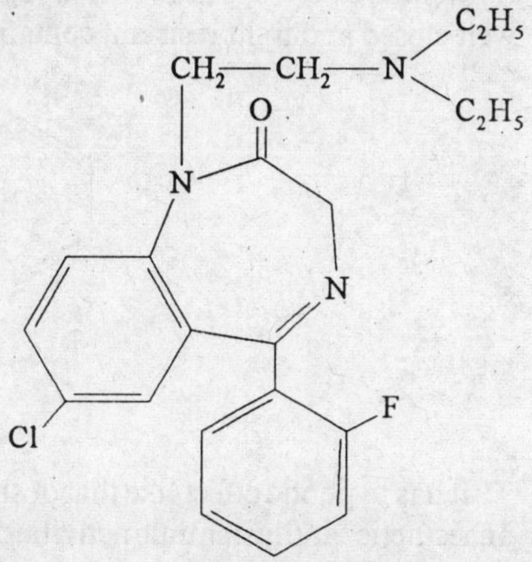

Flurazepam

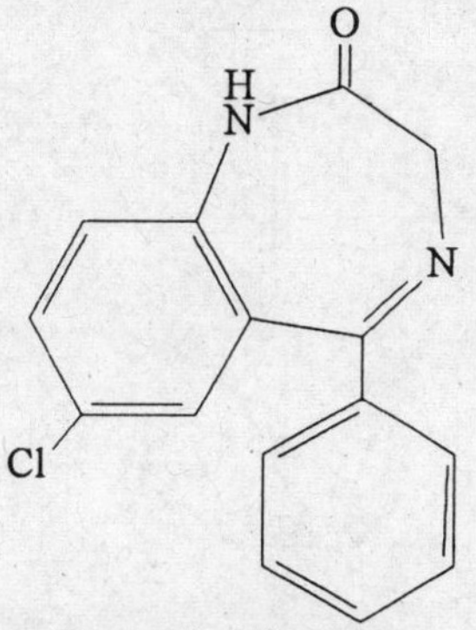

Desmethyldiazepam

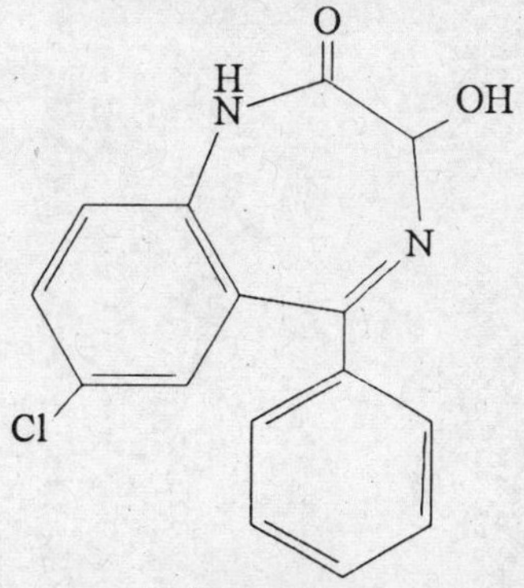

Oxazepam

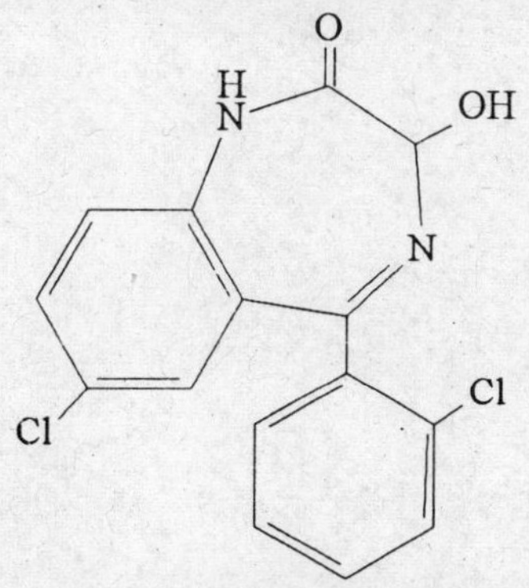

Lorazepam

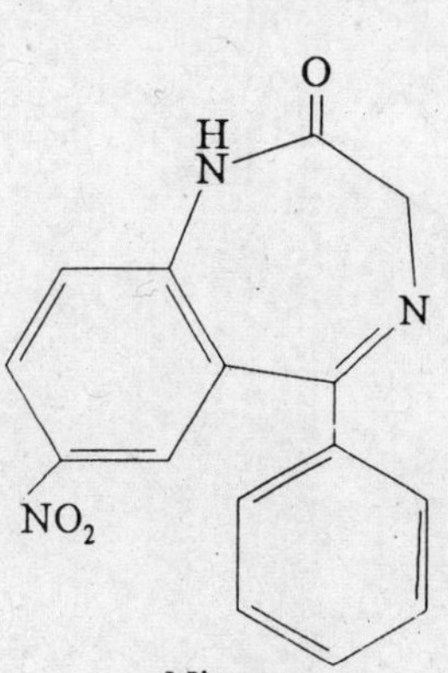

Nitrazepam

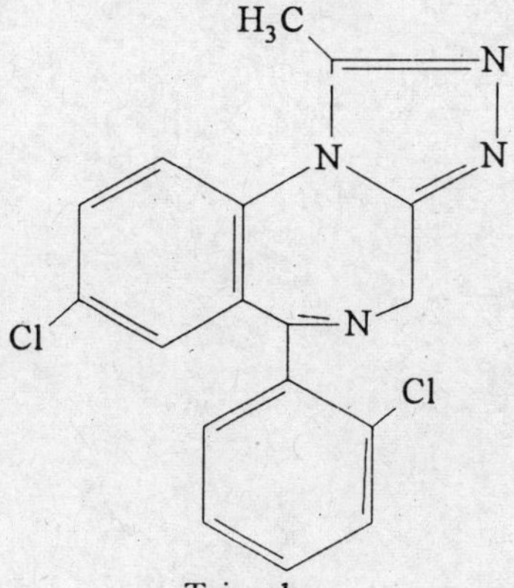

Triazolam

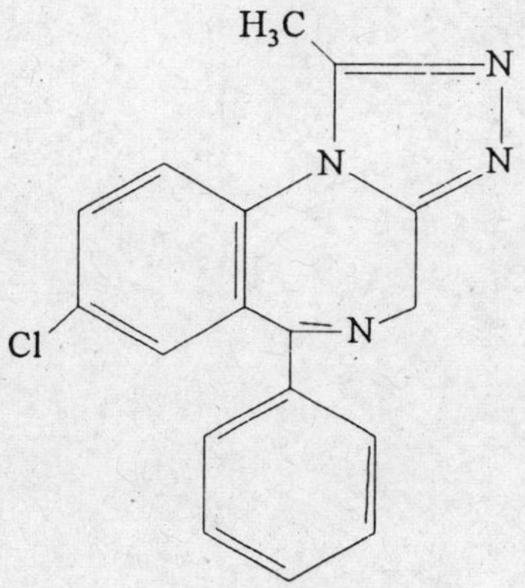

Alprazolam

Diazepam, Flurazepam and Nitrozepam are very effective hypnotics. These are given orally and are absorbed and distributed evenly in body. These are metabolized in the liver. They are relatively non-toxic and hence are preferred more than to barbiturates. These drugs produce near normal sleep pattern. An overdose causes respiratory depressions.

(*e*) **Inorganic Ions**

Bromides : Bromides are not much in use now a days because of its adverse side effects such as conjunctivitis, headache, gastric distress, neurological disturbances etc.

(*f*) **Miscellaneous Agents**

Methaqualone : It is a white crystalline powder which is insoluble in water. It should be stored in well closed and light resistant containers. Chemically it is 2-methyl-3-0-tolylquinazolin-4 one.

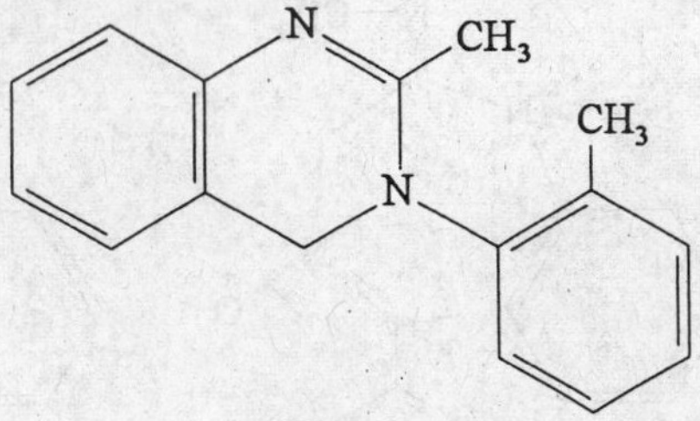

N CH_3 CH_3 N

Methaqualone

It has hypnotic effect like that of short acting barbiturates. These drugs also have antituisive, local anaesthetic, antihistaminic activities. It is generally administered orally in tablet forms.

Chapter 23

Antibiotics

After discovery of Pencillin in 1940, as a valuable medicinal preparation the word antibiotic came into wide use in science and everyday life. The term antibiotic is defined as a drug derived from living matter or micro-organism which either prevents the growth of other micro-organism or destroy them. Antibiotics are obtained from micro-organism such as fungi, actinomycetes, bacteria etc., The first well authenticated report on an antibiotic was made by Pasteur who found that certain aerobic non-pathogenic bacteria inhibited the growth of anthrax bacilli.

All the chemical substances obtained from living cells cannot be considered as antibiotics as the antibiotics have to satisfy certain conditions such as :

(*i*) It should be effective at low concentration.
(*ii*) It should antagonise growth or kill one or more species of the micro-organism.
(*iii*) It should not have significant toxic side effects.
(*iv*) It must be effective against a pathogen.
(*v*) It should be stored for a long time without appreciable loss of its activity.
(*vi*) The antibiotic should be completely eliminated from the body after its administration has been stopped.
(*vii*) It should be highly stable so that it can be isolated and processed into suitable forms of dosages which are readily absorbed.

Antibiotics are very important therapeutic agent and have been found to be clinically effective in many protozoan beacterial and fungal infections. The antibiotics are very specific in their action i.e. a given antibiotic is effective against certain types of micro-organisms only. The organisms susceptible to the inhibitory or lethal effect of an antibiotic are known as its specturm. Biological potency of antibiotics are normally expressed in conventional units contained in 1.0 ml or in 1.0 mg of a preparation. A unit of antibiotic activity is the minimum quantity of an antibiotic that will inhibit or slow down the growth of a certain quantity of cells of a standard strain of a test microbe contained in unit volume of a nutrient medium.

Classification of Antibiotics

Approximately 6000 antibiotic substances have been described by now and it would be very difficult to handle such great variety of antibiotics unless they are systematized properly.

There are several approaches to classification, based on the specturm of biological action of antibiotics or it may be based on the chemical structure of antibiotics.

I. Classification Based on the Specturm of Biological Action of Antibiotics

1. Antibiotics mainly effective against gram-positive bacteria.
   (*a*) Antibiotics used for systemic infection e.g. Penicillins, Fucidin etc.
   (*b*) Those employed topically e.g. Bacitracian.

2. Antibiotic mainly effective against gram-negative bacteria e.g. Streptomycin, Cycloserine etc.
3. Antibiotics effective against both gram-positive and gram-negative bacteria e.g. Ampicillin, Neomycin etc.
4. Antibiotics effective against gram-positive, ricketisae, chlamydia and also gram-negative bacteria e.g. Tetracyclines, Choloramphenicol etc.
5. Antibiotics effective against acid fast bacilli (M. tuberculosis) e.g. Streptomycin, Capreomycin etc.
6. Antibiotics effective against protozoa e.g. Tetracycline, Paromomycin.
7. Antibiotics effective against fungi e.g. Nystatin, Amphotericin B etc.
8. Antimalignancy antibiotics e.g. Actinomycin D, Mitomycin etc.

II. Classification According to Their Chemical Structures

Antibiotics can be distinguished as Antibiotics having acyclie structure.

1. **Acylie Structure.** This group of antibiotics comprises cylopentane, cyclohexane and cycloheptane derivatives, *e.g.* sarcomycin, actidione, thujic acid etc.

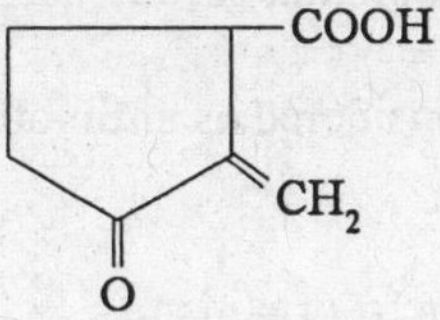

Sarcomy cin

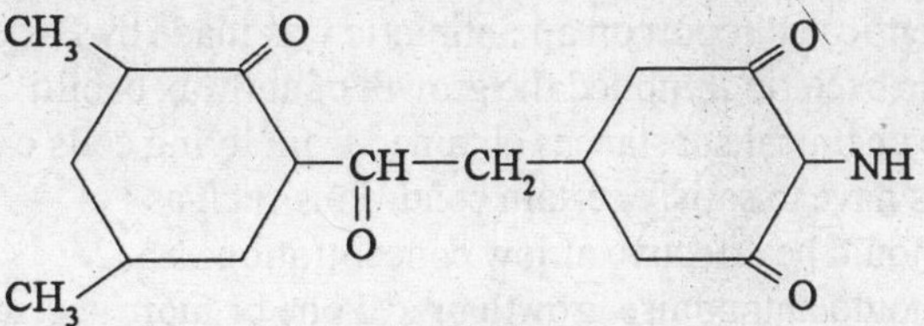

Actidione

2. **Tetracyclines.** This group includes compounds similar in their structure. The main representative is tetracycline.

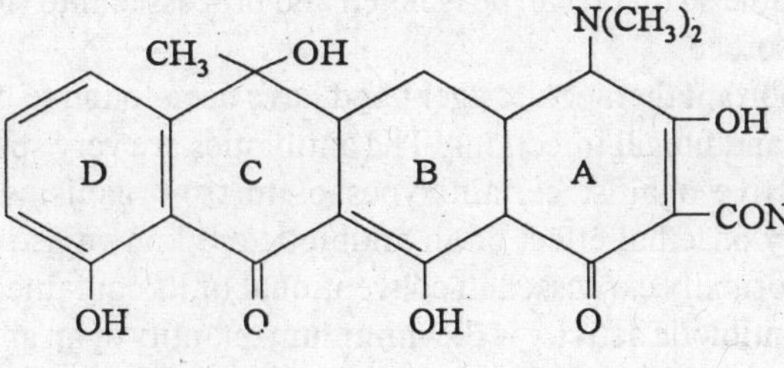

3. **Aromatic Antibiotics.** Antiobiotics included in this group are benzene derivatives e.g. gallie acid, chloramphenicols etc.

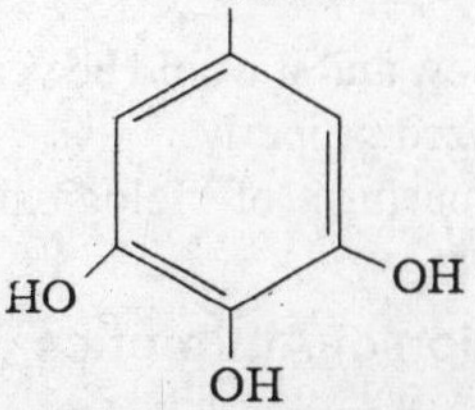

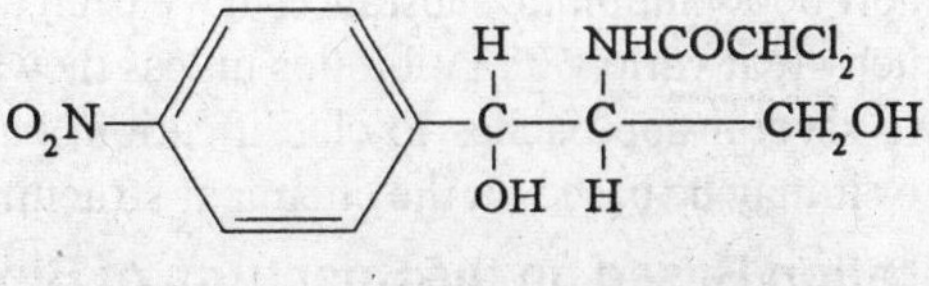

4. **Heterocyclic Oxy-Antibiotics.** This group includes many antibiotics containing either five-membered O-heterocycle compounds or six-membered O-heterycycle compounds e.g. Carlina oxide, Penicillic acid, Kojic acid, Novobiocin, cintrinin etc.

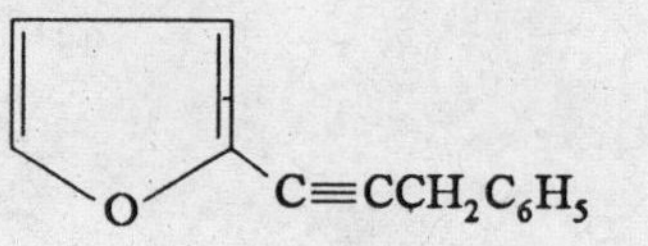

Carlina oxide

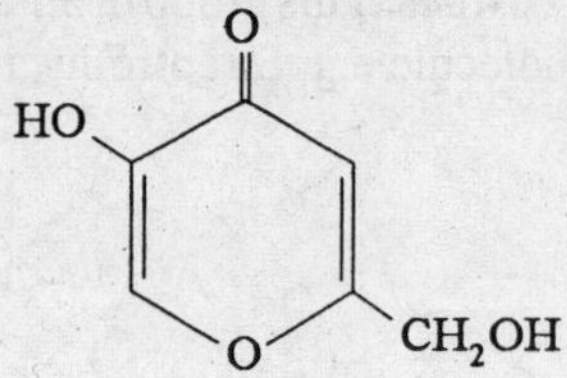

Kojic acid

5. **Antibiotics.** Antibiotics with several O-heterocycle rings e.g. Trichothecin.

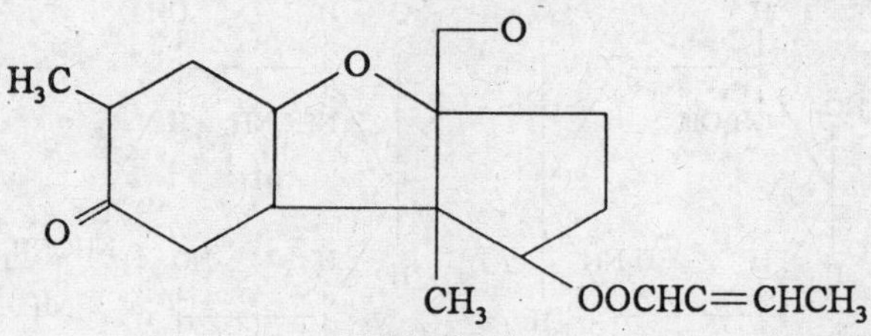

Trichothecin

6. **Macrolides.** The characteristic feature of these antibiotics are the presence of macrocyclic lactone ring connected with one or several carbohydrate residue such as amino sugars etc. e.g. erythromycin, methymycin etc.

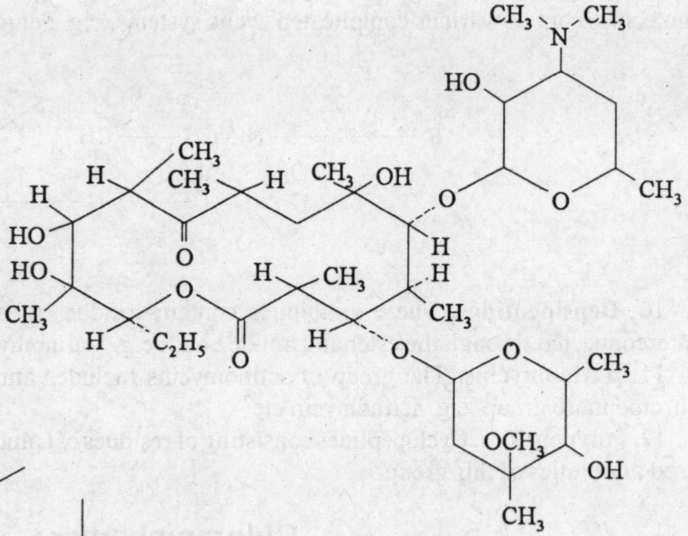

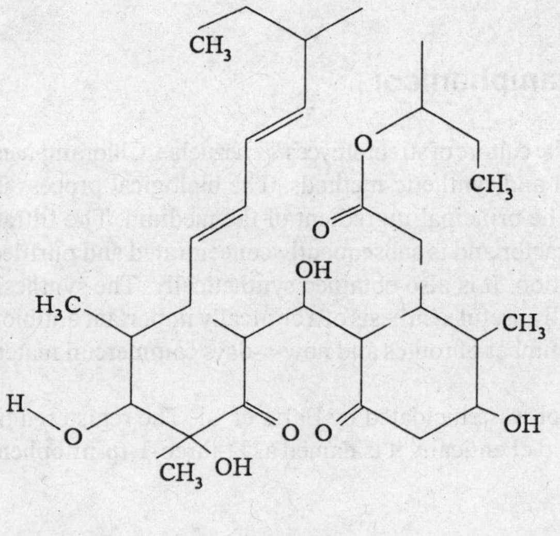

7. **Oligomycins.** This group includes compounds containing a conjugated diene system in their molecules. e.g. Oligomycins A, Chondamycin etc.

8. **Aminoglycosides.** This group of antibiotic substances includes compound containing glycoside bonds in their molecule e.g. Streptromycin, Gentamycins etc.

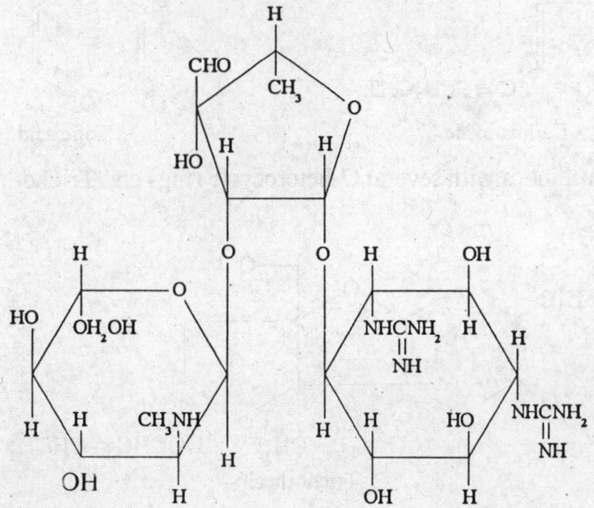

H
CHO CH_3 O
HO H H
H O O H OH
HO OH_2OH O $NHCNH_2$ H H
NH
H H CH_3NH H H HO $NHCNH_2$
H NH
OH H OH H

9. **Nitrogen containing Heterocyclic Compounds.** This group includes many antibiotics whose molecules contain various complicated cyclic systems. e.g. Penicillins, Proactinomycestes etc.

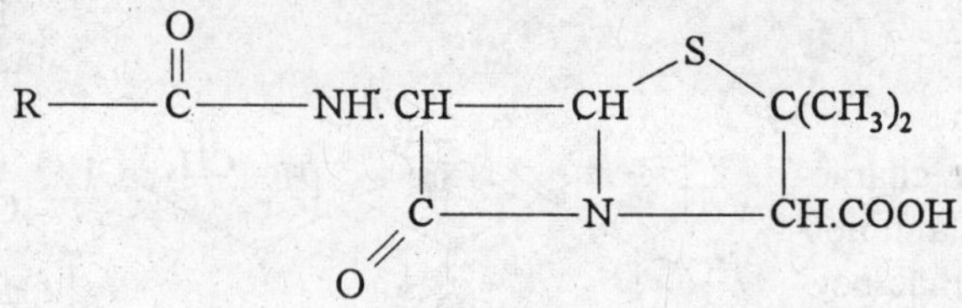

```
        O                         S
        ||                       / \
R ——— C ——— NH. CH ——— CH          C(CH3)2
                |        |           |
                C ——— N ——————— CH.COOH
               //
              O
```

10. **Depsipeptides.** These antibiotics contain residues of alpha-oxy and alpha-amino acids interconnected through the ester and amide bonds e.g. Valinomycin, amidomycin etc.

11. **Actinomycins.** The group of actinomycins includes antibiotics that contain the noxasine chromophore group. e.g. actinomycin etc.

12. **Polypeptides.** Cyclopeptides consisting of residues of L and D amino acids are the most widely used antibiotics of this group.

Chloramphenicol

It was discovered by Ehrlich et al. in 1947 from the culture of streptomyces Venezuelae. Chloramphenicol is produced commercially by both biological and synthetic methods. The biological process is an aerobic fermentation using wheat gluten as the principal ingredient of the medium. The filtrate is treated with amyl acetate in a centrifugal extractor and is subsequently concentrated and purified by means of vacuum evaporation and crystalization. It is also obtained synthetically. The synthesis of chloramphenicol constituted the first practically useful synthesis of a clinically important antibiotics. Chloramphenicol has been synthesized by a number of routes and now-a-days commercial materials are mostly produced by synthesis.

The chemical structure of chloramphenicol was elucidated by Bartz et. al. The registered trade name of chloramphenicol is chloromycetin and chemically it is named as D-threo-1-(p-nitophenol)-2-dichloro acctamido, propane 1, 3 diol.

$$O_2N-C_6H_4-\underset{OH}{\overset{H}{C}}-\underset{H}{\overset{NHCOCHCl_2}{C}}-CH_2OH$$

D-threochloramphenicol

$$O_2N-C_6H_4-\underset{H}{\overset{OH}{C}}-\underset{NHCOCHCl_2}{\overset{H}{C}}-CH_2OH$$

L-threochloramphenicol

The synthesized antibiotics is the D-threo form. The properties of chloramphenicol largely depend on the spatial configuaration of the molecule. The L-threo isomer is biologically inactive. It is highly active against a wide variety of gram-positive and gram-negative micro-organisms. Chloramphenicol acts by interfering with the synthesis of bacterial proteins hence it is bacteriostatic inaction.

Properties of Chloramphenicol

Chloramphenicol is a stable neutral compound, bitter in taste with a sharp melting point of 150.1°. It is soluble in many organic solvents but is sparingly soluble in water. It is optically active $(\alpha)^{25} = +19°$ in ethanol and –25.5° in ethyl acetate.

The molecualr composition of chloramphenicol $C_{11}H_{12}O_5N_2Cl_2$. It does not contain any free amino group or carbonyl group.

On acetylation by means of acetic anhydride in pyridine it gives diacetyl derivative due to the presence of two hydroxyl groups.

On hydrolysis with either acid or alkalis it gives dichloro acetic acid and an optically active base.

$$C_{11}H_{12}O_5N_2Cl_2 + H_2O \longrightarrow \underset{\text{Optically active base}}{C_9H_{12}O_4N_2} + \underset{\text{dichloro acetic acid}}{CHCl_2COOH}$$

Chloramphenicol on reduction with tin and hydrochloric acid followed by diazotisation and coupling with β-napththol gives anorange red precipitate due to the presence of an aromatic ring.

Chloramphenical has two asymmetric carbon atoms, therefore, it has four stereoisomers namely D(–) and L(+) threoisomers and D(–) and L(+) erythroisomers. Both isomers of erytho form are very toxic and are not used in medicine.

Spectrol analysis of chloramphenical suggests it to be a derivative of para substituted nitrobenzene.

Therapeutic Uses

Chloramphenicol was the first effective broad specturm antibiotic to be discovered. It is effective against both gram-positive, gram-negative and also against rickettsia. It inhibits the growth of staphylococcus, streptococcus, bacillus, vibrios etc. It is used in the treatment of typhoid fever (gram negative bacteria), pneumonia (gram-positive bacteria), rickettsic (epidermic typhus) dysentary, urinary tract infections, enteric fever, whooping cough etc. It also finds its use in the treatment of haemophilus influezae, meningitis, plague, syphilis, gnorrhoea etc.

This antibiotic is effectively absorbed from the intestinal tract and is diffused into the tissues. Being poorly water soluble its absorption depends to a great extent on its particle size.

Assay of Chloramphenicol

1. Chemical Method.

(*i*) **Ultraviolet Spectrophotometric Method.** It is a physical measurement of the absorption of an aqueous solution of chloramphenicol at 278mμ. This method is satisfactory for pure smaples of chloramphenicol and some of the pharmaceutical dosage forms.

This method is based on the basis of Beer-Lambert law according to this $E = \log_{10} = \frac{I_o}{I} = KCl$

Where E = Extinction
I_o = intensity of incident light
I = intensity of transmitted light
C = concentration of absorbing substance
K = a constant for the substance

Thus according to Beer Lambert law when extinction is plotted against concentration a straight line is obtained.

Method. Transfer Standard, Blank and sample solution to three 1.0cm. absorption cells respectively. Measure the absorbance at 278mμ, after setting the instrument to read zero absorbance at 278mμ with the blank. A standard curve of the absorbance is obtained by plotting the absorbance on the ordinate scale and concentration of chloramphenicol on the abscissa as shown in the Fig. 1.

The potency of the unknown can be determined from the graph.

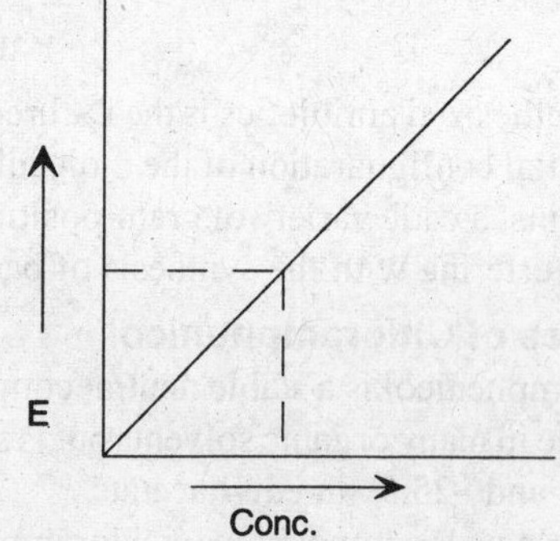

Fig. 1.

(*ii*) **The Micro-Biological Method :** This method is carried out by the cylinder plate diffusion technique using Sarcina lutea as the test organism as it is sensitive to chloramphnicol. The zone of inhibition exhibited by the drug against S. lutea is proportional to the concentration of the drug present. After incubation of the Agar plate for 16 to 18 hours at 32°C to 35°C. The concentration of the drug in the unknown sample can be calculated from the graph drawn with log conc. of the standard on X-axis and diameter zone of inhibition on the Y-axis.

Structure Activity Relationship of Chloramphenicol

The activity of a drug is closely related to its chemical structure. The physiological activity of a compound is associated with a particular structural unit or group called pharmacophore. If a drug having a particular pharmacophore group is found to be toxic then this group is modified using simple reactions to yield a physiologically active compound with less toxicity. Structure activity relationship of chloramphenicol is as follows.

1. A large number of analogies of chloramphenicol have been prepared but the antibacterial activity of the parent antibiotic is found to be more effective than other analogies.
2. If the phenyl ring of the chloramphenicol is replaced by a heterocyclic ring then the resulting antibiotic is found to be less effective than chloramphenicol.
3. The nitro group in the phenyl ring of chloramphenicol is replaced by cyanide group (—CN), amide(—$CONH_2$), primary amine (—NH_2), hydroxyl (—OH) and secondary amine (—NHR) group and their physiological activity is tested. It has been found out that the compound with these modification show antibacterial activity.
4. Di-chloroacetyl group of chloramphenicol is very important for its antibacterial activity. The dibromo acetyl group has only 80% of bacterial activity.
5. The nitro benzene radical of chloramphenicol may contribute to its bonemarrow toxicity though sometimes it has been found out that toxicity can also occur with analogues of chloramphenicol without the nitro group.
6. Chlarmphenicol sodium succinate has a high aqueous solubility in comparison to chloramphenicol base hence it is used for parental chloramaphnicol theraphy.

7. Chlormphenicol palmitate oral suspenson is taste less and is used for pediatric use.
8. The stereochemistry of the antibiotic plays an important role in its bacterial activity e.g. D-threo isomer is optically active whereas L-threo isomer does not show any physiological activity.
9. Any alteration in the nature of the side chain of chloramphenicol results in its deactivation.
10. Alteration in the aliphatic moiety of the molecule of chloramphenicol leads to diminished activity or loss of activity. Any alteration in the aromatic moiety do not effect its activity.
11. The propane-di-ol moiety of the antibiotic molecule is essential for its antibacterial activity.
12. Any structural change that effects the functional character of the hydroxyl group on carbon-3 atom or altered the length of the propane residue results in deactivation of the drug.
13. The hydroxamic acid analogue of chloramphenicol lacks antibacterial activity.
14. The bulky substituents reduce antibacterial activity.
15. Derivatives of chloramphenicol containing a wide range of p-substituents are physiologically active.

Penicillin

Penicillin was discovered by Alxender Fleming in 1920. It is very important and widely used antibiotic and was first extracted from the mould of Penicillin notatum. Later on a mutant of a related mould P. Chrysogenum was found to give the highest yield of penicillin and is now employed for commercial production of this antibiotic.

Penicillin is effective against gram-positive and gram-negative cocci and some gram-positive bacilli. It is bacteriostatic in action but when given in certain concetration it can act as a bactericidal agent. Pencillin inhibit the synthesis of the bacterial cell wall. It is most effective against activity multiplying organisms. Penicillin does not interfere with the tissue cell wall synthesis in humans and higher animals hence it is non-toxic to them even in large doses.

Penicillin is generally produced from a culture medium of corn steep, liquor, lactose and several inorganic salts. The medium is agitated by a stream of sterile air at 24°C. Under these conditions of agitation and aeration the mould grows throughout the bulk of the liquid, as globular pellicles consisting of mycelium. After fermentation the contents of the vessels are rapidly cooled and mycellium is filtered on a sterile rotatory filter. Then it is subjected to solvent extraction with butyl alcohol at controlled pH. Petroleum ether is added to the cold solvent extract and shaken with dilute sodium carbonate solution. The solution is adjusted to pH 6-7 and then rapidly evaporated to give sodium salt of penicillin which is dried in a high vacuum. It is made non-toxic, sterilised and free from pyrogens by allowing it to pass through asbestos which abosrbs micro-organism and pyrogens

Structure and Properties of Penicillin

The basic structure of the penicillin consists of a thiazolidine ring fused with a β-lactam ring. The two rings constitute the fundamental nueleus of all the penicillins, namely 6-aminopenicillinic acid. The molecular formula of penicillin is $C_9H_{11}N_2O_4$ SR. Depending on the nature of R there are six natural penicillin obtained from penicillin notatum, has the following structure.

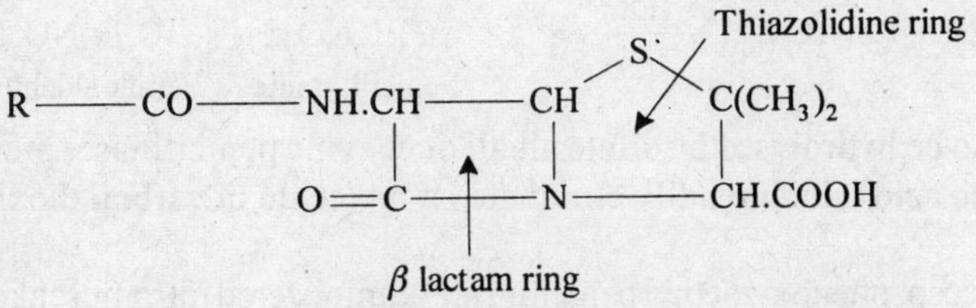

```
                                              Thiazolidine ring
                                           S
R —— CO —— NH.CH —— CH              C(CH3)2
                    |             |               |
              O=C ——— N ———— CH.COOH
                    ↑
              β lactam ring
```

6- Amino penicill-anic acid

Six natural penicillins which are derivatives of 6 aminopenicillanic acid are as follows (Table -1)

Table 1

S. No.	*Chemical Name of Penicillin*	*Other Conventional Penicillin*	*Structure of Radical R*
1.	Pent -2 enyl penicillin	F	$-CH_2-CH=CH\,CH_2\,CH_3$
2.	Benzyl penicillin	G	$-CH_2-C_6H_5$
3.	p-hydroxy benzyl penicillin	X	$-CH_2\,C_6H_4\,OH$
4.	n-heptyl penicillin	K	$-(CH_2)_6-CH_3$
5.	n-Amyl penicillin	Dihydro F	$-(CH_2)_4-CH_3$
6.	Phenoxy methyl penicillin	V	$-CH_2-O-C_6H_5$

The absolute configuration of pencillin is proved by X-ray analysis which shows that natural penicillin is D-isomer having the following configuration.

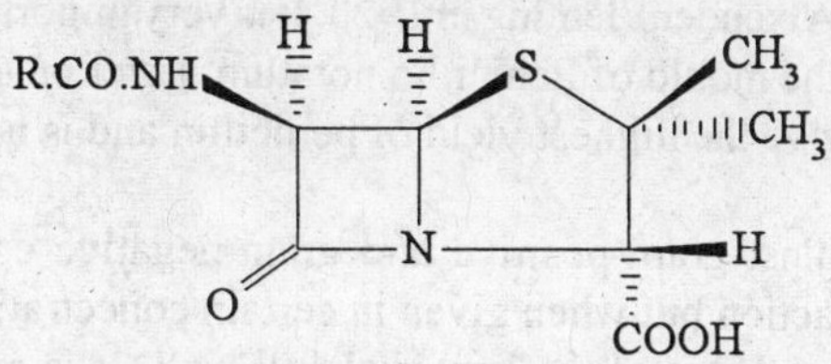

Absolute configuration of penicillin : Solid dark line indicate the groups in front of the plane, dotted lines indicate groups which are behind the plane, normal lines indicate the group along the plane of the paper.

Optically active monocarboxylic pencillins are all acids with a pk value of about 2.8 and are soluble in water and many solvents. They are unstable as free acids and also at pk above 8.5. The penicillin and their salts are deactivated by many reagents certain buffers are also added to medicinal pencillins preparation to counteract decomposition.

The penicillins are attacked by oxidizing and reducing agents but reaction with these reagents depends entirely on the individual compound.

The antibiotics containing unsaturated side chains are most readily hydrogenated and oxidized while benzyl penicillin is remarkably stable to oxidation even with chromium trioxide and hydrogen peroxide.

Halogens acts slowly upon the penicillins except those containing olefinic double bonds.

Under some circumstance the sulphide linkage of the thiazolidine ring is oxidized to sulphoxide or sulphone.

Penicillins are hydrolysed by hot dilute inorganic acids yielding carbondioxide, an amine known as penicillamine and an aldehyde called Penilloaldehyde

$$C_9H_{11}N_2O_4SR + 2H_2O \xrightarrow{HCl} \underset{\text{Penicill-amine}}{C_5H_{11}NO_2S} + \underset{\text{Penilloaldehyde}}{C_3H_4NO_2R} + CO_2$$

Penicillin can also be hydrolysed by dilute alkali or enzyme penicillnase, produces a dicarboxylic acid called pencilloic acid which readily eliminated a molecule of carbon dioxide to form penilloic acid.

Penicillin drugs except phenoxymethyl penicillin are administered intramuscularly or subcutancously in the form of solutions. The aqueous solution is prepared with a sterile isotonic solution of sodium chloride or with water. Phenoxymethyl penicillin is stable to the action of acids and therefore it can be taken orally in tablet form. (It does not decompose in the acidic medium of gastric juice).

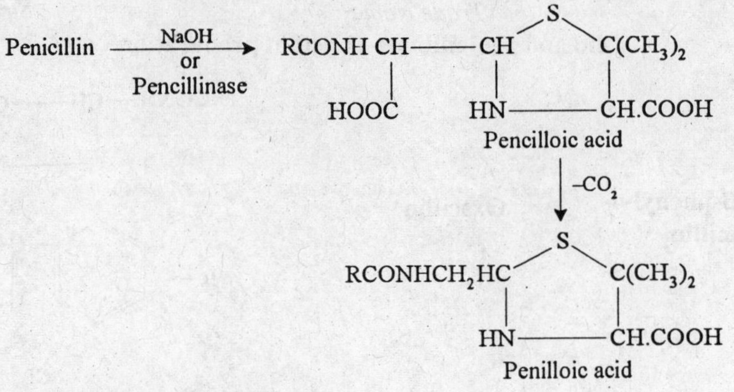

$$\text{Penicillin} \xrightarrow[\text{Pencillinase}]{\text{NaOH or}} \text{RCONH CH(HOOC)–CH(HN–)–S–C(CH}_3)_2\text{–CH.COOH}$$

Pencilloic acid

$$\downarrow -CO_2$$

$$\text{RCONHCH}_2\text{HC(HN–)–S–C(CH}_3)_2\text{–CH.COOH}$$

Penilloic acid

Semi Synthetic Production of Penicillin

Now a days the analogues of natural penicillin with some valuable properties are prepared by method which can be defined as a semisynthetic product as they are produced by the biological and chemical method.

The starting product for this synthesis is 6-amino-penicillanic acid. This acid is produced by biosynthesis in two ways :

(*i*) in development of the mould fungus penicillium chrysogenum and (*ii*) in the enzymatic hydrolysis of benzyl penicillin.

After the biosynthetic production of 6-amino pencillamine, an acid residue is introduced into it by the conventional methods of organic chemistry and a new penicillin is thus produced. All the drugs are white crystalline powder and except ampicillin dissolve well in water.

The semisythetic penicillins are listed as follows in Table 2.

Table 2

Penicillin	*Trade Name*	*Structure*
	Acid resistant preparation	
1. **Phenoxy methyl penicillin.** These are stable to acids and are very active in vitro against gram positive cocci. It is used in respitory tract, strep-to-coceal, pneumococcal infection.	Penicillin V	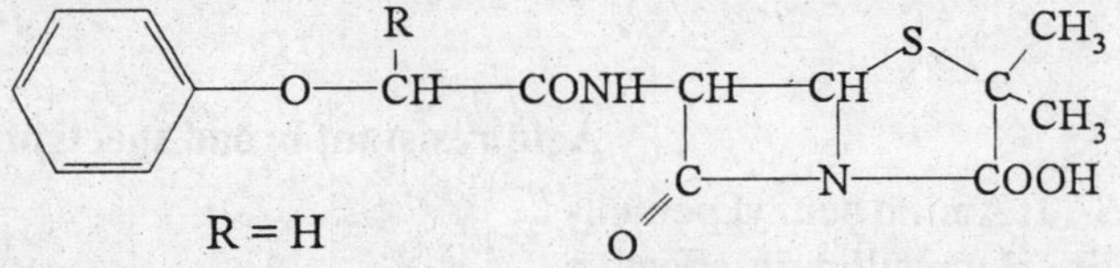 C_6H_5–O–CH(R)–CONH–CH–CH–S–C(CH_3)(CH_3)–COOH; C(=O)–N ring; R = H
2. **Phenoxy ethyl penicillin.** These are weakly bound to serum protien.	Phenethicillin	R = CH_3
3. **Phenoxy prophyl penicillin.** It is most easily absorbed.	Propicillin	R = CH_2 — CH_3
4. **Phenoxy prophyl penicillin.**	Phenbenicillin	R = C_6H_5
	Penicillinase resistant preparation	
1. **2, 6 dimethoxy phenyl penicillin.** It is resistance to be staphylococall penicillinase and are used for treating infection due to benzyl penicillin resistant organism.	Methycillin or Methicillin	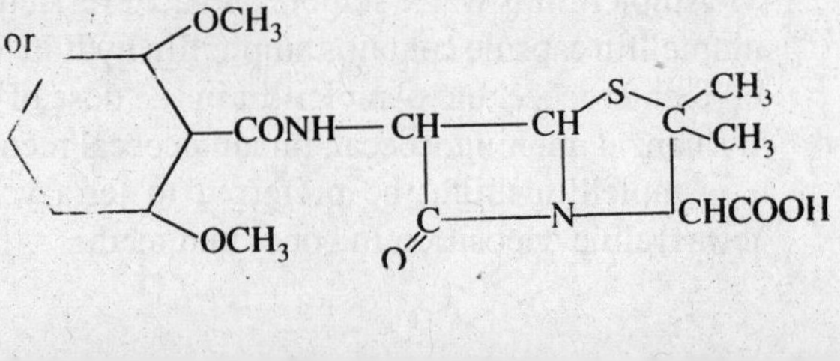 2,6-(OCH_3)$_2C_6H_3$–CONH–CH–CH–S–C(CH_3)$_2$–CHCOOH; C(=O)–N ring

Penicillin	*Trade Name*	*Structure*
Acid and penicillinase-resistant preparation		RCONH—CH—CH—S—C(CH_3)(CH_3); C(=O)—N—COOH
1. **5-methyl-3-phenyl-4-isoxy azolylpenicillin**	Oxacillin	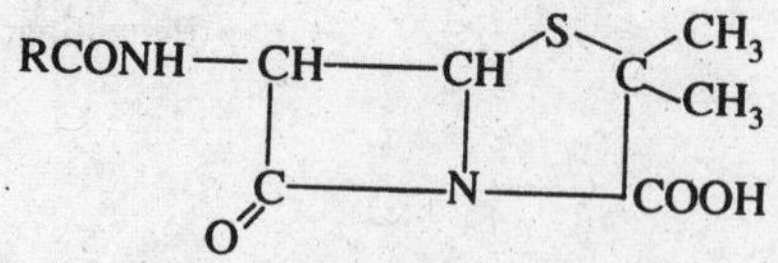 R= 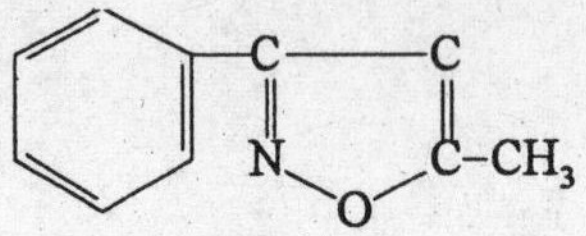
2. **0-chloro-phenyl-5-methyl-4-iso-oxasolyl.**	Cloxacillin	R= 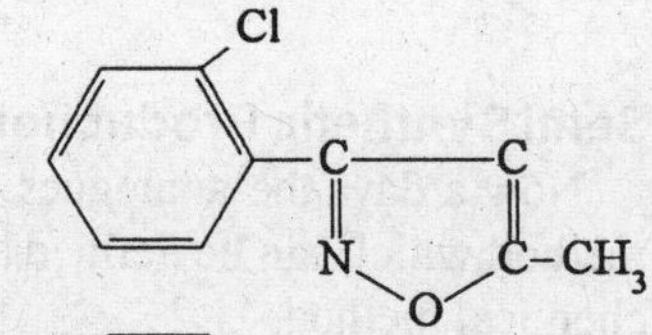 (Cl; C—C; N—O—C—CH_3)
3. **2-ethoxy-1-napthyl penicillin**	Nafcillin	R= 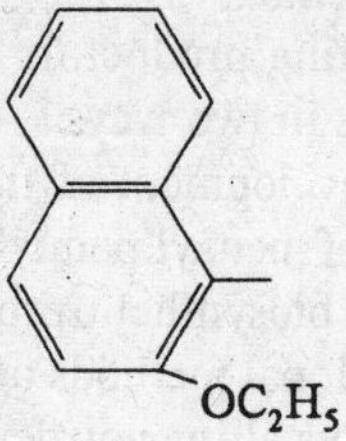 (OC_2H_5)
4. **Bipenyl-penicillin**	Diphenicillin	R= 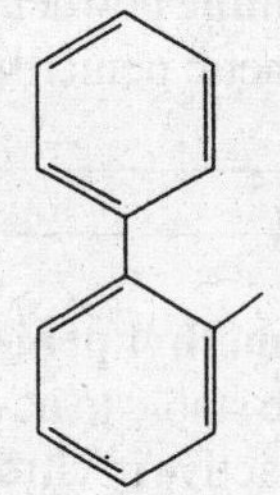
Acid resistant broad spectrum preparation		
1. **Amino benzyl penicillin.** Ampicillin is effective agent gram positive and gram negative organism. It is more effective than benzyl penicillin against a variety of gram negative bacteria.	Ampicillin	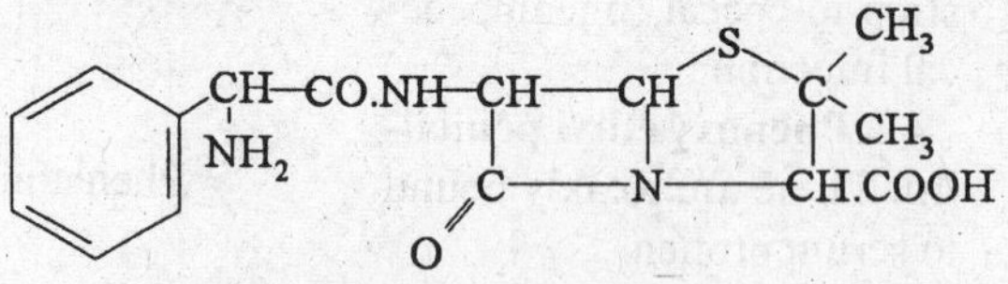 CH—CO.NH—CH—CH—S—C(CH_3)(CH_3); NH_2; C(=O)—N—CH.COOH

Ampicillin (Roscillin, Broacil)—This drug is effective against H influenzae, Neissera gonorrhoea, salmomella typhosa, E. coli, shigelle etc. The drug is moderaely effective against klebsiella. It is inactivated in presence of penicillinase.

Ampicillin is water soluble and acid resistant. It is readily absorbed on oral administration. The ampicillin capsule contains ampicillin hydrate which can be used against intestinal infections due E. coli, enterococci etc. Ampicillin in the dose of 6 to 12 g per day, is effective in the treatment of H. influenzal, meningococcal, pneumococcal meningitis.

Ampicillin might be preferred to tetracyclines in pregnant women and in infants to avoid tetracycline deposition in bones and teeth.

The mechanism of chemo-therapeutic action of penicillin is said to be both bacteriostatic as well as bactericidal. The bactericidal effect of penicillin does not resemble that of the antiseptics, the latter kill more readily as their concentration is increased while an optimum of bactericidal concentration is reached with penicillin and above that point its killing ability decreases. Pencillin is antibacterial against organisms mainly in their active growth phase and it interferes in some way with the synthesis or action of ribonucleic acid.

Therapeutic Uses of Penicillin

It is very effective against (*i*) pneumococcal infections like pneumonia, meningitis etc. (*ii*) Streptococcal infection like scarlet fever, pharyngitis etc., (*iii*) Staphylococcal infection like acute osteomyelitis etc., in these infections generally benzyl penicillin is used (*iv*) Meningococcal infections (*v*) Veneral diseases such as gonorrhoea, syphilis etc., (*vi*) Actinomycosis (*vii*) Rheumatic fever, in these infections mainly Benzyl penicillin is given. It is also used in the treatment of diphtheria, tetanus and gangrene. The antibiotic can be used topically, orally or parenterally. In oral therapy the dosage must be larger as the drug is in-activated to a significant extent by gastric acid. The stomach juices possess the enzyme penicillinase which hydrolyse penicillin by opening the lactam ring with the loss of activity when these enzyme are present in significant amount a resistance to penicillin is produced. The dry sodium or potassium salts are available in ampoules for sterile injection. The dry calcium salt can be applied topically as an ointment or as a powder. Though penicillin is a reasonably safe drug but some times it can produce some adverse reactions such as nausea, vomiting, anaphylaxis, allergic reactions etc.

Natural penicillins like benzyl penicillin have certain drawbacks such as they are inactivated by the gastric hydrochloric acid, their duration of action is short and are effective mainly against gram-positive organisms etc. Therefore attempts were made to synthesize penicillin free from these draw-backs. Penicillin chrysogenum which produces the natural penicillins through the formationof 6-aminopenicillianic acid nucleus before the attachment of the naturally occurring side chains can be substituted. Thus a variety of semisynthetic or synthetic penicillins are now produced. All these synthetic penicillins can produce allergic reactions similar to those caused by natural penicillins.

Assay of Penicillin

Chemical Method

Among all the chemical methods, Iodometric method is the most common one. This method was given by Alicino, Mindell et. al.

Procedure. Weigh approximately 30 mg of the sample and dilute with phosphate buffer solution (having pH at 6) to a concentration of approximately 1.2 mg/ml. Transfer 2 ml of this solution in to two 125 ml. iodine flask each. In one of the iodine flask add 2.0 ml. of IN sodium hydroxide and allow to stand at room temperature for 15 minutes. After this add 2 ml of 1.2 N HCl and 10 ml. of 0.01N iodine. Titrate the excess iodine with 0.01N sodium thiosulphate solution (which is standardised accurately against potassium iodate) using starch as an indicator. A blank titration is performed in the second iodine flask without adding sodium hydroxide and hydrochloride acid, add 10 ml of 0.01N iodine solution and immediately titrate against sodium thiosulphate solution using starch as an indicator. The difference in the titre values gives a measure of iodine reacted.

In the similar way prepare a standard solution of penicillin G and titrate against 0.01N iodine solution using starch as an indicator. The amount of 0.01N iodine absorbed by 1.0 mg of sodium penicillin G standard solution is known as factor 'F'.

The potency of the unknown drug in the given sample is then calculated as

$$= \frac{\text{difference in the titre value} \times \text{potency of working standard / mg}}{\text{mg. of sample in 2.0 ml} \times \text{F}}$$

The microbiological method is carried out by the same plate diffusion technique as described earlier. For penicillin the test organism is sarcina lutea and Micrococcus pyogenes as these are sensitive to penicillin.

Structure Activity Relationship of Penicillin

1. An intact β-lactam and thiazolidine fused ring system was believed to be essential for its activity though recent development indicates that certain modifications in the thiazolidine ring retains its biological activity.
2. The parent compound phenyl penicillin is less active than benzyl penicillin.

RCONH—CH—CH—S—C(CH$_3$)$_2$—CH(COOH)—N—C(=O)

R = C_6H_5—CH$_2$— Benzyl penicillin

R = C_6H_5— Phenyl pencillin

It has been found out that the introduction of substituents in the ortho position reduces its physiological activity due to steric hindrance. The first compound produced commercially was methicillin with methoxy substituents, followed by oxacillin, cloxacillin, nafcillin etc. (Table-2).

These compounds are not active against gram-negative organisms although some of them inhibit a number of the gram-negative lactamases.

3. Any substitution on the carbon atom of the hydrophobic side chain of benzyl penicillin with an amino (ampicllin) or carboxye (carbenicillin) group increase its activity.

RCONH—CH—CH—S—C(CH$_3$)$_2$—CH(COOH)—N—C(=O)

RCONHC—CH—S—C(CH$_3$)$_2$—CH(COOH)—N—C(=O)

R = C_6H_5—CH(NH$_2$)— Ampicillin

R = C_6H_5—CH(COOH)— Carbenicillin

4. The esters of penicillin antibiotics are not bactoristatic in vitro but have considerable activity in the living animal. The benzyl ester of benzyl penicillin is much more active than its parent compound.
5. The decarboxylation of the penicillins leads to complete loss of its bacteriostatic activity.
6. It has been found out that the activity of 6-amino penicillanic acid which is the fundamental nucleus of all penicillins, is 2000 times lower than that of benzyl penicillin.
7. The *p*-hydroxy benzyl penicillin has been coupled with a number of diazonium salts in neutral solution, to form azopenicillins, which shows the same activity as that of natural penicillins.
8. The introduction of certain functional groups such as cyano, methoxy, nitro, fluoro, bromo, iodo etc. at para position of the benzyl penicillin, enhances its activity. e.g.

p-methozyl benzyl (p-CH_3—O—$C_6H_4CH_2$—)

p-cyano benzyl (p-NC—C_6H_4 CH_2—)

p-iodo benzyl (p-Br C_6H_4 CH_2—)
p-chloro benzyl (p-Cl C_6H_4 CH_2—)
p-fluoro benzyl (p-F C_6H_4 CH_2—)
p-nitro benzyl (p-O_2N C_6H_4 CH_2—)
penicillin etc.

The substitution of fluorine at meta position shows maximum activity.

9. An introduction of β-bromoallyl thiomethyl group (CH_2== C Br CH_2 SCH_2—) in the natural penicillin results in considrable increase of the physiological activity.
10. The substitution of 3-carboxyl group of the thiazolidine ring results in biologically active compounds e.g. pivampicillin, talampicillin, bacampicillin etc.

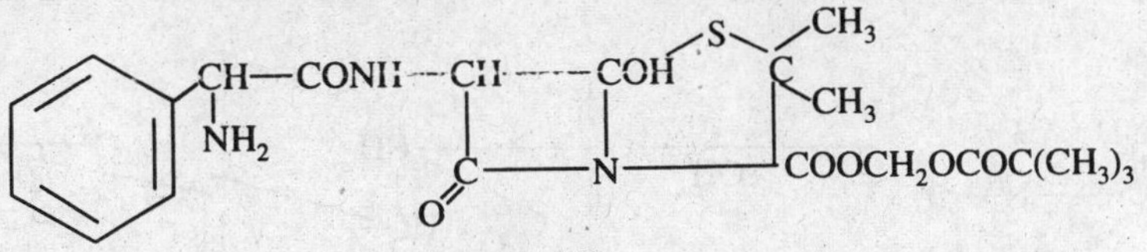

Pivampicillin

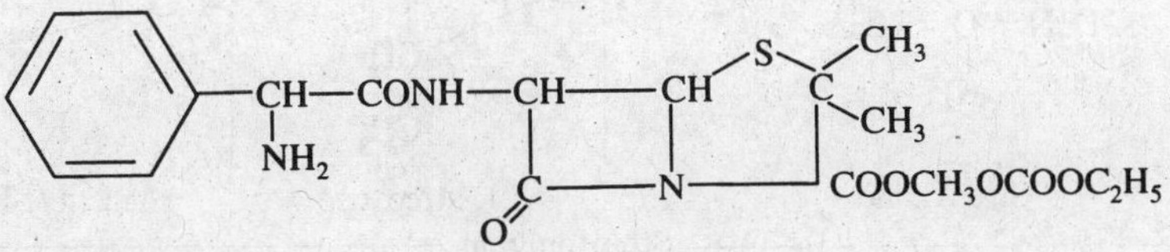

Becampicillin

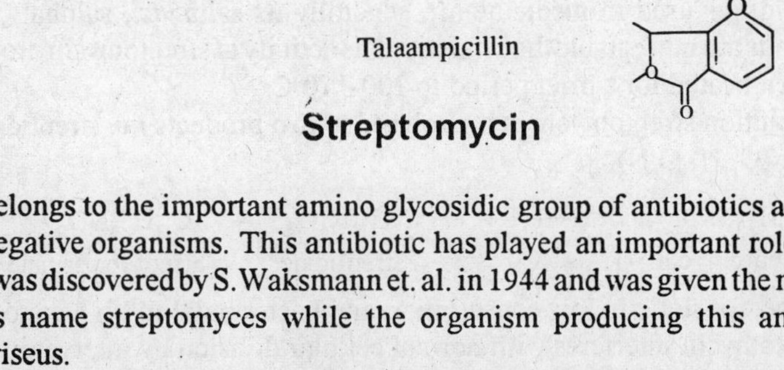

Talaampicillin

Streptomycin

Sterptomycin belongs to the important amino glycosidic group of antibiotics and are very effective against gram-negative organisms. This antibiotic has played an important role in the treatment of tuberculosis. It was discovered by S. Waksmann et. al. in 1944 and was given the name of streptomycin from the genus name streptomyces while the organism producing this antibiotic was named streptomyces griseus.

Streptomycin is produced by the submerged cultures. This culture medium contains protein materials such as soya been meal, carn steep liquor, cottonseed meal or meat extract. The culture solution is kept in large vats at 24-28°C when the growth of micro-organism begins, maximum yield is achieved after three to five days. It is found that the yield of the antibiotic increases if the culture solution is irridated by X-ray or ultra-violet light. After separating the mycelium and other metabolic products and wastes, the streptomycin is obtained from the filtrate either by absorption on charcoal or base exchange resins. The pure form of the streptomycin is isolated as sulphate or crystalline double salt of calcium chloride.

The complete sterile drug is obtained by dissolving crystalline streptomycin in distilled water to yield 25% solution which should be made free from undesiderable impurities like heavy metals, colour etc. by passing through seitz filter and then it is freeze-dired.

Structure and Properties

It is a colourless solid, dissolves readily in water and is insoluble in organic solvents such as alcohol, ether, chloroform etc. It is a leavorotatory, strong base. Its basic nature is attributed to the presence of two guanidine residues and N-methyl group of the saccharide component in the molecule. It is more stable than streptomycin and its molecular formula is $C_{21}H_{39}N_7O_{12}$. The structure of streptomycin is as follows.

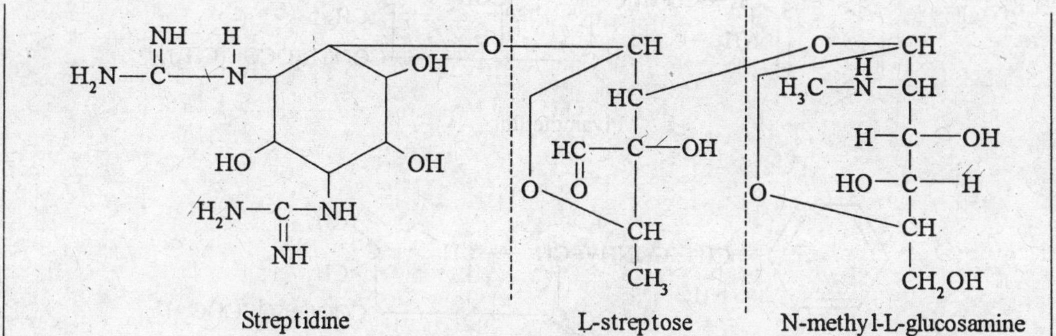

Streptidine L-streptose N-methyl-L-glucosamine

Streptomycin

Streptomycin is made up of three basic units namely streptidine, L-streptase and N-methyl-L-glucosamine. Streptidine unit is made up of two guanido groups. Streptidine is the common part of all the streptomycin molecules and it is joined by a glycoside bond to the disaccharide streptobiosamine consisting of L-streptose and N-methyl-L-glucosamine.

In the solid state, streptomycin salts show fluourescence in ultra violet light, streptomycin can form complexes with salts of some bivalent metals. This antibiotic possesses high hydrophilic property due to the presence of large number of hydroxyl and amine groups present in the molecule. An aqueous solution of the base has pH of 12.0. Its aqueous solution are the most stable at a temperature below 28° and a pH of 7.8. The activity of the antibiotic sharply drops in an acidic medium (at a pH about 6). A weakly alkaline medium is most favourable for its antibiotic action.

The streptomycin drugs used in medicine are generally its salts *viz.*, sulphats, hydrochloride, phosphates etc. The salts are more stable than its base. The activity of streptomysin drops sharply when stored at 50°C or when heated for a brief period to 100-110°C.

In strongly acid solution streptomycin is hydrolysed to two products *i.e.* streptidine ($C_8H_{39}O_4N_6$) and streptobiosamine ($C_{13}H_{23}O_9N$).

$$\underset{\text{Streptomycin}}{C_{21}H_{39}N_7O_{12}} \longrightarrow \underset{\text{Streptidine}}{C_8H_{18}N_6O_4} + \underset{\text{Streptobiosamine}}{C_{13}H_{23}NO_9}$$

Streptomycin is bacteriostatic in low concetration and bactericidal in high concentration. It has been found that streptomycin interferes with normal cellular division by intervening in ribonucleic acid metabolism. It blocks the reaction preceding the formation of more nucleotides.

Therapeutic Uses of Streptomycin

The most important use of streptomycin is in the treatment of tuberculosis. It is also useful in treating infection of E. coli, H. influenzae, plague, respiratory tract infections, meningitis, bacteremia, peritonitis etc.

Streptomycin drugs may show some adverse reactions like local irritation, skin rashes, renal irritation etc.

Besides streptomycin many other antibiotics are isolated from the genus streptomyces *e.g.* Neomycin obtained from strepto,yces fraclia, Kanamycir obtained from streptomuces kanamycetieus, gentamincin obtained from micromonospora purpuse paramonycin obtained from streptomyces remosus etc. These drugs have structural resemblance to streptomycin and are used as effectively against streptomycin infections, H. influenzae, staphylococci infections, Neomycin and gentamincin and are used in the form of ointment for the treatment of various skin diseases.

Assay of Streptomycin

(*i*) **Chemical Method.** This method is given by Donal C Grovel and William Randal.

Principle : Streptomycin reacts with oxidised alkaline nitropruside to produce an orange coloured derivative whose extinction is measured in the visible region.

Reagents : 10% sodium nitropruside, 10% potassium ferricyanide solution, 10% sodium hydroxide solution.

Preparation of Sodium Nitropruside Solution

It is prepared by mixing equal volume of sodium nitropruside, potassium ferric cyanide and sodium hydroxide in the same order. A deep red colour changing to yellow green colour solution is obtained in about 15 minutes at room temperature. Dilute 1.0 ml of this solution to 100 ml with water. A freshly prepared solution should be used.

Preparation of Standard Solution

Prepare an aqueous standard solution of streptomycin dissolving 1.0 mg/ml of streptomycin or dihydrostreptomycin.

Distribute 1.0 ml, 2.0 ml. 3.0 ml, and 4.0 ml of this standard solution in four different tubes and make up the solution to 10 ml with distilled water.

Preparation of the Blank Solution

Use 10 ml of distilled water as blank.

Sample which is to be assayed.

Dilute the sample to 250 ml and then take 0.5 ml of this solution and make up to 10.0 ml (*i.e.* 1ml of sample in 20 ml of water).

Method. Take three test tubes and pour 10 ml of blank, standard and sample respectively in the test tubes, add 10 ml of oxidised niropruside solution to each test tube and allow the colur to develop at room temperature for 3 minutes, Read at 490nm after 10 minutes in a spectrophotometer.

Calculations. Sample used contains (*i*) .0 g or 1000 mg in 250 ml solution. 1ml of this solution contains = 4 mg of sample 0.5 ml of this solution = 2 mg of sample 0.5 ml of the standard solution = 2 mg of sample.

(*ii*) Microbiological method is carried out by the same plate diffusion technique. The test organism is Bacillus subtilis and Klebsiella prieumonial. These organism are sensitive to streptomycin.

Structure Activity Relationship of Streptomycin

1. The antibiotic activity of streptomycin largely depends on the hydrogen ion concentration (pH) of the medium and its composition. The efficiency of streptomycin activity decrease in an acid medium while its biological activity is maximum in an alkaline medium.
2. Catalytic hydration of streptomycin gives dihydrostreptomycin ($C_{21}H_{41}O_2N_7$) which shows same bioligical activity as that of streptomycin. Dihydrostreptomycin is more stable than streptomycin.
3. If streptomycin is treated with carbonyl reagents (hydroxylamine, phemylhydrazine), it reacts with its aldehyde group resulting in complete loss of activity.
4. The reduction of the aldehyde group in streptomycin to a primary alcohol group in dihydrostreptomycin, does not abolish antibiotic activity of the dihydrostreptomycin.
5. The presence of D-Mannose sugar unit in streptomycin (Mannosidostreptomycin) makes the drug less active than streptomycin.

6. It has been found out that streptidine unit has no antibiotics properties.

7. Introduction of cysteine molecule $\left(\begin{array}{l}\text{SH}\quad\ \ \text{NH}_2\\ |\qquad\quad\ |\\ \text{CH}_2\text{—CH—COOH}\end{array}\right)$ in streptomycin results in its deactivation. The deactivated product on reduction with Raney nickle gives streptomycylamine which exhibit antibacterial activity against gram positive and gram negative micro-organism.

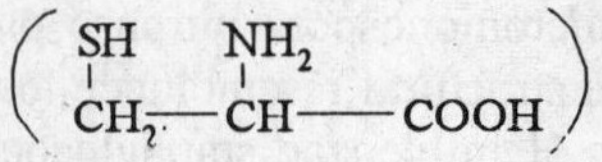

Tetracycline

These antibiotics have the broadest antimicrobial spectrum of all. These are obtained from a new species of actinomyces named streptomyces aureofaciens.

Tetracyclines comprise a group of antibiotics characterised by hydronaphthacene skeleton, the three most important members of the group are tetracycline, chloroteyracycline (aureomycin) and oxytetracycline (terramycin).

The structure of terramycin was the first of this group to be elucidated. The molecular formula was shown to be $C_{22}H_{24}N_2O_9$. The compound is found to contain eight active hydrogen atom dimethyl amino and a carbonamide group and a *c*-methyl group. Diazomethane formed a dimethyl ether showing the presence of two acidic hydroxyl group and the other two hydroxy groups are alcoholic. the structure of terramycin may be written as

$C_{18}H_9O_4NMe_2$
$CONH_2$
two OH group (acidic)
two OH group (alcoholic)

Aureomycin is obtained from streptomyces, aureofaciens, terramycin is isolated from streptomyces rimosus. Antibiotics have similar chemical and physical properties as well as antibacterial spectra. Tetracylines have the following structure.

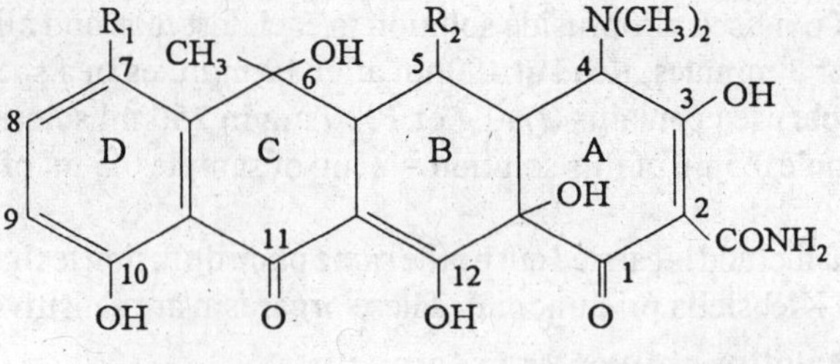

Tetracycline	R_1 = H,	R_2 = H
Chlorotetracycline	R_1 = Cl,	R_2 = H
Oxytetracycline (terramycin)	R_1 = H,	R_2 = OH

All the tetracyclines are produced by actinomycetes *viz*. streptomyces aureofacines, streptomyces rimosus, streptomyces aureomycin. Numerous methods have been derived for the isolation of tetracyclines. They may be preceptitated from filtered culture medium at pH 8-10 in presence of divolent cations or as a salt with cationic suface active agents such as calcium mixed barium magnesium salt, at pH 1 to 1.5 as the chelate compounds with ferric ion. Crude tetracyclines may be obtained by ion exchange chromatography or via their salts from which they are regenerated by mineral acids.

Properties

The moelcular formula of tetracycline is $C_{22}H_{24}N_2O_8$. Tetracycline drugs are yellow, odourless crystalline powers powder with a bitter taste. They dissolve sparingly in water except for tetracycline and oxytetraycline. In storage the drugs gradually turn dark because they are oxidized by the oxygen in the air. This occurs most readily in aqueous alkaline solutions which rapidly darken in the air

All the tetracycline are amphoteric in nature and forms crystalline salt with alkalies and acids. The basic properties of tetracycline is due to the presence of diethyl amino group $-N(CH_3)_2$ and the acidic properties are due to the phenol hydroxyl group (ring D).

Each tetracycline has three ionisable groups with *pka* values ranging from 3.3 to 9.7.

The common chromophoric system gives rise to a characteristic spectra which extend to visible region and imparts yellow colour to all the compounds of this group.

These form chelates with many metal ions and this property is responsible for their antibacterial activity. The phenol hydroxyl group in a tetracycline gives a colour reaction with ferric chloride and forms an azo dye when coupled with diazonium salts.

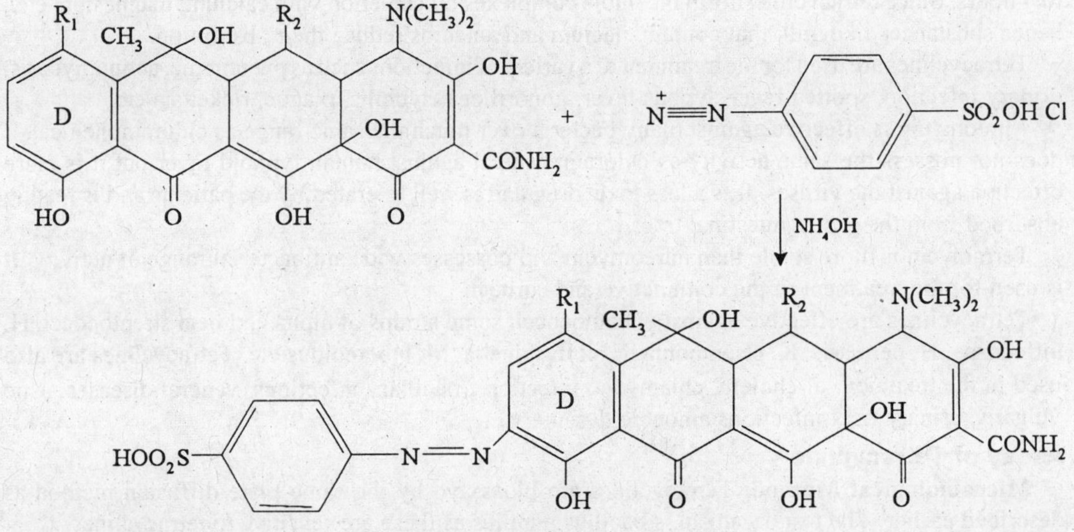

R1 CH3 OH R2 N(CH3)2 OH D OH CONH2 HO O OH O + $\overset{+}{N}\equiv N$–C6H4–SO_2OH Cl

$\xrightarrow{NH_4OH}$

HOO_2S–C6H4–N=N– [R1 CH3 OH R2 N(CH3)2 OH D OH CONH2 OH O OH O]

Azo-dye

Fusion of aureomycin with sodium hydroxide gives 5-chlorosalicyclic acid whereas oxytetracycline gives unsubstituted salicylic acid under the same condition. Less vigorous degradation of both antibiotics lead to similar products.

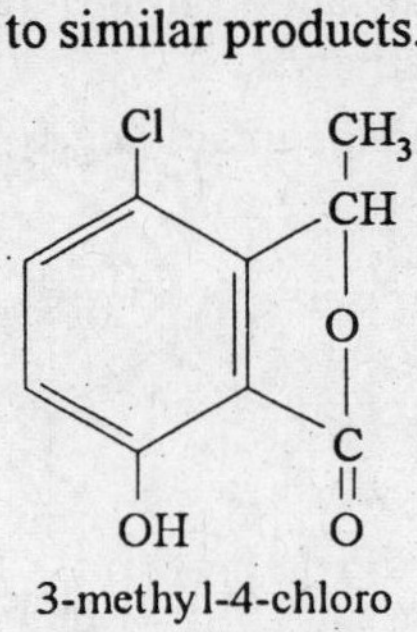

Cl CH3 CH O C O OH

3-methyl-4-chloro 7-hydroxy phthalide fom aureomycin

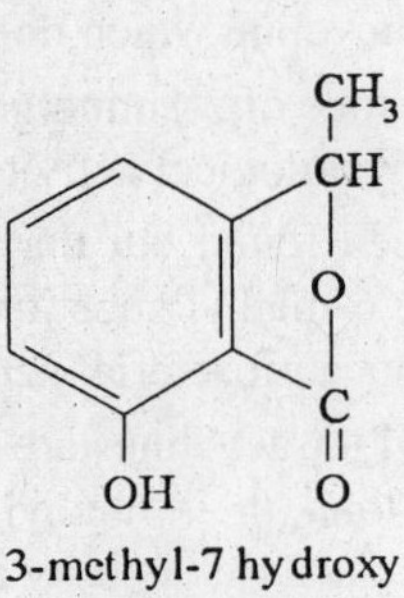

CH3 CH O C O OH

3-methyl-7 hydroxy phthalide from terramycin

After hydroxyl groups have been blocked by methylation, the oxidation of aureomycin (chloro tetracycline) with potassium permanganate gave the methyl derivative of a phenolic lactone, a phthalide containing one atom of chlorine at C-4, Tetracyclines may undergo reversible isomerization giving rise to new tetracycline isomer.

All the tetracyclines gives fluorescene in presence of ultra violet light which is employed to identify them.

Another property of the tetracyclines is their decompositon by an alkali with the formation of isotetracyclines whose colour is more intense than that of the antibiotic itself.

Therapeutic Uses

Tetracyclines are broad spectrum antibiotic which is effective against gram positive and gram negative micro-organisms, acid fast bacilli, spirochetes, rickettisae etc.

These drugs are bacteriostatic in action and inhibit certain essential enzyme systems of the bacterial cell.

Protein synthesis by ribosomes is immediately inhibited after treatment of susceptible organism with tetracycline. It chelates with certain cations like calcium, magnesium etc. and this property is considered very important for the action of these antibiotics.

These drugs are mainly absorbed from the duodenum and small intestine. Oral administration of a single therapeutic dose of tetracycline and oxy tetracycline, peak plasma level is reached within 3 to 4 hours. Since tetracyclines form insoluble complexes by cheletion with calcium, magnesium etc., hence substances like milk that contain clacium and antacids reduce their absorption.

Tetracyclines are used for the treatment of a variety of infections such as pneumonia, actinomycosis, urinary infection, spotted fever, typhus fever, gonorrhoea, syphilis, plague, rickettsia etc.

Aureomycin is effective against many bacteria over much the same range as chloramphenical, it does not possess the same activity as chloramphenical against human typhoid fever but it is more effective against the viruses. It is a less toxic drug and is well tolerated by the patients and is readily absorbed from the gastro intestinal tract.

Terramycin is more stable than aureomycin and possesses wide antibacterial range of activity. It is used for the treatment of the conjunctive and corneal.

Tetracyclines are effective against pneumococci, some strains of alpha and beta streptococci, H. influenzae, H. pertussis, K. prieumonia, E. coli, shigella, M. tuberculous etc. Tetracyclines are also used in the treatment of cholera, chlamydia, infections, bacillary infections, veneral diseases, acne vulgaris, urinary tract infections amoebic dysentery etc.

Assay of Tetracycline

Microbiological Method. Tetracyclines are bioassayd by the same plate diffusion method as described earlier. The test organism is bacillus pumilus as these are sensitive to tetracyclines.

Structure Activity Relationship of Tetracyclines

1. The stereochemistry of the ring system and the nature of the substituents are very important for its biological activity *e.g.*, if the configuration of the 4-dimethylamino group is reversed, it forms 4-epitetracycline which does not show any antibacterial activity.
2. The presence of 2-amino group,4-dimethylamino group and the intact ring system is essential for its physiological activity.
3. It has been found out that 7-chloro, 5-hydroxy, 6-hydroxy, 6-methyl substituents are not very necessary for its antibacterial activity.
4. Demethyl tetracyclines are more stable to acids and alkalies than their methyl homologus and shows same antibacterial activity as that of tetracycline.

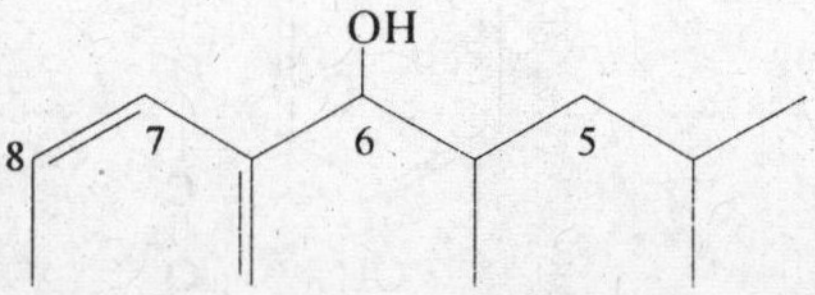

OH 8 7 6 5

De-methyl tetracycline

5. It has been observed that 4-epi, 5-epi, 5-α and 11α-dihydro derivatives of tetracycline are biologically inactive.
6. The 4-de dimethylamine derivative of tetracycline has some residual activity.
7. The antibiotic property of the tetracycline are attributed to their ability to inhibit the protein synthesis of micro-organism by removing the essestial trace metals by chelation. This is supported by the fact that magnesium ions inhibit the activity of tetracyclines and aureomycins.
8. Many derivative of tetracycline namely 7-chloro-6-demethyl-6-deoxy tetracycline hydrochloride, 7-bromo-6-demethyl deoxy tetracycline sulphate, 9-amino 6-demethyl-6-deoxy tetracycline hydrochloride have been found to be more active than the parent compounds.
9. The antimicrobial activity of some deiviatives of tetracycline against staph, aureus are as follows :

Antibiotics	*Activity%*
(*i*) Chlorotetracyclines	100
(*ii*) Demethyl chlorotetracycline	75
(*iii*) Tetracycline	25
(*iv*) Oxytetracycline	24

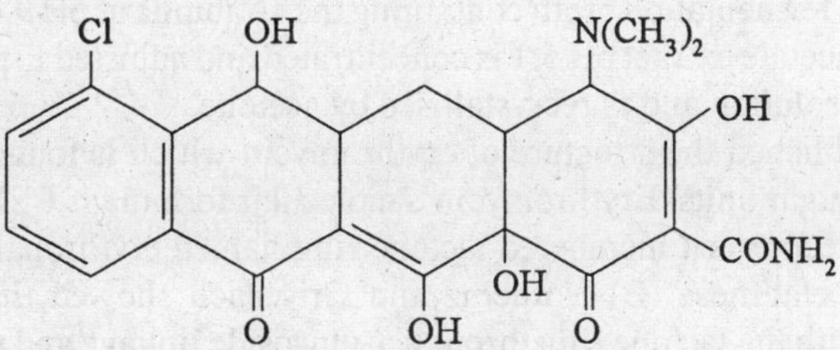

Cl OH $N(CH_3)_2$ OH $CONH_2$ OH O OH O

De-methyl chlorotetracycline

10. Introduction of Bromine at position-7 gives bromo tetracyclines, which is found to be six times more biologically active than tetracyline.

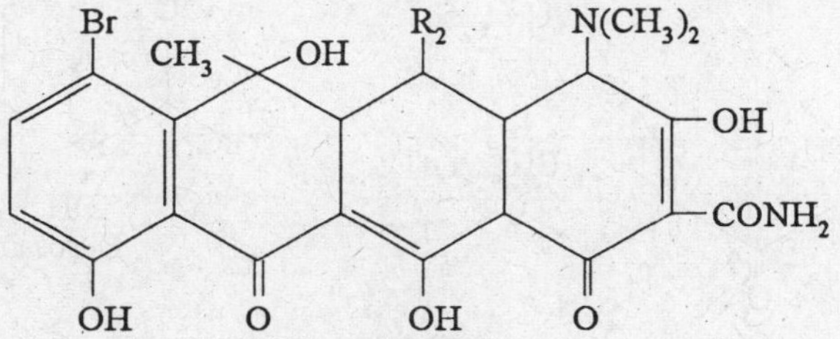

Br CH_3 OH R_2 $N(CH_3)_2$ OH $CONH_2$ OH O OH O

Bromo tetracycline

11. Methacycline which is obtained by modifying the molecule of 6-demethyl tetracycline, is a derivative of oxytetracyclines. Methacycline is a valuable antibiotic and is more physiologically active than parent compound.

Macrolides

Woodwards gave the name of macrolides to those compounds obtained as antibacterial products of metabolism of various streptomycetes with weakly alkaline properties and a high molecular weight. Macrolide antibiotics contain a large lactone ring (12-16 atoms) having a few double bonds and no nitrogen atoms in additions the ring is substituted with one or more sugar residue, some of which may be amino sugars. There are at least nine groups of macrolides depending on the nature and relative arrangement of their sugar moietes *e.g.* erythromycin magnamycin, methymycin, speranycin, oleandomycin, pickromycin etc. The presence of sugar moiety is essential for the biological activity of the macrolides.

The macrolides are normally active against gram-positive bacteria but some of them are effective against gram-negative bacteria and fungi. These are mainly bacteriostatic in action although in very high concentration they have bactericidal tendency. Macrolides inhibit the bacterial protein synthesis both in vivo and vitro.

Macrolides are divided into two main groups. One of the group is known as erythromycin group which comprises of erythromycin oleandomycin, magnamycin and some others and the other group comprises of levorim, filipin, pimoricin etc. Erythromycin is the most important member of the macrolidis antibiotics.

Erythromycin

It was discovered in 1952 by Meguire et. al. It is obtained from streptomyces erythreus belonging to the Actinomycetaceae family.

It is obtained from fermentation medium containing soyabean and glucose. Erythromycin is extracted by filtering the fermentation broth containing the antibiotic at pH 9.4 and by agitating with amyl acetate. The amyl acetate extract pH 5.1 is concentrated and adjusted to pH 11.0. The antibiotic is extracted from above solution and is recrystallised by acetone.

Clark et. al. has established the structure of erythromycin which is found to be a poly hydroxy lactone containing two sugar units. Erythromycin's molecular formula is $C_{37}H_{67}NO_{13}$. It contains an aglycone part consists of fourteen membered lactone ring named erythronalide which is joined to desosamine (A) and to cladinose (B) Hofheinz and Grisebach showed that both cladinose and desosamine are linked with the lactone ring through β-glucoside linkage and proposed the following structure.

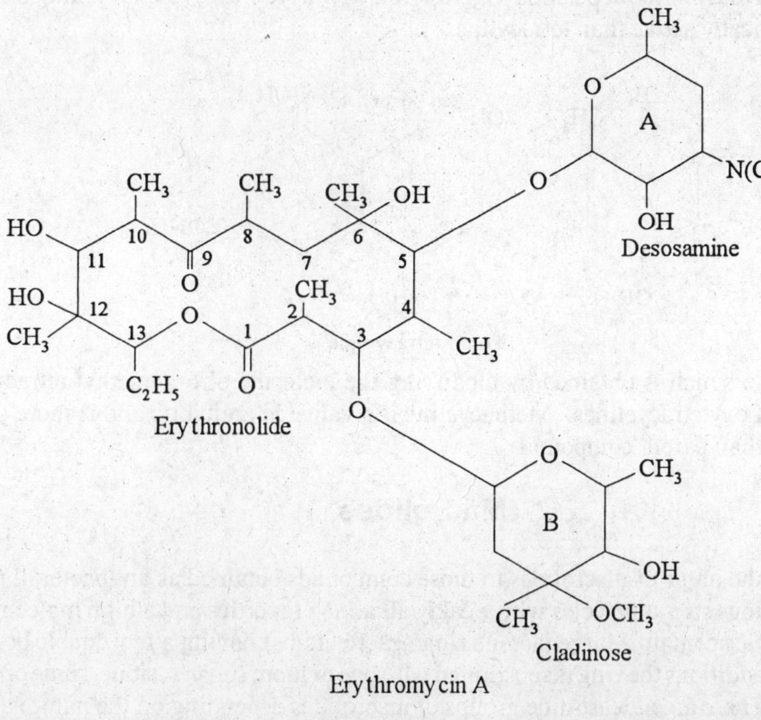

Erythronolide

Desosamine

Cladinose

Erythromycin A

(Designating the desosamine radical as R_1. erythronolide as R, the cladinose radical as R_2 the formula of erythromycin can be represented as R_1—O—R—O—R_2). The structure of erythromycin B and C has also been established. Erythromycin B differs from A in that containing hydrogen instead of OH group at position-12 and erythomycin C differs from A in that it contains the residue of a neutral carbohydrate mycarose instead of cladinose.

Erythromycin is a complex basic compound and its solubility in water amounts to about 2mg/ml. It is freely soluble in organic solvents such as alcohol, acetone, chloroform etc.

Acid hydroloysis of crystalline erythromycin gives cladinose ($C_8H_{16}O_4$) erythrolosamine ($C_{29}H_{49}O_8N$) and a molecule of water.

Hydrolysis of erythromycin with 6N hydrochloric acid gives dimethyl amine hydrochloride (C_2H_8NHCl) and desosamine.

Reduction of erythromycin with sodium borohydride gives dihydroerythromycin ($C_{37}H_{69}O_3N$).

Acylation of erythromycin with benzyloxy carbonyl chloride eliminates one N-CH_3 group and form Ortho N-dicarbobenzyloxy des-N mehtylerythromycin.

Therapeutic Uses

Erythromycin is now widely used in medicine. It is effective against gram-positive organism and some of the gram-negative organisms. Depedning upon the concentration of the antibiotics it can act as bacteriostatic or bacteriocidal in action. It is effective against streptococci, staphylococci, pneumococci, some strains of H. influenzae, tonsilitis, diphtheria, burns, sepsis wound infection, rickettsiae etc. It is also effective against penicillin resistance staphylococci and is useful in the treatment of amoebiasis and syphilis.

Erythromycin is mainly absorbed from the small intestine. The drug is partially destroyed by the gastric juice hence it has to be adminstered in the form of enteric coated tablets.

Erythromycin is an effective substitute for penicillin though its antibacterial action is weaker than that of benzyl penicillin. It is particularly valuable in patients who are allergic to penicillin. The drug can be used for prophylaxis, rheumatic fever, diphtheria etc.

Assay

Ford et. al. developed a chemical method of assaying erythromycin which is as follows:

Take 5.0 ml of solution of erythromycin (56 mg/ml and add 5.0 ml of 27 N sulphuric acid. After 30 minutes optical density was determined at 485 mm band in a 1.0 cm layer and compared with distilled water. Plot a standard curve using solution of pure preparation of erythromycin from which the results are read.

Structure Activity Relationship of Erythromycin

1. Dimethyl amino group plays an important role in the biological activity of the erythromycin as des-N-methyl erythromycin possess only about 5% of the biological activity of erythromycin.

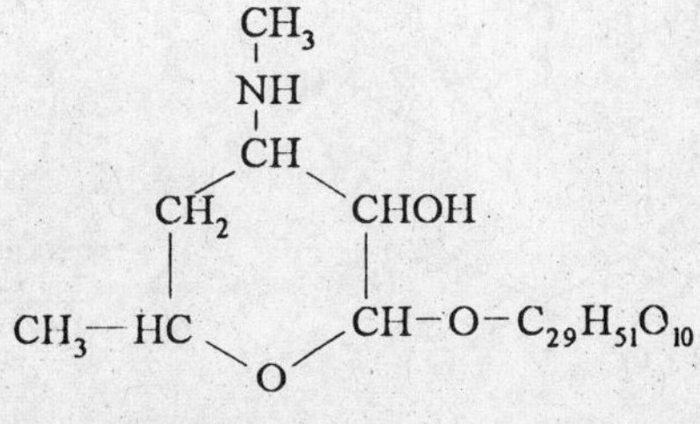

CH_3

NH

CH

CH_2 CHOH

CH_3—HC CH—O—$C_{29}H_{51}O_{10}$

O

des-N-methyl erythromycin

2. Substitution of one N-methylgroup by hydrogen, ethyl, propyl *n*-butyl, isobutyl or γ-phenyl propyl radical, deprives erythromycin of biological activity.

3. Activity of erythromycin increases ten fold when the pH of medium is raised by one in the range of pH between 5.5 and 3.5.
4. The presence of compound like sodium chloride, semicarbazide, urea has no effect on the activity of erythromycin.
5. The sugar residues of erythromycin are essential for its antibacterial activity when this antibiotic is devoid of one or more of their normal sugars they show diminished or altered activity.

Rifamycin

Rifamycin was isolated from the Streptomyces mediterranei by sensi et. al. in 1959.

There are five types of rifamycin namely A, B, C, D, and E. The mixture of antibiotics is quite unstable and is known as rifamycin complex. Its most stable component is rifamycin B.

Rifamycin is obtained from a medium containg 2.0% corn steepliquor, 0.9% calcium carbonate, 0.6% ammonium sulphate, 0.5% soyabean meal, 0.1% magnesium sulphate, 5.0% glucose and 0.3% potassium hydrogen phosphate, the strain of streptomyces mediterranei contain all the five rifamycins. If 0.2 percent of solution of sodium diethyl barbiturate is added to the nutrient medium, almost pure rifamycin is synthesized.

Rifamycins are complex macro cycli antibiotics and has the following structure.

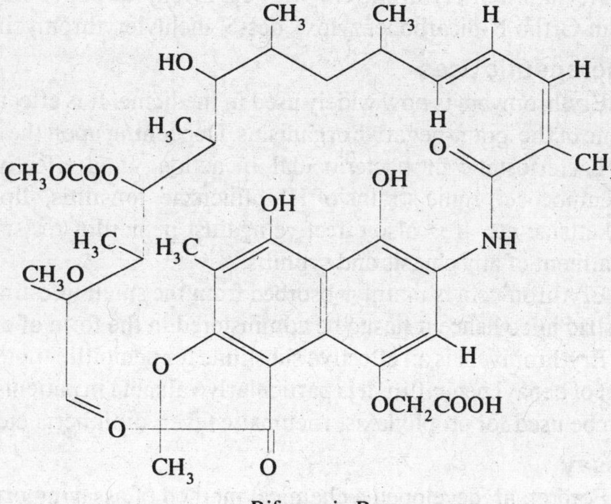

Rifamycin B

Properties

Rifamycin inhibit DNA dependent RNA polymerase of mycobacteria. It may inhibit the growth of certain viruses also.

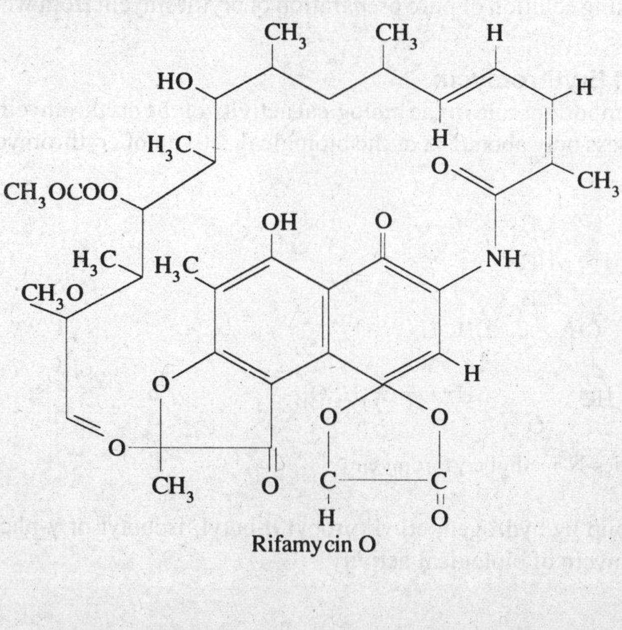

Rifamycin O

This drug is a zwitter ion and is soluble in organic solvent.

When rifamycin B is a arated it is converted into a new derivative, rifamycin S.

The intermediate product formed during conversion of Rifamycin B to Rifamycin S is known as rifamycin O which can be converted back to rifamycin B by some reducing substances eg. ascorbic acid.

Rifamycin S is highly active against gram-positive bacteria and mycbacterium tuberculosis. It is easily reduced to give rifamycin SV.

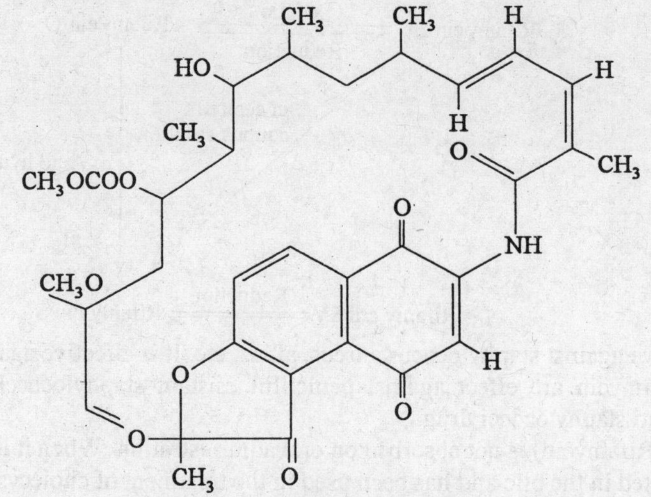

Rifamycin S

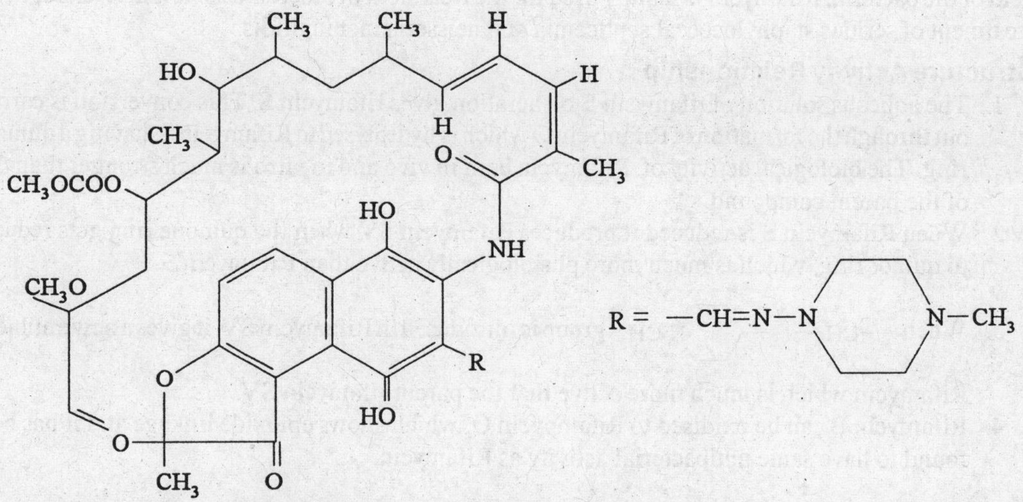

Rifamycin SV

$R = -CH{=}N-N\langle\rangle N-CH_3$

When one hydrogen atom of rifamycin SV is replaced by

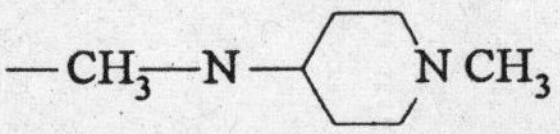

$-CH_3-N-\langle\rangle NCH_3$

a semi synthetic refamycin drug is obtained known as rifampicin.

The relations between various forms of rifamycin can be shown as as given on page 265.

The structure of all four rifamycin has been established.

It has been suggested that acetic acid and propionic acid are the basis of their biogensis.

Rifamycin SV in the most active and valuable antibiotic. It inhibit the growth of gram-positive forms of bacteria that are resistant to other antibiotics such as penicillins, erythromycin etc. The toxicity of rifamycin SV is low. These antibiotics inhibit the bacterial RNA polymerase.

Therapeutic Uses

It inhibit the growth of most gram-positive bacteria such as Escherichia coli, Kelebasiella etc.

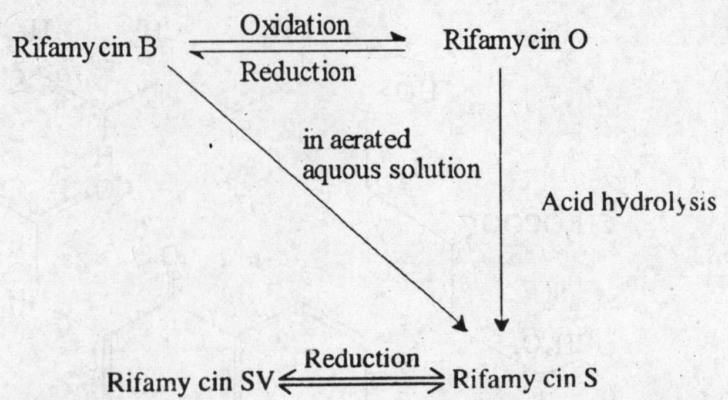

Rifamycin is active against staphylococus aureus, albus, etc. It is effective against My-cobacterium tuberculosis. Rifamycin are effect against penicillin esistant staphylococci. Rifamycin is used specially as an anti staphylococci drug.

Rifamycin B (Rifamycin) is not absorbed on oral administration. When it is given intravenously it is mainly excreted in the bile and has been used in the treatment of cholecystitis.

Rifamycin is well absorbed from the gastrointestinal tract and it block the biosysthesis of nucleic acid of the bacteria. Rifamycin is mainly used for the treatment of tuberculosis. It is also useful in the treatment of serious staphylococcal septicemia and neisseria meningitidis.

Structure Activity Relationship

1. The aqueous solution of rifamysin B on aeration gives rifamycin S. This conversion is carried out through the formation of Rifamycin O which is hydrolysed to Rifamycin S, having a quinone ring. The biological activity of Rifamycin both in vivo and in vitro is much stronger than that of the parent compound.
2. When Rifamycin S is reduced it produces Rifamycin SV when the quinone ring gets reduced to quinol ring, which is much more phsiologically active than Rifamycin S.
3. When $—CH_3—N\langle\quad\rangle N\,CH_3$ group is introduced in Rifamycin SV, it gives a new antibiotic Rifamycin which is much more active that the parent rifamycin SV.
4. Rifamycin B can be oxidised to Rifampycin O, which shows epioxide linkage and it has been found to have same antibacterial activity as Rifamycin.

Chapter 24

Relationship Between the Chemical Structure and Pharmacological Activity

Natural products are a rich source of organic substances which are used for the production of many drugs. These organic substances were isolated and subjected to detailed investigations for their structure and physiological activity of a compound is associated with a particular structural unit or a functional group known as pharmacophore group. It was observed that if the pharmacophore group is removed then the compound loses its activity. If the compound having a particular pharmacophore group is found to be toxic, then this group is somewhat modified by using simple reactions to yield a new compound which is physiologically more active or is less toxic or both. Though there is no specific relationship between the chemical structure and biological activity of the various groups yet some physiological effect has been found to be associated with a particular structural unit or group. The relationship between the chemical structure of organic substances and their pharmacological actions are illustrated as follows:

1. Effects of Unsaturation

It has been observed that unsaturated compounds, both aliphatic and aromatic are considerably more active than saturated ones *e.g.* Trichloroethylene is more active soporific agent (agent producing deep sleep) than Trichloroethane.

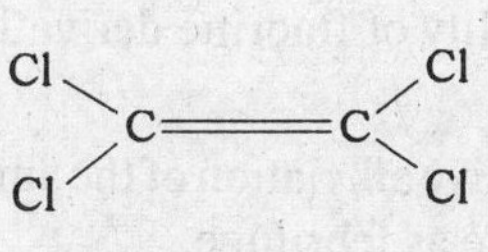

Trichloroethylene 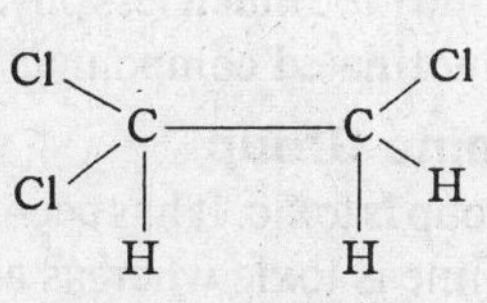 Trichloroethane

It has been observed that unsaturated compounds are more toxic than corresponding saturated compounds *e.g.* Propanol-1 ($CH_3CH_2CH_2OH$) is a non-poisonous compound having mild narcotic property where as allyl alcohol $CH_2{=}{=}CHCH_2OH$ has strong poisonous properties.

2. Effect of Chain Length

It has been found out that the activity of a compound decreases if there is an increase in the number of carbon atoms in a chain *i.e.* if the number is more than 5 to 6 carbon atoms in a chain. With an increase in the number of carbon atoms in a chain, the solubility of substances lowers and they are unable to get into the blood and fail to reach the corresponding centres and thus do not cause the required effect.

3. Effect of Isomerism

Structural isomers have shown marked differences in their pharmacologic activity. This phenomenon is illustrated in ortho, meta and para derivatives of aromatic compound, *e.g.* o-hydroxy benzoic acid is a physiological active compound whereas both para and meta isomers are physiologically inactive.

Stereoisomerism also plays an important role in deciding the pharmacological activity of organic substances. Both geometrical and optical isomers show different physiological activity. It has been established that levorotatry (–) isomers are often more active than dextrorotatory (+) ones. *e.g.* (–) nicotine is twice as active as (+) form.

4. Effect of Halogens

It has been observed that both aliphatic and aromatic series are more active than the parent hydrocarbons, however, halogenation increases toxicity also. Depending on the electrochemical nature, halogens exert a pronounced effect on the physiological behavior of the organic compounds. The negative halogen *i.e.* halogen substituted in non-conjugated positions usually increases the activity of the compounds, although it increases toxic properties also, but it is not so pronounced and hence halogen substitution is used as a mean of increasing the activity and widening the margin of safety in a given series.

The introduction of positive halogens as in α-halogenocarbonyl compounds, acid chlorides, etc. decreases the toxicity of the compounds. Those compounds which have available halogens such as chloramines are strong antiseptics, depending upon the percentage of hypohalous acid liberated on hydrolysis.

Among halogens Cl, Br, I show pronounced effect on the biological behavior of the compound for example, hypnotic properties generally decrease with increase in atomic weight whereas the antiseptic properties increase. Among $CHCl_3$, $CHBr_3$ and CHI_3 the hypnotic activity decreases where as antiseptic activity increases from chloroform to iodoform.

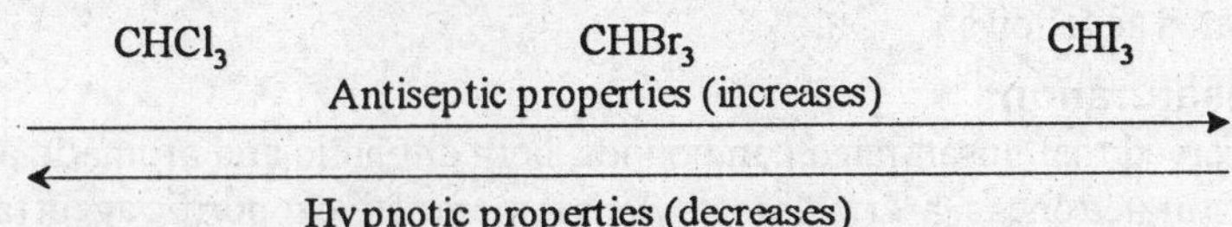

Fluorine is found to be much less physiologically active than the corresponding other halogens. The less activity of fluorinated compounds is due to the stability of fluorine derivatives.

5. Effect of Amino Group

The amino group is toxic. It has been found that successive alkylation of the amino group decreases toxicity *e.g.* Aniline is toxic whereas acetanilide finds use as febrifuge.

Sulphonation and carboxylation also decrease the pharmacological effect of the amino group *e.g.* Aniline is poisonous in nature while *p*-aminobenzoic acid is a component of vitamin B complex.

The introduction of a second amino group increases the toxicity, *e.g.* all the three phenyldiamines are more poisonous than aniline.

6. Effect of Nitro and Nitrite Compound

It has been observed that introduction of a nitro group into aromatic compounds increases their toxicity, *e.g.* nitronaphthol, notrobenzene are more toxic than parent hydrocarbons.

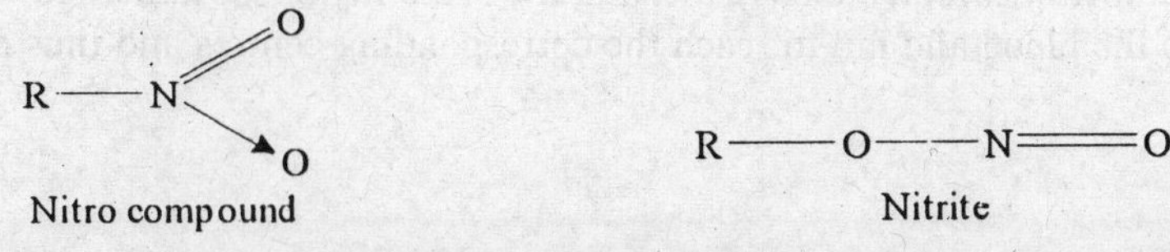

Nitro compound   Nitrite

Nitrite group is isomeric with nitro group but is different in action from that of nitro group e.g. aliphatic nitrates have dilating effect on blood vessels whereas nitro compounds have no action and thus alipahtic nitrites are used to lower blood pressure, this property increases from methyl to amyl nitrite.

7. Effect of Nitrile (—CN) Group

The parent compound of this group is HCN, which is a strong poison. The introduction of nitrile group gives rise to two series of compounds, the nitriles (RNC), both are poisonous.

In aliphatic series, the lower nitriles have been found to be more poisonous than the higher nitriles. Isonitriles are also very poisonous and paralyse the respiratory systems.

The presence of cyanide ion in some inorganic compounds also exerts more poisonous effect, *e.g.* potassium thiocynate and sodium nitroprusside are poisonous substances.

8. Effect of Acidic Group

The introduction of an acidic group such as carboxyl, sulphonic sharply lowers the biological potency of an active compound or a poisonous drug, *e.g.* phenol is poisonous whereas benzene sulphonic acid is not poisonous.

It has been observed that nitro, amino and hydroxy compounds becomes less active, or completely inactive pharmacologically, *e.g.* Nitrobenzene is poisonous but nitrobenzoic acids are harmless. Similarly, amines are toxic whereas amino acids are food stuffs.

However, it has been found out that the biological properties which have been lost by the entrance of acidic group can be restored by esterification, *e.g.* the poisoning action of tyrosine is restored in its ester derivative.

OH

$CH_2.CH(NH_2).COOH$

Tyrosine
(harmless)

NH_2

COOR

p-aminobenzoic esters
(poisonous)

The acylation of basic compounds by means of organic acids reduces the basicity and physiological action of the compound and this property is of great importance in the preparation of synthetic drugs.

The benzoylation increases the pharmacological activity of the various compounds, *e.g.* ecgonine methyl ester is inactive while its benzoyl derivative cocaine is an important local anesthetic.

CH_2—CH—$CHCOOCH_3$
$N.CH_3$ CHOH
CH_2—CH—CH_2

Ecogonine methylester
(inactive)

CH_2—CH—$CH.COOCH_3$
$N.CH_3$ $CH.O.COC_6H_5$
CH_2—CH—CH_2

Cocaine
(active)

9. Effect of Aldehyde and Ketone Groups

Aldehydes are usually more reactive than Ketones and consequently it exhibit more intense biological activity, *e.g.* Formaldehyde is a potent antiseptic. The higher members have the combined property of the aldehyde and alkyl groups. The introduction of hydroxyl groups in the aldehyde molecule decreases the biological activity of the compound and can also make it medically inert.

It has been observed that the pharmacological properties of Ketones are similar to that of corresponding secondary alcohols, *i.e.* Ketones in general possess narcotic action, *e.g.* acetophenone, a mixed Ketone is a strong hypnotic, α, β unsaturated ketones possess diuretic activity.

Aldehyde and Keto groups occur in such a wide variety of compounds that generalizations are not possible.

10. Effect of Hydroxyl Group

The introduction of hydroxyl group into an aliphatic compound decreases its physiological properties. This decrease is roughly proportional to the number of hydroxyl groups introduced into the aliphatic compound, *e.g.* *n*-propanol is more active than glycerol, similarly Hexanol is more active than Sorbital etc.

Sometimes the presence of a hydroxy group makes the compound lose its particular physiological activity, *e.g.* caffeine is physiologically active, whereas hydroxy caffeine does not show any physiological activity.

The isomeric alcohols which have the same number of carbon atoms usually exhibit a drop of activity, and the order of activity is T > S > P.

Teritiary alcohols are the most active because of their ability to form carbonium ions.

The physiological activity generally decreases by etherification and esterification. The hydroxyl group itself in such compounds does not have physiological action of its own but it anchors the molecule on a reactive position of cell chemical.

The introduction of a hydroxyl group in an aromatic compound increases the physiological activity of the compound, *e.g.* phenol is more toxic and a stronger antiseptic than benzene, salicylic acid is a stronger antifungal and antirheumatic agent than the parent compound, benzoic acid.

In contrast to the aliphatic compounds, introduction of a hydroxyl group in aromatic nucleus increases the physiological activity of the compound, *e.g.* phenol is more toxic and a stronger antiseptic than benzene. Introduction of more hydroxyl groups in aromatic nucleus increases the toxicity, *e.g.* resorcinal and pyrogyllol are more toxic than phenol.

11. Effect of Alkyl Group

Whenever an active hydrogen of a drug is substituted by an alkyl group the following changes have been observed:

(*i*) In some compounds, biological activity decreases on alkylation *i.e.* the physiological activity of the alkylated compound is less than that on nonalkylated derivatives, *e.g.* convulsive properties of ammonia is decreased by the introduction of methyl groups and thus trimethylamine is free from all such effects. Similarly, the convulsive properties of aniline are diminished when the hydrogen atoms of aniline are replaced by alkyl groups.

Other examples in which the introduction of alkyl group lower the biological activity are as follows :

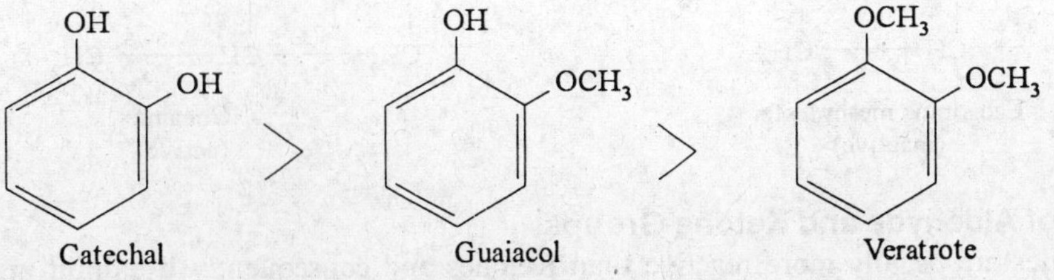

OH, OH — Catechal > OH, OCH_3 — Guaiacol > OCH_3, OCH_3 — Veratrote

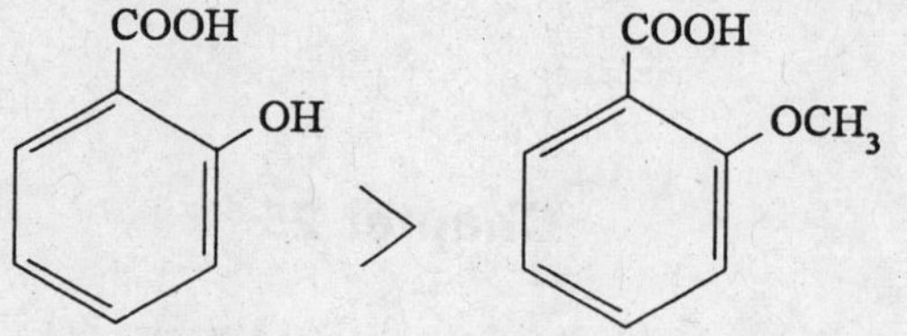

COOH, OH (benzene ring) > COOH, OCH_3 (benzene ring)

Salicylic acid Methoxy benzoic acid

(*ii*) In some compounds the toxicity increases on alkylation, *e.g.* ethers (R—O—R) are more toxic than ammonia, some simple alkylamines have been found to be more toxic than ammonia.

(*iii*) In some compounds, the alkylation of a carboxyl, hydroxy and amino group helps to bring out certain masked properties, *e.g.* cocaine is a strong anaesthetic, while its analogue acid is inert, antipyrine is a strong antipyretic while its analogue having only one methyl group is inert.

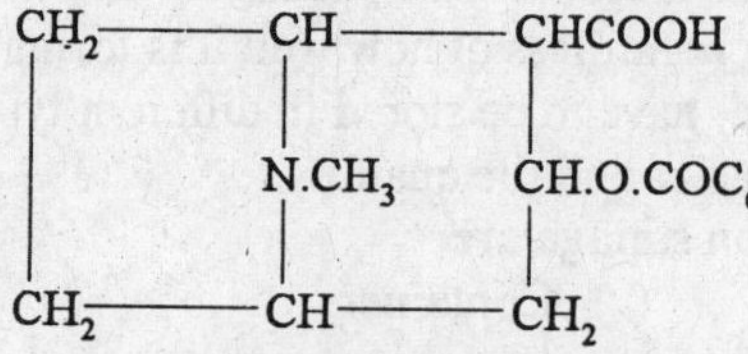

CH_2—CH—CHCOOH | $N.CH_3$ | $CH.O.COC_6H_5$ | CH_2—CH—CH_2

Analogue acid of cocaine (inactive)

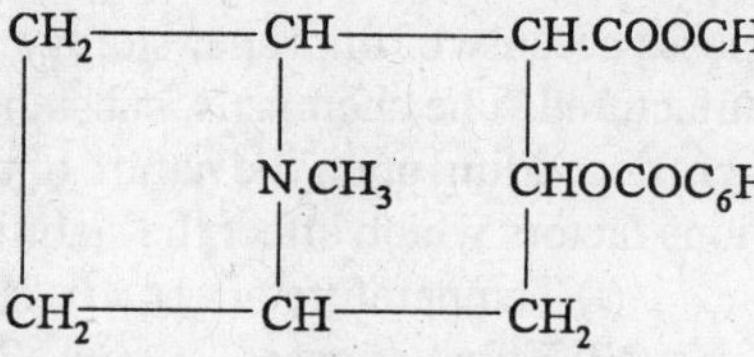

CH_2—CH—$CH.COOCH_3$ | $N.CH_3$ | $CHOCOC_6H_5$ | CH_2—CH—CH_2

Cocaine (active)

The size of alkyl group also shows marked effect on the biological activity, *e.g.* the ethyl group has more marked effect than that of methyl group, *e.g.* diethyl ketone is a stronger hypnotic than acetone, replacement of methyl group of acetophenone ($C_6H_5COCH_3$) by ethyl group forms $C_6H_5COC_2H_5$, which has more hypnotic effect.

Similarly, Dulcin having an ethyl group is about 200 times as sweet as its methyl analogue, which is tasteless.

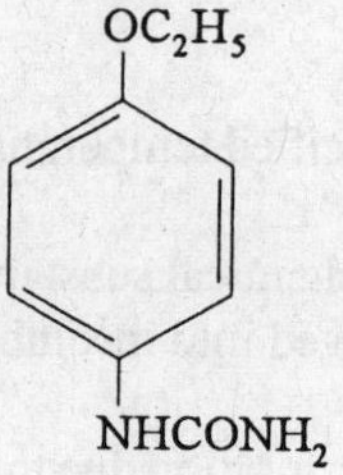

OC_2H_5 (benzene ring) $NHCONH_2$

Dulcin (sweet)

OCH_3 (benzene ring) $NHCONH_2$

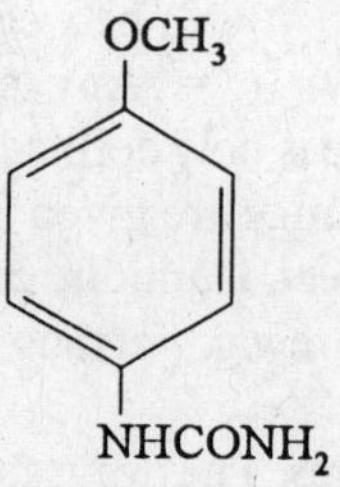

Methyl analogue (tasteless)

Chapter 25

Storage of Pharmaceutical Substances

A drug product may become useless and even hazardous if it reacts with packaging material or decomposes because of improper storage. This may happen regardless of how well it is formulated or manufactured. The chemicals, substances when prepared, have to be stored in different types of containers depending upon the nature of the material, batch size and the quantity.

Various factors which affect the stability of medication on storage are :

(*i*) Temperature effect (*ii*) Humidity effect (*v*) Containers
(*iii*) Effect of gases (*iv*) Effect of light

(*i*) **Temperature Effect.** High temperature accelerates the rate of chemical reaction.

High temperature may accelerate physical changes also *e.g.* colour fading, sublimation, solvent loss, sedimentation of ointment and suspensions etc., cracks in tablets, especially sugar coated.

Similarly extremely low temperature also affects crystallisation, phase separation of emulsion etc.

Temperature definitions for storage and packaging are as follows :

Cold	=	Any temperature not exceeding 8°C.
Refrigerator	=	A cold place in which the temperature is between 2°C and 8°C.
Cool	=	Any temperature between 8°C to 15°C.
Room temperature	=	The temperature prevailing in a working area. Controlled room temperature is a temperature maintained thermostatically between 15°C to 30°C.
Excessive heat	=	Any temperature above 40°C.

Room temperature is now defined as a cool place. Some specified temperature range required for storage of different drugs are given in Table 1.

(*ii*) **Humidity Effect.** Humidity potentiates hydrolysis and chemical substances undergo changes due to air and moisture *e.g.* Ferrous sulphate slowly gets changed into insoluble Ferric oxide by air and moisture.

(*iii*) **Effect of Gases.** The drugs may undergo oxidative deterioration due to atmospheric oxygen. This oxidation can be prevented by adding an appropriate antioxidant to the pharmaceutical preparations. *e.g.* Sodium meta bisulphate is used as an antioxidant for aqueous solutions like Procaine and Adrenaline injections.

Other gases can also affect the storage of drugs *e.g.* chlorine, ammonia, and gasoline engine exhaust can have deleterious effect of flavours and aroma of the drugs.

(*iv*) **Effect of Light.** Light rays, especially ultra violet rays, can have an adverse effect on drugs. *E.g.* surgical solution of chlorinated soda rapidly deteriorates upon exposure to light and heat, Vitamin B_{12}, adrenaline etc. are also affected by light, pure chloroform readily forms traces of poisonous phosgene by a reaction, catalysed by light, formation of phosgene can be prevented by the addition

Table 1

	Drug	Storage Requirements
1.	Aminosalicylic Acid	Preserve in tight, light-resistant containers at a temperature not exceeding 30°C.
2.	Cod Liver Oil	It is preserved in tight containers, from which air has been expelled by production of vacuum or by an inert gas.
3.	Calcium Hydroxide Solution	It is preserved in well-filled tight containers at a temperature not exceeding 25°C.
4.	Ethinylestradiol	It is preserved in tight, non-metallic, light-resistant containers.
5.	Formaldehyde Solution	It is preserved in a well-closed container preferably at a temperature not below 15°C.

of 1 or 2 per cent of ethanol. Sunlight also affects a large number of chemicals *e.g.* bismuth carbonate gets blackened on exposure to sunlight.

The light sensitive materials should be stored in light resistant containers like darkened glass vessels or metal containers to prevent photochemical decomposition.

(*v*) **Containers.** The container is the device that holds the drug and may have a direct contact with the drug. The "immediate container" is that which is in direct contact with the drug at all times. the 'closure' is that part of the container that can be opened and closed to facilitate the access of content and their removal. The possibility of reaction between container and contents can be very harmful, hence every precaution should be taken to avoid such reactions, *e.g.* creams and ointments are liable to react with metal surfaces, such as salicylic acid ointment must not be packed in metal tubes unless the containers are lacquered internally to inhibit reaction.

Alkali sensitive materials, especially which are sterilized by autoclaving, should be packed in glass ampoules. These glass ampoules or containers should comply the test for hydrolytic resistance. Generally the glass containers are of three grades I, II and III. The type I is made of Borosilicates with high hydrolytic resistance. These are more inert than soda lime glass. Type II also have high hydrolytic resistance due to surface treatment of the glass. These are generally made of soda lime glass that has been de-alkalized. Type III glass has limited hydrolytic resistance than glasses of I and II types. These are made of untreated soda lime glass and are suitable for storing anhydrous solid parental drugs.

All glass containers used for injectable preparations should comply with the test for hydrolytic resistance.

Plastic containers require careful evaluation and should be free from undesirable properties which affect the safety of the drugs. The plastic materials used for packaging are mainly made of ethylene, polypropylene, polystyrene, polyvinylchloride (either plasticised or unplasticised) etc. along with appropriate additives like lubricants, plasticisers, antioxidants etc.

Containers should provide specified degree of protection according to the following official definitions :

(*i*) **Light-resistant container.** It protects the contents from the effect of incident light by virtue of the specific property of material of which it is composed of including any coating applied to it. A clear, colourless or translucent container can be made light-resistant by means of opaque enclosure and in such cases opaque covering is needed until the contents have been used up. Sometimes opaque capsule shells are used for protection from light *e.g.* opaque capsule shells are used to protect chlordiazepoxide from decomposition from light.

(*ii*) **Well-closed container.** A well-closed container protects the contents from outside impurities, customary handling, storage and distribution.

(*iii*) **Tight containers.** A light container protects the content from contamination by extraneous liquids, solids, vapours and also protect the loss of drug due to evaporation, deliquiscence, efferves-cence, etc. under ordinary conditions of handling, shipment, storage and distribution.

(*iv*) **Hermetic container.** These types of containers protect the drug from the influence of air or any other gas under the ordinary conditions of storage, handling and distribution.

(*v*) **Single close containers.** This type of containers are closed in such a manner that the contents cannot be removed without destruction of the 'closure' of the container and contents are intended to be used fully immediately after it is opened *e.g.* glass ampoules used as single dose injectable material.

(*vi*) **Multiple dose container.** These types of containers permit the withdrawal of successive portions of the contents without damaging the closure of the container and also without changing the strength, quality or purity of the remaining portion of the drug.

(*vii*) **Aerosol container.** It is a package whose contents are expelled through an open valve by means of internal pressure of the material contained within. The material is generally a pressurized self-propelling material, powered by means of liquefied or compressed gas.

In addition to these general terms used to restrict storage environment the Indian Pharmacopoeia also states specific restrictions in certain individual drug monographs applicable to particular drug products.

Encapsulation

The process of enclosing in a capsule or a membrane is known as encapsulation. Capsule is a unit dosage form in which drug is enclosed in either a hard or soft water soluble shell or container.

There are mainly two types of capsules, *i.e.* hard gelatin capsules and soft gelatin capsules. There are many advantages in capsule form of dosage as they are tasteless shells for drugs. They are easier to administer. They are economically produced in large quantities and the shells are physiologically inert and get easily digested in gastro-intestional tract.

Hard Gelatin Capsules

The modern method for the industrial manufacture of hard gelatin capsules is undertaken on fully automatic machines housed in sophisticated air-conditioned buildings. As a result, they are manufactured by only a small number of specialist companies who supply empty capsule shells to the pharmaceutical industry to fill them with their own products. These are used specially for administration of solid medicines. As the shell takes several minutes to dissolve the taste of the unpalatable drug is effectively masked. The hard gelatin shell is made up of methyl cellulose, polyvenyl alcohol or gelatin and platicizer like glycerin and sorbital and they consist of cylindrical body and a capsule. Both of them have hemispherical ends.

Sometimes the capsules are coated with an anti-microbial, to reduce any tendency to mould growth on storage.

The prefabricated cylindrical shells of the capsules are filled by volume. The desired charge or quantity of drug is diluted with lactose until it is just fit with the body of capsule.

The filling of the capsule is mechanised which can be of two types, partly or fully.

The partly mechanised process consists of a horizontal circular die plate, into which the empty capsules are fitted by hand, the drug is then fed by the hopper into the body of the capsule. As the die plate rotates each empty capsule is filled as it comes under the hopper. The capsule is then capped by hand. In case of fully mechanised process the machine automatically opens the empty capsules, fills them up and finally discharge the complete sealed capsules, example, chloramphenicol capsules, diphemylamine capsules, etc.

Soft Gelatin Capsules

These capsules are used for powers, non-aqueous liquids or pastes.

These capsules are available in various sizes and shapes such as speherical, cylindrical with hemispherical ends, or oval in shape, *e.g.* medical preparation for local purposes, ophthalmic products, volatile liquid drugs, etc.

Soft capsules are prepared from gelatin sheets or gelatin mass.

In rotary die process the gelatin mass and the liquid to be filled are fed continuously. This machine is designed in such a manner that it can continuously and automatically produce finished capsules from the raw material.

The gelatin mass from an overhead tank is converted into two continuous ribbons and are brought together between two rotary dies. Simultaneously, the product to be filled in the capsules from the reservoir is added. The pump action is so adjusted that is corresponds with the arrival of the die pockets at the injection sites. The force of the material being filled, forms the ribbons into half capsules, which are then seated together by heat and pressure of the rotating dies.

The capsules formed are washed thoroughly and dried.

The following Figure 1 shows capsule filling machine and Urschel capsule slitter.

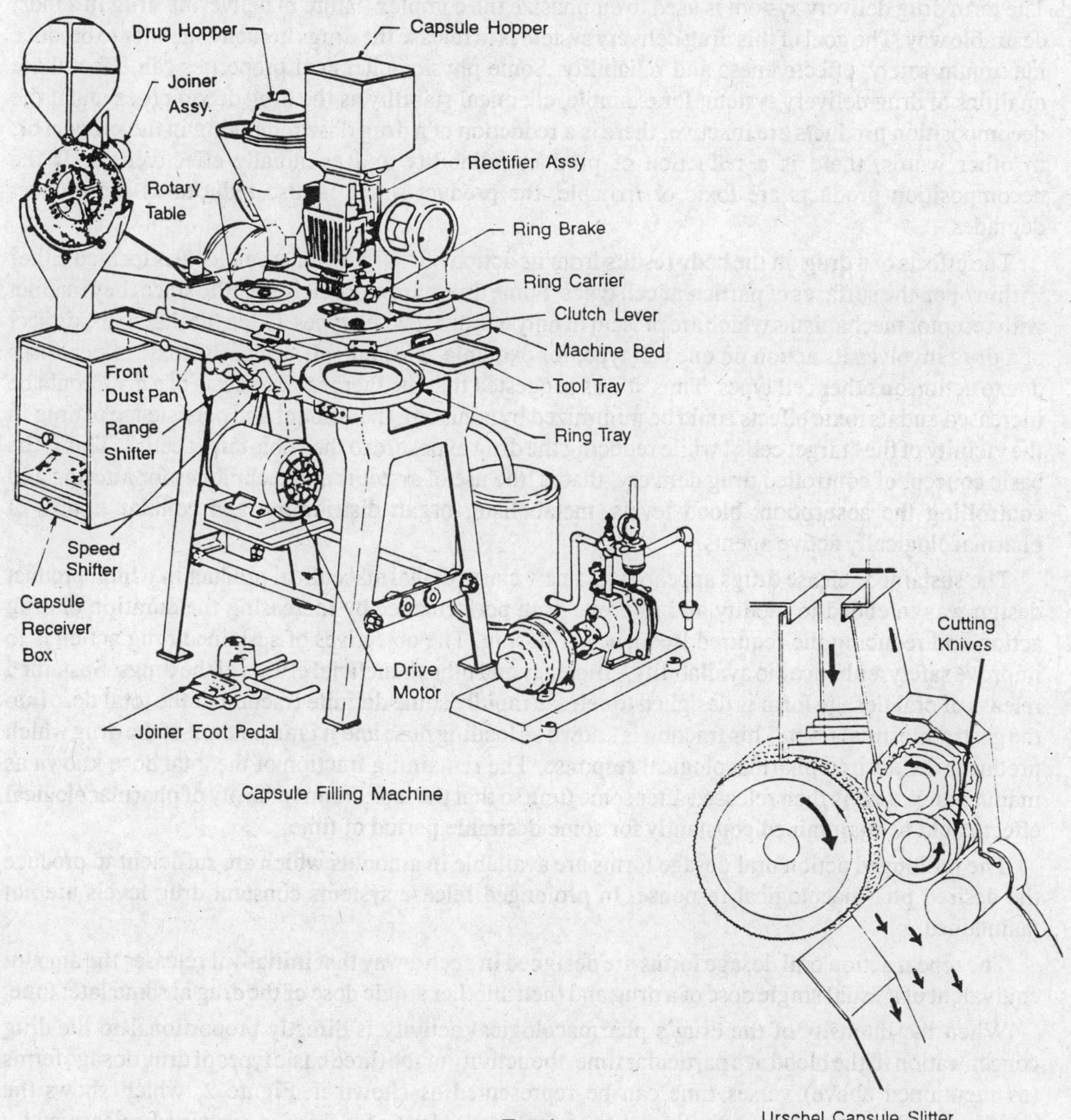

Fig. 1

Source: *Multiparticulate Oral Drug Delivery*, Issac Ghebre Selarie, Marcel Dekker Inc., NY, 1994

Finishing of Capsules

After filling and sealing and locking of capsules, these are subjected to the treatment of finishing. Finishing process is a must to remove the undesired matter adhering to the capsules, hence, this process clears the capsules and gives a better appearance to the capsules on a piece of cloth containing small amount of inert oil. Capsules are brushed with soft rotating brushes for better finishing.

Products filled into gelatin capsules have to comply with the requirements of the official compendia. There are approximately 30 pharma copies in the world and out of these 27 have monograph or set limits for capsule products.

Drug Delivery Systems, Sustained Release of Drugs

The term drug delivery system is used to emphasize the complex nature of delivering drug in a most desirable way. The goal of this drug delivery system is to release the drugs in such a manner to produce maximum safety, effectiveness and reliability. Some physico-chemical properties can effect these qualities of drug delivery system, for example, chemical stability as the drug decomposes and if the decomposition products are inactive, there is a reduction of a drug dose remaining in the product or, in other words, there is a reduction of product reliability and eventually effectiveness. If the decomposition products are toxic or irritable, the product safety is also reduced as the product degrades.

The effects of a drug on the body results from its action on "target" macromolecules located either within or on the surfaces of particular cell types. Some drugs are extremely specific since they interact with receptor mechanisms which are present in only one or a few cell types. Often the therapeutic effect of a drug involves its action on one cell type, for example, a tumor cell, while the toxic effect arises due to action on other cell types. Thus, it was suggested that the therapeutic effects of a drug could be increased and its toxic effects could be minimized by adjusting the amount and persistence of drug in the vicinity of the "target cells" while reducing the drug exposure to the "non-target cells". This is the basic concept of controlled drug delivery, that is, the use of systems and techniques for altering and **controlling the absorption, blood levels, metabolism, organ distribution and cellular uptake of** pharmacologically active agents.

The sustained release drugs appeared as a new class of pharmaceutical product in which product design was intended to modify and improve drug performance by increasing the duration of drug action and reducing the required frequency of dosing. The objectives of sustained drug action is to improve safety, enhance bio availability, efficiency and other beneficial effects of the drugs. Sustained release of oral dosage form is designed to release rapidly some definite fraction of the total dose into the gastrointestinal tract. This fraction is known as loading dose and it is an amount of the drug which produces the desired pharmacological response. The remaining fraction of the total dose known as maintenance dose is then released after some time so that the maximum intensity of pharmacological **effect could be maintained constantly for some desirable period of time.**

The prolonged action oral dosage forms are available in amounts which are sufficient to produce the desired pharmacological response. In prolonged release systems constant drug levels are not maintained.

The repeat action oral dosage forms are designed in such a way that initially it releases the amount equivalent of a usual single dose of a drug and then another single dose of the drug at some later time.

When the intensity of the drug's pharmacological activity is directly proportional to the drug concentration in the blood at a particular time, the activity of the three basic types of drug dosage forms (as mentioned above) verses time can be represented as shown in Figure 2, which shows the relationship between drug activity and time for single dose of a drug, a sustained release and a prolonged action formulation.

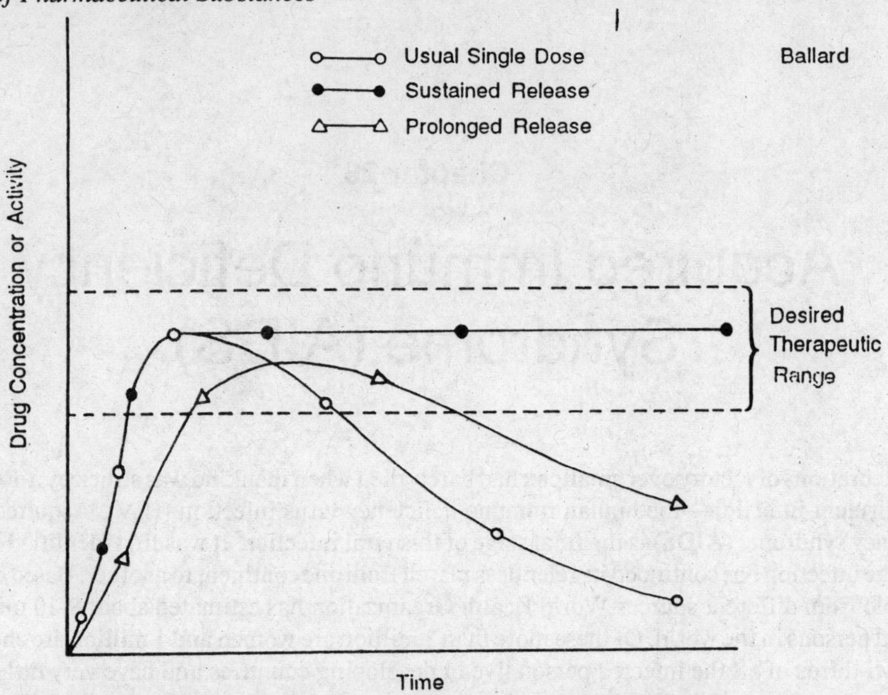

Fig. 2 Relation between drug concentration or activity and time

Source : J.R. Robinson, *Sustained and Controlled Release Drug Delivery Systems* Marcel, Dekker Inc. New York. 1978.

Chapter 26

Acquired Immuno Deficiency Syndrome (AIDS)

Our celebrations of victory over smallpox had barely died when mankind was struck by a deadlier and more virulent infection—the human immuno deficiency virus infection (HIV). Acquired immuno deficiency syndrome (AIDS) is the final stage of this viral infection. It was first identified in 1981 in USA, the infection has continued its relentless march from one continent to another. Based on the date available from different sources World Health Organisation has estimated about 8-10 million HIV infected persons in the world. Of these more than 3 million are women and 1 million are children and over two-thirds of all the infected person live in developing countries and have very little access to health care.

AIDS is caused by human immunine deficiency virus (HIV) which is an RNA retro virus, which produces the enzyme reverse transcriptase inside the cell of the infected person. This enzyme transforms viral RNA to DNA and this new DNA is known as provirus which is incorporated into the host cell DNA. The host cell then produces new copies of the virus that pass out into tissue fluid and blood and infect other host cells. When infected host cells divide, copies of the provirus are integrated into the DNA of daughter cells and thus spreads the disease within the body.

The HIV virus invades the T-lymphocyte cells and possibly neuroglial cells in the brain. When these cells are infected their number is reduced and this causes suppression of T-cell immunity and consequently there is wide spread infection often by microbes of relatively low pathogenicity. The immunity of the infected individual decreases due to HIV virus. A schematic diagram of the structure of the most dreaded virus is shown in Fig. 1.

HIV has been isolated from semen, cervical secretions, lymphocytes, plasma, cerebrospinal fluid, tears, saliva, urine and breast milk. The secretions known to be especially infectious are semen, cervical secretions, blood and blood products.

This infection generally spreads through the following sources :

1. By sexual intercourse.
2. By infected blood, blood products, donated semen and organs.
3. By contaminated needles used.
4. From an infected mother to her child, infection can spread possibly by breast milk or through the placenta before birth or while the baby is passing through the birth canal.

The presence of antibodies to HIV indicates that the individual has been exposed to the virus but not that a naturally acquired immunity has developed. All those who have antibodies in their blood do not develop AID although they may spread the infection to others. One of the methods to detect AID infection is ELISA technique, which was developed by Engvall and Perlmann in 1971. This method employs the antigen-antibody reaction to estimate minute quantities of hormones and other substances in biological fluids.

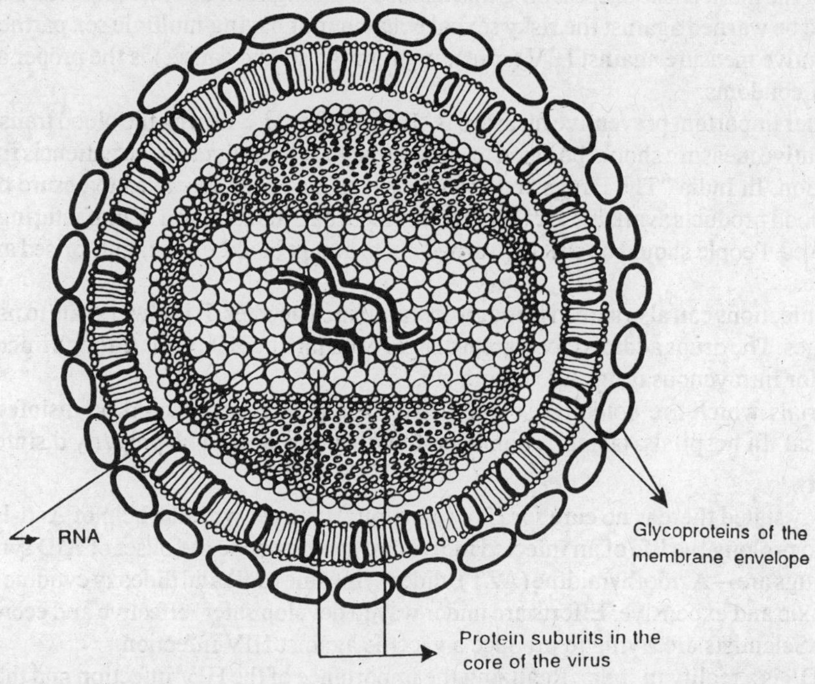

Fig. 1 Schematic Structure of HIV virus.

General symptoms of AIDS

A few weeks after the infection, there may be an acute influenza-like illness with no special features, followed by a period of three or more years without any symptoms when AID develops the main complications are wide spread in recurrent infections and tumours. Some other symptoms are as follows :

(*a*) Pneumonia commonly caused by Pneumocystis carini or by any other microbes may be involved.

(*b*) Persistent nausea, diarrhoea and loss of weight due to recurrent infection of the alimentary tract by a wide variety of microbes.

(*c*) Recurrent meningitis, encephalitis brain abscesses etc. which may be caused by microbes or possibly by HIV. It may be followed by deterioration in neurological functions, characterised by forgetfulness, loss of concentration, confusion, limb weakness etc.

(*d*) Wide spread eruptions *e.g.* eczema, impetigo, psoriasis, wearts etc. may be found in the body.

(*e*) Non-infective enlargement of lymph nodes may be found.

(*f*) Malignant tumours also may occur.

(*g*) The interaction, between HIV and other infectious agent is of great health concern. The most significant may be M-tuberculosis. It has been found out that with dual infection of HIV and TB develop clinical Tuberculosis more rapidly than persons without HIV infection. The epidemic of HIV has triggered off a secondary epidemic of tuberculosis world wide.

Prevention

It is presumed that HIV was introduced in India somewhere in the early 1980s and an epidemic spread of the virus started in 1985 onwards. There is as such no cure for AID. Some anti-HIV drugs have shown encouraging results, but these are expensive and can cause severe adverse reaction. The HIV infections can only be stopped by preventive measures such as mass communication, information and educating the people about the dreaded disease. Some useful preventive measures are:

(*i*) One of the main cause of spreading this disease is through sexual intercourse and hence people should be warned against the risky sexual behaviour of having multiple sex partners. The only preventive measure against HIV (spreading through sexual causes) is the proper and effective use of condoms.

(*ii*) Another important preventive measure is "Blood safety" *i.e.* during the blood transfusion every preventive measure should be taken to ensure that the blood given to the patient is free from HIV infection. In India "The Drugs and cosmetics Act" has been amended to ensure that all blood and blood products available are HIV free and also that, the standard manufacturing practice are followed. People should be strictly warned not to accept blood from unauthorised and unknown sources.

(*iii*) HIV infections can also spread through contaminated needles. It is always safe to use disposable syringes. The drugs addicts face a great danger of acquiring HIV virus through infected syringes used for intravenous drugs like heroin etc.

(*iv*) Materials which are potential source of accidental infection should be disinfected prior to disposal. In hospitals, labour room, operation theatres, should be properly disinfected.

Treatment

As already stated there is no cure for HIV infection. However with the help of Anti-HIV drugs, it is possible to prolong the life of an infected individual and postpone the onset of AID symptoms. The anti-HIV drugs are—Azidothymidine (AZT), dideoxyinosine (ddI) and dideoxy cytidine (ddc). These drugs are toxic and expensive. Efforts are under way to develop safer, effective and economical anti-HIV drugs. Scientists are trying to produce a vaccine against HIV infection.

Now AIDS is a reality in India. Realizing the importance of the HIV infection and the crucial role it would play in shaping the health of the people of India. The government. has set up many research foundations such as National Aids Committee, a board for AIDS control formed from officials of Ministry for health etc. The current dimensions and future implications of AIDS are frightening. Like science, the community too should gear itself to meet the challenge, only through a spirited and committed response by the community can we hope to emerge victorious against this deadly viral disease, AIDS as we have against small pox.

Appendix I

UNITS AND THEIR ABBREVIATIONS

LENGTH

m	=	meter
cm	=	centimeter, 10^{-2} m
mm	=	millimeter, 10^{-3} m
μm	=	micrometer, 10^{-6} m
nm	=	nanometer, 10^{-9} m
pm	=	picometer, 10^{-12} m

VOLUME

l	=	liter
dl	=	deciliter, 10^{-2} l
ml	=	milliliter, 10^{-3} l
μl	=	microliter, 10^{-6} l

MASS

kg	=	kilogram, 10^{3} g
g	=	gram
mg	=	milligram, 10^{-3} g
μg	=	microgram, 10^{-6} g
ng	=	nanogram, 10^{-9} g
pg	=	picogram, 10^{-12} g
grs	=	gross
oz	=	ounces

CONCENTRATION AND CHEMICAL ACTIVITY

Osm	=	osmole, osmolar
mOsm	=	milliosmole, milliosmolar, 10^{-3} Osm
M	=	mole, molar
mM	=	millimole, millimolar, 10^{-3} M
Eq	=	equivalent
mEq	=	milliequivalent, 10^{-3} Eq

TIME

hr	=	hour
min	=	minute
sec	=	second
msec	=	millisecond, 10^{-3} sec
μsec	=	microsecond, 10^{-6} sec

ELECTRICAL

v	=	volt
mv	=	millivolt, 10^{-3} v
μv	=	microvolt, 10^{-6} v
f	=	farad

ENERGY

cal	=	calorie
Cal	=	Calorie, 10^{3} cal
kcal	=	kilocalorie, 10^{3} cal

FREQUENCY

Hz	=	Hertz (cycles per second)

A HANDY METRIC CONVERSION TABLE

To change	*To*	*Multiply by*
Ounces (oz)	grams (g)	28
Pounds (lbs)	kilograms	0.45
Pints (pt)	liters (l)	0.47
Gallons (gal)	liters (l)	3.8
Fahernheit temperature (°F)	Celsius temperature (°C)	5/9 after subtracting .32

TABLE OF MEASURES

The International Metric System

Listed here are all the base units of the International System and all the prefixes, plus the most common derived units that have special names and other common units that are not part of the International System but are used with it. The International System is also called "the SI", for its official name, *Système international d'unités* (International System of Units).

SI base units

name	*symbol*	*quantity*
meter	m	length
kilogram	kg	mass
second	s	time
ampere	A	electric current
kelvin	K	thermodynamic temperature
mole	mol	amount of substance
candela	cd	luminous intensity

Common SI derived units with special names

name	*symbol*	*quantity*
hertz	Hz	frequency
pascal	Pa	pressure, stress
watt	W	power, radiant flux
volt	V	eletric potential, electromotive force
gram	g	mass (= 1/1000 of a kilogram; note that base unit is kilogram not gram)

SI prefixes

name	*symbol*	*multiplying factor**
exa-	E	$\times 10^{18}$
peta-	P	$\times 10^{15}$
tera-	T	$\times 10^{12}$
giga-	G	$\times 10^{9}$
mega-	M	$\times 10^{6}$
kilo-	k	$\times 10^{3}$
hecto-	h	$\times 10^{2}$
deca-	da	$\times 10$
deci-	d	$\times 10^{-1}$
centi-	c	$\times 10^{-2}$
milli-	m	$\times 10^{-3}$
micro-	μ	$\times 10^{-6}$
nano-	n	$\times 10^{-9}$
pico-	p	$\times 10^{-12}$
femto-	f	$\times 10^{-15}$
atto-	a	$\times 10^{-18}$

Common units used with the SI

name	*symbol*	*quantity*
liter	l or L	volume or capacity (= 1 dm^3)
degree Celsius	°C	temperature (=1 k; 0°C = 273.15 K)
hectare	ha	area (= 10,000 m^3)
ton	t	mass (= 1,000 kg)
nautical mile	M	distance (navigation) (= 1,852 m)
knot	kn	speed (navigation) (=1 M/h)
standard atmosphere	atm	atmospheric pressure (= 101.3 kPa)

*$10^2 = 100$, $10^2 = 1000$; $10^{-1} = 0.1$, $10^{-32} = 0.01$. Thus, 2 km = 2 × 1000 = 2000 m and 3 cm = 3 × 0.01 = 0.03 m. Any prefix may be used with any base or derived unit (e.g. mm = millimeter, MHz = megahertz). Some common prefixes are also used with the liter (*e.g.* ml = mililiter)

Common Conversion Factors

1 centimeter = 0.39 in	1 gram = 0.04 oz	1 square centimeter = 0.16 sq in
1 meter = 39.4 in	1 kilogram = 2.20 lb	1 square meter = 1.20 sq yd
1 kilometer = 0.62 mi	1 ton = 1.10 short tons	
	1 liter = 0.88 qt	
	1 cubic centimeter = 0.06 cu in	
	1 cubic meter = 1.31 cu yd	

Length

inch		(2.54 cm)
foot	12 in	(0.3048 m)
yard	3 ft	(0.9144 m)
rod	5½ yd	(5.029 m)
statute mile	1,760 yd	(1.6093 km)

Volume or Capacity

Liquid Measure

fluid ounce		(29.57 ml)
pint	16 fl oz	(0.4732 l)
quart	2 pt	(0.9464 l)
gallon	4 qt	(3.7854 l)

Dry Measure

pint		(0.5506 l)
quart	2 pt	(1.1012 l)
peck	8 qt	(8.8098 l)
bushel	4 pecks	(35.2390 l)

Area

square inch		(6.452 cm^2)
square foot	144 sq in	(929.03 cm^2)
square yard	9 sq ft	(0.8361 m^2)
acre	4,840 sq yd	(0.4047 ha)
square mile	640 acres	(2.590 km^2)

Weight

Avoirdupois Weight

ounce		(28.3495 g)
pound	16 oz	(453.59 g)
hundredweight	100 lb	(45.36 kg)
ton	20 short hundredweight	(907.18 kg)

Troy Weight

grain		(0.0648 g)
carat	3.086 grains	(200.00 mg)
pennyweight	24 grains	(1.5552 g)
ounce	20 pennyweights	(31.1035 g)
pound	12 oz	(373.24 g)

Appendix II

SUMMARY OF THE DRUGS

General Clinical Tests

Test Name and Method	*Specimen Required*	*Normal Range*	*Interpretation*
1. Sugar (*a*) Fohlin & Wu	Serum	80–120 mg/100 ml	*Elevated in* diabetes mellitus, hyper-tHyroidism.
(*b*) Nelson Somoygi	Serum	65–110mg/100 ml	Hyperpituitarism, hepatic diseases.
(*c*) O-Toluidine	Serum (used to diagnose)	60–95/100 ml	*Decreased* in adrenal insufficiency, functional hypoglycemia, hypopituitarism.
Sugar (a) Clinistix StrIp Test Procedure—Strip test paper impregnated by glucose oxidase, O-Tolidine, vegetable peroxidase, dipped in urine-turns blue if sugar is present in urine.	Urine		This test is sensitive to as little as 0.1% glucose in urine presence of sugar or glucose in urine indicates diabetes mellitus.
(b) Benedict's Test Procedure—5 drops of urine and a tablet containing copper sulphate and sodium hydroxide—ppt. shows the presence of sugar.	Urine		Green ppt—0.25% of glucose in urine. Yellow through orange to brick red colour ppt.—2% or more glucose in urine.
2. Cholesterol (a) Sackett's Method (Calorimetric estimation)	Serum	150–250mg/100ml	*Elevated in* familial hypercholesterolemia, hypothyroidism, chronic hepatitis obstructive jaundice etc. *decreased in* acute hepatitis, hyperthyroidism, acute infections, anaemia, malnutrition.

Test Name and Method	Specimen Required	Normal Range	Interpretation
3. Sodium (flame photometry used to detect change in water and salt balance in the body)	Serum	130–154mEq/litre	*Elevated in* dehydration hyperaldosteronism, carticosteroid excess, coma, diabetes insipraus *decreased in* renal insufficiency, acute or chronic diarrhoea, edema intestinal obstruction.
	Urine	120 mEq/litre	*Elevated in* aldosteronism, alkalosis, fever. *Decreased in* renal diseases.
4. Potassium	Serum	3.6–5.6 mEq/litre	*Elevated in* Addison's disease, adrenal and renal insufficiency, acute bronchial asthma, *decreased in* starvation severe vomiting, metabolic alkalosis, familial peridic paralysis.
	Urine	25–100 mEq/litre	Elevated in aldosteronism, alkalosis, fever.
5. Chloride	Serum	98–108 mEq/litre	Elevated in renal insufficiency, nephrosis, dehydration, overtreatment of saline solution. Decreased in gastrointestional diseases, over treatment with diuretics, metabolic alkalosis.
	Urine	170–250 mEq/litre	Decreased in Pneumonia, nephritis associated with edema.

Appendix II contd.

Cardiovascular Drugs

(Drugs used for the treatment of diseases of heart and blood vessels)

Kinds of Drugs	*Chemical Structure*	*Therapeutic Uses*	*Adverse Effects*
1. Cardiotonic Drugs (*i*) Cardiacglycosides *e.g.* digitalis (*a*) Digoxin (digoxigenin-aglycone D-digitoxose-sugar) (*b*) Digitoxin (digitoxigenin-aglycone D-digitoxose-sugar)	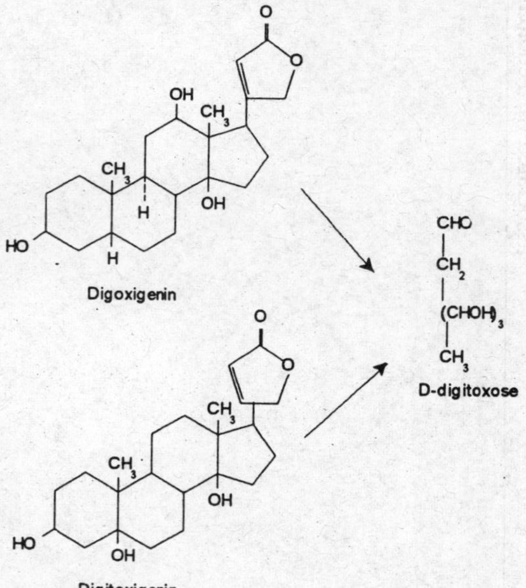 Digoxigenin Digitoxigenin D-digitoxose: CHO–CH_2–$(CHOH)_3$–CH_3	1. Used in the treatment of congestive cardiac failure. Significantly improves cardiac output. 2. Used as a diuretic.	Headache, nausea and vomiting, anorexia, etc.
2. Antiarrhythemic Drugs Drug that depresses myocardium (*i*) Quinidine	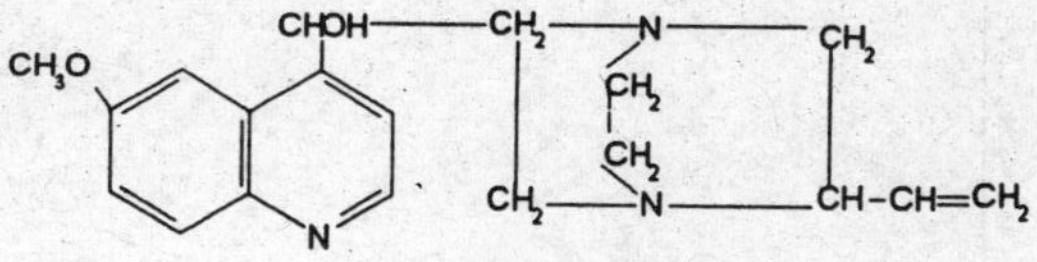 Quinidine	1. It is a valuable drug in the management of most cardiac arrhythmias. 2. Used for the treatment of paroxysmal ventricular tachycardia. 3. Used for the treatment of arrhythmia in patients with hypertension.	Nausea, vomiting, diarrhoea, headache, etc.

Kinds of Drugs	*Chemical Structure*	*Therapeutic Uses*	*Adverse Effects*
(*ii*) Procainamide	$H_2N-C_6H_4-C(=O)-NHCH_2CH_2N(CH_2CH_3)_2 \cdot HCl$ Procainamide	1. Prevents arrhythmias under cyclopropaine anesthesia. 2. Useful in the treatment of junctional and ventricular arrhythmias.	Hypotension, vomiting, headaches, etc.
β-adrenergic drug (*iii*) Propanolol-hydrochloride	$OCH_2CHCH_2NHCH(CH_3)_2$ (OH) on naphthalene .HCl Propanolol	1. Used as an antiarrhythmic drug. 2. Used in angina pectoris. Used in hypertension.	Fatigue, muscle cramps, lethargy, hallucinations, etc.
(*iv*) Alprenolol (*v*) Oxprenalol	$OCH_2CHCH_2NHCH(CH_3)_2$ (OH), R .HCl $R-CH_2=CHCH_2-$ Alprenolol $R-CH_2=CHCH_2O-$ Oxprenalol		
(*vi*) Pinadolol (*vii*) Practolol	$OCH_2CH(OH)CH_2NHCH(CH_3)_2$ on indole (H, N) Pindolol $OCH_2CHCH_2NHCH(CH_3)_2$ (OH), $NHCOCH_3$ Practolol		Same as Propanolol
Cholinergic drugs antagonists of calcium (*viii*) Verapamil hydrochloride		1. Used in migraine and cluster headaches. 2. It is an important calcium antagonists. 3. Used in the treatment of paroxysmal supraventricular tachycardia. 4. Used in vasospastic angina.	Hypotension, causes low heart rate.

Kinds of Drugs	*Chemical Structure*	*Therapeutic Uses*	*Adverse Effects*
3. **Hypotensive or Antihypertensive Drugs** (*i*) Clonidine-hydrochloride	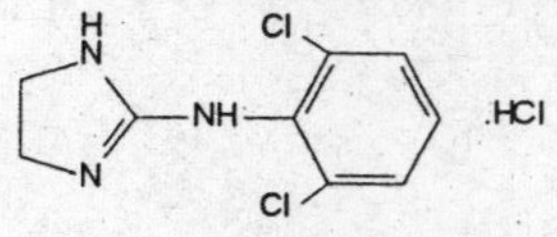 .HCl Clonidine	Used as hypotensive drug. Stimulates the alpha adrenergic receptors in the brain.	Dry mouth, drowsiness, etc.
(*ii*) Alpha Methyldopa (Aldomet)	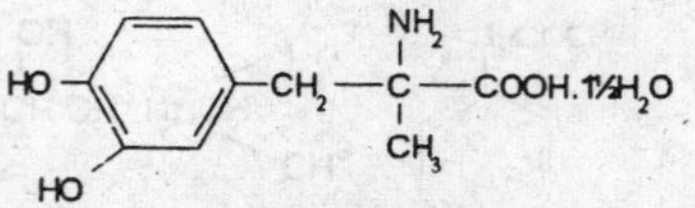 HO–C$_6$H$_3$(HO)–CH$_2$–C(NH$_2$)(CH$_3$)–COOH.1½H$_2$O	Used as hypotensive drug.	Drowsiness, depression.
Ganglion Blockers (*iii*) Pentoliniumtartrate	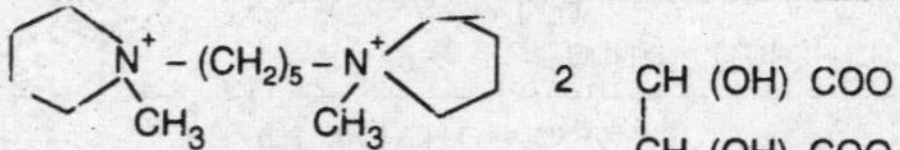 N$^+$(CH$_3$) – (CH$_2$)$_5$ – N$^+$(CH$_3$) 2 CH (OH) COO / CH (OH) COO Pentoliniumtartarate	Used as hypotensive drug.	Faintness, weakness, nausea and vomiting.
Adrenergic Blockers (*iv*) Guanethidine Sulphate	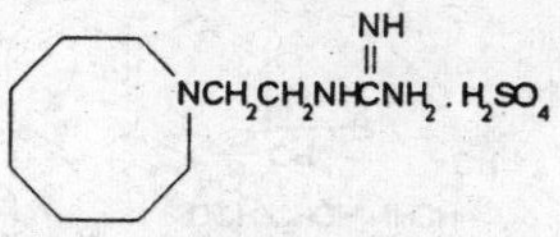 NCH$_2$CH$_2$NHC(=NH)NH$_2$. H$_2$SO$_4$	Used as hypotensive drug in the treatment of moderate to severe hypertension.	Diarrhoea, muscle aching, etc.
(*v*) Reserpine	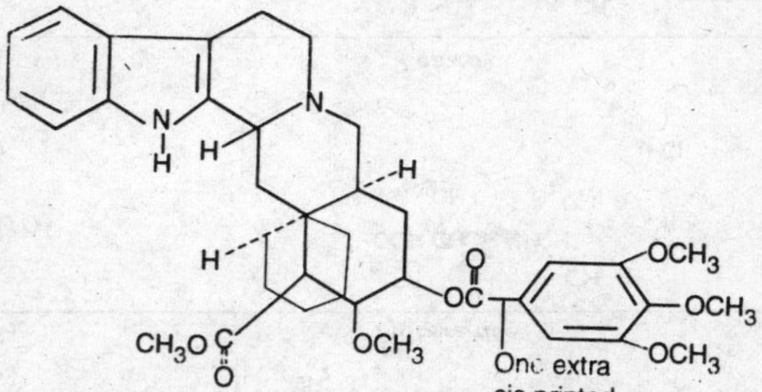 OCH$_3$, OCH$_3$, OCH$_3$, CH$_3$OC(=O), OCH$_3$ Onc extra cis printed	Used in treating hypertensive emergencies.	Nasal stuffiness, gastric hyperacidity, mental depression, etc.
Beta Blockers (*vi*) Phenoxybenzamine	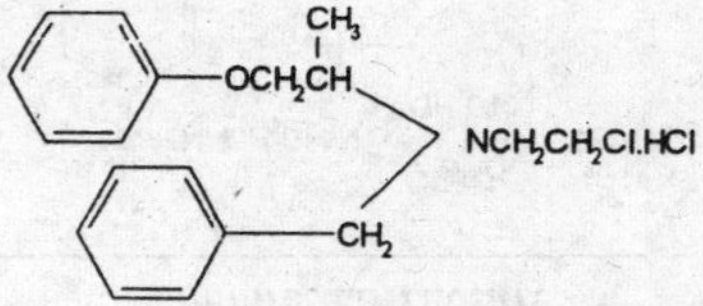 C$_6$H$_5$–OCH$_2$CH(CH$_3$)–NCH$_2$CH$_2$Cl.HCl, –CH$_2$–C$_6$H$_5$	Used in the treatment of peripheral vascular disorders. Control hypertension.	Depression.

Kinds of Drugs	Chemical Structure	Therapeutic Uses	Adverse Effects
4. Antianginal Agents (*i*) Nitrites *e.g.* (*a*) Glyceryltrinitrate	CH_2ONO_2 | $CHONO_2$ | CH_2ONO_2	1. Used in the treatment of angina pectoris. 2. Effective vasodilators. 3. Used in the treatment of congestive heart failure due to acute myocardial infarction.	Headache.
(*b*) Anyl Nitrite		1. Helps in diagnosis of congenital and rheumatic valcular disease. 2. It is an antianginal drug.	Headache, hypotension.
(*c*) Isosorbide Dinitrate	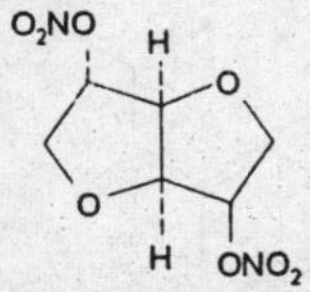 O_2NO, H, O, O, H, ONO_2	1. Used in congestive heart failure. 2. Used in the treatment of angina pectoris.	
(*ii*) Dipyridamole	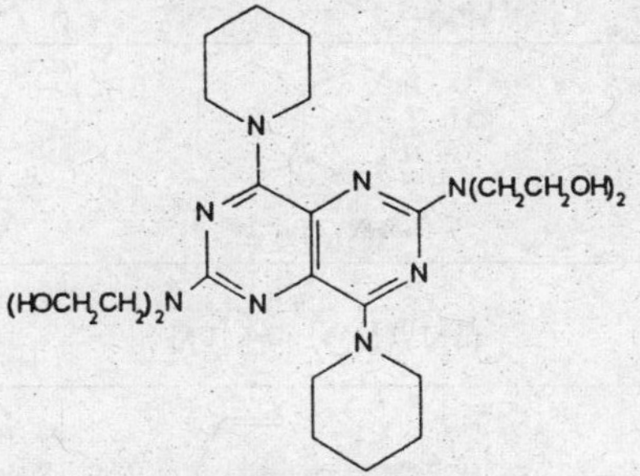 N, N, N, $N(CH_2CH_2OH)_2$, N, $(HOCH_2CH_2)_2N$, N, N Dipyridamole	1. It is used in the treatment of angina pectoris. 2. Used as antithrombotic drug.	
5. Vasodilators Alpha adrenergic blocking agent (*i*) Tolazoline hydrochloride	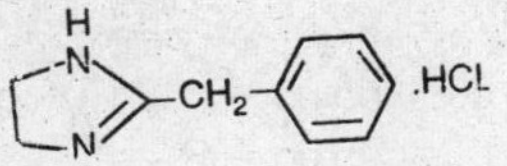 H, N, N, CH_2, .HCl	Used in the treatment of peripheral vascular disorders.	

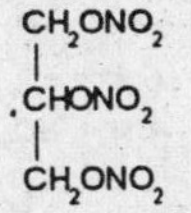

Kinds of Drugs	Chemical Structure	Therapeutic Uses	Adverse Effects
Beta Adrenergic Stimulant (*ii*) Isosuxusprine-hydro-chloride		1. Used in peripheral vascular diseases. 2. Used in the treatment of threatened abortion or premature labour.	
(*iii*) Sodium Nitroprusside	$Na_2 Fe (CN)_5 NO$	1. Used as hypotensive drug. 2. Used in acute myocardiac infarction.	
(*iv*) Hydrallazine hydro-chloride		Used as vasodilator in hyper-tension.	Headache, palpitations.
(*v*) Papaverine		1. It has shown coronary dilation in animal experiments. 2. It is used in arterial embolism (heart disease).	
(*vi*) Nicotinic acid	Vitamin B complex	Weak vasodilator. Large doses induce hypotension.	Large doses cause flushing and burning of skin and dizziness.
(*vii*) Inositol Nicotinate		1. Used as vasodilator. 2. Used in the treatment of peripheral vascular disorders. 3. It is used to reduce hyper-cholesterolacmia.	Vasoconstriction.
(*viii*) Methyl nicotinate		Used as vasoditor in cream and ointments for relieving pain ir. muscular rheumatism lumbago, etc.	

Sulphonamides

Name of the Drug	*Chemical Structure*	*Therapeutic Uses*	*Adverse Effects*
1. Sulphanilamide	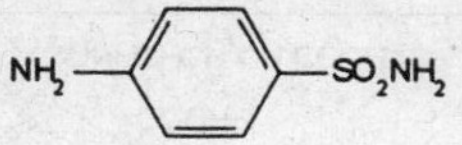 NH_2 — SO_2NH_2	Used in the treatment of pneumococci, streptococci, meningocci, urinary tract infections.	Systemic side effects such as nausea, vomiting, exfoliative dermatitis, urinary tract disturbances, etc.
2. Sulphadiazine	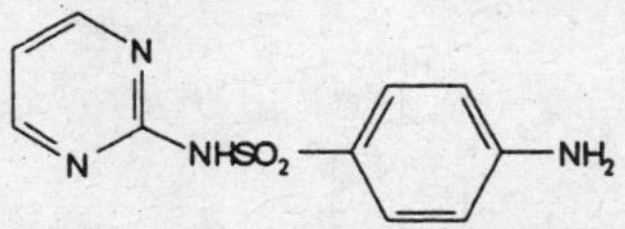 N, N, $NHSO_2$ — NH_2	Used for systemic infections in the treatment of meningococcal meningitis.	Same as above.
3. Sulphapyridine	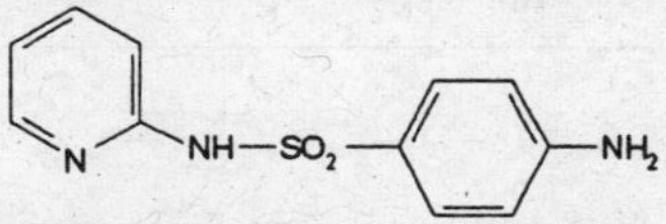 N, $NH—SO_2$ — NH_2	Used in the treatment of pneumonia causing bacteria, meningococcal, streptococcal infections.	Same as above.
4. Sulphathiazole	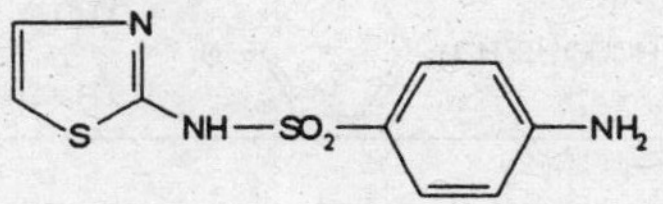 N, S, $NH—SO_2$ — NH_2	1. Used in staphylococcal infection. 2. Used in bubonic plague.	Same as above.
5. Sulphafurazole	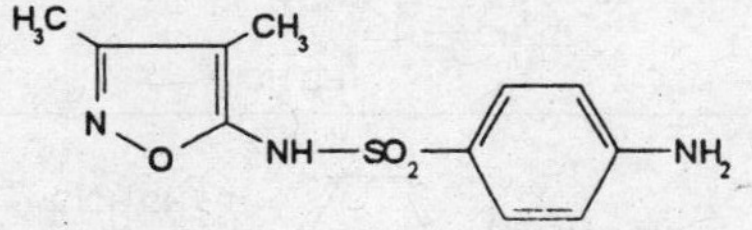 H_3C, CH_3, N, O, $NH—SO_2$ — NH_2	Used in urinary tract infection.	Same as above.
6. Prontosilsoluble	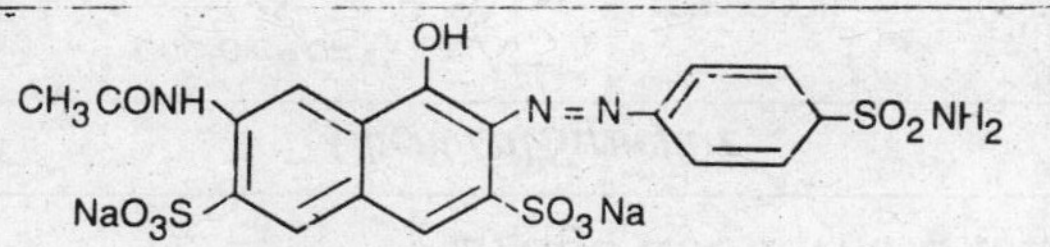 OH, CH_3CONH, N = N, SO_2NH_2, NaO_3S, SO_3Na	Used in streptococcal sore throat. Streptococcal fever and acute endocarditis.	Same as above.

Antineoplastic Drugs
Drugs used in the treatment of Cancer

Name of the Drug	*Chemical Structure*	*Therapeutic Uses*	*Adverse Effects*
1. Alkylating Agents (*i*) Chlorambucil	$ClCH_2CH_2$, $ClCH_2CH_2$ N — (benzene ring) — $(CH_2)_3COOH$	Used in the treatment of chronic lymphocytic leukaemia.	Gastrointestional irritation, bone marrow depression with leukopenia.
(*ii*) Mustinehydrochloride	$ClCH_2CH_2$, $ClCH_2CH_2$ $\overset{+}{N}$(H) — CH_3 Cl^-	1. Used in lymphoma. 2. Used in carcinoma of the lung and other solid tumours.	Nausea and vomiting.
(*iii*) Melphalan	$ClCH_2CH_2$, $ClCH_2CH_2$ N — (benzene ring) — $CH_2CHCOOH$ (NH_2)	1. Used in the treatment of multiple mycloma. 2. Used in the treatment of the carcinoma of the breast and ovary. 3. Used in auto immune disorders.	No acute toxicity.
2. Antimetabolites (*i*) Methotrexate	(pteridine ring: N, N, N, N; NH_2, NH_2) — CH_2 — N(CH_3) — (benzene ring) — CONHCH(COOH)CH_2CH_2COOH	1. Used in acute lymphatic leukaemia. 2. Useful in choriocarcinoma.	Megalobastic anaemia, intestional liver damage etc.
(*ii*) Azathioprine	(imidazole ring: N, N—CH_3, NO_2) — S — (purine ring: N, N, N, NH)	Useful in rheumatoid arthritis, renal disorders, etc. In tissue transplants.	

Name of the Drug	*Chemical Structure*	*Therapeutic Uses*	*Adverse Effects*
(*iii*) Fluorouracil	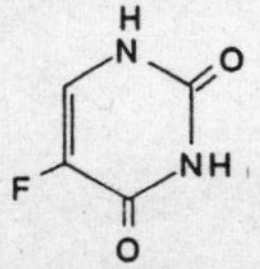	1. Used in malignant neoplasms of breast, liver etc. 2. Used topically for treating actinic keratoses and tumours of skin.	Nausea, gastrointestinal ulcers, bone marrow-depression.
3. Plant products (*i*) Vinblastine	Structure of Vinblastine 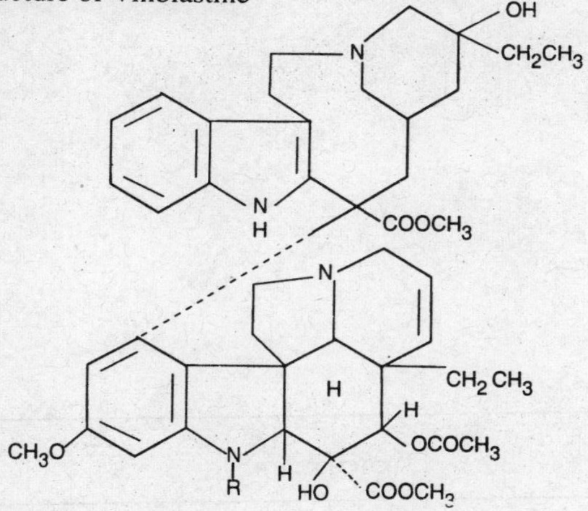	1. Used in Hodgkin's disease. 2. Used in choriocarcinoma.	Vomiting, mental depression, peripheral neuropathy etc.
(*ii*) Vincristine	R = CH_3 Vinblastine R = CHO Vincristine	Used in acute lymphatic leumaemia in children.	Vomiting, depression, peripheral neuropathy.
4. Hormones (*i*) Hydrocortisone		Useful in treating hypercalcenia caused by certain tumours.	Hypertension, fluid-retention, etc.
(*ii*) Prednisone		Useful in treating hypercalcenia caused by certain tumours.	Hypertension, fluid-retention, etc.
(*iii*) Estrogens		Useful in treating carcinoma of breasts.	Fluid retention, uterine bleeding.
(*iv*) Progestins		Used in breast cancer.	Occasional fluid retentions.

Name of the Drug	*Chemical Structure*	*Therapeutic Uses*	*Adverse Effects*
(*v*) Androgens		Used in chronic lymphocytic leukemia.	Fluid retention, cholestatic jaundice.
5. Antibiotics (*i*) Actinomycin-D		Used in Wilm's tumour. Hodgkin's disease.	
6. Radioactive isotopes		Used in thyroid cancer.	
7. Miscellaneous Agents (*i*) Hydroxy Urea	$HONHCONH_2$	Tumours of ovary.	Bone marrow depression.
(*ii*) Urethane	$H_2NCOOHCH_2CH_3$	Used in the treatment of mycloid leukaemia.	

Diabetes and Hypoglycemic Drugs

Name of the Drug	*Chemical Structure*	*Therapeutic Uses*	*Adverse Effects*
1. Insulin	M.wt.-6000 It consists of two amino acid chain. A and B joined by two (—S—S—) linkage, chain A contains 21 amino acids and B contains 30 amino acids, all in a known sequence.	1. Used in the treatment of diabetes mellitus. 2. Glucose insulin drip is used for the treatment of hyperkalaemia. 3. Insulin tolerance test with plasma GH is used as a diagnosistic test for hypopituitarism. 4. Glucose insulin potassium drip is used to treat mycardial infarction.	Some allergic reaction.
2. Oral-Hypoglycaemic Drugs (*i*) Sulponylureas	$R_1-SO_2NHCONH-R_2$		
Some important drugs are: (*a*) Tolbutamide	$CH_3-C_6H_4-SO_2NHC(=O)-NH(CH_2)_3CH_3$	Most commonly used drug for the treatment of juvenile diabetes.	Skin rashes.
(*b*) Chlorpropamide	$Cl-C_6H_4-SO_2NHC(=O)NHCH_2CH_2CH_3$	1. Used in diabetes insipidus. 2. Used as effective antidiuretic.	Excess dose may increase the risk of jaundice, nausea, allergy etc.
(*c*) Glibenclamide	$(5\text{-}Cl, 2\text{-}OCH_3)C_6H_3-C(=O)-NH-CH_2-CH_2-C_6H_4-SO_2NHC(=O)-NH-C_6H_{11}$	Useful drug for treating diabetes in patients who can not take insulin.	
3. Biguanides (*a*) Phenformin	$(R_1)(R_2)N-C(=NH)-NH-C(=NH)-NH_2$ Or $NH_2-C(=NH)-NH_2$ $C_6H_5-CH_2CH_2NHC(=NH)NH-C(=NH).NH_2.HCl$	Controls all types of diabetes mellitus.	Unsafe in patients with renal or heart disease because of danger of lactic acidosis.
(*b*) Metformin	1, Idimethyl biguanide	It can be used for a long time for treating diabetes.	Same as that of Phenformin.

Anti-Convulsant Drugs

Name of the Drug	*Chemical Structure*	*Therapeutic Uses*	*Adverse Effects*
1. Barbiturates (*i*) Phenobarbitone	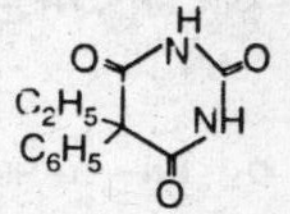 H, N, O, O, NH, O, C_2H_5, C_6H_5	Used in any type of epilepsy except petit mal.	Depression, drowsiness.
(*ii*) Methylphenobarbitone	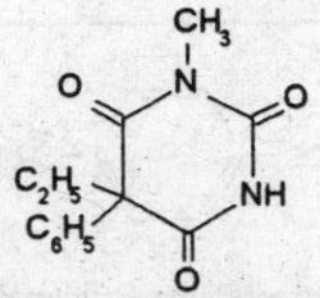 CH_3, N, O, O, NH, O, C_2H_5, C_6H_5	Anticonvulsant drug.	Depression, drowsiness.
(*iii*) Amylobarbitone	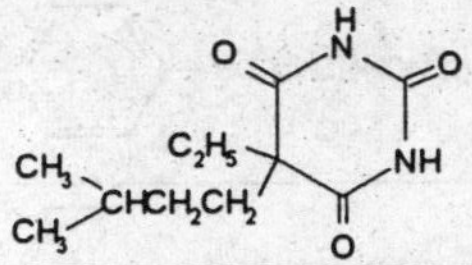 H, N, O, O, NH, O, C_2H_5, CH_3, CH_3, $CHCH_2CH_2$	Used as a sedative and as an anticonvulsant drug.	Depression, drowsiness.
(*iv*) Pentobarbitone Sodium	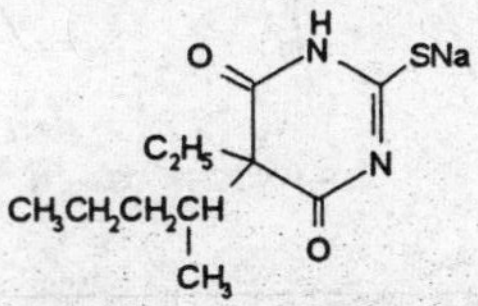 H, N, SNa, O, N, O, C_2H_5, $CH_3CH_2CH_2CH$, CH_3	Used as an anticonvulsant drug.	Depression, drowsiness.
(*v*) Secbutobarbitone	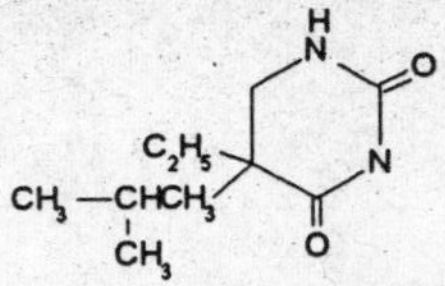 H, N, O, N, O, C_2H_5, CH_3-CHCH_3, CH_3	Anticonvulsant Drug.	Depression, drowsiness.
(*vi*) Quinalbarbitone	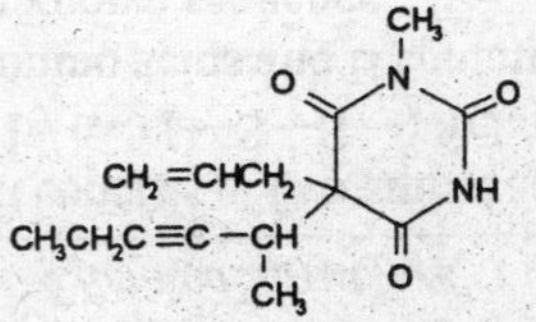 CH_3, N, O, O, NH, O, $CH_2=CHCH_2$, $CH_3CH_2C\equiv C-CH$, CH_3	Anticonvulsant Drug.	Depression, drowsiness.

Name of the Drug	*Chemical Structure*	*Therapeutic Uses*	*Adverse Effects*
2. Hydantions (*i*) Phenytoin	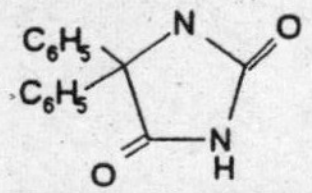	1. Used in major epilepsy. Also used in temporary and focal epilepsy. 2. Used as cardiac arrhythmias.	Nervousness, drowsiness, rash etc.
3. Oxazolidinediones (*i*) Troxidone (Trimethadione)	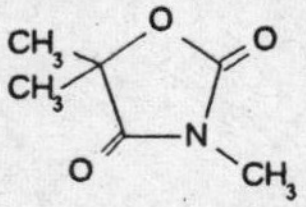	Used in petit mal epilepsy and psychomotor epilepsy.	Drowsiness, blurring of vision, skin rashes, dermatitis, etc.
(*ii*) Paramethadione	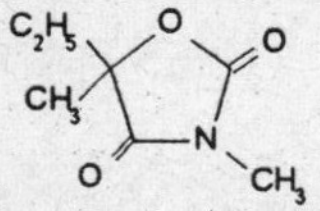	Used in petit mal epilepsy and psychomotor epilepsy.	Drowsiness, blurring of vision, skin rashes, dermatitis, etc.
4. Acetylurea Derivative (*i*) Phenacemide	$NH_2-CO-NH-CO-CH_2-C_6H_5$	Used in psychomotor epilepsy.	Hepatitis, headache, dermati tisete.
5. Succinimides (*i*) Ethosuximide	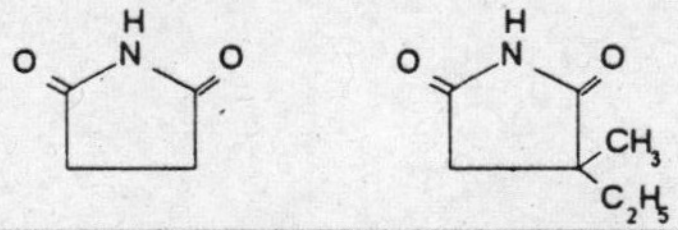	Used in petit mal epilepsy.	Nausea, vomiting, drowsiness, etc.
(*ii*) Methosuximide		Used for temporal lobe epilepsy.	Nausea, vomiting, drowsiness, etc.
6. Miscellaneous Drugs (*i*) Diazepam	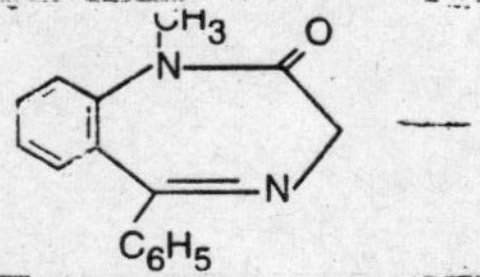	Used in petit mal, temporal lobe epilepsy.	Drowsiness.
(*ii*) Sodium Valproate	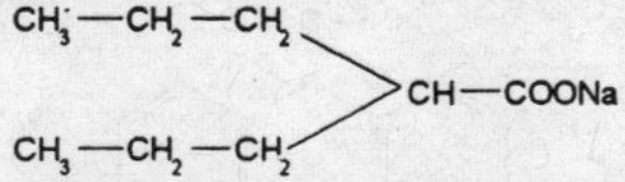 $(CH_3-CH_2-CH_2)_2CH-COONa$	Used in petit mal, major and temporal lobe epilepsy.	Drowsiness.
(*iii*) Sulthiamine	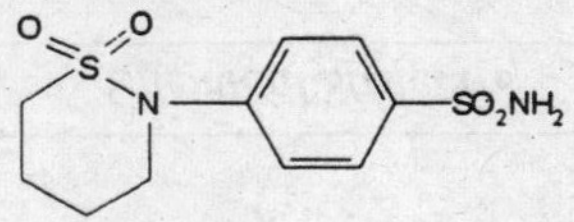	1. Used in temporal lobe epilepsy. 2. Used in major epilepsy and myodonic seizures.	Blurring of vision, weight loss, psychotic reaction, kidney damage.

Name of the Drug	Chemical Structure	Therapeutic Uses	Adverse Effects
(iv) Carbamazepine	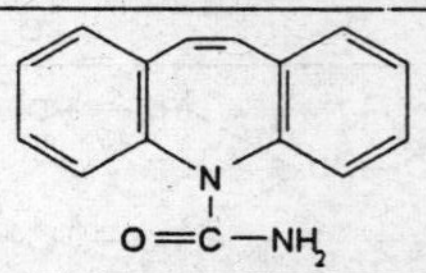 N $O=C-NH_2$	Used in grand mal, psychomotor, temporal lobe and major epilepsy. Also used as analgesic to control pain in nerves.	Drowsiness, blurred vision, bone marrow, depression, etc

Analgesics, Antipyretics and Anti-inflammatory Agents

Name of the Drug	Chemical Structure	Therapeutic Uses	Adverse Effects
A. **Narcotic Analgesies** Naturally Occuring (*i*) Morphine	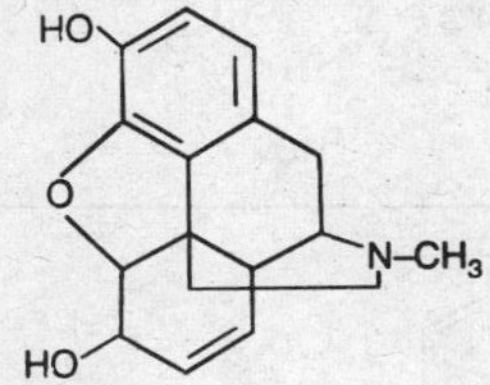 HO, O, HO, N–CH_3	1. Can relieve all types of pain (Analgesic) and is used to reduce pain in conditions such as acute myocardial infarction, fractures of bones, burns, pleurisy, etc. 2. It is used before surgery. 3. It reduces post-operative pain.	Causes addiction leading to acute Morphine poisoning. Dryness of mouth, mental clouding, dysphoria, vomitting headache, constipation.
Derivatives of Morphine Heroin Apomorphine	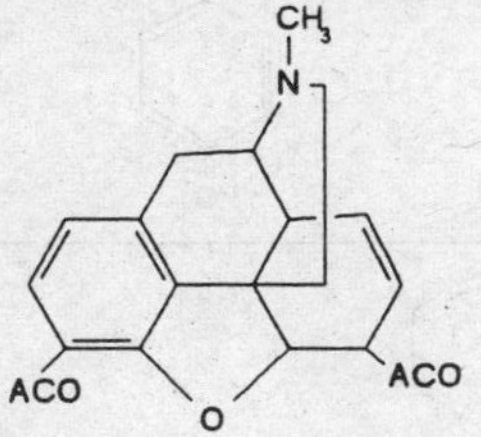 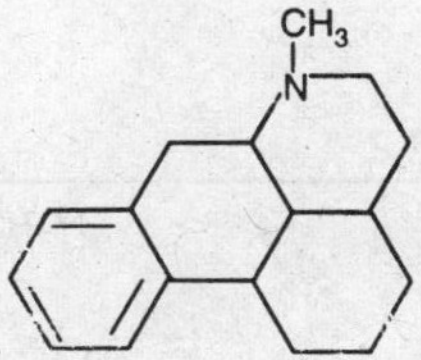 CH_3, N, ACO, O, ACO Heroin CH_3, N Apomorphine	1 Powerful Analgesic produces Euphoria. 1. Emetic, stimulant of chemoreceptor trigger zone.	Causes severe addiction.
Synthetic (*ii*) Pethidine (*iii*) Methadone	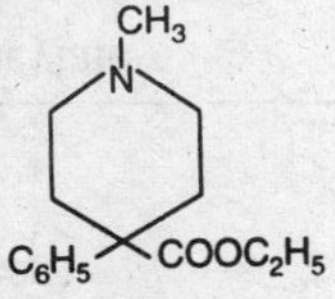 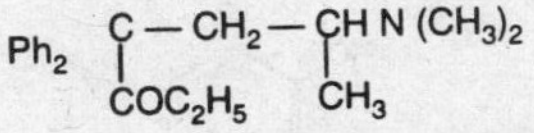 CH_3, N, C_6H_5, $COOC_2H_5$ (*ii*) Ph_2 C — CH_2 — CH N $(CH_3)_2$, COC_2H_5, CH_3 (*ii*) (This structure is wrong)	1. Strong Analgesic, used in the treatment of shock, burns, myocardial infarction. 1. Stronger Analgesic than Morphine, relives viceral pain.	Respiratory depressant. Respiratory depressant.
(*iv*) Morphinan (*v*) Ben Zomorphan Compounds (Phenazoline, Pentazoline)	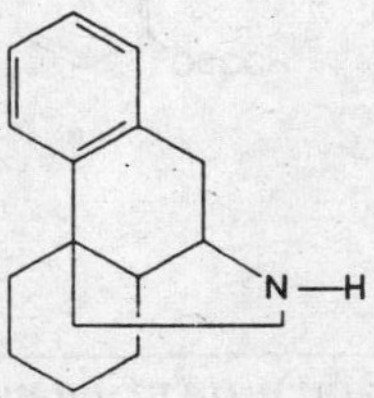 N—H	1. Used as analgesics and is stronger than Morphine, well-absorbed in the gastrointestinal tract. 1. More potent than Morphine.	

Name of the Drug	Chemical Structure	Therapeutic Uses	Adverse Effects
B. Non-Narcotic Analgesics			
1. Salicylic Acid Derivatives			
(*i*) Aspirin	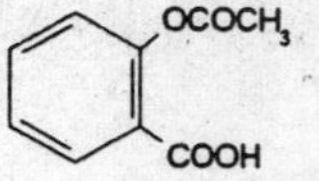 $OCOCH_3$, $COOH$	1. Mild Analgesic, anti-inflammatory, anti-rheumatic antipyretic.	Causes disturbance in gastro intestinal tract. Large dose can cause hepatic damage.
(*ii*) Methyl Salicylate (oil of wintergreen)	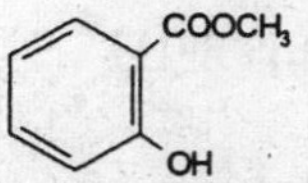 $COOCH_3$, OH	1. Topical application. 2. Flavoring agent. 3. Used as Analgesic in sciatica, rheumatic, lumbago.	Same adverse effect as that of Aspirin.
(*iii*) Sodium Salicylate	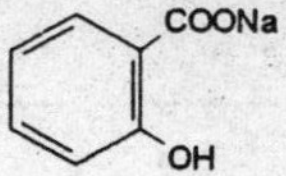 COONa, OH	1. Integumental pain and acute rheumatic fever.	Same adverse effect as Aspirin.
Salicin	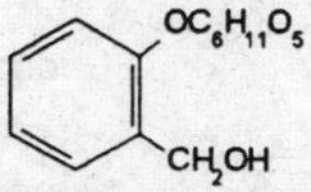 $OC_6H_{11}O_5$, CH_2OH	2. Anti-inflammatory, rheumatic pain and fever.	Same adverse effect as Aspirin.
2. Para Amino Phenol Derivatives			
(*i*) Paracetamol	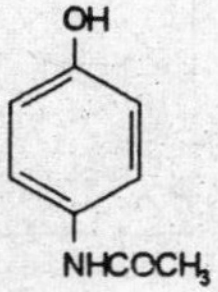 OH, $NHCOCH_3$	1. Antipyretic and Analgesic.	Excessive sweating, nausea, vomiting etc.
(*ii*) Phenacetin	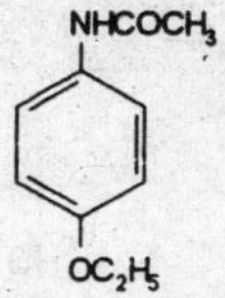 $NHCOCH_3$, OC_2H_5	1. Relief of integumental pain.	More toxic than Aspirin.
3. Pyrazole Derivatives			
(*i*) Phenylbutazone	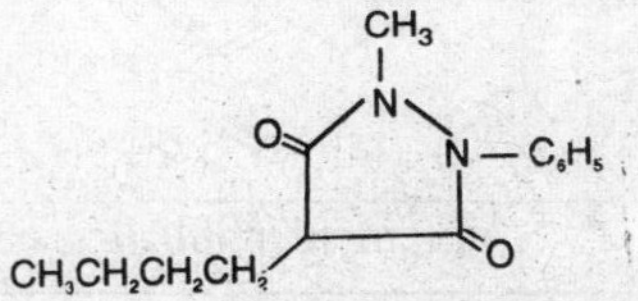 CH_3, N, O, N — C_6H_5, O, $CH_3CH_2CH_2CH_2$	1. Relieving pain in acute gout. 2. Treatment of spondylitis, rheumatoid arthritis.	Various side effects, hence it should not be used routinely as an analgesic or anti-pyretic.
(*ii*) Oxyphenbutazone		1. Relieving pain in acute gout. 2. Treatment of spondylitis, rheumatoid arthritis.	Gastric irritation.

Name of the Drug	Chemical Structure	Therapeutic Uses	Adverse Effects
(*iii*) Phenazone	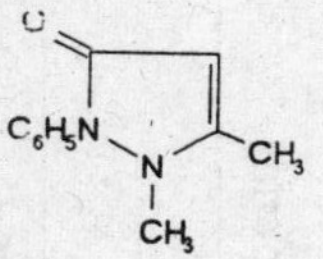	1. Potent Analgesic and anti-pyretic action.	
(*iv*) Analgin	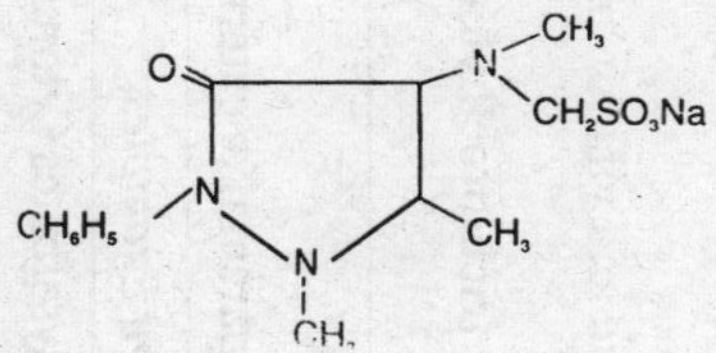	1. Potent analgesic and anti-pyretic.	
4. Indole Derivatives and Aryl Acetic Acid Derivatives			
(*i*) Indomethacin	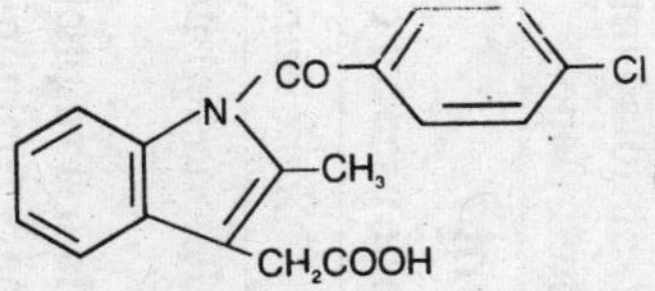 indomethacin	1. Inflammatory, Analgesic, antipyretic action. 2. Effective in rheumatic disorders and treatment of Gout.	Nausea, vomitting, skin rashes, diarrhoea, etc.
(*ii*) Sulindac	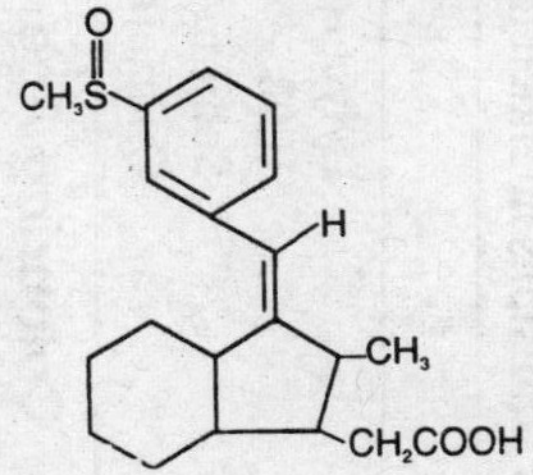 Sulindac	1. Has longer duration than Indomethacin. 2. Used in the treatment of rheumatic and Musculo-skeletal disorders.	Headache, blurring of vision, psychotic disturbances. Giddiness, skin rashes.
5. Miscellaneous Drugs			
(*i*) Ibuprofen (Burfen)	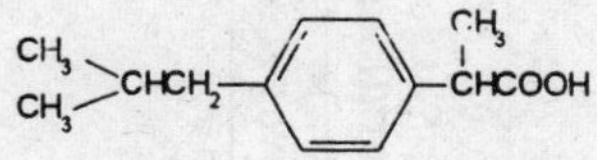	1. It has Analgesic and weak anti-inflammatory action. 2. Used in rheumatoid disease.	Headache.
(*ii*) Ketoprofen	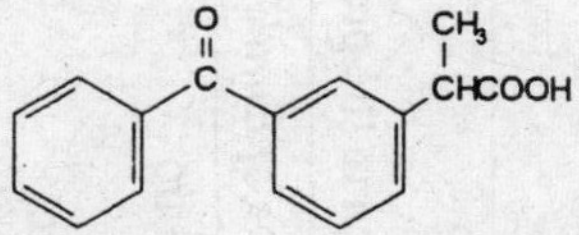	1. Anti-inflammatory, analgesic and anti-pyretic action. 2. More potent than Brufen.	
(*iii*) Naproxen	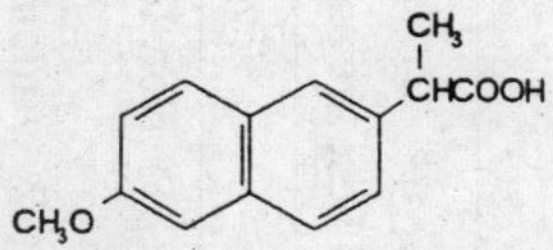	1. Rheumatoid arthritis.	Gastro-intestinal adverse effects.
(*iv*) Tromaril		1. Analgesic, anti-pryetic anti-inflammatory.	

ORGANIC PHARMACEUTICAL AIDS

Preservatives

Antimicrobial compounds capable of inhibiting microbial growth in preportions.

Name of Important Preservatives	Uses
1. Benzoic acid (C_6H_5COOH)	Food preservative Ointment for fungus infection of the skin.
2. Hydroxy benzoates (HOC_6H_4COOK)	Food preservative. Used in cosmetics and in pharmaceutical preparations.
3. Sodium benzoata (C_6H_5COONa)	Food peservative.
4. Benzyl esters of benzoic acid	Used in pharmaceutical preparations.
5. Chlorinated tertiary alcohols *e.g.*, Chlorbutol ($C1_3CC(CH_3)OH$)	Used in dye drops and injections.
6. Phenyl mercuric nitrate ($C_6H_5HgNO_3$)	Used in contraceptive preparations.
7. Phenyl mercuric hydroxide (C_6H_5HgOH)	Same as Phenylmercuric nitrate.
8. Benzalkonium Chloride	Used in eye drops.
9. Methyl propy1 Paraben	Used in pharmaceutical preparation.
10. Cetrimide	Used in various injections.

Antioxidant

Agents which prevents atmospheric oxidation and thus prevents spoilage of various preparations.

Name of Important Antioxidants	Uses
1. Gallic acid and its ethyl, propyl, octyl, dodecyl derivatives	Used to stabilize compounds like ether and paraldehyde.
2. Butylated hydroxy anisole	Used in food oils, fatty acid and vitamin preparation.
3. Thymol	Used as antioxidant to trichloroethylene and tetrachlorethylene.

Emulsifying Agents

These are used in the preparations of emulsions intended to be used for oral, parental or external use.

Type of Emulsifying Agents	Some Important Examples
1. Non-Ionic	Lylyceryl monostetrate Polysorbates Cetomacrogal 1000
2. Anionic	Sodium lauryl sulphate
3. Cationic	Quarternary Ammonium Compounds

Colouring Agents

These are employed for the purpose of imparting colours.

Name of Colouring Agents	Source and Uses
1. Natural colouring agents (*i*) Chlorophyll, β-Carotene, Alizarin, Indigo	Obtained from plants. Used to colour lotions, cosmetics and other preparations.
(*ii*) Cochineal	Obtained from dried female insects knows as Dactylopus Coccus Costa.
(*iii*) Caramel	Obtained by burning sugar. Used to colour syrups.
2. Synthetic colouring agents	
(*i*) Amaranth	Synthetic ago dye. Used for colouring medicines.

Ointment Bases

Types of Ointment Bases	Various Preparations	Properties
1. Hydrocarbon bases (not absorbed by the skin, generally immiscible in water).	Obtained from petroleum known as Paraffins it can be of three types:	
	(*i*) Soft paraffin	Semisolid, white or yellow coloured, soft greasy, odourless. Used for colourless medicament.
	(*ii*) Hard paraffin	Odourless, colourless, translucent wax, insoluble in water.
	(*iii*) Liquid paraffin	Transparent, colourless, odourless oily liquid used as antioxidant.
2. Absorption Bases (hydrophyllic base, absorbs water)	(*i*) Non-emulsified base (penetrate the skin) *e.g.*,	
	(*a*) wool fat	Absorb 50% weight of water. Used in eye ointments.
	(*b*) Hard fat	Mixture of mono-, di- and tri-glycerides of saturated fatty acid. Used for perparation of suppositories.
	(*c*) Bees wax	Waxy esters mainly myricyl-palmitate. Used as an ingredient of paraffin ointment.
	(*d*) Hard wax (Theobroma oil)	Yellowish white hard wax, it is a mixture of glyceryl esters of stearic, palmitic oleic acids used as suppository base.
	(*e*) Wool Alcohol	Saponified wool grease of sheep, golden brown coloured, solid. Used in making wool ointments and water-in-oil emulsion.
	(*ii*) Emulsions (absorb water)	70% wool fat and 30% water
	(*a*) Hydrous wool fat (lanolin)	Used as emollient in several ointment bases.
	(*b*) Oily cream base	Mixture of equal parts of ointment and water, hydro-philic in nature, adhere well to skin.
3. Water Miscible Bases (can be washed easily with water)	(*i*) Cetomacrogol emulsifing wax	Mixture of soild aliphatic alcohols. Used to improve emollient properties of paraffin ointments.
	(*ii*) Polyethlene glycols	Wax-like solids, soluble in water. Used in the preparation of aquous creams.
4. Water Soluble Basis (anhydrous, hydrophilic)	(*i*) Macrogels	Polyethlene glycols. Used in suppositories, pessaries cosmetic creams, etc.

Antiseptics and Disinfectants

Name of the Drug	*Chemical Structure*	*Uses*
1. Phenols and related Compounds Phenols (carbolic acid)	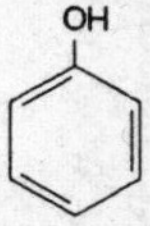	1. It is a potent bactericidal and fungicidal agent. 2. It is general protoplasmic poison which injures the microorganisms and the cells of the body tissue.
Different concentration of phenol (*i*) Liquified Phenol	80% w/w/ of phenol and water.	Effective against gram positive and gram negative bacteria.
(*ii*) Dilute Solution of Phenol	0.2 to 1.0%	Bacteriostatic in action, stronger concentration is bactericidal in action.
(*iii*) Liquified Phenol	20% solution	1. Used in dentistry as anaesthtic agent. 2. Used as dressings for small wounds.
(*iv*) Phenol Glycerin	16% w/w of phenol in glycerin.	Used as ear drops and mouth wash.
Important Derivatives of Phenol (*a*) Methylated (*i*) Cresol (ortho, para, meta)	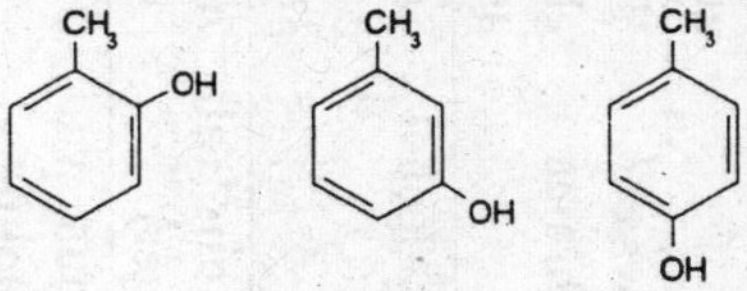 Ortho Meta Para	Used as general disinfectant.
(*ii*) Lysol	Saponified preparation of cresol with vegetable oil.	Used as disinfectant for domestic and hospital use.
(*iii*) Thymol	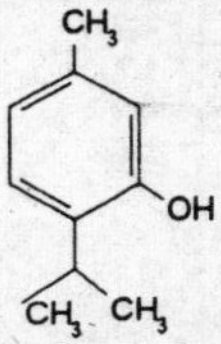	1. Used as an antiseptic and antifungal agent. 2. Used as antiseptic lotions, mouth washes, pain balm toothache drops, etc. 3. Used as an antioxidant.
(*b*) Chlorinated (*i*) Chloroxylenol	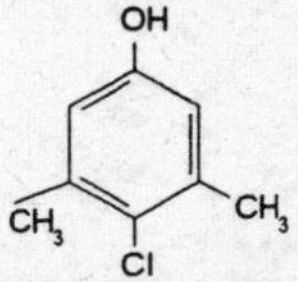	1. It is used in antiseptic creams. 2. Used as surgical dettol for disinfecting surgical instrument.

Name of the Drug	*Chemical Structure*	*Uses*
(*ii*) Chlorocresol	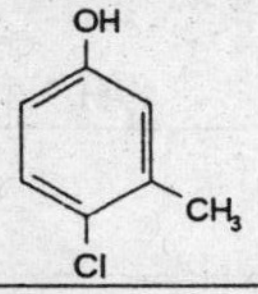	1. It is a potent bactericide. 2. Used in eye drops, paste, creams. Aquous injections, etc.
(*iii*) Hexachlorophane	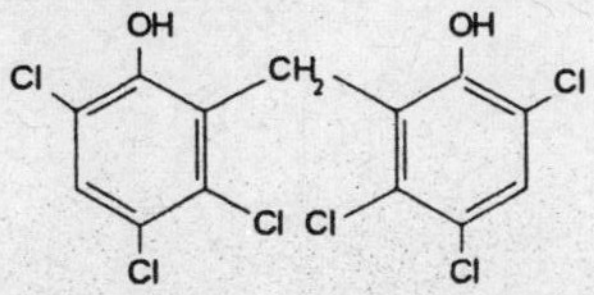	1. It is bacteriostatic in action. 2. Used in medicated soaps, tooth-pastes, deodorants, etc. 3. Used topically in the treatment of dermatitis, impetigo, carbuncles.
(*iv*) Chlorhexidine	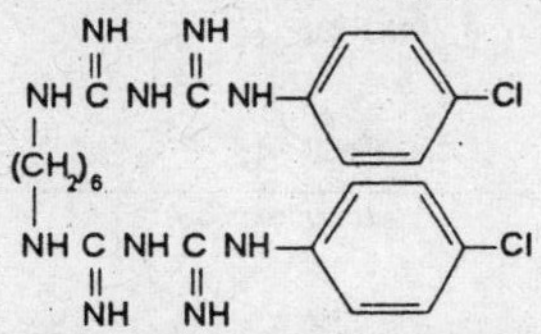	1. It is used in the preparation of antiseptic creams like savlon. 2. It is used in dusting powders, creams, ointments, etc.
2. Halogen Compounds (*i*) Chloramines (*a*) Chloramine-T	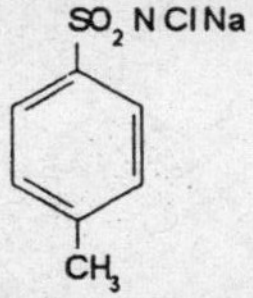	1. Used as disinfectant lotion in diseases like measels, scarlet fever. 2. Used as a mouth wash.
(*b*) Dichloramine-T	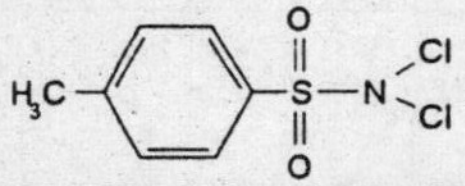	Used as disinfectant for dressing wounds.
(*ii*) Halozone	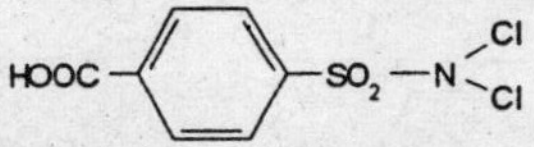	Mainly used for purification of drinking water.
(*iii*) Bleaching powder		Used for purification of drinking water.
3. Dyes (*i*) Crystal Violet	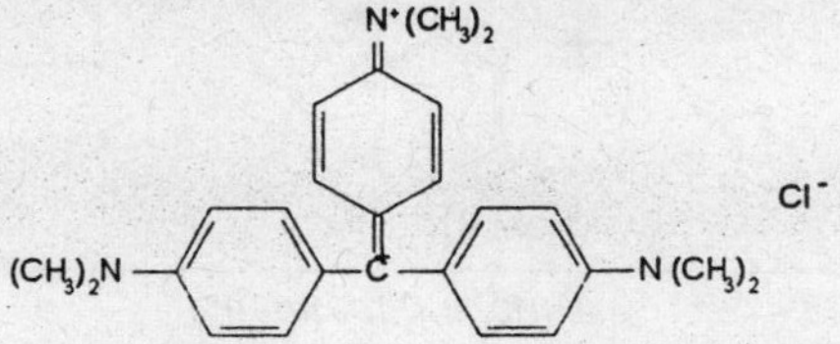	1. It is a potent antiseptic. 2. Used topically in the treatment of burns, boils, skin infections. 3. Used as an anthelmentic agent.

Name of the Drug	*Chemical Structure*	*Uses*
(*ii*) Brilliant Green	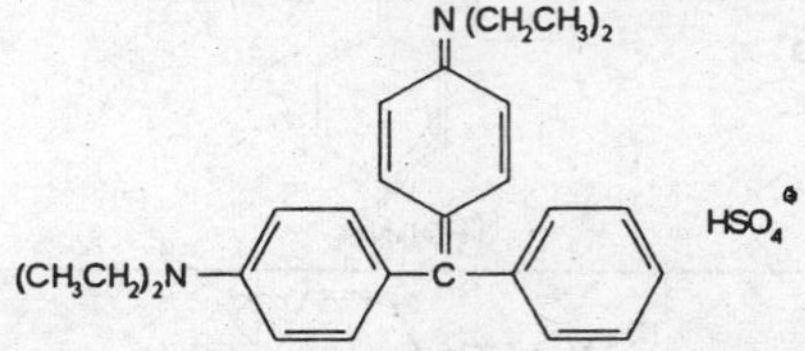 $(CH_3CH_2)_2N$–…; $N(CH_2CH_3)_2$; $HSO_4^{\ominus}$	1. Used for sterilising skin. 2. Used in mycotic infection of the lung by aerosol inhalation.
(*iii*) Methylene Blue	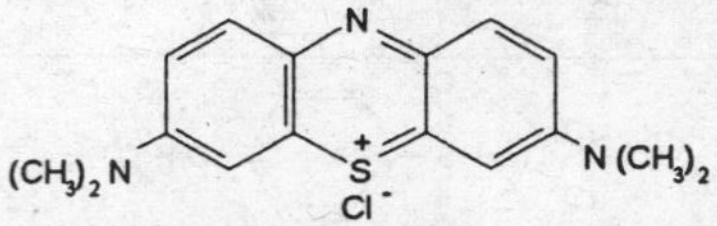 $(CH_3)_2N$; $N(CH_3)_2$; Cl^-	1. Used as an antidote in the treatment of cyanide poisoning. 2. Used as an antiseptic.
(*iv*) Mercurochrome	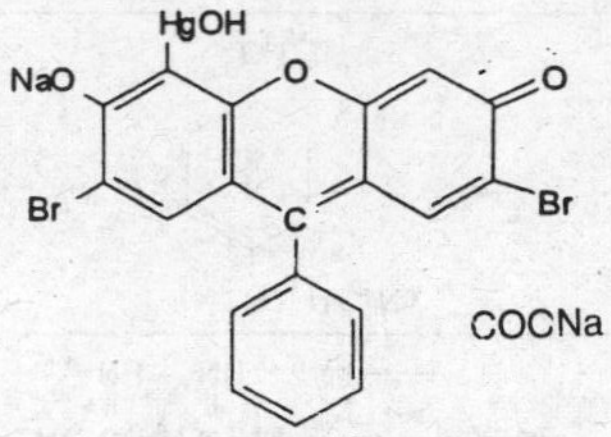 HgOH; NaO; Br; Br; O; COCNa Mercurochrome	1. Used as an antiseptic for skin.
4. Organic Mercurials (*i*) Thermersol	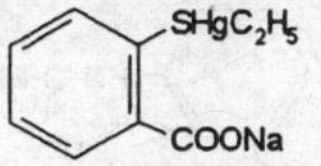 $SHgC_2H_5$; COONa	1. Used for disinfecting skin before surgical operations. 2. Used for sterilizing surgical instruments.
(*ii*) Nitromersol	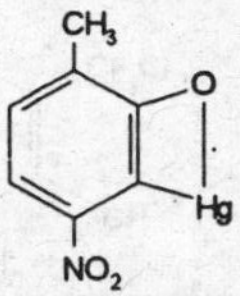 CH_3; O; Hg; NO_2	1. It is powerful antiseptic. 2. Used in eye ointment.
(*iii*) Phenylmercuric Nitrate		1. Used as an antibacterial and antifungal agent. 2. Used as a preservative for injections.
5. Formaldehyde and Its Derivatives (*i*) Formaldehyde	HCHO	1. Highly effective against several bacteria, viruses and fungi. 2. Used as preservative for pathological specimen.

Name of the Drug	*Chemical Structure*	*Uses*
(*ii*) Gluteraldehyde	$H-\overset{O}{\overset{\Vert}{C}}-(CH_2)_3-\overset{O}{\overset{\Vert}{C}}-H$	1. It is a potent bacteriocide and fungicide. 2. Used for sterilizing surgical instruments.
6. Nitrofuran Derivatives (*i*) Nitrofurazone	O_2N–(furan ring, O)–$CH=NNHCONH_2$	1. It is an effective antiseptic against both gram positive and gram negative organisms. 2. It is used as an antiseptic in burns, ulcers, skin diseases.
(*ii*) Nitrofuroxime	O_2N–(furan ring, O)–$CH=NOH$	Used as an antiseptic in vaginal infections.
7. Cationic Surface Active Agents (*i*) Benzalkonium Chloride	$C_6H_5-CH_2-\overset{CH_3}{\underset{CH_3}{N^{\oplus}}}-R\ \ Cl^{\ominus}$ $R = C_8H_{17}$ to $C_{18}H_{37}$	1. Used as detergent, karatolytic and as an emulsifying agent. 2. Used as an antiseptic for wounds , burns and in obstetrics. 3. Used as a disinfectant in baby diapers and surgical instruments. 4. Used as preservative for eye drops.
(*ii*) Cetrimide	$[CH_3(CH_2)_{13}N^+(CH_3)_3]Br$	1. Used as preservative for eye drops. 2. Used as an antiseptic for sterilizing surgical instruments, wounds, burns, etc.
(*iii*) Cetylpyridinium Chloride	(pyridinium ring)$\overset{+}{N}(CH_2)_{15}CH_3\ \ Cl^{\circ}$	1. Used as an antiseptic and germicide. 2. Used in lozenges and mouth wash. 3. Used for cleaning skin before an operation.
(*iv*) Domiphen Bromide	Dodecyldimethy 1-2-phenoxyethyl ammonium bromide.	1. Used as an antiseptic detergent. 2. Used in surgical dressings. 3. Used as lozenges for disinfecting mouth and throat.
(*v*) Dequalinium Chloride	H_2N–(quinolinium, CH_3)–$\overset{+}{N}-(CH_2)_{10}-\overset{\oplus}{N}$–(quinolinium, CH_3)–NH_2 $\ 2Cl^{\ominus}$	1. Used as an antibacterial and antifungal agent. 2. Used as an antiseptic for throat infections.

Anaesthetics

Name of the Drug	*Chemical Structure*	*Advantages*	*Disadvantages*
I. General Anaesthetics 1. Volatile Anaesthitic (*i*) Ether	$C_2H_5OC_2H_5$	1. It induces surgical anaesthesia without any preanaesthetic medication. 2. It produces muscular relaxation. 3. It does not modify blood pressure. 4. It is easier to administer. 5. It is economical and quite stable.	1. The induction is slow. 2. It boils at low temperature. 3. Recovery is slow. 4. Nausea and vomiting appear. 5. Ether convulsions may occur in children.
Derivatives of Ether (*a*) Methylpropyl Ether (*b*) Vinyl Ether	Methyl-n propyl ether $CH_2{=}CH{-}O{-}CH{=}CH_2$	1. Has less adverse effects. 2. It is more potent than ether. 3. Useful in dental procedures. 4. Used for minor operation of shorter duration.	1. It damages lever. 2. It is flammable.
(*ii*) Chloroform	$CHCl_3$	1. Induction is prompt. 2. Vapours are not flammable. 3. Used for obstetrical anaesthesia.	1. It has lever and cardiac toxicity.
(*iii*) Halothane	$F_3C{-}CH(Br)(Cl)$	1. It is not flammable and does not irritate the respiratory passage. 2. It is a potent anaesthetic. 3. Post operative vomiting is low. 4. It relaxes brochial muscle. 5. It is used during plastic surgery.	1. It causes cardiovascular depression, hence blood pressure decreases. 2. It depresses respiration. 3. It is poor analgeric. 4. It has to be given through a special apparatus. 5. It casuses hepatic damage.
(*iv*) Trichloro-ethylene	$H(Cl)C{=}C(Cl)_2$	1. It is a potent analgsic. 2. It is used for short surgical procedures. 3. Used as an analgesic for migraine pain, angina pectrois child brith, etc.	1. It breaks down in presence of soda lime. 2. It produces poor muscle relaxation. 3. It is hepatotixic and produce teachypnoea.

Name of the Drug	*Chemical Structure*	*Advantages*	*Disadvantages*
(*v*) Ethylchloride	CH_3CH_2Cl	1. It can be used as local anesthetic for minor surgeries. 2. As general anaesthetic it produces anaesthesia within 1 to 2 minutes. 3. Recovery is quick.	1. It may damage liver. 2. It does not produce adequate muscle relaxation. 3. Its margin of safety is low. 4. It is difficult to maintain steady depth of anesthesia.
Gaseous Anaesthetics (*vi*) Nitrous Oxide	N_2O	1. One of the safest anaesthetic agent. 2. Non-inflammable and non-irritating. 3. Rapid induction and rapid recovery. 4. Pronounced analgesic action.	1. It is not a potest anaesthetic. 2. May produce violent excitation Hypoxia. Increases cardiac irregularties.
(*vii*) Cyclopropane	CH_2 / CH_2—CH_2 (ring)	1. It is a potent anaesthetic and has fast induction. Maintains blood pressure. 2. It does not produce irritation of respiratory passage. 3. Good muscle relaxant. 4. Valuable anaesthetic for major thoracic and abdominal surgeries. 5. Does not produce visceral damage.	1. It may produce cardiac irregularities. 2. Depresses respiratory centres. 3. It increase capillary oozing. 4. Anaesthetic mixture of cyclopropane and oxygen is explosive.
2. Non-volatile Anaesthetics (*i*) Thiopental Sodium	C_2H_5, $CH_3CH_2CH_2CH(CH_3)$ on C; CO—N=C—SNa; CO—NH-	1. It is non-explosive. 2. Its induction is rapid. 3. Tendency of nausea, vomiting and excitement is less. 4. Does not effect respiration. 5. No post-anaesthetic complications.	1. Needs constant supervision to avoid an overdose. 2. Depresses vasomotor centre and myocardium. 3. Does not produce muscular relaxation. 4. Sometimes it may produce fatal reaction like cessation of respiration.
(*ii*) Methohexitone	H N, ONa, O, N, O; $CH_2=CHCH_2$, $CH_3CH_2CH_2CH(CH_3)$	Similar like that of Thiopental Sodium.	Similar like that of Thiopental Sodium.

Name of the Drug	*Chemical Structure*	*Advantages*	*Disadvantages*
(*iii*) Propanidid	Benzene ring bearing $CH_2COOHCH_2CH_2CH_3$, OCH_3 and $OCH_2CON(C_2H_5)_2$	1. It is a short acting anaesthetic. 2. It is an analgesic.	1. Fall in blood pressure, respiratory depression. 2. Produces nausea, vomiting, headaches, excessive salivation.

The Esters

Name of the Drug	*Chemical Structure*	*Advantages*	*Disadvantages*
II. Local Anaesthetics 1. Cocaine	Bicyclic ring (CH_2–CH_2, N–CH_3 bridge, CH, CH_2) bearing $COOCH_3$ and CH–O–C(=O)–C$_6$H$_5$	1. It is a surface anaesthetics. 2. It stimulates central nervous system.	1. It is a drug of addiction. 2. It dilates the pupils and raises the blood pressure. 3. Produces euphoria and fatigue.
2. Benzocaine	Benzene ring bearing $COOC_2H_5$ and NH_2 (para)	1. Used in throat lozenges. 2. Used for relieving pain of gastric ulcers. 3. It is a surface anaesthetic.	
3. Procaine	Benzene ring bearing $COOCH_2CH_2N(C_2H_5)_2$ and NH_2 (para)	1. It has vasodilator property. 2. It is non-irritant and non-habit forming drug.	It is poorly absorbed from mucous membrane.
4. Amethocaine	Benzene ring bearing $COOCH_2CH_2N(CH_3)_2$ and $NHCH_2CH_2CH_2CH_3$ (para)	1. It is used topically as eye drops. 2. It is used as spinal and nerve block injections.	It is not suitable for operations of shorter duration.

Name of the Drug	*Chemical Structure*	*Advantages*	*Disadvantages*
5. Proxymetacaine	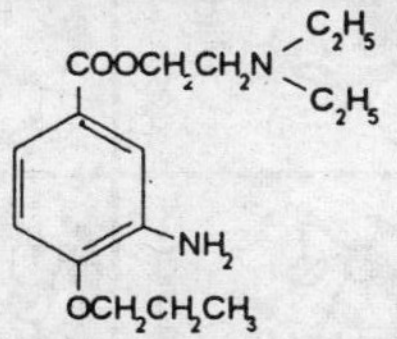 $COOCH_2CH_2N(C_2H_5)_2$; NH_2; $OCH_2CH_2CH_3$	1. It is a potent anaesthetic. 2. Used as topical local anaesthetic in opthalmic preparations.	

The Amides

Name of the Drug	*Chemical Structure*	*Advantages*	*Disadvantages*
1. Lignocaine	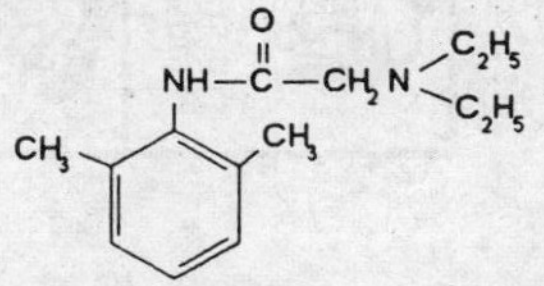 $NH-C(=O)-CH_2N(C_2H_5)_2$; CH_3; CH_3	1. It is resistant to hydrolysis. 2. It has quick set of action. 3. It is good analgesic agent and is used in dental surgeries.	It causes drowsiness.
2. Cinchocaine hydro-chloride	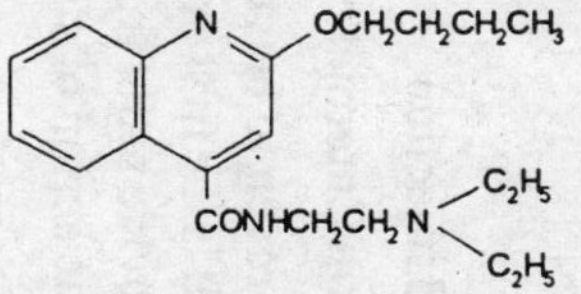 N; $OCH_2CH_2CH_2CH_3$; $CONHCH_2CH_2N(C_2H_5)_2$	1. It can be used locally in the ointment form. 2. It can also be used as solution form as injection for spinal alaesthesia.	

Appendix III

Important Drugs in Pharmaceutical Chemistry

1. Aspirin

Available as : Tablets and syrup.

Sold as : Asabuf, Disprin, Micropyrin, Winsprin etc.

Used in the Treatment of : It is used in the treatment of common headaches, muscular aches. It provides relief from cold and flu symptoms, relief of sore throat. Rheumatoid Arthritis and acute rheumatic fever.

Possibly Effective : It is proven to be effective to a certain extent in the prevention of strokes and heart attacks when used is very low doses [300 mg a day].

Common Side Effects : It can cause dizziness, nausea and vomiting skin rashes, stomach irritation and ringing in the ears as illustrated above.

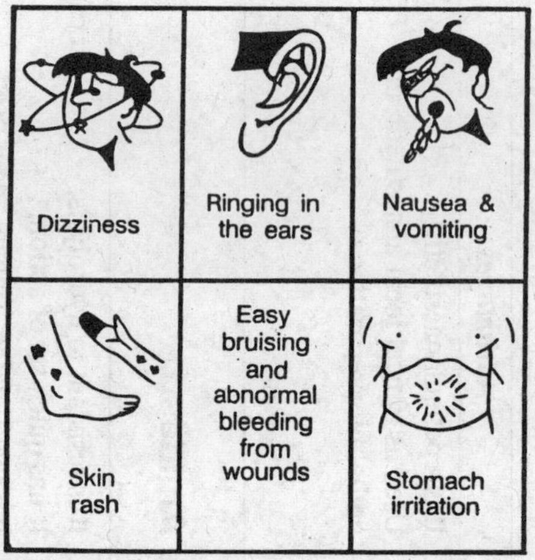

Warning : Aspirin can produce abnormal bleeding tendencies in some patients. Alteration of the colour of the stools to a black or dark nature indicates bleeding within the stomach.

2. Chloramphenicol

Available as : Syrup, capsules and injection.

Sold as : Biomycetin, Oteck, Paraxin Enterojet etc.

Used in the Treatment of : It is proven to be very effective in curing Typhoid and it is the first drug of choice. It is also proven to provide relief from infections that are resistant to the action of simpler antibiotics.

Common Side Effects : It can cause headaches, sore mouth, rashes mild depression, diarrhoea, nausea and vomiting.

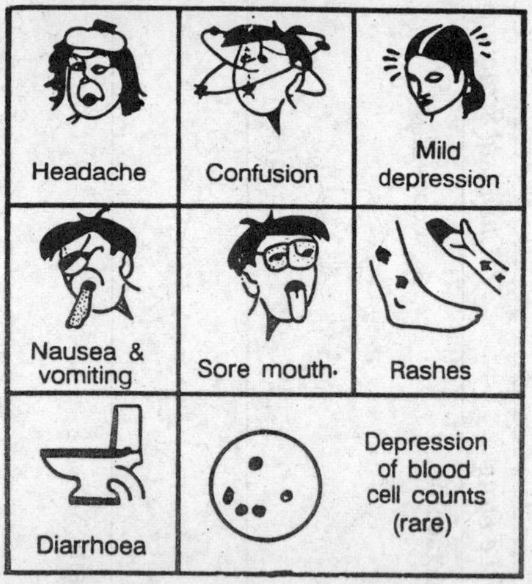

Warning : Since this can cause serious side effects, such as depression of blood cell counts, the drug should be used with caution and only under close medical supervision. In rare cases this can cause blood toxicity which can be fatal. Patients on this drug must have their blood counts periodically checked.

3. Paracetamol

Available as : Tablets, syrup and injection.

Sold as : Calpol, Metacin, Crocin, Dolapar, Zymol Combiflam etc.

Used in the Treatment of : It is a very effective antipyretic drug, used in reduction of fever, to provide relief from muscle and joint pains, colds, flu symptoms, common headaches and also to provide relief from painful mensuration.

Common Side Effects : Side effects of paracetamol are usually very rare and even if present, they are very mild.

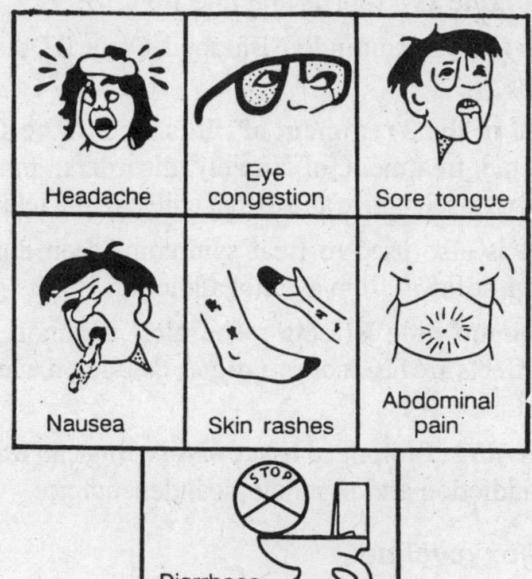

4. Trimethoprin/Sulfamethoxazole

Available as : Tablets, syrup and injection.

Sold as : Actiprin, Bacteriam, Esprin etc.

Used in the Treatment of : It is used to provide relief from common and simple infections of the urinary tract, throat, ear and chest. It is used to provide relief from exacerbations of chronic bronchitis.

It is proven to be very effective in curing "Shigeela" diarrhoea caused by a bacterium.

Common Side Effects : Side effects of the drug are rare and usually mild. The common side effects are headache, eye congestion, nausea, sore tongue and abdominal pain.

5. Ibuprofen

Available as : Tablets and syrup.

Sold as : Anaflena, Brufen, Ibudix etc.

Used in the Treatment of : This is a popular pain reliever, and used in the treatment of common arthritic problems like spondylitis rhueumatoids etc. It is used to provide relief from symptoms of diseases such as pain and inflammation. It is also used to provide relief from painful menstruations.

Common Side Effects : Some of the mild side effects of this drugs are that it causes headaches, skin rashes, nausea, dizziness and blurred vision.

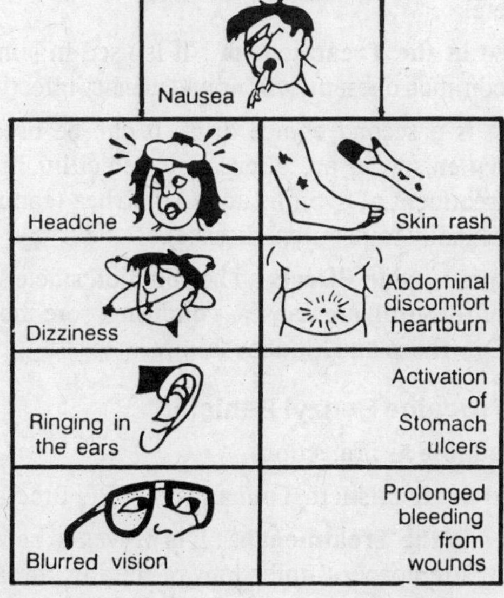

6. Gentamycin

Available as : Injection.

Sold as : Bactigen, Garamycin, Genticyn, Primicen etc.

Used in the Treatment of : It is generally required for the treatment of severe infections that will not respond to less powerful antibiotics.

It is proven to be effective in curing serious infections involving the urinary tract, the lungs, the bones, the nervous system etc.

Common Side Effects : The most common and serious side effects of these drugs are damage to the kidney and damage to the nerve of hearing.

Warning : Gentamycin is a very powerful antibiotic that has a great potential for serious side effects.

The damage to the inner ear and the kidney damage that occur due to the intake of Gentamycin are both irreversible.

7. Diazepam

Available as : Tablets and injection

Sold as : Butacontindon, Butapud, Conaril Dolagin etc.

Used in the Treatment of : It is used in the short time treatment of anxiety disorders, muscle spasms. It is also and to that epilepsy and tetanus.

It is also used to treat symptoms seen during withdrawal from alcohol addiction.

Common Side Effects : The most common side effects are headaches, fatigue, confusion, nausea etc.

Warning : Prolonged usage of this drug can lead to addiction and psychological dependence.

8. Doxycycline

Available as : Capsules.

Sold as : Akini, Alcyclin-O, Deklin, Ebdox Leen etc.

Used in the Treatment of : It is used in simple, common chest infections and urinary infections.

It is a second choice drug; it can be used in patients who are allergic to penicillin in the treatment of Syphilis and Gonorrhea (common sexually transmitted disease).

Common Side Effects : The common side effects of this drug are nausea, vomitting, sore mouth, Diarrhoea and kindney failure.

9. Procaine Benzyl Penicilin

Available as : Injection.

Sold as : Dicristicin, Oninamycin, PPF, Procillin Forte.

Used in the Treatment of : It is proven to be very effective in the treatment of streptococeal infections of the ear, throat and lungs. It is used to treat common pneumonia, common skin infections such as boils etc.

It is also used to treat common and sexually transmitted diseases like Syphilis and Gonorrhoea.

Common Side Effects : The most common side effects of this drug are skin rashes, anxiety, confusion, agitation, weakness and a feeling of "impending doom."

Warning : The use of this drug can produce an allergic shock state known as "anaphylanis", this is rare, and it is very serious and can even result in death if not treated promptly.

10. Penicillin-V

Available as : Tablets and syrup

Sold as : Gystapin-V, Fenocin, Kaypen Pensol-V, Veripen.

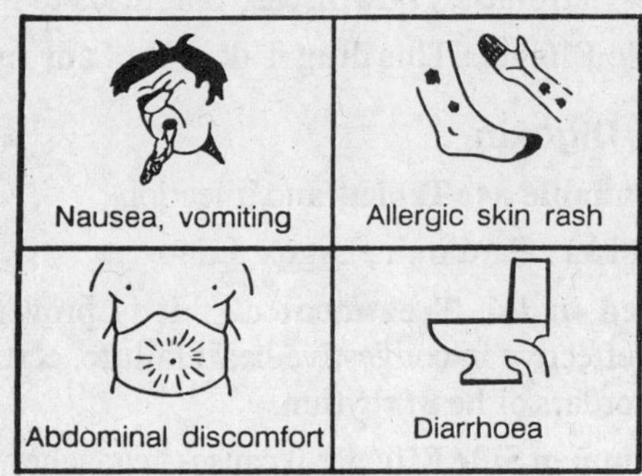

Used in the Treatment of : It is used in the treatment of steep-to-coccal throat infections, pneumonia, common skin infections like boils etc. It is also used to treat mild ear infections.

Common Side Effects : The most common side effects of this drug are nausea, vomiting, allergic skin rashes, diarrhoea and abdominal discomforts.

Warning : The use of drug can produce an allergic shock state known as "anaphylanis". Though this condition is rare but is very serious and can result in death if not treated promptly.

11. Erythromycin

Available as : Capsules, syrup and injections.

Sold as : Ettocin, E-mycin, Ero-B, Erthrocin, Ethimycin etc.

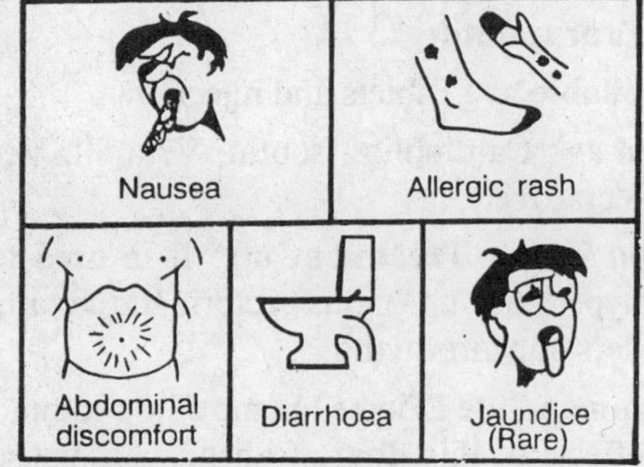

Used in the Treatment of : It is proven effective in the treatment of Mycoplasneapruenconia—a rare form of lung infection and in the treatment of intestinal amoebiasis.

Erythromycin is of great value as an alternative to penicillin if patient is allergic to penicillin.

It is also used in the treatment of Syphilis and Gonorrhea—a common sexually transmitted disease.

Common Side Effects : The side effect of this drug are simple respiratory infections and skin infections. It also causes nausea, diarrhoea and allergic rashes.

12. Tetracycline

Available as : Capsule and injection.

Sold as : Alcyclin, Hostacydine, Reoinen Tetramac, U-cycline etc.

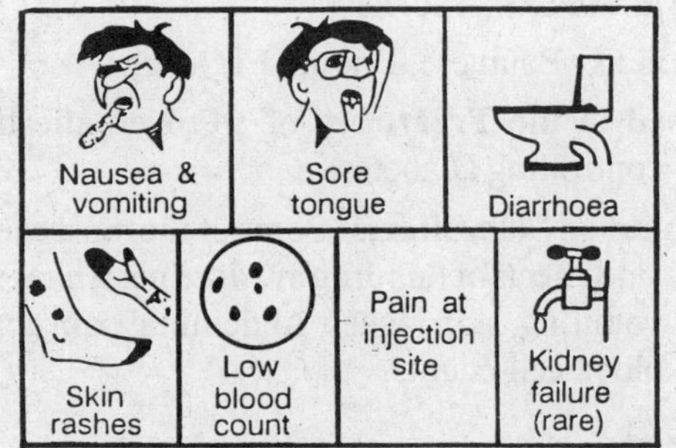

Used in the Treatment of : It is used in simple, common, chest infections, urinary infections, and it is used along with other standard drugs in the treatment of intestinal amoebiasis. It is also used in the treatment of common sexually transmitted diseases like Gonorrhea and Chancroid.

It is a drug of second choice and can be used in patients who are allergic to penicillin in treating sexually transmitted diseases.

Common Side Effects : The most common side effects are nausea, sore tongue, diarrhoea, skin rashes and low blood count. In rare cases it might even cause kidney failure.

13. Ranitidine

Available as : Tablet and injection.

Sold as : Aciloc, Histac, Ranitin, Rantac, Ultac, Rintid etc.

Used in the Treatment of : Ulcers of the stomach, erophageal reflux, and in curing Zollinger-Ellison syndrome (a rare disease that leads to the production of very large amounts of stomach acid).

Side Effects : This drug is devoid of any major side effect and if any they are usually mild.

14. Digoxin

Available as : Tablets and injection.

Sold as : Cardionin, Digox, Lanoxin.

Used in the Treatment of : It is proven to be effective in congestive heart failure, certain disorders of heart rhythm.

Common Side Effects : It causes headaches, nausea and vomiting and general weakness. One of the most serious side effect of this drug is that it decreases or increases the heart beat to a very great extent which can prove lethal.

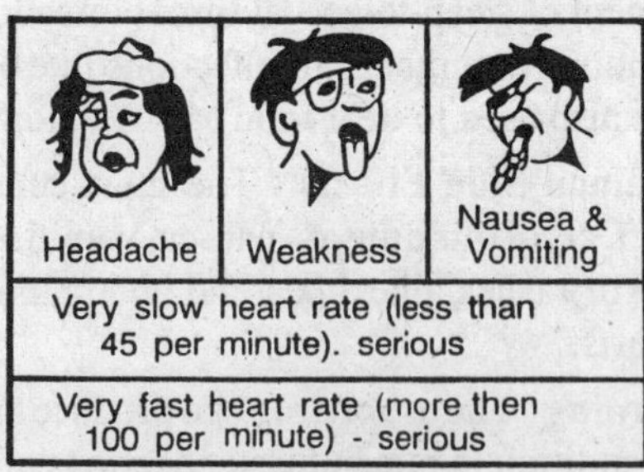

15. Verapamil

Available as : Tablets and injection.

Sold as : Cardioplin, Isoptin, Veramil, Veraptin, Veratril etc.

Used in the Treatment of : It is used to treat hypertension, Angina pectoris and certain disorders of heart rhythm.

Common Side Effects : Some of the common side effects of this drug are headache, nausea and vomiting, fatigue, dizzeness, low pressure (Low B.P.), constipation and in some cases very slow heart rate (less than 45/min) which is quite serious.

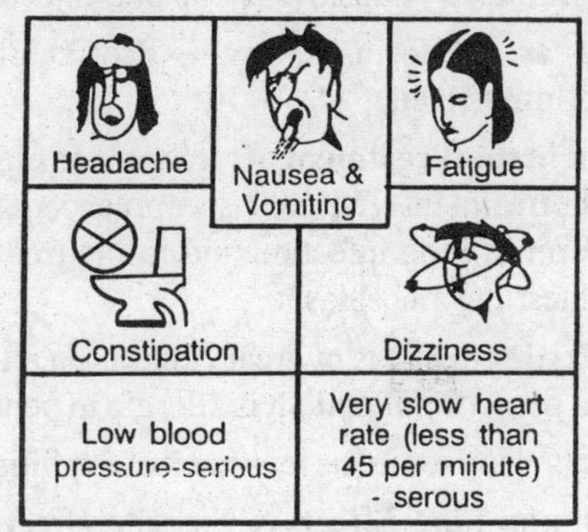

16. Glibenclamide

Available as : Tablets.

Sold as : Batanese, Daonil, Erglucon

Used in the Treatment of : Proven effective in controlling Diabetes.

Common Side Effects : Some of the most common side effects of this drug are dizziness, nausea and vomiting, skin rashes, Abdominal pain, and low blood cell counts.

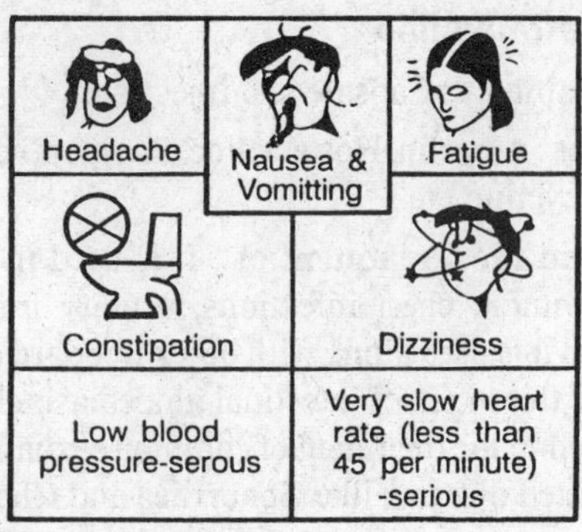

17. Cephalexin

Available as : Capsule and syrup.

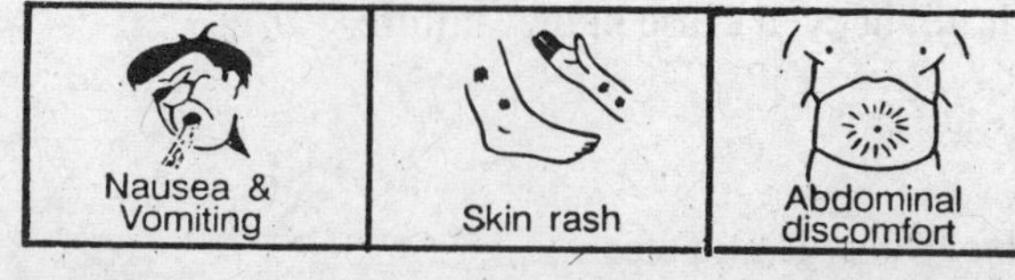

Sold as : Alcephin, Cephxin, Sponidex etc.

Used in the Treatment of : It is proven to be very effective in treatment of simple common lung infections, in treatement of middle ear infections, urinary infections and also in the treatment of common skin and soft tissue infections.

Common Side Effects : Some of the side effects are nausea, skin rashes, diarrhoea and abdominal discomfort.

Warning : It is to be used with caution in patients who are allergic to penicillin.

18. Mebendazole

Available as : Tablets.

Sold as : Besantin, Ebea, Kit-kat, Metrazole, STA-500 etc.

Used in the Treatment of : It is used to treat worm infestations of the intestinal tract.

Common Side Effects : It causes mild abdominal pain and diarrhoea.

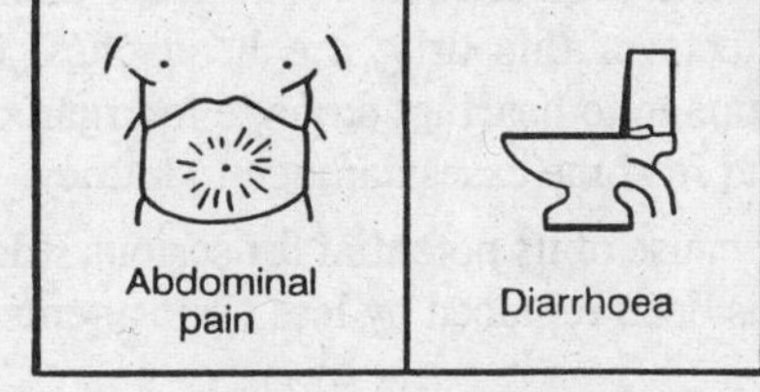

Warning : Not to be used in pregnant women, it can produce harmful effect and malformations on the developing foetus.

19. Chloroquine

Available as : Tablets, syrup and injection.

Sold as : Amicline, Ciplaguin, Enaguin, Resodin

Used in the Treatment of : It is proven to be effective in the treatment of acute attacks of common varieties of malaria and amoebiases.

Common Side Effects : It causes headaches, blurred vision, nausea, itching diarrhoea and abdominal cramps.

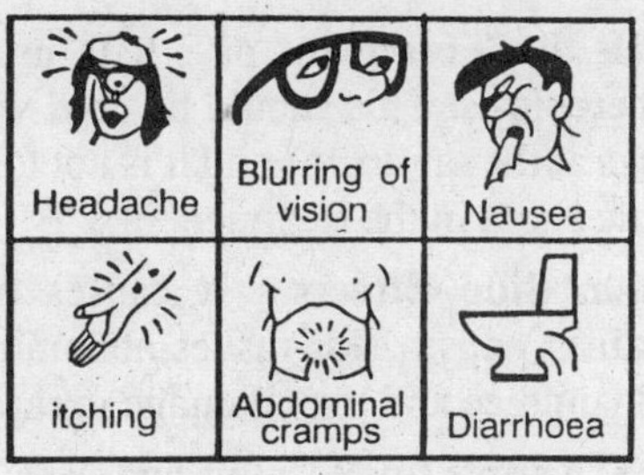

20. Diphenyldantoin/Phenytoin

Available as : Tablets, injections.

Sold as : Corutal, Dilartin, Epiliptin, Garoin etc.

Used in the Treatment of : All varieties of epilepsy.

Common Side Effects : It causes, nausea, headache, dizziness and confusion.

21. Metronidazole

Available as : Tablets, syrup and injections.

Sold as : Alfunnet, Anuvan, Dinite, Furan, Furzol, Unimezol etc.

Used in the Treatment of : It used to treat amoebiasis, Genital infections caused by Trichomonas, Anarobic infections caused by bacteria belonging to the group known as Bacteroides.

Common Side Effects : It causes, nausea, vomiting, abdominal cramps and skin flushing may occur.

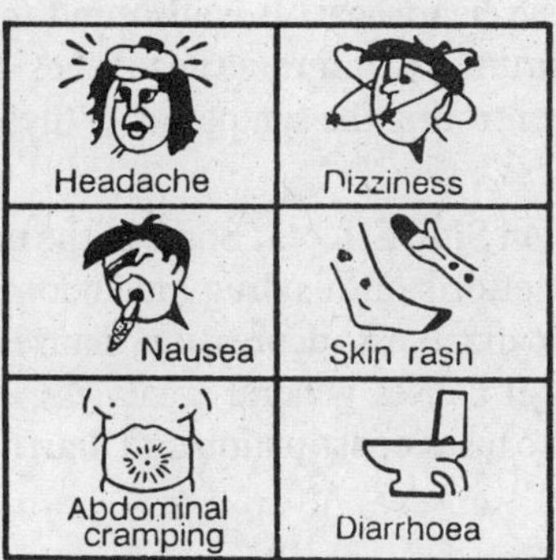

Warning : Patient's advised not to take alcohol while on this drug.

22. Streptomycin

Available as : Injections.

Sold as : Ambitrya - S, Imosec, Sugacin etc.

Used in the Treatment of : It is proven effective in treatment of tubuculosis, Pleague, Tularenia (a very rare disease) and also in the infections of the heart valves and linings.

Common Side Effects : The most common side effects of this drug are headaches, dizziness, damage to hearing, damage to organ of balance and in some cases damage to kidney.

Because of its potential for serious side effects it has been replaced by less toxic agents

Headache	Dizziness
Damage to hearing	Damage to organ of balance
Incoordination	Kidney damage

23. Rifampicin

Available as : Capsules and syrup.

Sold as : Alkil, Anticox, Arzide, Conid Rifadin, Tibicin etc.

Used in the Treatment of : It is used to treat Tuberculosis. This should be used with 2, or 3 drugs at the same time and it is not to be used as a sole agent in the treatment of T.B.

Common Side Effects : It causes headaches, dizzines, nausea, skin rashes, abdominal cramps and jaundice and liver damage in rare cases.

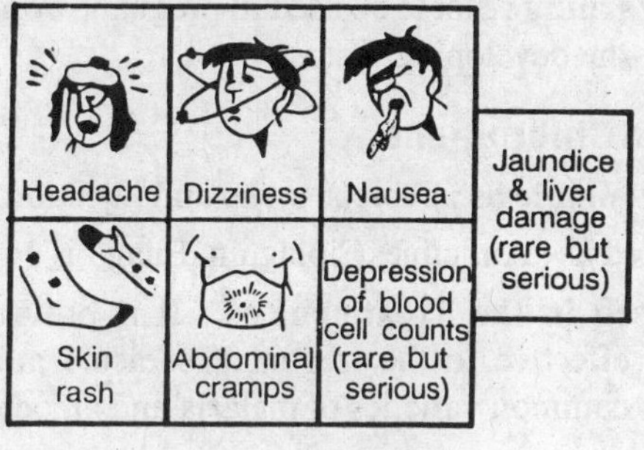

Note : The urine, stools, saliva and sweat of patients taking this drug may become stained red orange.

24. Propranolol

Available as : Tablets, injection.

Sold as : Beta-xipamid, Betazene, Ciplan, Pronol etc.

Used in the Treatment of : It is proven to be effective in treating hypertension (high blood pressure), angina pectoris, certain irregularitics of the heart beat. It is also and to prevent the recurrence of migraine headaches. It is also used in controling the symptoms of thyroid harmone excess.

Common Side Effects : Some of the most common side effects of this drug are abdominal cramps, light dizziness, depression, nausea and vomiting. It causes general weakness, sometimes it also causes constipation and diarrhoea.

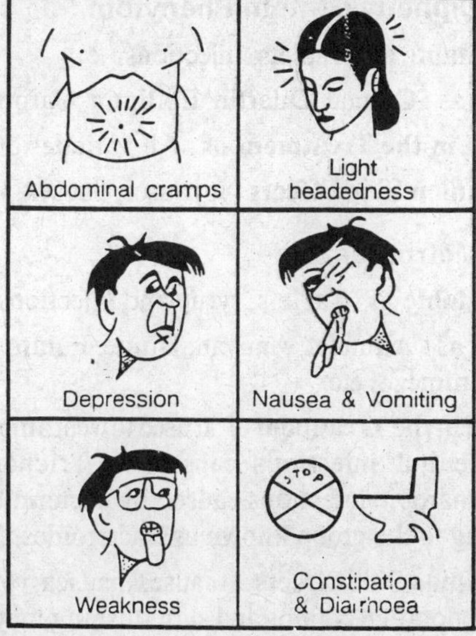

25. Furosemide.

Available as : Tablets and injections

Sold as : Aquamide, Duiral, Lasix, Spiromide, Terbolan etc.

Used in the Treatment of : It is proven to be effective in treating hypertension and diseases that are characterised by retention of excess water in the tissues (edema) such as heart failure.

It is also used to treat kidney failure and liver disorders.

Common Side Effects : Some of the common side effects caused by this drug are headaches, dizziness, nausea and vomiting, loss of hearing, diarrhoea, skin rashes, cramping,. It also causes general weakness.

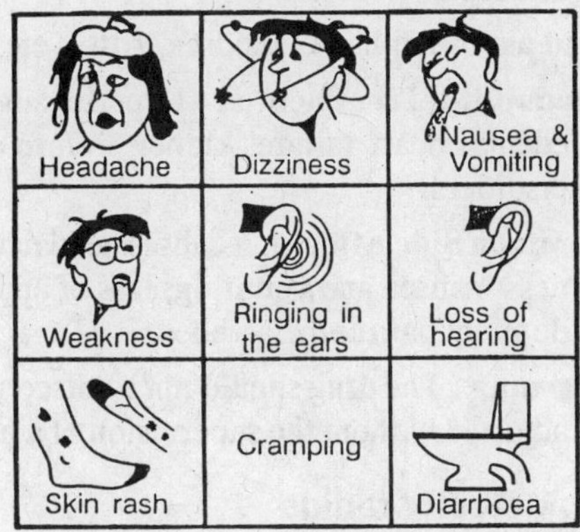

26. Ethambutol

Available as : Tablets and capsules.

Sold as : Bidlebutol, Combutol, Etibi, Etinol, Iditol etc.

Used in the Treatment of : It is found to be effective in the treatment of tuberculosis.

Common Side Effects : Some of the common side effects of this drug are headaches, dizziness, numbness of fingers and toes, nausea, abdominal pain and skin rashes. It causes decrease in visual acuity and it is quite serious.

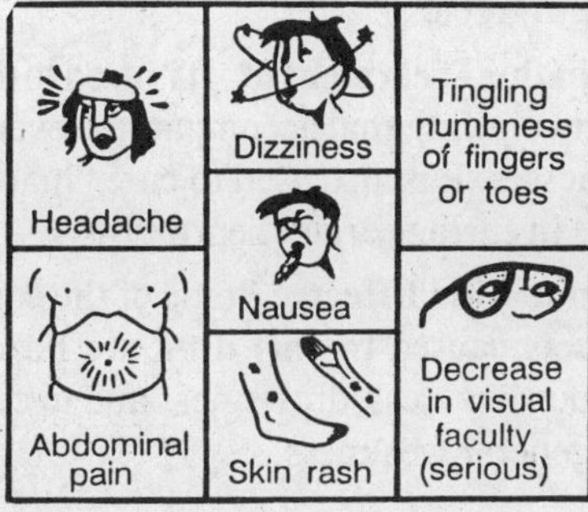

27. Isonicotinic and Hydrazide [INH]

Available as : Tablets, syrup.

Sold as : Anticox, Inapas, Rifakin etc.

Used in the Treatment of : Tuberculosis.

Common Side Effects : It causes nausea, skin rashes, depression of blood cell counts, vision problems, numbness of toes and fingers and in rare cases liver damage (serious).

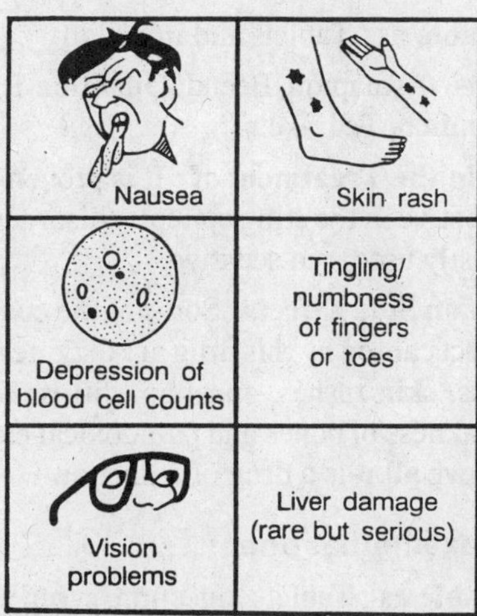

28. Diethylcarbamazine

Available as : Tablets, syrup.

Sold as : Barocide, Citrazan, Hetrazan etc.

Used in the Treatment of : Filarial infections, tropical eosinophilia and also to treat Onchocerciasis (a rare worm infection).

Common Side Effects : It causes headaches, abdominal cramps and dizziness.

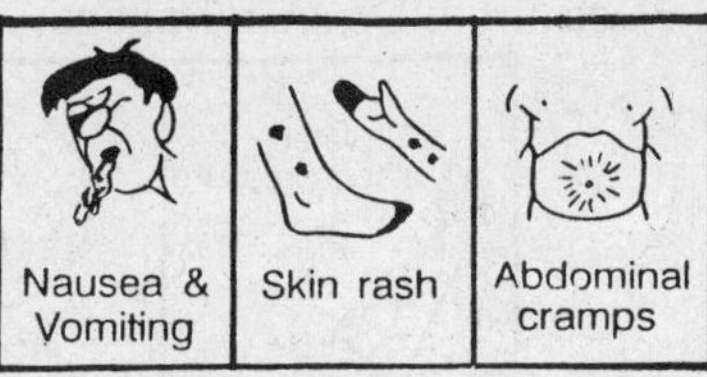

29. Hydrochlorthiazide

Available as : Tablets.

Sold as : Adelphane - Esidrex, Altrip etc.

Used in the Treatment of : Hypertension, to cure edema, heart failure, kidney failure and liver disorders.

Common Side Effects : It causes headaches, dizziness, nausea and vomiting, loss of appetite, abdominal pain and diarrhoea.

Warning : The drug should not be discontinued or adjusted without the supervision of a physician.

Headache
Dizziness
Nausea & vomiting
Loss of appetite
Abdominal pain
Diarrhoea
Allergic skin rash

30. Metoclopranide

Available as : Tablets and injections

Sold as : Artivon, Emaxol, Emenil, Tonid, Reglan, Regurgit etc.

Used in the Treatment of : It is used to treat nausea and vomiting that accompanies several kinds of illnesses. It is also used to cure "hiabris hernia" and in curing certain heartburns.

Common Side Effects : Some of the common side effects caused by this drug are headaches, fatigue, dizziness, drowsiness and in certain cases of general weakness.

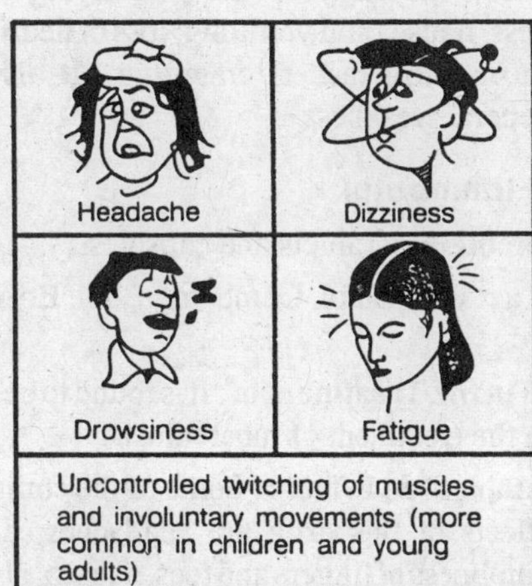

31. Phenobarbitone

Available as : Tablets and injection

Sold as : Asmapion, Beetal, Dilantion-P, Garodin, Luminal, Tedral etc.

Used in the Treatment of : It is proven to be very effective in the sting epileptic disorders and it is mostly used as a sedative.

Common Side Effects : Some of the common side effect caused by this drug are dizziness, drowsiness, skin rashes, anaemia, diffuse body pain, weakness of bones and paradoxical excitability. Above all it is a drug of addiction.

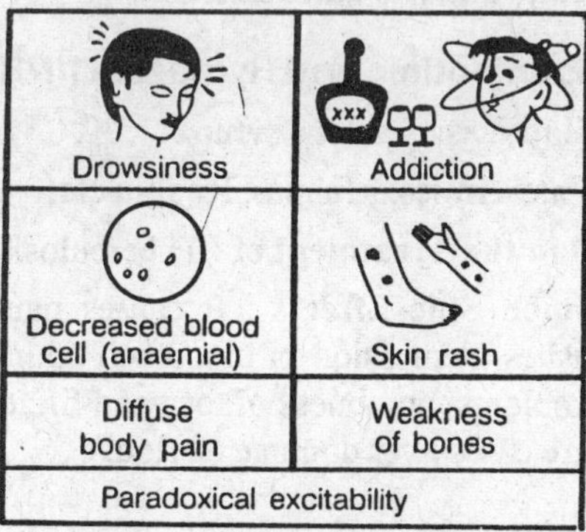

32. Dexamethasone

Available as : Tablets, injection, syrup and ointment.

Sold as : Decadron, Dexabu, Dexona, Mycidex, Topicasone etc.

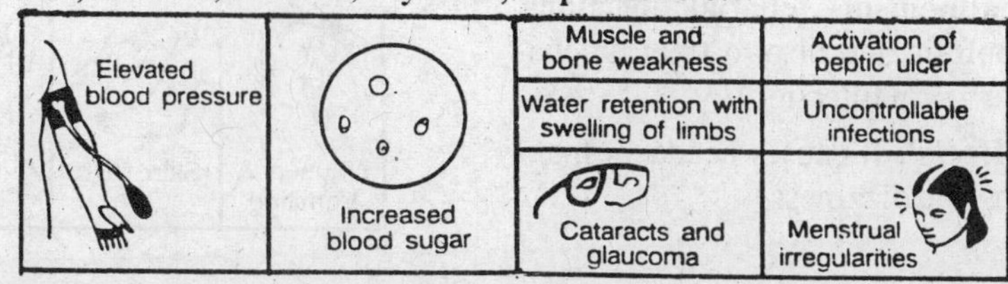

Used in the Treatment of : It finds usage in various disorders of skin, joints, eyes, lungs, blood and bowel as well as in the treatment of allergies and in cancer.

Common Side Effects : When improperly used, side effects are often severe and include, elevated blood pressure, increased blood sugar, uncontrollable infections, cataracts and glaucoma.

33. Pyrazinamide

Available as : Tablets.

Sold as : Alzide, Lynamida, Tricox etc.

Used in the Treatment of : It is a drug of second choice in the treatment of tuburculosis.

Common Side Effects : It causes jaundice, nausea, joint pains, burning urination and joint pains and it worsens diabetic.

Jaundice	Nausea & vomiting
Burning urination	Joint pains
Worsening of diabetic state	

34. Salbutamol

Available as : Tablets, injection and inhalant.

Sold as : Aspul, Bataphil, Butasol etc.

Used in the Treatment of : It is used in the treatment of Bronchise asthma.

Common Side Effects : It causes headache, muscle cramps, palpitations and tremors.

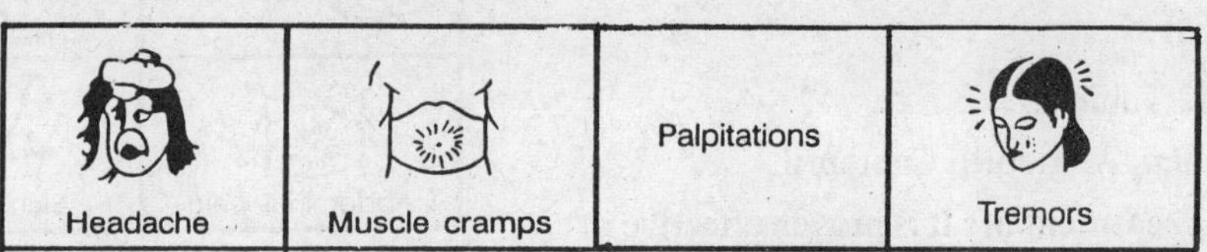

35. Nitroglycerin

Available as : Tablets, injection and ointment.

Sold as : Nitrodura, Nitramack and Nitroparli.

Used in the Treatment of : It is used to treat angina pectoris.

Common Side Effects : It causes headaches, dizziness, skin rash, palpitations and flushing of skin.

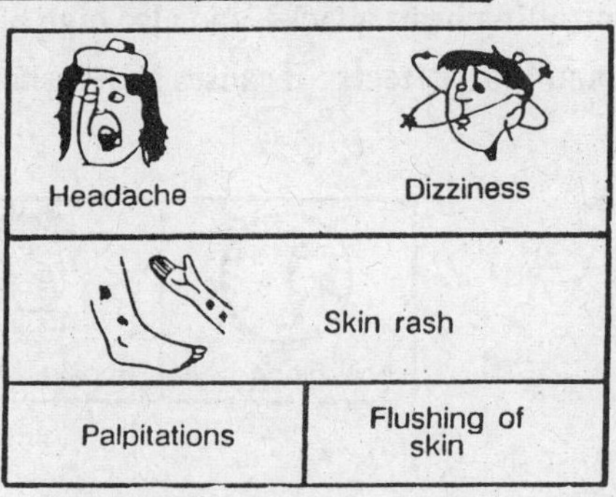

36. Promethazine

Available as : Tablets, injection and syrup

Sold as : Avonial, Phenergan, Tixylix etc.

Used in the Treatment of : It is proven to be effective in treatment of allergic disorder, as a sedative in adults and children, to increase the effectiveness of other pain killers, to prevent vomiting and also to prevent motion sickness.

Common Side Effects : It causes blurred vision, dizziness and dryness of mouth.

37. Theophyline

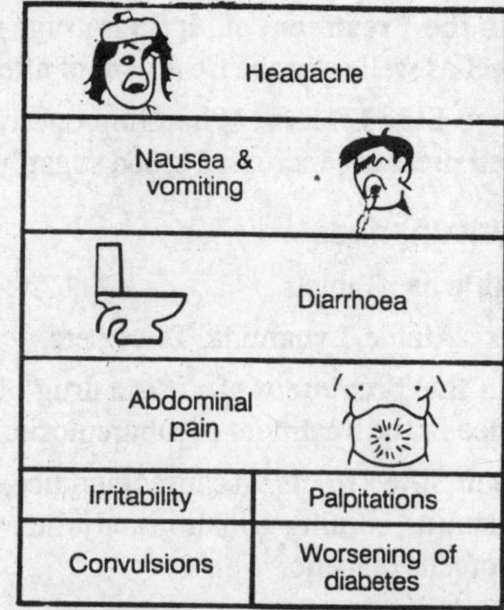

Available as : Tablets, injection and syrup.

Sold as : Asbion, Bcontyl, Tranol, Vert etc.

Used in the Treatment of : It proved to be effective in bronchial asthma.

Common Side Effects : It causes headache, nausea and vomiting, diarrhoea, abdominal pain, irritability, palpitations, convulsions and worsening of diabetes.

38. Nifedipine

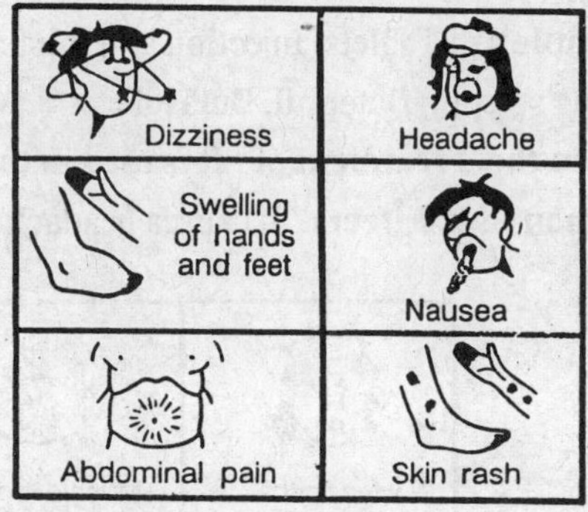

Available as : Tablets and Capsules.

Sold as : Anginal, Calbloc, Calcigard, Depin, Nifedin, Neifalac etc.

Used in the Treatment of : Used in treatment of high blood pressure, angina pectoris and certain disorders of heart rhythm.

Common Side Effects : It causes headache, dizziness, nausea, abdominal pain and skin rashes.

39. Captopril

Available as : Tablets.

Sold as : Aceten, Angiopril, Captopril.

Used in the Treatment of : It is proven effective in controlling heart attacks, and also high blood pressure.

Common Side Effects : It causes skin rashes, dizziness, impaired taste and low blood cells.

40. Carbamazepine

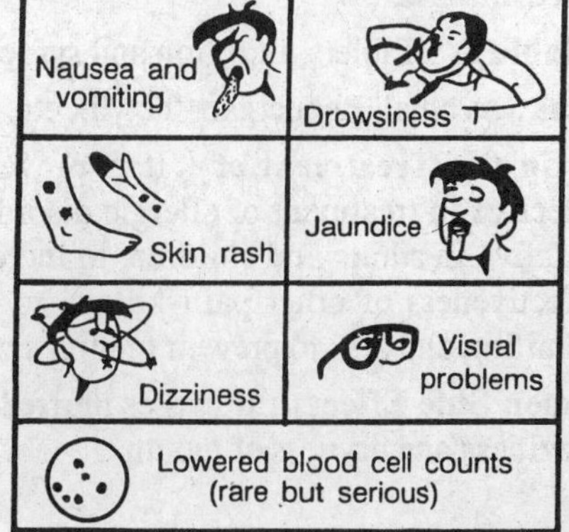

Available as : Tablets.

Sold as : Carbadac, Cartatol, Epitol, Mazetol etc.

Used in the Treatment of : It is proven to be effective in treatment of epiletic disorders, Trigeminal neuralgia (a condition characterised by periodic attacks of severe pain in the region of the face).

41. Diphenylhydantoin (Phenytoin)

Available as : Tablets and injections.

Sold as : Comital, Dilantin, Epilepten, Eptoin, Garoin.

Used in the Treatment of : It is proven to be effective in curing epilipsy.

Common Side Effects :

Dizziness	Headache	Vomiting	Depression of blood counts (rare)
Confusion	Skin rash	Liver damage (rare)	

References

1. Charles R. Craig, Robert E. Stitzel, *Modern Pharmacology*, 3rd Edition, Little Brown and Company, Boston, 1990.
2. Corwin Hansch, Peter G. Sammer, John B. Taylor, Peter D.K. Kennewell, *Comprehensive Medicinal Chemistry*, Pergmon Press, Great Britain, 1990.
3. Bertram G. Katzung, *Basic and Clinical Pharmacology*, Lange Medical Publications, Los Atos, California, 1982.
4. James H. Jandl, George Richards Minot, *Text Book of Hematology*, Little Brown and Company, Boston, 1987.
5. William F. Ganong, Jack and DeLoris Lange, *Review of Medical Physiology*, San Francisco, Appleton and Lange Norwalk, Connecticut, 1993.
6. Goodman, Gilmann's *The Pharmacological Basis of Therapeutics*, 6th Edition, Macmillan Publishing Co., New York, 1975.
7. Arthur C. Guyton, *The Textbook of Medical Physiology*, Mississippi, Saunders Company, 1986.
8. Paul F. Gringer, Robert J. Panzer, Philip Greenland, *Clinical Diagnosis and The Laboratory Tests*, Yearbook Medical Publisher, Chicago, London, 1986.
9. Jacques Wallach, *Interpretation of Diagnostic Tests*, Little Brown and Company, Boston, 1992.
10. Monica Cheersbrough, *Medical Laboratory Manual for Tropical Countries*, Austin Son Ltd., England, 1981.
11. C.L. Ghai, *A Textbook of Practical Physiology*, 3rd Edition, Amritsar, Jaypee Brothers, Medical Publishers, New Delhi, 1987.
12. Donald C. Grove, William A. Randall, *Assay Methods of Antibiotics*, New York, 1955.
13. Ruth R. Levine, *Pharmacology Drug Actions and Reactions*, 4th Edition, Little Brown and Company, Boston, 1990.
14. Howard C. Ansel, Nicholas G. Popovich, Loyd V. Allen, *Pharmaceutical Dosage Forms and Drug Delivery Systems*, Williams and Wilkins, Philadelphia, 1955.
15. Corwin, Peter G. Samnes, John B. Taylor, *Comprehensive Medicinal Chemistry*, Rhone Poulene Ltd., Pergamonn Press, Great Britain, 1990.
16. R. Craig, Robert E. Stitzel, *Modern Pharmacology*, 4th Edition, Little Brown and Company, Boston, 1994.
17. Wesley G. Clark, C. Craig Barter, Alice R. Johnson, *Goth's Medical Pharmacology*, 13th Edition, Mosby Yearbook Publications, 1988.
18. A.V.S.S. Rama Rao, Dr. A. Surya Lakshmi, *A Textbook of Biochemistry For Medical Students*, 7th Edition, VBS Publishers Distributors Ltd., New Delhi, 1994.
19. Sarada Subrahmanyan, K. Madhavankulty, *Textbook of Human Physiology*, 4th Edition, S. Chand & Company Ltd., New Delhi, 1995.
20. David T. Plummer, *An Introduction to Practical Biochemistry*, 2nd Edition, Tata Mcgraw Hill Publishing Company Ltd. New Delhi, 1985.

21. S.P. Singh, *Manual of Biochemistry*, CBS Publishers and Distributors, Delhi, 1988.
22. R.C. Gupta, S. Bhargava, *Practical Biochemistry*, CBS Publishers and Distributors, New Delhi, 1985.
23. Harold Varley, *Practical Clinical Biochemistry*, 4th Edition, Arnold-Heinemann, New Delhi, 1976.
24. Eric E. Conn, P.K. Stumpf, *Outlines of Biochemistry*, 4th Edition, Wiley Eastern Limited, New Delhi, 1987.
25. Lubert Stryer, *Biochemistry*, 2nd Edition, CBS Publishers and Distributors, New Delhi, 1986.
26. Frances Talaska, Fish Bach, *A Manual of Laboratory Diagnostic Tests*, 3rd Edition, J.B. Lippincott Company, Philadelphia, 1988.
27. Norbert W. Tietz, *Clinical Guide to Laboratory Tests*, W.B. Saunders Company, 1990.
28. Parveen Kumar, Michael Clark, *Clinical Medicine*, 3rd Edition, Bailliere, Tindall, 1994.
29. Alex C. Sonnen Wirth, Leonard Jarett, *Gradwohl's Clinical Laboratory Methods and Diagnosis*, 7th Edition, Vol. II, C.V. Mosby Company, 1980.
30. Mario Werner, *Handbook of Clinical Chemistry*, 3rd Edition, Vol. II, CRC Press, Florida, 1985.
31. Butter Worths, *Medical Dictionary*, ELBS, Anchor Brendon Ltd., Great Britain, 1986.
32. R.S. Satoskar and S.D. Bhandarkar, *Pharmacology and Pharmacotherapeutics*, Vol. I and II, Popular Prakashan, Bombay, 1980.
33. Harold I. Kaplan, Benjamin J. Sodock, *Pocket Handbook of Psychiatric Drug Treatment*, Williams and Wilkins, Baltimore, 1993.
34. Jerrold G. Bernstein, *Handbook of Drug Therapy in Psychiatry*, 3rd Edition, John Wright, Boston, 1984.
35. Marcus A. Krupp, Milton J. Chatton, *Medical Diagnosis and Treatment*, Lange Medical Publications, California, 1978.
36. G.R. Chatwal, *Pharmaceutical Chemistry*, Vol. II, 1st Edition, Himalaya Publishing House, Bombay, 1991.
37. Joseph R. Robinson, *Sustained and Controlled Release Drug Delivery System*, Marcel Dekker, Inc., New York, 1978.
38. Gilbert S. Banker, Christopher T. Rhodes, *Modern Pharmaceutics*, Marcel Dekker, New York, 1979.
39. Ronald Borchardt, Arnold Repta, Valentino Stella, *Directed Drug Delivery*, Humana Press, Clifton, New Jersey, 1985.
40. R.L. Juliano, *Drug Delivery Systems*, Oxford University Press, New York, Oxford, 1980.
41. Michael H. Rubinstein, *Pharmaceutical Technology–Controlled Drug Release*, Vol. I, Ellis Horwood Limited, John Wiley and Sons, New York, 1987.
42. Eugene Rosenberg, Irun R. Cohen, *Microbialbiology*, Saunders College, Philadelphia, 1983.
43. B.I. Duerden, TMS Reid, J.M. Jewsbury, D.C. Turk, *A New Short Textbook of Microbial and Parasitic Infection*, Hodder and Stoughton, London, 1988.
44. Avram Goldstein, Lewis Aronow, Sumner M. Kalman, *Principles of Drug Action–The Basis of Pharmacology*, Harper and Row, New York, 1968.
45. Michael G. Gerald, *Pharmacology, An Introduction to Drugs*, Prentice Hall Inc., Englewood Cliffs, New Jersey, 1974.
46. Goldstein, A., L. Aronow and S.M. Kalman, *Principles of Drugs Action*, Hoeber Medical Division, Harper and Row, New York, 1968.
47. DiPalma, *Drill's Pharmacology in Medicine*, 4th Edition, McGraw Hill, New York. 1971.
48. Ariens, E.J., *Molecular Pharmacology*, Academic Press, New York, 1964.
49. Cambell, G.D., *Oral Hypoglycemic Agents*, Academic Press, New York, 1969.
50. T.S. Casaky, Byron A. Barnes, *Cutting's Handbook of Pharmacology–The Actions and Uses of Drugs*, Appleton–Century–Crofts/Norwalk, Connecticut, 1984.

51. Melmon, K.L., Morrelli, H.F., *Clinical Pharmacology: Basic Principles in Therapeutics*, 2nd Edition, Macmillan Publishing Co. Inc., New York, 1978.
52. White, A., Handler, P., Smith, E.L., Hills, R.L., Lehman, I.R., *Principles of Biochemistry*, 6th Edition, McGraw Hill Book Co., New York, 1978.
53. John H. Block, Edward R. Roche, Taito O. Soine, Charles O. Wilson, *Inorganic Medicinal and Pharmaceutical Chemistry*, Lea and Febiger, Philadelphia, 1974.
54. William O. Foye, *Principles of Medicinal Chemistry*, Lea and Febiger, Philadelphia, 1981.
55. Leslie G. Chatten, *Pharmaceutical Chemistry*, Vol. 2, Marcel Dekker, New York, London, 1969.
56. Joseph L. Rabinowitz, Ralph M. Myerson, *Topics in Medicinal Chemistry*, Vol. I, Interscience Publishers, Division of John Wiley and Sons, New York, London, Sydney, 1967.
57. Samuel H. Yalkowsky, Anthony A. Sinkula, Shri C. Valvani, *Physical Chemical Properties of Drugs*, Marcel Dekker Inc., New York and Basel, 1980.
58. J. Thuro Carstensen, *Theory of Pharmaceutical Systems*, Vol. II, Academic Press, New York, London, 1973.
59. P.G. Sammes, *Topics in Antibiotic Chemistry*, Vol. I, Ellis Horwood Limited, Division of John Wiley and Sons, Inc. New York, 1977, Vol. 6, 1982, Vol. 5, Vol. 2, 1978.
60. Mary Griffiths, *Introduction to Human Physiology*, 2nd Edition, Macmillan Publishing Co., New York, 1981.
61. J. Robert McClintic, *Basic Anatomy and Physiology of the Human Body*, 2nd Edition, John Wiley and Sons Inc., Canada, 1975.
62: Robert L. Vick, *Contemporary Medical Physiology*, Addison Wesley Publishing Company, Inc., California, 1984.
63. Arthur C. Guyton, *Physiology of the Human Body*, 5th Edition, W.B. Saunders Company, Philadelphia, 1979.
64. Judy L. Hubbard, Derek J. Mechan, *Physiology For Healthcare Students*, Churchill Livingstone, Edinburgh, London, Melbourne and New York, 1987.
65. Myer Freed, The Association of Vitamin Chemists, Inc. *Methods of Vitamin Assay*, 3rd Edition, Interscience Publishers, New York, London, Sydney, 1966.
66. Issac Ghebre Sellassie, *Multiparticulate Oral Drug Delivery*, Marcel Dekker Inc., New York, Basel, Hong Kong, 1994.
67. Manzur-Ul-Haque Hashmi, *Assay of Vitamins in Pharmaceutical Preparations*, John Wiley and Sons, London, New York, Sydney, Toronto, 1973.
68. Frances K. Widmann, *Clinical Interpretation of Laboratory Tests*, 8th Edition, F.A. Davis Company, Philadelphia, 1979.
69. Robert S. Galen, Leslie Brennam, *A Laboratory Diagnosis and Patient Monitoring Clinical Chemistry*, Medical Economics Company, New Jersey, 1981.
70. Alex Kaplan, Laverne L. Szabo, Kent E. Opheim, *Clinical Chemistry Interpretation and Techniques*, 3rd Edition, Lea and Febiger, Philadelphia, 1988.

Glossary

Anaerobic (*i*) Growing most vigorously in an atmosphere which is free from oxygen. (*ii*) Able to sustain life without free oxygen.

Arteries Vessels carrying blood from the heart to the different parts of the body.

Arterioles Tiny arterial branches

Aldosteronism A condition in which the blood contains abnormally high level of aldosterone.

Abrasion A scraping away of a portion of skin or of a mucous membrane as a result of injury.

Anorexia Loss of appetite. Seen in malaise. Commencement of all fevers and illnesses, also in disorders of alimentary tract.

Arrhythimas Lack of rhythm in heart action.

Anopheles A genus of mosquitoes belonging to the family culicidae, order Diptera.

Aphrodisiac An agent which stimulates sexual desire.

AtrophyA wasting due to lack of nutrition of any part.

Ataxia (*i*) Disorder or irregularity. (*ii*) Muscular incoordination that manifested. When voluntary muscular movements are attempted.

Beriberi A deficiency disease associated with malnutrition.

Bile A secretion of the liver.

Bilirubin The orange coloured or yellowish pigment in bile.

Colocynth [Gr. kolokynthe, fruit of citrullus colocynthis] Dried pulp of unripe colocynth fruit (family cucurbitaceae).

Colon That part of a large intestine which extends from the ileocolic junction in the right iliac fossa to end in the pelvis by becoming the rectum in front of the third saeral vertebra.

Convulsion Paroxysms of involuntary muscular contractions and relaxations.

Coronary Encircling, as the blood vessels that supply blood directly to the heart muscles.

Catarrhal Simple inflammation of mucous membrane.

Corrosive Producing corrosion *i.e.* slow disintegration of something by a destructive agent.

Demulcent An agent that will soothe the part or soften the skin to which applied.

Diaphoretic An agent which increases perspiration.

Diarrhoea Frequent passage of watery bowel movements.

Diataxia Ataxia of both sides of body.

Dilatation Expansion of an organ or vessel.

Distention enlarged.

Diuretic Secretion and passage of abnormally large amounts of Urine.

Dosage The amount of medicine to be administered to a patient at one time.

Douche A current of vapour or stream of hot or cold water directed against a part.

Dysarthria Difficult and defective speech due to impairment of the tongue or other muscles essential to speech.

Dysphoria Exaggerated feeling of depression and unrest without apparent cause.

Dyspnea Difficult breathing.

Dyspraxia Pain in performing any function.

Edema A local or generalized condition in which the body tissues contain an excessive amount of tissue fluid.

Emetic An agent that produces vomiting.

Emollient An agent that will soften and soothe the part when applied locally.

Epilepsy Recurrent transient attacks of disturbed brain function.

Epinephrine [Gr. epi upon + nephros, kidney] $C_9H_{13}NO_3$. This substance and norepinephrine are the two active hormones produced by the adrenal medulla.

Emphysema Distention of tissues by gas.

Hallucination False perception having no relation to reality and not accounted for by any exterior stimuli.

Hematology The science concerned with blood and the blood forming tissues.

Hemolysis The destruction of red blood cells with the liberation of hemoglobin which diffuses into the fluid surrounding them.

Hemostatic Checking hemorrhage.

Hyperkalemia Excessive amount of potassium in the blood.

Hyperthyroidism A condition caused by excessive secretion of the thyroid glands.

Ileum Lower three fifths of the small intestines from the jejunum to the ileocecal valve.

Impermeable Not allowing passage of fluid or air.

Ischemia Local and temporary anemia due to obstruction of the circulation to a part.

Jejunum The second portion of the small intestine extending from the duodenum to the ileum.

Jalap [Jalapa or Xalapa, a Mexican city where the drug was exported] The tuberous root of Exogonium punga or Ipomoea purga (family convolvulaceae) used as cathartic.

Lavage Washing out of a cavity.

Leukemia A chronic or acute disease of unknown etiological factors characterized by unrestrained growth of leukocytes and their precursors in the tissues.

Leukopenia Abnormal decrease of white blood corpuscles.

Lobar Preumonia Inflammation of one or more lobes of the lungs.

Malignant Growing worse, resisting treatment, said of cancerous growths.

Microflora The microscopic flora of a region.

Myocardium The middle layer of the walls of the heart, composed of cardiac muscle.

Myocardial ischemia Inadequate circulation of blood to the myocardium, usually as a result of coronary artery disease.

Myopathy Any disease or abnormal condition of striated muscles.

Nephritis Inflammation of the kidney.

Organelle Any cell, organ or specialized part of a protozoan, having some individual function such as that of movement, reproduction or metabolism.

Parasympathetic Craniosacral division of the autonomic nervous system.

Parenteral Introduced subcutaneously, intravenously or by any route other than by way of the digestive tract.

Paroxysm (*i*) A sudden spasm or convulsion of any kind. (*ii*) Sudden emotional state. (*iii*) A sudden periodic attack.

Peripheral Situated at the periphery.

Post synaptic Pertaining to the area on the distal side of a synaptic cleft.

Prophylaxis (*i*) Observance of rules necessary to prevent disease. (*ii*) In dentistry cleaning the teeth's surface.

Proteolytic Proleens are reduced to absorble form.

Protozoa The phylum of the animal kingdom which includes the simplest animals. Most are unicellular.

Reticular Resembling a network.

Spasm An involuntary sudden movement or convulsive muscular contraction.

Sputum Substance expelled by coughing or clearing the throat.

Tachycardiac Afflicted with tachycardia

Tachypnea Abnormal rapidity of respiration.

Tetany A nervous affection characterized by intermittent tonic spasms which are usually paroxysmal.

Toxicity Degree of being poisonous.

Urticaria A vascular reaction of the skin characterized by the eruption of pale evanescent wheals, which are associated with severe itching.

Vascular Composed of blood vessels.

Varicose veins Enlarged twisted veins.

Vermifuge Agent for expelling intestinal worms.

Wilson's disease A rare disease of degeneration of carpus striatum and cirrhosis of the liver characterized by tremulous distortion of the muscles, dysarthria, dysphagia, emotionalism.

MADRAS UNIVERSITY

B.Sc. Degree Course

Branch IV—CHEMISTRY

APPLICATION ORIENTED SUBJECT

PHARMACEUTICAL CHEMISTRY

SYLLABUS

Unit I :

1.1. Introduction : (10) : Importance of Chemistry in pharmacy. Important terminologies used, their meaning–molecular pharmacology, pharmacodynamics, pharmacophore, metabolites, virus antimetabolites, bacteria, fungi, actinomycetes.

Names of drugs : code No., chemical proprietary, trivial, trade. Non proprietary names, synonyms (meaning only) storage of drugs–different temperature conditions.

Mechanism of action of drug types.

Assay-Biological, chemical immunological–statement only. Mechanism : Metabolism of drugs and their effect on pharmacological activity. Absorption of drugs. Drug delivery system, sustained release of drugs.

1.2. Physiological effects of different functional groups in drugs (2)

1.3. Testing of potential of drugs and their side effects.

Indian medicinal plants and trees–adathode, tulsi, thoothuvalai, shoeflower, fia, neem, mango, kizhanelli ocimum, grass and greens (2).

1.4. Causes and symptoms of Common diseases—tuberculosis, asthma, jaundice, piles, leprosy, epilepsy, typhoid, malaria, cholera, filaria. Diagnostic tests for sugar, salt and cholesterol in blood and urine. Detection of poisons and antidotes for them. First aid in emergencies. Use of plaster of Paris (6).

1.5. Important inorganic compounds of Al, P, As, Hg and Fe. Their uses.

UNIT II :

2.1. ANTIBIOTICS : Definition. Structure, uses of chloramphenicolam picillin, streptomycin, tetracycline, rifamycin Macrolides–Erythromycin–proporties and uses.

Structural features – SAR functional group responsible for drug action, structural modification that changes the potency of the above drugs. Conditions for their use as therapeutic agents. Fields of application (6).

2.2. SULPHONAMIDES : Substiutents in the amide group. General properties and drug action. Preporation and uses of sulphadiazin, sulphapyridine, sulphathiazole, sulphafurozole and prontosil (4).

2.3. VITAMINS : Classification. Role in Chemotherapy. Uses. Deficiency and symptoms. Estimation of Vitamins A, B_1 B_2 and C.

2.4. ANTINEOPLASTIC DRUGS : Causes for cancer. Antineoplastic agents–cytotoxic, antimetabolites, plant products, hormones (3).

2.5. HYPOGLYCEMIC DRUGS : Diabetes–types–Causes. control symptoms. Control, Insulin – preparation, uses. Oral Hypoglycemic agents. Sulphonyl ureas (3).

2.6. ANTICONVULSANT AGENTS (2) : Barbiturates, hydantoins, Oxazolidenediones, succinimides, Miscellaneous drugs.

UNIT III :

Antipyretic, analgesics, anti-inflammatory agents (8)

3.1. Classification : Action of analgesics. Narcotic analgesics : Morphine and its derivatives. SAR. synthetic analgesics–pethidine and methadones.

Salicylic acid and its derivatives, indolyl derivatives, aryl-acetic acid derivatives, pyrazole, *p*-aminophenol derivatives–mechanism of action.

3.2 ANTISEPTICS AND DISINFECTANTS (8) : Definition. Standardization of disinfectants. Use of phenols, dyes, chloramines, chlorohexadiene, Organomercurials. Dequalinium chloride, formaldehyde, cationic surface active reagents, Chloraminet – nitrofurazone.

Distinction between antiseptics and disinfectants.

3.3. ANAESTHETICS (8) : Definition. Classification. Uses of volatile anaesthetics–nitrous oxide, ethers, cyclopropane, chloroform, halothane, trichloroethlene, ethyl chloride–storage, advantages and disadvantages. Intravenous anaesthetics–thiopental-sodium, methohexitone, propanidid.

Local anaesthetics : Requisites. Uses of ethers–cocaine, benzocaine, procaine, amethocaine, proxymetacaine. Amides–lignocaine, cinchocaine hydrochloride.

UNIT IV :

4.1. ALKOLOIDS (15) : Sources, isolation and purification, colour reactions and detection–general.

Quinine and morphine – sources, extraction, structure, important features and SAR.

Tranquilizers, sedatives, hypnotics, psychedelic drugs.

4.2 Organic pharmaceutical aids : role as preservatives, antioxidants, colouring, flavouring, sweetening, emulsifying agents, stabilising and suspending agents. Ointment bases. solvents (4).

4.3. Minerals : Biological role of salts of Na, K and Ca, trace elements Cu, Zn and I. Deficiency. Sources (3)

4.4 Diagnostic agents: Organic types) for different actions, examples (2).

UNIT V :

5.1. BLOOD (10) : Composition. Analysis of blood sample – grouping, Rh factor. Tests for urea, bile carbonyl compounds, serum + protein in a sample. Physiological function of plasma protein. Role of blood as oxygen carrier. Structure of heme. Clotting mechanisms. Factors involved. Blood pressure–normal, low and high – causes and control, anemia causes, detection. Antianemic drugs.

5.2. HAEMATOLOGICAL AGENTS (8) : Coagulants and anticoagulants. coagulants – Vitamin K, Protamine, sulphate, dried thrombin, proteins, amino acids, anti coagulants— coumarins, indane dioals, citric acid, 2-sulphonyloacids, quinoxaline, throm lodyn, haemostatics, amino-caproic acid, tranexamic acid, Anemia : Causes, detection, antianemic drugs.

5.3. CARDIO VASCULAR DRUGS : Cardiac glycosides, antiarrhythemic drugs, antihypertension drugs, antianginal agents, Vascodilators, lipid. One example for each (4).

5.4. AIDS—Causes. HIV virus. propagation; Prevention, treatment.

(For Candidates admitted from 1992-93)

(Non-Semester—Third Year)

B.Sc. Degree Examination, March 1995

Core Course—Application Oriented Subject

Branch IV—Chemistry

PHARMACEUTICAL CHEMISTRY

Time : Three hours Maximum : 100 marks

Passing minimum : 40 marks

SECTION A — (25 × 1 – 25 marks)

Answer ALL questions

Choose the correct answer from the following : (15 × 1 = 15)

1. Functional group present in the antipyretic phenacetin is

(*a*) Ester (*b*) Ketone

(*c*) Phenol (*d*) Amide

2. Heterocycle present in caffeine is

(*a*) Purine (*b*) Pyrimidine

(*c*) Pyridine (*d*) Piperidine

3. Interaction of urea and melanoyl chloride leads to the formation of

(*a*) Atropine (*b*) Barbutyric acid

(*c*) Insulin (*d*) Procainamide.

4. Functional group present in prontosil is

(*a*) Ester (*b*) Alcohol

(*c*) Azide (*d*) Azo.

5. In SRA sulfonamide drugs mimic

(*a*) PABA (*b*) Folic acid

(*c*) Vitamin A (*d*) DOPA.

6. Chloroquin is a / an

(*a*) Antimalarial (*b*) Antipyretic

(*c*) Analgesic (*d*) Sedative.

7. Cholesterol belongs to a class of organic compounds known as

(*a*) Terpenoid (*b*) Steroid

(*c*) Polyphenolics (*d*) Alkaloids.

8. Haemoglobin contains

(*a*) Al (*b*) Cu

(*c*) Fe (*d*) Mg.

9. Sulfa drugs can be classified as

(*a*) Antibiotics (*b*) Analgesics

(*c*) Antidiabetics (*d*) Vermicides.

10. One of the following is an example for anaesthetic

(*a*) Quinine (*b*) Paracetamol

(*c*) Trichlorethylene (*d*) Lysergic acid.

11. Starting material for industrial synthesis of ether is

(*a*) Chloroform (*b*) Methanol

(*c*) Propanol (*d*) Ethanol.

12. Organism which is responsible for the causes of malaria

(*a*) Stephelococus (*b*) Plasmodium

(*c*) Erythroxyl (*d*) Streptomyces.

13. Azidothiamidine (AZT) acts against a virus which causes

(*a*) Smallpox (*b*) Lepracy

(*c*) Chicken pox (*d*) AIDS.

14. Antibiotic, streptomycin is employed for the treatment of

(*a*) Tuberculosis (*b*) Polio

(*c*) Rickets (*d*) Brain fever.

15. Functional group present in the antibiotic penicillin G is

(*a*) Lactam (*b*) Lactone

(*c*) Acetal (*d*) Nitro.

16. State whether the following statements are True or False. If false correct the statement. $(5 \times 1 = 5)$

(*a*) Penicillamine gives indigo colour reaction with neutral ferric chloride.

(*b*) Trichloroethylene is used as a bactericide.

(*c*) Nitrogen oxide in a liquid at room temperature.

(*d*) Paracetamol is used for its antipyretic and analgesic properties.

(*e*) Industrial synthesis of phenacetin starts from phenol.

17. In each of the following fill in the blanks with correct answers :

(*a*) Drugs which cause depression of motor functions of voluntary muscles are known as ____________.

(*b*) An example for anticonvulsant is ____________.

(*c*) Nicotine, an alkaloid, is found in the plant ____________.

(*d*) Chloramphinicol acts agains gram ____________. and bacteria.

(*e*) One of the aminoacid units present in penicilline is

SECTION B — (5 × 6 = 30 marks)

Answer any FIVE questions

All questions carry equal marks.

18. Name three Indian medicinal plants and drugs which are obtained from them.

19. How is penicillin G industrially obtained? What are the analogies of penicillin which are commonly used as drugs?
20. Outline the preparation of sulfafurazole.
21. What are the causes of anaemia and diabetes? Give the name and structure for one drug used for their treatment.
22. Describe briefly the mode of action of synthetic analgesic such as methadone.
23. What are disinfectants? give the structure and industrial synthesis of one of them.
24. Give the structure of LSD. How is it prepared from Lysergic acid? How is it assayed?
25. State briefly about the vairous uses of organic acids in pharmaceutial preparations.

SECTION C—(3 × 15 = 45 marks)

Answer ALL the questions

All questions carry equal marks.

26. (*a*) Discuss briefly with suitable examples, elementary basis for designing a drug with specific biological response with least side effects.

Or

(*b*) Write short notes on :

(*i*) Metal chelates and drugs.

(*ii*) Tranquilizers.

27. (*a*) How are anti malarials classified? Give one example each with detailed structure. Write SAR relationship of any one drug.

Or

(*b*) What are anaesthetics? Give two examples with structure. Explain four stages of anaesthesia.

28. (*a*) Explain briefly the general method for extraction of alkaloids from plant sources by taking one example. What are the colour reactions that nicotine is expected to give?

Or

(*b*) What are anticoagulant and coagulant drugs? Give an industrial synthesis of Coumarin.

(For candidates admitted from 1992-93)

(Non-Semester—Third Year)

B.SC. DEGREE EXAMINATION, MARCH 1996

Core Course—Application Oriented Subject

Branch IV—Chemistry

PHARMACEUTICAL CHEMISTRY

Time : Three hours Maximum : 100 marks

SECTION A—(25 × 1 = 25 mark)

Answer ALL the questions

Choose the correct answer from among the following (15 × 1 = 15)

1. Chloramphenicol is also known as

(*a*) tetracycline (*b*) chloromycetin (*c*) penicillin (*d*) streptomyces venezulae.

2. Mechanistic action of drugs is evident from

(*a*) metabolism of drugs (*b*) structures of drugs (c) medicinal use of drugs (*d*) none of these.

3. Bone fracture is cured by plastering with

(*a*) $CaSO_4\frac{1}{2}\ H_2O$ (*b*) $CaSO_4.\ 2H_2O$

(c) $MgSO_4\frac{1}{3}\ H_2O$ (*d*) $MgSO_4.7H_2O$.

4. SAR stands for ·

(*a*) structure and reactivity (*b*) stereochemistry and reactivity

(*c*) stereochemistry and reationship (*d*) structure activity relationship.

5. Which is NOT true of sulpha drugs?

(*a*) they are sulphanilamides (*b*) they are effective chemotherapeutic agents

(*c*) they are antibiotics (*d*) they are useful in the cure of bacterial infections.

6. The structure of prontosil is

(*a*) H_2N–⟨O⟩–SO_2NH–(isoxazole ring: N, O)

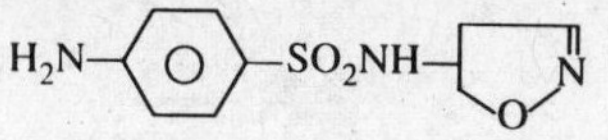

(*b*) H_2N–⟨O⟩–N = N–⟨O⟩–SO_2NH_2 (with NH_2 on the first ring)

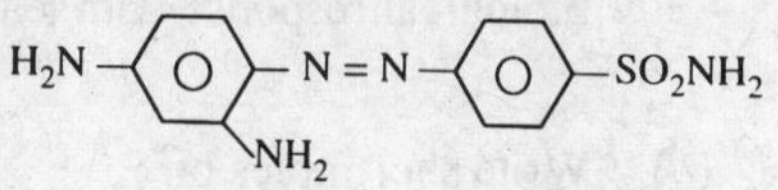

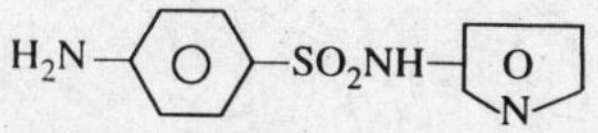

(*c*) H_2N–⟨O⟩–SO_2NH–(ring: O, N)

(*d*) H_2N–⟨O⟩–SO_2NH_2

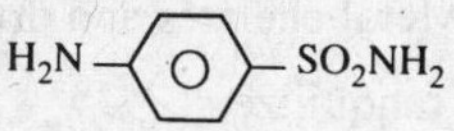

7. Which is NOT true of antiseptics?

(*a*) they are chemical agents which would prevent the growth of micro organisms in living tissues

(*b*) brilliant green and crystal violet are good examples (*c*) they are used on inanimate things

(*d*) they are used on animate things.

8. The one which is used as anaesthetic is

(*a*) ether (b) vinyl ether

(*c*) halothane (*d*) all the above.

9. Heroin is

(*a*) morphine (*b*) codeine

(*c*) diacetyl morphine (*d*) 3-ethyl ether of morphine.

10. Alkaloids develop a creamy yellow precipitate on treatment with

(*a*) Wagner's reagent (*b*) Maeyer's reagent

(*c*) Dragen dorffs reagent (*d*) $FeCl_3$.

11. The functional groups present in quinine are

(*a*) –OH, –O–, $-NCH_3$ (*b*) –OMe, –CHOH–, $-CH=CH_2$

(*c*) –OH, $-CHOH_2$, $-CH=CH_2$ (*d*) –CHOH, $-CH=CH_2$.

12. The one that does NOT serve as an antioxidant is

(*a*) cochineal (*b*) propyl gallate

(*c*) thymol (*d*) butylated hydroxyanisole.

13. Fluid fraction of blood is known as

(*a*) plasma (*b*) erythrocyte

(*c*) platelet (*d*) leukocyte.

14. Blood is grouped as

(*a*) A (b) B (c) O (d) all the above.

15. Methadone is used as a/an

(*a*) cardiovascular drug (*b*) antipyretic

(*c*) analgesic (*d*) anticancer agent.

16. State whether each of the following statements is True or False. If false, give the correct statement.

(*a*) Anaemia is caused by deficiency of Vitamin A

(*b*) The penicillins are all strong monobasic acids.

(*c*) The chemical name of halothane is 2-brom–2-chloro- 1, 1, 1-trifluoro ethane.

(*d*) Quinine is isolated from poppy.

(*e*) Thiopentone sodium is a thiobarbiturate.

17. Fill in the blanks with correct answer in each of the following (5 × 1 = 5)

(*a*) The assy of penicillin is based on the fact its decomposition product consumes a quantity of ———— .

(*b*) Sulphonamide drugs are derivative of ————.

(*c*) is the structure of————.

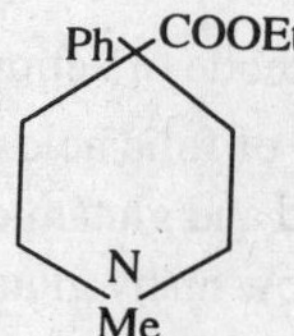

(*d*) ———level rises in the case of diabetics

(*e*) ——— mechanism is triggered by the release of fibrin and fibrinogen.

SECTION B — (5 × 6 = 30 marks)
Answer any FIVE questions
Each answer should not exceed 300 words.

18. Illustrate with suitable examples how drugs are named.

19. Write a brief note on first aid for accidents.

20. Describe the synthesis of chloramphenicol.

21. Give the preparation of

(*a*) sulphadiazine (*b*) L – (+) – ascorbic acid.

22. What are narcotic analgesics? Write down the structure of morphine. How can it be converted into codeine? Name a semi synthetic morphine derivative.

23. Write a note on tranquilisers.

24. Describe the synthesis of thiamine.

25. Describe how high and low blood pressure cases can be managed.

SECTION C— (3 × 15 = 45 marks)

Answer ALL the questions.

26. (*a*) (*i*) Write a note on important terminologies used in pharmaceutical chemistry.
(*ii*) How can diabetes be detected and cured ?
(*iii*) Give a brief account of mercury based drugs. (3 × 5)

Or

(*b*) Give the synthesis of :
(*i*) Paracetamol. (*ii*) Phenacetin (*iii*) Prontosil (*iv*) Vitamin B_2. (3 + 3 + 4 +5)

27. (*a*) (*i*) Indicate how the potency of a drug can be enhanced or retarded by structural modifications.
(*ii*) Explain the action of analgesics with a suitable illustration.
(*iii*) What is LSD? Draw its structure. (6 + 5 + 4)

Or

(*b*) Give a detailed account of extraction methods, colour reactions and detectıon of alkaloids. (15)

28. (*a*) (*i*) Give a simple method of estimation of urea explaining the principle involved in the same.
(*ii*) Explain the following :
Role of blood as oxygen carrier; Clotting mechanism (8 + 7)

Or

(*b*) (*i*) Give an example, mode of action and use of a member of cancer drugs.
(*ii*) Draw the structure of folic acid and mark out the 2-amino-4-hydroxypteridine, *p*–amino-benzoic acid and glutamic acid moleties in the structure. What is its source? What is its use? How many glutamic acid residues are there in the structure? (7 + 8)

(For candidates admitted from 1992-93)

(Non-Semester—Third Year)

BSc. DEGREE EXAMINATION, SEPTEMBER 1996

Core Course—Application Oriented Subject

Branch IV—Chemistry

PHARMACEUTICAL CHEMISTRY

Time : Three hours Maximum : 100 marks

SECTION A— (25 marks)

Answer ALL the questions

All questions carry equal marks

Choose the correct answer from among the following: (15 × 1 = 15 marks)

1. The elements present in chloramphenicol are :
(*a*) C, H only (*b*) C, H, N. O only
(*c*) C, H, O, N, Cl (*d*) C, H, N, Cl.

2. The potency of a drug
   (*a*) is based on its successful clinical trial
   (*b*) does not depend on its widespread use
   (*c*) does not depend on the approval of the state Govt.
   (*d*) can not be predicted.
3. Diabetes is a condition of
   (*a*) deficiency of red cells in blood (*b*) raised level of glucose in blood
   (*c*) decreased level of glucose in blood (*d*) decreased level of glucose in urine.
4. Which of the following is referred to as queen of drugs?
   (*a*) streptomycin (*b*) sulphadiazine
   (*c*) pencillin (*d*) chloromycetin.
5. Morphine belongs to the category of
   (*a*) narcotic analgesic (*b*) totally synthetic analgesic
   (*c*) sulpha drug (*d*) anaesthetic.
6. Aspirin is
   (*a*) methyl salicylate (*b*) Sodium salicylate
   (*c*) acetyl salicylic acid (*d*) a *p*-aminophenol derivative.
7. Heroin is a derivative of
   (*a*) caffeine (*b*) pethidine
   (*c*) morphine (*d*) diacetyl morphine.
8. Chemical agents that prevent the growth of micro organisms in living tissues are called
   (*a*) disinfectants (*b*) anaesthetics
   (*c*) antiseptics (*d*) none of these
9. Which of the following is NOT an anaesthetic?
   (*a*) chloroform (*b*) vinyl ether
   (*c*) halothane (*d*) chlorophene.
10. Quinine is obtainable from
   (a) opium (b) betel nut
   (c) tobacco (d) cinchona
11. SAR is
   (*a*) structure and reactivity (*b*) structure activity relationship
   (*c*) related to clinical trial of a drug (*d*) none of these.
12. Burnol is
   (*a*) an acidine-dye based ointment (*b*) an antineoplastic agent
   (*c*) an antidiabetic agent (*d*) a triphenylmethane dye.
13. The metal atom present in haem is
   (*a*) Mg (*b*) Co (*c*) Fe (*d*) Mn

14. Which of the following is a cardiovascular drug?

(*a*) terramycin (*b*) streptomycin

(*c*) digitoxin (*d*) methadone.

15. Plasmaprotein is helpful

(*a*) as a nutritive agent

(*b*) in the maintenance of osmotic pressure

(*c*) in the mechanism of blood clotting

(*d*) in all the above.

16. State whether the following are True of False. If false give the corrected statement. ($5 \times 1 = 5$)

(*a*) Accident victims are to be given first aid.

(*b*) Sodium salt of *p*-aminophenyl arsonic acid on heating with chloroacetamide in aqueous solution affords carbarsone.

(*c*) The structure of sulphapyridine is

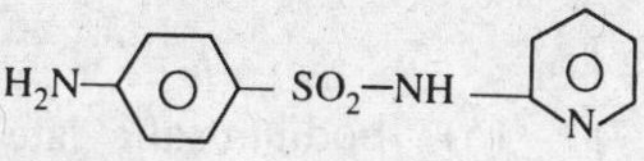

H_2N—(O)— SO_2—NH —(O, N)

(*d*) Pentothal sodium is a long acting anaesthetic

(*e*) LSD is a hallucinating agent.

17. Fill in the blanks in each of the following with correct answers.

(*a*) ——— is a good antacid.

(*b*) Benzoyl cyanide when reacted with dichloroethyl methyl amine and subjected to hydrolysis and esterification affords ———.

(*c*) Deficiency of vitamin A leads to ———.

(*d*) Alkaloids give pale yellow precipitate with ——— reagent.

(*e*) Release of fibrin from fibrinogen triggers the mechanism of ———.

SECTION B— (5 × 6 = 30 marks)

Answer any FIVE questions

All questions carry equal marks.

18. Give a brief account of various assaying methods.

19. Explain how phosphorus, mercury and iron-based drugs are medicinally important.

20. Give the preparation of :

(*a*) Sulphanilamide (*b*) Sulphafurazone.

21. Give a brief account of cationic surfactive agents.

22. Draw the structure of quinine and mark out the chiral centres on the same. How is it related to cichonine? What are the medicinal uses of quinine?

23. What is the normal blood pressure? How are high and low blood pressure patients treated?
24. Trace the importance of trace elements Cu and Zn.
25. How is diabetes caused? How is it treated?

SECTION C — (3 × 15 = 45 marks)

Answer ALL questions.

All questions carry equal marks

26. (*a*) How are the following detected in blood and urine?

(*i*) glucose (*ii*) lithium (*iii*) cholesterol (3 × 5)

Or

(*b*) With any two drugs of your choice, explain the concept of SAR. (15)

27. (*a*) Give the synthesis of :

(*i*) Vitamin C (*ii*) chloramphenicol

(*iii*) DL-penicillamine. (3 × 5)

Or

(*b*) Explain the role of preservatives, colouring agents, sweetening agents and suspending agent's with suitable examples. (3+4+4+4)

28. (*a*) (*i*) Name any two antipyretic agents and draw their structures. (6)

(*ii*) How are anaesthetics classified? Give an example for each category. How is amithocaine prepared? (9)

Or

(*b*) (*i*) How is urea estimated? (8)

(*ii*) Discuss the mode of action of anti-cancer drugs. (7)

(For candidates admitted from 1993-94)

(Non-Semester—Third Year)

BSc. DEGREE EXAMINATION, SEPTEMBER 1997

Core Course—Application Oriented Subject

Branch IV—Chemistry

PHARMACEUTICAL CHEMISTRY

Time : Three hours Maximum : 100 marks

SECTION A— (25 × 1 = 25 marks)

Answer ALL the questions

All questions carry equal marks

Choose the correct answer from among the following. (15 × 1 = 15 marks)

1. The study of pharmacological action of a drug is called

(*a*) Pharmacokinetics (*b*) pharmacodynamics

(*c*) molecular pharmacology (*d*) pharmaceutical chemistry.

2. Ferrous sulphate is used as
   (*a*) haematinic (*b*) antiseptic
   (*c*) antacid (*d*) antiemetic.
3. Which of the following statements is true with respect to the assay of antibiotics by biological method?
   (*a*) it is more accurate and more reliable
   (*b*) it is less accurate but more reliable
   (*c*) it is less accurate and less reliable
   (*d*) it is more accurate but more complicated.
4. Antineoplastic agents are used in the treatment of
   (*a*) hypertension (*b*) malaria
   (*c*) diabetes (*d*) cancer.
5. Sulphadiazine is a derivative of
   (*a*) pyridine (*b*) pyrazole
   (*c*) pyrimidine (*d*) pyrazine.
6. Deficiency of Vitamin B causes
   (*a*) beriberi (*b*) scurvy
   (*c*) night blindness (*d*) all of these.
7. Which is not true about pethidine?
   (*a*) it is an ester (*b*) it is a base
   (*c*) it is an antiseptic (*d*) it is an analgesic.
8. Which one of the following is an anaesthic?
   (*a*) nitric oxide (*b*) nitrous oxide
   (*c*) nitrogen dioxide (*d*) ammonia.
9. Halothane contains
   (*a*) three chlorine atoms (*b*) two bromine atoms
   (*c*) one bromine, one chlorine and three fluorine atoms
   (*d*) six chlorine atoms.
10. Which is not true of quinine?
    (*a*) it is an antimalarial (*b*) it has a methoxy group
    (*c*) it is an alcohol (*d*) it has three nitrogens.
11. Vannilin is used as a
    (*a*) sweetening agent (*b*) flavouring agent
    (*c*) colouring agent (*d*) none of these.
12. Which one of the following is an antioxidant?
    (*a*) ethyl gallate (*b*) benzoic acid
    (*c*) saccharin (*d*) burnt sugar.

13. Which one differs from the other three?
    (*a*) vitamin K (*b*) dried torombin
    (*c*) dicamarol (*d*) protamine sulphate.
14. Which is used in the treatment of anemia?
    (*a*) vitamin K (*b*) folic acid
    (*d*) sulphonyl ureas (*d*) none of these.
15. Alkaloid used as antihypertensive agent is
    (*a*) reserpine (*b*) morphine
    (*c*) quinine (*d*) nicotine
16. State whether each of the following statement is True of False. If false, give the correct statement (5 × 1 = 5)
    (*a*) Yellow mercuric oxide is an antiseptic
    (*b*) Chlorambucil is a nitrogen mustard.
    (*c*) Halothane is a less potent anaesthetic than chloroform.
    (*d*) All cinchona alkaloids contain methyoxyl group.
    (*e*) A person with blood pressure of 80/60 mm has low blood pressure.
17. Fill in the blanks with correct answer in each of the following : (5 × 1 = 5)
    (*a*) The active constituent in adathoda is ___________.
    (*b*) The first chemical compound used as drug is ___________.
    (*c*) Cyclopropane is stored in ___________.
    (*d*) Benzoic acid is a ___________.
    (*e*) ___________ are employed in the treatment of severe haemorrhage.

SECTION B— (5 × 6 = 30 marks)

Answer any FIVE questions

All questions carry equal marks

18. Write a note on storage of drugs
19. Discuss briefly the estimation of Vitamin A_1.
20. Discuss the use of sulphonyl ureas in controlling diabetes.
21. Discuss any two derivatives of salicylic acid used as analgesics.
22. Write a note on intravenous anaesthetics.
23. Give a brief account of sedatives.
24. Discuss briefly the biological role of Ca and I_2.
25. What is AIDS ? How can it be prevented?

SECTION C— (3 × 15 = 45 marks)

Answer ALL questions

All questions carry equal marks

26. (*a*) Discuss the factors controlling the absorption of drugs in stomach and liver.

Or

(*b*) Give an account of the medicinally imprtant compounds of aluminium and phosphorus.

27. (*a*) What is chloramphenicol? Discuss its antibiotic activity. Discuss also the effects of structural modifications of chloramphenical on its antibotic activity.

Or

(*b*) What is a local anaesthetic? What are the requirements of a good local anaesthetic? Discuss any two derivatives of benzoic acid used as local anaesthetics.

28. (a) What is a diagnostic agent? What are the three types of diagnostic agents? Explain the use of any two organic diagnostic agents.

Or

(*b*) Discuss any three types of cardiovascular drugs with examples.

(For candiates admitted from 1993-94)

(Non-Semester—Third Year)

B.Sc. DEGREE EXAMINATION, APRIL 1998.

Core Course—Application Oriented Subject

Branch IV—Chemistry

PHARMACEUTICAL CHEMISTRY

Time : Three hours Maximum : 100 marks

Passing Minimum : 40 Marks

SECTION A — (25× 1 = 25 marks)

Answer All the questions

All questions carry equal marks

Choose the correct answer from among the following : (15 × 1 = 15)

1. The group responsible for the pharmacological action of a drug is called

(*a*) pharmacophore (*b*) chromophore

(*c*) medicophore (*d*) none of these.

2. Antimetabolites are used in the treatment of

(*a*) malaria (*b*) cancer

(*c*) diabetes (*d*) typhoid.

3. Aluminium hydroxide is used as

(*a*) antiseptic (*b*) antacid

(*c*) coagulating agent (*d*) emulsifying agent.

4. Commercial source of insulin is the pancreas of

(*a*) pig (*b*) horse

(*c*) goat (*d*) monkey.

5. Which has sulphur ?

(*a*) Vitamin B_2 (*b*) Vitamin C

(*c*) Vitamin B_1 (*d*) All of these.

6. Which is not true of ethosuximide?
   (*a*) it is a five-membered ring compound
   (*b*) it is an anticonvulsant agent
   (*c*) it does not have an asymmetric carbon
   (*d*) it has one ethyl group.
7. Which is not true of thiopentone sodium?
   (*a*) it has two asymmetric carbons (*b*) it is an intravenous anaesthetic
   (*c*) it has one nitrogen (*d*) it has sulphur.
8. Aspirin is
   (*a*) acetylsalicylic acid (*b*) methyl salicylate
   (*c*) ethyl salicylate (*d*) none of these.
9. Which one differs from the others?
   (*a*) methidine (*b*) morphin
   (*c*) cocaine (*d*) paracetamol
10. Which element is essential for the growth of bones and teeth?
    (*a*) iodine (*b*) calcium
    (*c*) zinc (*d*) iron.
11. Which one is a sweetening agent?
    (*a*) ethyl acetate (*b*) vannilin
    (*c*) methyl salicylate (*d*) saccharin.
12. Which is not true of LSD?
    (*a*) it is a tertiary amine (*b*) it is an amide
    (*c*) it is a secondary amine (*d*) it is a bicyclic compound.
13. Which is a coagulant?
    (*a*) Dicumarol (*b*) Vitamin K
    (*c*) Phenindione (*d*) Warfarin sodium.
14. Which one is an antiarryhythmic drug?
    (*a*) quinidine (*b*) reserpine
    (*c*) morphine (*d*) none of these.
15. In pharmacy nitroglycerine is used as
    (*a*) vasodilator (*b*) antianemic drug
    (*c*) antihyper tension drug (*d*) none of these.
16. State whether each of the following statements is True or False. If false correct the statement. (5 × 1 = 5)
    (*a*) Sustained release formulations are intended for the maintenance of required blood level of the drug for longer duration.
    (*b*) Prontosil is a dye.

(*c*) Ether is used as a local anaesthetic.

(*d*) Morphine is a secondary amine.

(*e*) Vitamin K is an anticoagulant.

17. In each of the following fill in the blanks with correct answers : (5 × 1 = 5)

(*a*) Iron compounds are used as__________.

(*b*) Streptomycin is used in the treatment of __________.

(*c*) Structure of indomethacin is __________.

(*d*) Barbital is __________.

(*e*) AIDS is caused by the virus __________.

SECTION B— (5 × 6 = 30) marks)

Answer any FIVE questions

All questions carry equal marks.

18. Discuss the use of adathoda and tulsi in pharmacy.
19. discuss any two inorganic compounds of mercury used in pharmacy.
20. How is sulphathiazole prepared? What are its uses?
21. What are antineoplastic agents? Discuss any two plant products used as antineoplastic agents.
22. What is salol? How is it prepared? What are its uses?
23. Write a note on cationic surface active agents.
24. Discuss the biological role of zinc. Mention any two sources of zinc.
25. What are the pressure values of normal, low and high blood pressures?

SECTION C— (3 × 15 = 45 marks)

Answer ALL questions

All questions carry equal marks

26. (*a*) Discuss the mechanism of metabolism and absorption of drugs.

Or

(b) What is ampicillin? Why it is a better antibiotic than pencillin? Discuss any two structural modifications made on ampicillin and their effects on its activity.

27. (*a*) Discuss any three halogen containing compounds used as anaesthetics.

Or

(*b*) How is quinine extracted? Discuss its structure.

28. (*a*) What is anemia? What are its causes? How can it be detected and cured?

Or

(*b*) (*i*) What are diagnostic agents? Discuss any two organic diagnostic agents used in pharmacy. (10)

(*ii*) What are cardiac glycosides? What are their uses? Give one example.

(For candidates admitted in 1992-93)

(Non-Semester—Third Year)

B.Sc. DEGREE EXAMINATION, APRIL 1998

Core course—Application Oriented Subject

Branch IV — Chemistry

PHARMACEUTICAL CHEMISTRY

Time : Three hours Maximum : 100 marks

Passing Minimum : 40 marks

SECTION A— (25 × 1 = 25 marks)

Answer ALL the questions.

All Questions carry equal marks.

Choose the correct answer from among the following : (15 × 1 = 15)

1. Plaster of Paris is

   (*a*) $CaSO_4$ (*b*) $Ca(OH)_2$ (*c*) $CaSO_4 \frac{1}{2} H_2O$ (*d*) $CaSO_4 Ca(OH)_2$.

2. Which ore is used in pharmacy?

   (*a*) Arsenic oxide (*b*) Lead arsenate (*c*) Calcium arsenate (*d*) None of these.

3. Anemia is treated using

   (*a*) Vitamin B_{12} (*b*) Ferrous gluconate (*c*) Folic acid (*d*) All of these.

4. Which is not true about streptomycin?

   (*a*) It is used in the treatment of tuberculosis (*b*) It is an aldehyde

   (*c*) It is a base (*d*) It is a macrocyclic compound.

5. Vitamin B_2 is called as

   (*a*) Ascorbic acid (*b*) Pyridoxine (*c*) Riboflavin (*d*) None of these

6. Which one of the following has an azo group?

   (*a*) Sulphadiazine (*b*) Prontosil (*c*) Streptomycin (*d*) None of these.

7. Thymol is

   (*a*) 2-isopropyl-5-methylphenol (*b*) 3-methyl-6-isopropylphenol

   (*c*) 3.4-dimethylphenol (*d*) None of these

8. Which is not an anaesthetic?

   (*a*) Cyclopropane (*b*) Nitrous oxide (*c*) Formaldehyde (*d*) Vinyl ether.

9. Which is not true of methadone?

   (*a*) It is a ketone (*b*) It is an analgesic

   (*c*) It is a tertiary amine (*d*) It does not have an assymmetric carbon.

10. Barbituric acid is also called as

    (*a*) Veronal (*b*) Vannilin (*c*) Veratric acid (*d*) Barbital.

11. Which is used as a colouring agent?

    (*a*) Methyl orange (*b*) Amaranth (*c*) Congo red (*d*) Alizarin.

12. Chloroambucil is a/an

    (*a*) Antineoplastic agent (*b*) Sedative

    (*c*) Hypoglycaemic agent (*d*) None of these

13. Vitamin used in the treatment of anemia is

    (*a*) D (*b*) B_{12} (*c*) C (*d*) A.

14. Normal blood pressure is
    (*a*) 120/80 (*b*) 150/80 (*c*) 120/60 (*d*) 80/60
15. Deficiency of iodine causes
    (*a*) Cancer (*b*) Anemia (*c*) Thyroid (*d*) Diabetes
16. State whether each of the following statement is True or False. If false, give the correct statemtnt : (5 × 1 = 5)
    (*a*) Cold is caused by bacteria. (*b*) Pencillin is a broad spectrum antibiotic
    (*c*) Phenol can be used as an antiseptic and a disinfectant.
    (*d*) Vannilin is used as a flavouring agent. (*e*) Vitamin K is an anticoagulant.
17. Fill in the blanks with correct answers in each of the following : (5 × 1 = 5)
    (*a*) Structure of chloropropamide is ________.
    (*b*) Deficiency of vitamin C causes ________.
    (*c*) Chloramin T is the sodium salt of ________.
    (*d*) Petroleum jelly is used as an ________.
    (*e*) The heterocyclic system present in haeme is ________.

SECTION B — (5 × 6 = 30 marks)

Answer any FIVE Questions

All questions carry equal marks

18. Write a note on first aid for accidents.
19. Discuss any two compounds of iorn used pharmacy.
20. Discuss the mechanism of action of sulpha drugs.
21. Outline the synthesis of chloramphenicol.
22. What is a local anaesthetic? What are its requirements?
23. Discuss the use of diethyl ether as an anaesthetic.
24. What is an antioxidant? Discuss the use of antioxidants in pharmacy with suitable examples.
25. Discuss briefly the various blood groups and Rh factor.

SECTION C— (3× 15 = 45 marks)

Answer ALL Questions

All questions carry equal marks

26. (*a*) Discuss the testing of potential drugs using experimental animals.

    Or

    (*b*) (*i*) Discuss the effects of structural modifications on the potentiality of penicillin.
    (*ii*) Outline the synthesis of vitamin A.
27. (*a*) What is an analgesic? Give an account of the salicylic acid derivatives used as analgesics.

    Or

    (*b*) What is quinine? What are its uses? How its activity is influenced by structural modifications?
28. (*a*) Discuss the physiological functions of plasma proteins.

    Or

    (*b*) (*i*) Discuss with examples any two kinds of cardiovascular drugs.
    (*ii*) What is diabetes? How it is caused? How can it be controlled?

Time : Three hours Maximum : 100 marks

SECTION A = 25 marks)
Answer ALL questions.
All questions carry equal marks.

Choose the correct answer from among the following: (15 × 1 = 15)

1. Iron compounds are used in the treatment of
(*a*) anemia (*b*) diabetes (*c*) tuberculosis (*d*) all the above.
2. Which is the major site of absorption of most drugs?
(*a*) Kidney (*b*) Stomach (*c*) Small intestine (*d*) Liver.
3. Which one is caused by virus?
(*a*) Typhoid (*b*) Tuberculosis (*c*) Small pox (*d*) None of these.
4. Which is not true of tetracyclines?
(*a*) They are antibiotics (*b*) They have four rings
(*c*) All of them have phenolic group (*d*) All of them have chlorine.
5. Which has a seven-membered ring?
(*a*) Chloropropamide (*b*) Tolbutamide (*c*) Tolazamide (*d*) None of these
6. Which is an antineoplastic agent?
(*a*) Chlorambucil (*b*) Vitamin A (*c*) Ripamycin (*d*) None of these
7. Which is not true of morphine?
(*a*) It has a phenolic group (*b*) It is an ether
(*c*) It is a base (*d*) It is a tetracyclic compound.
8. Nitroburazone is
(*a*) Semicarbazone of 5-nitro-2-buraldehyde (*b*) hydrazone of 5-nitro-2-buraldehyde
(*c*) phenyl hydrazone of 3-nitro-2-buraldehyde (*d*) none of these.
9. Which is not a tranquilizing agent?
(*a*) Reserpine (*b*) Promethazine (*c*) Chloropropamide (*d*) Ampicillin.
10. Benzocaine is
(*a*) ethyl 4-aminobenzoate (*b*) methyl 4-amino benzoate
(*c*) ethyl 4-nitrobenzoate (*d*) none of these.
11. Which is an antioxidant?
(*a*) 2,6-di-t-butyl-4-methylphenol (*b*) 4-chlorophenol
(*c*) $CHCl_3$ (*d*) All the above.
12. Which should not be used as a solvent in pharmaceutical preparations?
(*a*) Water (*b*) Glycerine (*c*) Methanol (*d*) Propylene glycol.
13. Anemia is diagnosed by estimation of
(*a*) blood iron content (*b*) blood haemoglobin
(*c*) blood cyanocobalamin (*d*) blood vitamin B_{12}.
14. Which is an antihypertension agent?
(*a*) Papaverine (*b*) Morphine (*c*) Quinine (*d*) None of these.
15. Which is not true of dicumarol?
(*a*) It is an anticoagulant (*b*) It is an enol
(*c*) It has no nitrogen (*d*) It is a pentacyclic compound.

16. State whether the following statements are True or False. If False correct the statement. (5 × 1= 5)

(*a*) An increase in the rate of metabolism of a drug increases its efficiency.

(*b*) Chloramphenicol has a nitrogroup.

(c) Pethidine is an ester.

(*d*) All alkaloids do not contain nitrogen.

(*e*) Nitroglycerine is an antianginal agent.

17. Fill up the blanks with suitable answers. (5 × 1 = 5)

(*a*) $Al(OH)_3$ is used as an __________.

(*b*) Night blindness is caused by the deficiency of ________.

(*c*) Acetyl salicylic acid is known as __________.

(*d*) Diazepam has __________ nitrogens.

(*e*) Calibrate is a __________ agent.

SECTION B — (5 × 6 = 30 marks)

Answer any FIVE questions.

All questions carry equal marks.

18. Explain sustained release of drugs.
19. Discuss any two compounds of mercury used in medicine.
20. How is insulin prepared. How it is used in controlling diabetes?
21. Explain how phenol can be used as an antiseptic as well as a disinfectant.
22. Write a note on psychedelic drugs.
23. Write a note on ointment bases.
24. What is normal blood pressure? Explain high and low blood pressures.
25. Write a note on AIDS.

SECTION C — (3 × 15 = 45 marks)

Answer ALL questions.

All questions carry equal marks.

26. (*a*) Give an account of the uses of tulsi, neem and kizhanelli in medicine.

Or

(*b*) Give the synthesis and uses of any three sulpha drugs.

27. (*a*) (*i*) How is vitamin C estimated ?

(*ii*) Discuss briefly the use of cationic surface active agents as antiseptics.

Or

(*b*) What is a local anaesthetic? What are the requirements of a good local anaesthetic? Discuss any three local anaesthetics.

28. (*a*) Give an account of colouring, flavouring and agents used in pharmacy sweetening.

Or

(*b*) What are anticoagulants? Discuss any three type of anticoagulants with examples.

Time : Three hours Maximum : 100 marks

SECTION A — (marks)

Answer ALL questions.

All questions carry equal marks.

Choose the correct answer from the following:

(15 × 1 = 15)

1. Phenformin is used in the treatment of
   (*a*) tuberculosis (*b*) cancer (*c*) diabetes (*d*) all of these.
2. Which is not true of Chlorambuchil?
   (*a*) It is a carboxylic acid
   (*b*) It is tertiary amine
   (*c*) It is used in the treatment of leukaemia
   (*d*) It is optically active.
3. Which is not a local anaesthetic?
   (*a*) Cocaine (*b*) Benzocaine (*c*) Quinine (*d*) Procaine.
4. Which is true of thiomersal ?
   (*a*) It contains sulphur (*b*) It contains mercury
   (*c*) It is an antiseptic (*d*) All of these.
5. Which has a fluorine atom?
   (*a*) Diazepam (b) Oxazepam (c) Flurazepam (d) None of these.
6. Peopyl gallate is used as
   (*a*) sweetening agent (*b*) flavouring agent (*c*) antioxidant (*d*) none of these.
7. Which is not true of tranexamic acid?
   (*a*) It is a haemostatic (*b*) It is a cyclohexane derivators
   (*c*) It is a tertiary amine (*d*) It is an amino acid.
8. Halothan is
   (*a*) 1–bromo–1–chloro ethane (*b*) Chloroethane
   (*c*) 1–bromo–1–chloro–1–fluoroethane (*d*) 2-bromo–2–chloro– 1, 1, 1, –trifluoro ethane
9. Which can show optical activity?
   (*a*) **Vitamin K_1** (*b*) **Vitamin K_2 (30)** (*c*) **Vitamin K_2 (35)** (*d*) None of these
10. Which is not true?
    (*a*) Vimcristine is an aldehyde (*b*) Vimcristine is an alkaloid
    (*c*) Vimblastine is an amine (*d*) None of these
11. Kizhanelli is used in the treatment of
    (*a*) jaundice (*b*) typhoid (*c*) tuberculosis (*d*) none of these
12. Aluminium hydroxide gel is used is to
    (*a*) antiseptic (*b*) antacid (*c*) antibiotic (*d*) all of these.
13. The number of cysteine residues in insulin is
    (*a*) six (*b*) four (*c*) eight (*d*) none of these.
14. The compound which was used first in chemotherapy is
    (*a*) prontosil (*b*) pencillin (*c*) quinine (*d*) none of these.

15. Deficiency of Vitamin B_1 causes
 (*a*) beriberi (*b*) rickets (*c*) night blindness (*d*) none of these.
16. Fill in the blanks with suitable answers: (5 × 1 = 5)
 (*a*) Active principle in adathoda is ________ .
 (*b*) Iodine is administered as ________ .
 (*c*) Deficiency of calcium causes ________ .
 (*d*) Heparin is used as ________ .
 (*e*) Glycerine trinitrate is used as a ________ .
17. State whether the following statements are True or False. If False correct them. (5 × 1 = 5)
 (*a*) Nightblindness is caused by the deficiency of vitamin C.
 (*b*) Propane is a gaseous anaesthetic.
 (*c*) Benzalkonium chloride is used as an antiseptic.
 (*d*) All alkaloids have oxygen atom.
 (*e*) Clofibrate is an amide.

SECTION B — (5 × 6 = 30 marks)

Answer any FIVE questions.

All questions carry equal marks.

18. Explain the storage conditions of drugs with examples.
19. Write a note on the compounds of iron and mercury used in medicine.
20. Give the structure, properties and uses of erythromycin.
21. Describe briefly the structural elucidation of chloramphenicol.
22. What is thiopental sodium? Discuss its properties and uses.
23. Write a note on psychedelic drugs.
24. Write a note on coumarins used as anticoagulants.
25. What are diagnostic agents? Give examples. What are their uses?

SECTION C — (3× 15 = 45 marks)

Answer ALL questions.

All questions carry equal marks.

26. (*a*) Discuss the metabolism of drugs and its effect on pharmacological activity.

Or

 (*b*) What are anticonvulsant agents? Discuss any three kinds of anticonvulsant agents with examples.
27. (*a*) Write the structure of morphine. Discuss the analgesic activity of morphine and its derivatives.

Or

 (*b*) How is quinine isolated? Discuss the structure activity relationship in quinine.
28. (*a*) What is anemia? How is it detected and cured?

Or

 (*b*) Write notes on the following: (5 + 5 +5)
 (*i*) biological action of sodium
 (*ii*) nitrofurazone
 (*iii*) HIV virus.

Time : Three hours Maximum : 100 marks

SECTION A — (25 marks)

Answer ALL questions.

All questions carry equal marks.

Choose the correct answer among the following:

(15 × 1 = 15)

1. The formula of ammoniated mercury is
   (*a*) $Hg(NH_3)_2Cl_2$ (*b*) $HgNH_3$ (*c*) $HgNH_2Cl$ (*d*) $Hg(NH_2)_2$.
2. Which is a source for vitamin B_2?
   (*a*) Egg (*b*) Milk (*c*) Green vegetables (*d*) All of these
3. Which has a seven-membered ring?
   (*a*) Tolbutamide (*b*) Tolazamide (*c*) Chloropropamide (*d*) None of these.
4. Which is not true about paramethadione?
   (*a*) It has an asymmetric carbon (*b*) It is an anticonvulsant drug
   (*c*) It has NH group (*d*) It has three oxygens.
5. Which is not an ester of *p*-amino benzoic acid?
   (*a*) Proxymetacaine (*b*) Procaine (*c*) Benzocaine (*d*) None of these.
6. Nitroburazone is the semicarbazone of
   (*a*) 5–nitro-2–puraldehyde (*b*) 2–puraldehyde
   (*c*) 3–nitro–2–puraldehyde (*d*) 5–nitro–3–puraldehyde.
7. Which is not a pyrazole derivative?
   (*a*) Analgin (*b*) Phenazone (*c*) Benorylate (*d*) None of these.
8. Normal level of Zn in blood is
   (*a*) 100 mg/100 ml (*b*) 900 μg/100 ml (*c*) 0.3mg/kg (*d*) 20mg/100ml.
9. Copper is an essential constituent of
   (*a*) cytochrome (*b*) catalase (*c*) ascorbic acid oxidase (*d*) all of these
10. Which is a diagnostic agent?
    (*a*) Sodium diatrizoate (*b*) Congo red (*c*) Crystal violet (*d*) None of these.
11. Which is a preservative?
    (*a*) Sodium benzoate (b) Cinnamic acid (c) Phenoxy ethanol (d) all of these.
12. Sodium lauryl sulphate is used as
    (*a*) suspending agent (*b*) stablising agent (*c*) emulsifying agent (*d*) none of these
13. Which is used in the treatment of anemia?
    (*a*) Ferrous fumarate (*b*) Folic acid (*c*) Cyanocobalamin (*d*) all of these.
14. Which is not true about clopibrate?
    (*a*) It is a lipid lowering agent (*b*) it is an easter
    (*c*) It is an acidic compound (*d*) it is a neutral compound.
15. Which is a vasodilator?
    (a) Papaverine (b) Morphine (c) Quinine (d) None of these
16. Fill in the blanks suitably: (5 × 1 = 5)
    (*a*) Kozhanelli is used in the treatment of ______________.
    (*b*) ______________ is used in the treatment of typhoid.

(*c*) Sorbitol is used as a ________________.

(*d*) AIDS is caused by ______________ virus.

(*e*) Chemical name of profen is ______________ .

17. State whether the following statements are True or False. If False correct them : (5 × 1 = 5)

(*a*) Aluminium phosphate is a slow acting antacid.

(c) Phenopormin has six nitrogens.

(d) Quinine is an antitubercular compound.

(e) Drugs which prolong the coagulation of blood are called coagulants.

SECTION B — (5 × 6 = 30 marks)

Answer any FIVE questions.

All questions carry equal marks.

18. Give the medicinal uses of adathoda and thoothuvalai.

19. Write a note on storage of drugs.

20. Give an account of the properties and uses of streptomycin.

21. Give the preparation and uses of sulphadiazin.

22. Discuss the effect of any three structural modifications on the analgesic activity of morphine.

23. Give an account of the derivatives of phenol used disinfectants.

24. Write a note on transculizers.

25. What are antihypetension drugs and antianginal agents? Give one example for each.

SECTION C — (3× 15 = 45 marks)

Answer any FIVE questions.

All questions carry equal marks.

26. (*a*) Explain the terms molecular pharmacology, pharmacodynamics, pharmacophore, metabolite and antimetabolite.

Or

(*b*) What are antineoplastic agents? Discuss briefly any three kinds of antineoplastic agents.

27. (*a*) Give an account of volatile anaesthetics.

Or

(*b*) How are alkaloids isolated and purified? Discuss the color reactions of alkaloids.

28. (*a*) Discuss the causes and control of low and high blood pressure.

Or

(*b*) Write notes on the following :

(*i*) Drug delivery system

(*ii*) Biological role of iodine

(*iii*) Haemostatics.

UNIVERSITY OF MADRAS

B.Sc. - BRANCH IV CHEMISTRY

(Effective from the academic year 2002-2003 and there after)

1. That in the Regulations relating to B.Sc. Degree course in Appendix -A scheme of Examinations core courses, under 4, Branch – IV Chemistry, the following amedments be adopted :

scheme of Examinations

branch – IV – Chemistry

		Duration	Max. marks
First Year :			
Paper I	General Chemistry I	3 Hrs.	100
Practical I:	Inorganic qualitative analysis and preparation	3 Hrs. (Record - 5 + 35)	40
Second Year :			
Paper II	General Chemistry II	3 Hrs.	100
Practical II	Volumetric Estimation	3 Hrs. (Record - 10 + 30)	40
Third Year :			
Paper III :	Inorganic Chemistry	3 Hrs.	100
Paper IV :	Organic Chemistry	3 Hrs.	100
Paper V :	Physical Chemistry	3 Hrs.	100
Paper VI :	Analytical Chemistry		
Application Oriented Subject		3 Hrs.	100
Paper VII :	Applied Chemistry	3 Hrs	100
Practicals			
Practical III:	Physical Chemistry	3 Hrs. (Record- 10 + 30)	40
Practical IV:	**Gravimetric Estimation**	**3 Hrs.** (Record -	

3. Practical chemistry - S. Sundaram- 3 Volumes- S. Viswanathan
4. Vogel's text book of practical organic chemistry- Longman

Application Oriented Subject :

Applied Chemistry

1. A textbook of pharmaceutical chemistry-Jayashree Ghosh-S.Chand.
2. Pharmaceutical chemistry-S. Lakshmi Sultan Chand.
3. Pharmacology and pharmatherapetics - R.S. Satoskar.
4. Medicinal chemistry-Asutosh Kar-New Age.
5. A textbook of synthetic drugs-O.D. Tyagi-Anmol publications.
6. Introduction to biological chemistry - J. Awapara-Prentice Hall

7. A textbook of biochemistry-Ambika.S]
8. Biochemistry-A.L. Lehinger
9. Essentials of biological chemistry-James Fanley-East West press.
10. Outline of dairy technology-Sukumar De
11. Principles of dairy chemistry-Robert Jenness
12. Indian diary products-K.S. Rangappa and K.T. Acharya
13. Polymer chemistry-M.G. Arora-Anmol publications-New Delhi
14. Textbook of polymer science-F.W. Billmeyer-New Age International
15. Polymer chemistry an introduction-M.P. Stevens-Oxford
16. Chemical technology of leather-ISI
17. Fundamentals of leather science-Woodroffe
18. Tanning process-Crthman
19. Publications of CLRI-Chennai
20. Nature and properties of soils-Harry, O.Buckman
21. Soil sciences-A. Sankara